Gas Hydrates

Geological Society Special Publications
Series Editors
A. J. Fleet
A. C. Morton
A. M. Roberts

It is recommended that reference to all or part of this book should be made in one of the following ways:

Henriet, J.-P. & Mienert, J. (eds) 1998. *Gas Hydrates: Relevance to World Margin Stability and Climate Change*. Geological Society, London, Special Publications, **137**.

Long, D. Lammers, S. & Linke, P. 1998. Possible hydrate mounds within large sea-floor craters in the Barents Sea. *In:* Henriet, J.-P. & Mienert, J. (eds) *Gas Hydrates: Relevance to World Margin Stability and Climate Change*. Geological Society, London, Special Publications, **137**, 223–237.

GEOLOGICAL SOCIETY SPECIAL PUBLICATION NO. 137

Gas Hydrates

Relevance to World Margin Stability and Climate Change

EDITED BY

J.-P. HENRIET

Renard Centre of Marine Geology
University of Gent
Belgium

J. MIENERT

GEOMAR
Kiel
Germany

1998

Published by

The Geological Society

London

THE GEOLOGICAL SOCIETY

The Society was founded in 1807 as The Geological Society of London and is the oldest geological society in the world. It received its Royal Charter in 1825 for the purpose of 'investigating the mineral structure of the Earth'. The Society is Britain's national society for geology with a membership of around 8500. It has countrywide coverage and approximately 1500 members reside overseas. The Society is responsible for all aspects of the geological sciences including professional matters. The Society has its own publishing house, which produces the Society's international journals, books and maps, and which acts as the European distributor for publications of the American Association of Petroleum Geologists, SEPM and the Geological Society of America.

Fellowship is open to those holding a recognized honours degree in geology or cognate subject and who have at least two years' relevant postgraduate experience, or who have not less than six years' relevant experience in geology or a cognate subject. A Fellow who has not less than five years' relevant postgraduate experience in the practice of geology may apply for validation and, subject to approval, may be able to use the designatory letters C Geol (Chartered Geologist).

Further information about the Society is available from the Membership Manager, The Geological Society, Burlington House, Piccadilly, London W1V 0JU, UK. The Society is a Registered Charity, No. 210161.

Published by The Geological Society from:
The Geological Society Publishing House
Unit 7, Brassmill Enterprise Centre
Brassmill Lane
Bath BA1 3JN
UK
(*Orders*: Tel. 01225 445046
Fax 01225 442836)

First published 1998

British Library Cataloguing in Publication Data
A catalogue record for this book is available from the British Library.

ISBN 1-86239-010-X
ISSN 0305-8719

Typeset and printed by Alden Group, Oxford, UK.

Distributors
USA
AAPG Bookstore
PO Box 979
Tulsa
OK 74101-0979
USA
(*Orders*: Tel. (918) 584-2555
Fax (918) 560-2652)

Australia
Australian Mineral Foundation
63 Conyngham Street
Glenside
South Australia 5065
Australia
(*Orders*: Tel. (08) 379-0444
Fax (08) 379-4634)

India
Affiliated East-West Press PVT Ltd
G-1/16 Ansari Road
New Delhi 110 002
India
(*Orders*: Tel. (11) 327-9113
Fax (11) 326-0538)

Japan
Kanda Book Trading Co.
Cityhouse Tama 204
Tsurumaki 1-3-10
Tama-Shi
Tokyo 0206-0034
Japan
(*Orders*: Tel. (0423) 57-7650
Fax (0423) 57-7651)

Contents

Gas Hydrates: the Gent debates. Outlook on research horizons and strategies

J.-P. HENRIET[1] & J. MIENERT[2]

[1] *Renard Centre of Marine Geology, Geological Institute, University of Gent, Krijgslaan 281-S8, B-9000 Gent, Belgium*
[2] *GEOMAR, Research Centre for Marine Geosciences, Wischhofstr. 1–3, D24148 Kiel, Germany*

In 1811 Sir Humphry Davy, who gained fame for both his research on the methane-laden atmospheres in British coal mines and his synthesis of various new elements and compounds, witnessed the first chlorine hydrate crystallizing. At that time he probably did not imagine that 185 years later methane hydrates would fuel heated debates under the gothic vaults of a former Dominican monastery in Gent.

Natural gas hydrates have come a long way. From a mere chemical curiosity they proved, as early as the mid-1930s, to be a nuisance for the natural gas industry. Its impact increased in the 1970s, with even the largest pipelines from offshore or arctic fields or the wells from high-pressure underground storage facilities becoming clogged by hydrate plugs. As todays hydrocarbon industry makes a strategic move towards the continental slope, oceanic gas hydrates are increasingly recognized as a major potential hazard for the stability of offshore structures in various deep-water hydrocarbon provinces.

Beyond these direct interferences with man's industrial ventures, gas hydrates are gradually moving onto the foreground of global climate debates. If present estimates of methane hydrate volumes stored in the oceanic margin sediments are substantiated, natural gas hydrates represent, under the present climatic and oceanographic conditions, by far the largest mass of organic carbon stored in a potentially dynamic reservoir of the globe's carbon cycle. Their stability is controlled to a large extent by the temperature regime of the oceans and by pressure conditions on the seabed, both of which are directly linked to sea-level changes.

This vast quantity of 'frozen' greenhouse gas may have played a significant role in the global symphony of ice ages and the dramatic climatic shifts which characterize the Late Cenozoic world. And it may still play a role today, perhaps hardly noticed against the background of the present world's CO_2-driven climate machine. But we cannot exclude the possibility that any major modification of the ocean's dynamic thermal structure in a warming world might unleash vast amounts of methane from its seabed reservoir. Such a release would potentially contribute to the already anticipated evolution of the Earth's atmosphere, with methane at sometime in the foreseeable future replacing carbon dioxide in its prime role as a global warming agent.

Is such a scenario possible, even plausible? Can we quantify it and introduce it into our models? Does it imply a fundamental change in our modelling approaches, and should the now familiar coupled ocean–atmosphere models give way to coupled seabed–ocean–atmosphere models? Can the estimates of the scale of the hydrate reservoir be substantiated by any ground-truthing? How can the mechanisms of mass and energy transfer between the seabed, ocean and atmosphere, linked to hydrate growth and decay, be unravelled and quantified? And which strategy should be adopted for achieving such ambitious goals?

This profusion of questions, in the first instance, calls for dialogue and concerted efforts between various disciplines, from geophysics and geochemistry to ocean and atmospheric physics and chemistry, sedimentology, thermodynamics, ice core stratigraphy, biogeochemistry and many more fields of research. The questions prompt immediate dialogue between academic and industrial partners and call for sustained, long-term efforts, focused in their objectives but global in their relevance.

Such questions were also the subject of the workshop 'Gas hydrates: relevance to world margin stability and climatic change', organized in Gent in September 1996 under the auspices of the European Commission's Marine Science and Technology (MAST) Programme and the European Marine and Polar Science Boards (EMaPS).

This eminent patronage occurred because of the outstanding fundamental scientific relevance of this topic, which truly blends with far-reaching economical and societal considerations, and because the northwest European continental

HENRIET, J.-P. & MIENERT, J. 1998. Gas hydrates: the Gent debates. Outlook on research horizons and strategies. *In:* HENRIET, J.-P. & MIENERT, J. (eds) *Gas Hydrates: Relevance to World Margin Stability and Climate Change.* Geological Society, London, Special Publications, **137**, 1–8.

margin, Europe's new frontier for hydrocarbon resources, turns out to be a vast natural laboratory for the interaction between gas hydrates and slope stabilities, where some of the world's largest submarine slides have yet to be studied with advanced research. This is exactly the type of natural European laboratory which has the potential to lead to fundamental breakthroughs of global relevance.

With such outstanding natural laboratories along Europe's margins, research will transgress the European scientific and human dimensions. The interaction between oceanic conditions and the seabed along the North Atlantic margins is a basin-wide scientific challenge that calls for organization and, wherever possible, cooperation with U.S. and Canadian partners. On Europe's northern frontier, in the Arctic seas, and on its southern flank, in the eastern Mediterranean and Black Sea, hydrate research has been pioneered by Russian institutes, which are becoming Europe's active partners. Last but not least, Europe should not overlook the important momentum in fundamental and applied hydrate research in the Far East, particularly at Japanese institutes.

This volume draws on state-of-the-art hydrate research information from leading scientists, from the USA, Russia and the European Union. **Kvenvolden** (US Geological Survey in Menlo Park) and **Sloan** (Colorado School of Mines) set the stage for the volume. Both take the reader on a journey from the fundamentals of oceanic hydrates, or clathrates, to their vision of the relevance of hydrates as an energy resource, as a factor of margin stability and of climatic change. This journey benefits from both the affinities and the differences in perspective of both authors, partly due to the different 'cultural' environments from which they come. Kvenvolden builds upon USGS databases to highlight the geological occurrence and the resource dimension of gas hydrates in a global context, while Sloan utilizes his laboratory studies to introduce the reader to the fundamental physical and chemical properties of gas hydrates, their characterization and kinetics of formation and decay. The latter contribution sets the stage for the following chapter, which addresses some fundamental insights into hydrate generation and identifies challenges in modelling.

The following sections of this introductory paper summarize the outcomes of the round table debates which animated the Gent workshop, and introduce papers in this volume. It should be emphasized that where present-day forums on oceanic gas hydrates research currently encompass the trilogy 'resource–hazard–climate issue', as introduced by Kvenvolden and Sloan, the Gent workshop deliberately focused on slope stability and climatic issues. The contents and structure of this Special Publication reflect this focus.

The first round table discussion of the workshop focused on the challenges ahead in understanding the generation of oceanic gas hydrates and in assessing the physics and chemistry of oceanic hydrate-bearing sediments. Contributors to the discussion were Ben Clennell and Gabriel Ginsburg (panel members), Myriam Kastner (moderator), Hans Amann, Bill Dillon, Bilal Haq, Jean-Pierre Henriet, James Kennett, Keith Kvenvolden, Dave Long, Charles Paull, Carolyn Ruppel, Alister Skinner, Dendy Sloan, Tjeerd Van Weering and Warren Wood.

Analysis and modelling of hydrate formation

A prime question emerging from the debate is how we actually view the hydrate zone as a complex, multiphase system and in particular how we view multiphase fluid transport and hydrate generation in porous media. There are still vast opportunities offered to hydrate research to draw upon the extensive literature in related fields such as chemical engineering (chemical reactors) and various fields of civil engineering, and to design suitably controlled experiments. An important issue is the residence time of gas within the hydrate stability zone. Only a minor portion of the gas migrating through a hydrate stability zone seems to be actually trapped in hydrates. The kinetics of such entrapment needs clarification. However, it is clear that the host sediments are not merely behaving as a passive matrix in which hydrates accumulate. Surface forces on mineral particles may control the kinetics of hydrate formation and the stability of hydrates in the sediment pores. And it is the grain size which is likely to control the spatial distribution of hydrates: in coarse sediments, gas hydrates are often found to cement the grains throughout the matrix, whereas in finer sediments they frequently occur as discrete nodules, sheets or lenses.

Ginsburg focuses on the spatial variability of gas solubility in pore waters, and highlights the inhomogeneous distribution of hydrates within the hydrate stability zone. **Rempel & Buffett** venture to construct relatively simple models that quantitatively predict the development of hydrate layers, both for advection of gas from deeper reservoirs and for *in situ* biogenic production. The latter model meets Ginsburg's observation about a shallow region with elevated hydrate saturation. Natural gas hydrates however do not have a simple composition: they are solid solutions of water and a variety of gases such as CO_2, CH_4, N_2 and C_2H_6, and may contain significant concentrations of H_2S. **Bakker** develops and models stability conditions for fluid systems and gas clathrates with a complex composition, thoroughly testing them against experimental data. **Lu & Matsumoto** focus on the formation of CO_2 hydrates under varying pH conditions.

Exploration strategy and field evaluation methodology

The habitat of hydrates in sediments will control their geophysical response, whether electrical, seismic or thermal. The calibration of geophysical responses will require extensive ground-truthing through drilling and the recovery of undisturbed samples for analysis in the laboratory under the original ambient pressure. This calls for the development of a new generation of pressure core samplers and core logging devices, as well as instruments allowing imaging (NMR imaging, CAT-scanning) and controlled experiments inside pressurized cores. Only this type of approach will allow the proper calibration of downhole logging responses.

Booth *et al.* introduce a comprehensive first overview of the spatial distribution of hydrates in various 'reservoirs' and environments, as appraised from numerous well sites located around the world. **Goldberg & Saito** comment on the results of downhole logging on the ODP Blake Ridge sites, standing up for 'logging-while-drilling' in hydrate investigations.

In general, geophysics has the potential to achieve more than it has to date. While significant research is still needed to understand the true physical nature of the BSR (the 'bottom simulating reflector'), it is necessary to treat it on its own merits and, through the above calibrations and a move towards more high-resolution seismics, further focus on the most significant sedimentary section, which stretches from the BSR to the seabed.

Both **Hobro *et al.*** and **Tinivella *et al*.** demonstrate the power of seismic tomographic inversion studies for elucidating the velocity distribution above BSR horizons. **Mienert *et al.*** discuss high-resolution results obtained from ocean bottom seismometers and from modelling using trial and error ray tracing.

But how far can we go with ground-truthing? The exploration phase for oceanic hydrates is in its early stages. We are still seeking to identify the physical criteria which will allow mapping of hydrate reservoirs: 'BSR'-mapping is still the only, and debatable, strategy. There is an urgent need for ample drilling efforts and adequate sample recovery. In most cases targets will call for three-dimensional approaches. However, drilling is expensive, so a strategy is needed and site selection criteria have to be identified. Ground-truthing should focus on end members and type localities which span a wide spectrum of hydrate habitats. Research should proceed until each end member, in each type of hydrate accumulation, is categorized.

There was a consensus in the workshop debates that sites for investigation should ideally be selected where seeps are observed, and that such seeps need to be quantified, and qualified. This calls for monitoring of fluxes and for sample collection and *in situ* chemical analyses. It is important to assess how much gas is seeping from hydrate zones and non-hydrate zones, from focused vents and diffuse venting areas. How much is released as a steady state flow, and how much as episodic venting, inasmuch as the latter can be evaluated during times available for observation.

We need fresh observations of modern venting systems in order to try to understand their dynamics. We can only issue warnings today about trends in CO_2 levels in the atmosphere because of decades of observation. The basic technology is available, and improvements of instruments are straightforward. Even low-cost, long-term thermistor stations can teach us how temperature drives the system. If we are truly convinced that the issue is scientifically important, we must start monitoring stations in the deep sea. Over a time scale of a few decades, especially in areas prone to slope instabilities, an event will eventually happen that we will be able to record.

Major gas hydrate occurrences: case studies

A second debate at the workship ventured to target areas where international and multi-disciplinary research efforts, over a sustained period, may shed light on significant oceanic gas hydrate occurrences and quantify the elements needed for both resource assessment and climatic modelling. Contributors to the debate were Bill Dillon and Michael Ivanov (panel members), Jürgen Mienert (moderator), Mike Baillie, Angelo Camerlenghi, Gabriel Ginsburg, Myriam Kastner, Bilal Haq, Mike Helgerud, Jean-Pierre Henriet, Charles Paull, Dominique Raynaud, Carolyn Ruppel, Dendy Sloan and Warren Wood.

Introducing the evaluation of potential target areas raised the issue of gaps in our approaches: both topical gaps and scale gaps. We have already come a long way in bridging topical gaps, in particular between geophysical research and geochemical investigations, which have dominated the scene so far, by the study of chemical gradients over large sections, laboratory studies, well logging and thermal studies. But we still have scale gaps. Somewhere between the scale of molecules and logging scales we already seem to have a problem. What is going on within the pores? What is going on at a scale of centimetres to metres? This is where much more research involvement is required.

Such involvement will primarily require field exploration and experiments, initially using downhole experiments. We need to look at hydrates even before they are brought up the hole. For such purposes some field study sites should be set up and agreed upon by a wide research community. Places where high concentrations of gas hydrates are found and where the geological background is already well documented should be used for such experiments.

Still on the question of scales, but at the other end of the spectrum: as we move towards analysing the possible coupling between ocean and seabed through bottom currents, the transmission of thermal pulses from a warming or cooling atmosphere via the hydrosphere to the seabed and the resulting decay or growth of oceanic hydrates, the responses of slopes through giant slides or sequences of slides, we analyse basin-wide phenomena. The logical scale for studying the climatic role of hydrates and for making the link with thermal models of the ocean will be the oceanic basin, such as the North Atlantic, or the Arctic. In a climate perspective, the choice of study sites should also aim to clarify the behaviour at the basin scale.

In the Atlantic, following ODP Leg 164, the Blake Ridge is probably one of the best known hydrate areas. **Paull *et al.*** and **Thiéry *et al.*** report on the preliminary field data and some most interesting geochemical analyses of pore fluids. New ideas will be developed from the research that has been done so far and this is a sound reason for re-visiting these sites to check these ideas, to validate the models on the holes and to continue research, with high-resolution geophysics on the well sites and with a possible move towards *in situ* experimenting and monitoring. But for research within the context of European programmes, more proximal target areas will have to be identified.

Europe's Mediterranean flank and, in particular, the eastern Mediterranean and the Black Sea, offer attractive perspectives though specifically in relation to the geodynamic setting of convergent margins and to the particular oceanographic environment of confined seas. The association of seeps with mud volcanism and H_2S degassing is important, while there is an ample background database including multibeam, single- and multichannel seismic data. The Black Sea offers the possibility of a transect from the deep-sea to the shelf and even to the land, and the chance to observe a whole range of degassing processes and the distribution of hydrates as a function of water depth. Shallow and relatively near-shore sites would allow cost-effective re-sampling and monitoring.

De Lange & Brumsack report on gas hydrates from mud domes in the eastern Mediterranean and present what is probably the first estimate of the total amount of shallow methane associated with mud dome structures on the eastern Mediterranean Ridge. Their results, seen against a background of the recently discovered fast rise in bottom-water temperature in the Mediterranean, have a particular significance. **Woodside *et al.*** describe gas hydrates and mud volcanoes associated with the Anaximander Mountains offshore southwest Turkey, in the vicinity of recently discovered giant mass movements such as the Great Slide in the eastern Mediterranean. Deep fluid fluxes and hydrate accumulations associated with authigenic carbonates and bacterial mats are reported by **Ivanov *et al.*** and **Bouriak & Akhmetjanov** on the Crimean continental margin in the Black Sea.

The degassing potential of the eastern Mediterranean and the Black Sea brought human development into focus at the Gent workshop. The eastern Mediterranean, being the cradle of our literate civilisation, has generated a 5000-year long human record, which, on at least one occasion, underwent total collapse. Without directly suggesting a causal link, we should bear in mind that gas release, especially when toxic gases like H_2S are implied, may constitute a significant hazard.

If, however, we are looking for recent significant hydrate destabilization and seabed degassing related to temperature rise, we should probably focus on Europe's northernmost facade, the Arctic Seas. Deglaciation caused the sudden flooding of vast areas of Arctic shelves by water that was 30°C warmer than the formerly exposed substratum. If there is any place in the world where hydrate breakdown should have manifested itself, it should be in the high-latitude shelf seas.

Though **Long *et al.*** do not directly address the former issue, which was raised in the debate at the workshop, their observations in the Barents Sea still suggest that the present-day decay of hydrates near large sea-floor craters supplies widespread methane plumes which fluctuate in response to seasonal warming of the bottom water. Later in this book **Mienert *et al.*** interpret similar observations differently.

Authors like **Veerayya *et al.***, **Neben *et al.*** and **Delisle *et al.*** remind us of the significant hydrate potential of Indian and southeast Asian continental margins. The last paper models the recovery of a BSR after a thermal disturbance at the sea floor due to slumping, for instance. Such modelling seems to show that the repositioning of a BSR is a very slow process, taking place over thousands of years.

There was a consensus in the conclusions of the workshop debate that focusing on a few areas with a multidisciplinary approach is an essential objective. The selection of such areas however requires more preparation.

Relevance to margin stability and climatic change: research horizons

Delisle *et al.*'s exercise in the recovery of a thermal equilibrium in hydrate stability zones following a major disturbance of the seabed indirectly links with what certainly turned out to be the most stimulating and 'debatable' issue of the Gent workshop, i.e. the possible coupling of oceanic gas hydrates, margin instabilities and climate. This brought scientists from different backgrounds into direct discussion. The direct confrontation of ideas between atmospheric modellers and ice core scientists on one side and hydrate geochemists and geophysicists on the other side turned into a captivating experience, which left both parties with a taste for more.

Participants in this debate were Charles Paull, Dominique Raynaud and Robert Thorpe (panel members), Dendy Sloan (moderator), Hans Amann, Lars Berge, Angelo Camerlenghi, Bill Dillon, Bilal Haq, Myriam Kastner, Peter Kennett, Tom McGee, Walther Van Kesteren and Warren Wood.

It is widely recognized that the present-day, steady-state methane release from the ocean, including from oceanic hydrates, is not impressive and that much larger releases, which could affect climate, call for dramatic events such as giant slides which, by one mechanism or another may find their origin in the decay of hydrate horizons on the continental slope. It is already known, through extensive acoustic mapping, that vast stretches of the oceanic margins, in particular the North Atlantic, show evidence of major large-scale slides and slumps.

The basic mechanisms through which decaying hydrates may affect the stability of slopes are still poorly understood. Overpressure in the pores below a decaying hydrate base may account for a decrease in shear strength, but no experimental evidence for any overpressure under BSR horizons has been put forward to date. Any build-up of overpressure will depend upon the balance between hydrate decay and pressure dissipation through possible permeability barriers. Furthermore, and by analogy with the rationale behind the analysis of the generation of hydrates, we should not regard a slope sediment as a simple, passive matrix, in particular not along formerly glaciated margins. The nature of the clays significantly controls the stability of the sediments, and for some clays in particular the stability of the particle matrix is directly controlled by the sodium content of the pore water. Any freshening of the pore water, e.g. by hydrate decay, may decrease the sodium concentration and hence trigger slope instabilities through a possible 'quick clay' behaviour. The most dramatic examples of quick clay surges which have been documented occur in glacial sediments of the Scandinavian and Canadian coastal plains and valleys which were deposited in submarine environments and subsequently uplifted and then leached by freshwater infiltration.

Mienert *et al.* demonstrate that deep-water gas hydrate horizons are coincident with areas and depths of slope failures in continental margin sediments along the north-eastern Atlantic margin, in particular in the Storegga slide area. While across the Atlantic, in the Blake Ridge hydrate province, faulted and collapsed depressions clearly seem to root at the base of the hydrate stability zone and are interpreted by **Dillon *et al.*** in terms of overpressure of gassy sediments just below the zone of hydrate stability.

Another process which needs clarification is the fate of steady fluxes of methane from hydrate reservoirs to the seabed and from the seabed to the sea surface and atmosphere. We need to understand the process of transport of gas across geochemically active horizons like the sulphate reduction zone. And beyond the seabed, what is the fate of methane in the water column, as a function of discharge rate, bubble size and dynamics? Do we fully master the chemical kinetics, and have we a feeling of how much is subject to oxidation and take-up by the biosphere on the way to the surface? The absence of experts in this field was seen to represent a knowledge gap at the workshop.

If any significant slide-induced flux of methane from the ocean to the atmosphere is substantiated, the next step is to attempt to calculate and model the radiative forcing of potentially large and sudden releases of methane on climate change.

It is clear that we need to understand the relative sensitivity of the whole environment in which clathrates occur and decay, in which methane migrates, and the relative sensitivity through time. We have evidence of rapid major climate changes in the Quaternary and there is no answer so far as to what is forcing these rapid changes. We have evidence of major slide events in the Quaternary, and suspect a significant coupling, but have no real clues yet. But as the past is the key to the future, we have to scrutinize long-term records of slope instabilities and fast climate changes to gather such data.

Haq reviews the possible coupling mechanisms between seabed degassing and climatic changes and screens possible long-term climatic and slope-stability records. Stable isotope records suggest the Late Paleocene – Early Eocene warming peak may be a past analogy, which could offer further clues about the behaviour of gas hydrates and their contribution to global warming. He also digs into the more remote past, towards 'pre-psychrospheric' times.

A study of the stable isotope record of the Santa Barbara Basin off California (an oral communication by Peter Kennett *et al.* at the Gent workshop) based on recently drilled ODP holes has revealed rapid warming events which can be explained by seabed degassing, and that seem to be synchronous with warmings associated with Dansgaard-Oeschger events recorded in Greenland ice cores. One reason why clathrates have so far not emerged as a dynamic component of the climate machine is that for a long time there has been a sense that the deep seas beyond the continental shelves were relatively stable in terms of hydrographic conditions. One of the results coming out of ODP and in particular from the Santa Barbara data is that this is not true. At least in the Northwest Pacific, physical oceanographic conditions have been very unstable in Quaternary times, with bottom-water temperatures varying up to 3°C, which is very significant in terms of hydrate stability. Hydrates can thus be produced and released in a pumping type mechanism, that could affect global climate change. However the global relevance of results from one basin must be treated with caution until they can be substantiated by further data.

As emphasized by the climatologists and ice-core scientists at the workshop, what turns out to be a priority now, certainly if we are aiming to model a possible coupling between seabed and atmosphere, is the need to refine the sampling resolution and hence the time resolution in all these records and thus allow reliable correlations and identification of lead and lag mechanisms.

Thorpe *et al.* model a 'Catastrophic Hydrate Release' (CHR) of methane at the termination of the last glaciation as indicated by ice core results to test its potential impact on glacial climate. They come to the conclusion that methane alone could not trigger deglaciation. However, a combination of methane release, increase in carbon dioxide and changes in heat transport by the ocean – coupled with high climate sensitivity – could simulate changes of the observed magnitude. **Raynaud *et al.*** also confirm that present evidence from ice cores does not show evidence for CHR, but the resolution of the ice record is not refined enough to unequivocally resolve leads and lags between methane and carbon dioxide surges and temperature rises. According to Thorpe *et al.*, if the GRIP core sampling interval could be reduced to allow a resolution of around 50 years, then it would be possible to more fully test the CHR hypothesis.

Still a major issue for the climatic relevance of the world hydrate reservoir is the validation of its size, at a global scale. We will never achieve accurate estimates, but we need meaningful estimates. Can the Blake Ridge results be regarded as representative of other rich hydrate provinces, or is it an exception? This question leads us back to the exploration and ground-truthing debate, which is most significant for the estimation of the relevance of oceanic hydrates for the Earth's climate.

Beyond these scientific arguments, the climate debate has profound human dimensions of culture, communication and strategic priorities. As suggested by a climate modeller at the workshop, we might wonder why the climate modelling community has traditionally paid so little attention to the significance of gas hydrates, though there is increasing evidence of their huge potential. One reason is that the climate community is presently deeply concerned with trying to understand what is going on now and what is going to happen in the near future, and what we might be about to do to the climate system other than what we are doing at this time.

It is a matter of economic pressure, of population growth, of how much energy resources are going to be used, of how much CO_2 will be emitted. In this process, uncertainties are large – about emissions, about climate sensitivity, about the climate response to forcing by greenhouse gases – and it is generally felt that these uncertainties coupled together are very much greater than the potential uncertainties associated with the possible release of clathrate gases in the future. That is, to a large extent, why clathrate research has traditionally had a low profile.

Which strategy for international cooperation?

The final Gent debate focused on defining strategies for international and multidisciplinary research efforts. Participants were Jean Boissonnas (EC MAST), Bilal Haq and R. Heinrichs (NSF), Laurent d'Ozouville (EMaPS) (panel members), Jean-Pierre Henriet (moderator), Ray Cranston, Jean-Paul Foucher, Myriam Kastner, Jim Kennett, Tom McGee, Dominique Raynaud, John Roberts, Carolyn Ruppel and M. Veerayya.

The challenge of unveiling the possible coupling of the seabed and climate through gas hydrates is a topic of global cooperation across discipline boundaries. Over the next decade, the Ocean Drilling Program will include a significant component looking at gas hydrates, with one or more focused legs, but it is resource-limited. Other organizations are needed, other science structures that will enable the scientific community to get at the fully multidisciplinary and interdisciplinary nature of clathrates. The possible input of a Japanese drilling ship, currently at the planning stage, might broaden the scientific perspectives for hydrate research.

For the European Union, the Gent workshop coincided with the drafting of the 5th Framework Programme (5th FP) which will significantly differ in its structure and objectives from the previous programmes. Emphasis will be laid on industry-driven research, research that more closely answers the needs of society, and research on the environment and the resources of the living world. Clathrate research could fit into such schemes, provided scientists can convey the proper messages to their national authorities as the 5th FP is finalized.

But whatever the prospects may be, European Union support would necessarily focus on European margins. These include the Mediterranean Sea, and possibly the Black Sea. However, such priorities do not preclude international agreements, and talks have been opened with the US Administration to draft an official agreement which includes matters relating to marine science and technology. At present teams from the US, Canada and Australia can join EU teams on a project provided they bring their own funding. Cooperation schemes with Russian partners also exist and will be developed. Regarding cooperation with a country with advanced clathrate research like Japan, far-stretching bilateral agreements are already in place with the US and Canada respectively which focus on Pacific basins research and matters of technology. However cooperation with other partners, in particular along the Indian Ocean margins and in southwest Asian regions, should not be overlooked.

Still on the European scene, the European Board for Marine and Polar Science (EMaPS) has promoted communication (networking) among scientists and the case for a European strategy for ocean drilling, involving a new synergy with industrial partners. Two workshops have already been organized, one on scientific strategies in Europe for ocean drilling, including ODP and programmes like CORSAIRES or IMAGES, and another to identify new technologies for scientific drilling.

In terms of programmes, attention should also be paid to the US MARGIN initiative, on its way to becoming internationalized. The British and French continental margins communities have recently started their own programme. Such programmes include fluid flow as a broad general topic and encompass clathrate research.

Throughout these programmes, technology development for exploration, for ground-truthing and for monitoring needs to be a focal point. Communication of data is also important, as is sharing of equipment to collect new data. It is the conveners' hope that the Gent workshop and its debates will contribute to these objectives.

The Gent workshop 'Gas Hydrates: Relevance to World Margin Stability and Climatic Change' was supported by the EC MAST 3 Concerted Action CORSAIRES. Behind the convenors' names, a large team contributed to the success of the workshop and the editing of this volume. Special credit goes to Tine Missiaen, Maarten Vanneste, Marc De Batist, Marc Faure Didelle and Dries Declercq. Alister Skinner and Dan Evans from the British Geological Survey are gratefully acknowledged for a careful reading of this manuscript.

Professor Robert Kidd

Professor Robert Kidd was keen to be among the participants of the workshop 'Gas Hydrates: Relevance to World Margin Stability and Climatic Change'. Relevant work he intended to propose dealt with sediment instability on the margin of the Canaries, as part of the EC MAST2 'STEAM' project, and with methane driven mud volcano extrusion on the Mediterranean Ridge, as part of TREDMAR. He also planned a more general talk on JOIDES' interest in gas hydrates, based on ODP's Long Range Plan: a topic he already had brilliantly presented in front of the EMaPS Marine Board in Rotterdam, on 9 May 1996. Tragically, Rob passed away a few weeks later, in early June, aged 48. His manifold contributions to marine sciences will be sadly missed.

A primer on the geological occurrence of gas hydrate

K. A. KVENVOLDEN

US Geological Survey, MS999 Menlo Park, California 94025, USA

Abstract: Natural gas hydrates occur world-wide in polar regions, usually associated with onshore and offshore permafrost, and in sediment of outer continental and insular margins. The total amount of methane in gas hydrates probably exceeds 10^{19} g of methane carbon. Three aspects of gas hydrates are important: their fossil fuel resource potential; their role as a submarine geohazard; and their effects on global climate change. Because gas hydrates represent a large amount of methane within 2000 m of the Earth's surface, they are considered to be an unconventional, unproven source of fossil fuel. Because gas hydrates are metastable, changes of pressure and temperature affect their stability. Destabilized gas hydrates beneath the sea floor lead to geological hazards such as submarine slumps and slides, examples of which are found world-wide. Destabilized gas hydrates may also affect climate through the release of methane, a 'greenhouse' gas, which may enhance global warming and be a factor in global climate change.

Increasing interest in gas hydrates has spawned a wealth of ideas regarding their significance in the natural world. For example, in *Global Warming, The Greenpeace Report*, Leggett (1990) points out that gas hydrates may produce a potentially enormous 'greenhouse' feedback effect on future global climate. This positive feedback will result because of the large amount of methane (a 'greenhouse' gas) that could be released from destabilized gas hydrates. MacDonald (1990*a*), in his comprehensive review "The future of methane as an energy resource", emphasizes the importance of gas hydrates as a potential natural fuel resource of methane as well as of hydrogen which can be abstracted from the hydrate methane. Just what are these substances that have captured the attention of environmental groups and appear to be so important to aspects of future global economics?

Definition

Gas hydrates, also called gas clathrates, are naturally occurring solids composed of water molecules forming a rigid lattice of cages with most of the cages each containing a molecule of natural gas, mainly methane. Gas hydrates are essentially water clathrates of natural gas in which water crystallizes in the isometric crystallographic system rather than the hexagonal system of normal ice. Two structures, I and II, of the isometric (cubic) lattice are recognized in nature with structure I (Fig. 1) being most common. In structure I, the cages are arranged in body-centred packing and are large enough to include methane, ethane and other gas molecules of similar molecular diameters, such as carbon dioxide and hydrogen sulphide. In structure II, diamond packing is present resulting in some cages being large enough to include not only methane and ethane but also gas molecules as large as propane and isobutane (Sloan 1990).

The maximum amount of methane that can occur in a methane hydrate is fixed by the clathrate geometry. In a fully saturated structure I methane hydrate, 1 molecule of methane is present for every 5.75 molecules of water; in theory then, when appropriate hydrate expansion factors are considered, 1 m^3 of methane hydrate can contain up to 164 m^3 of methane gas at standard conditions (Fig. 2). Thus, gas hydrates in shallow reservoirs less than about 1.5 km beneath the surface can have more methane per unit volume than can be contained as free gas in the same space (Hunt 1979). This fact explains some of the serious interest in gas hydrates as a potential: (1) future energy resource; (2) submarine geohazard; and (3) factor in global climate change. Each of these potential aspects of gas hydrates will be addressed in this paper.

Controls

The occurrence of gas hydrates in nature is controlled by an interrelation between the factors of temperature, pressure and composition, as illustrated by means of a phase diagram (Fig. 3). In the pressure–temperature domain of methane hydrates, the position of the phase boundary is determined not only by the composition of the gas mixture, i.e. the presence of gases other than methane, but also by the ionic impurities in the water. Because the exact composition of gas and water in sediment pore spaces is usually not known, a pure methane–pure water system is commonly assumed to predict the depth and

KVENVOLDEN, K. A. 1998. A primer on the geological occurrence of gas hydrate. *In:* HENRIET, J.-P. & MIENERT, J. (eds) *Gas Hydrates: Relevance to World Margin Stability and Climate Change*. Geological Society, London, Special Publications, **137**, 9–30.

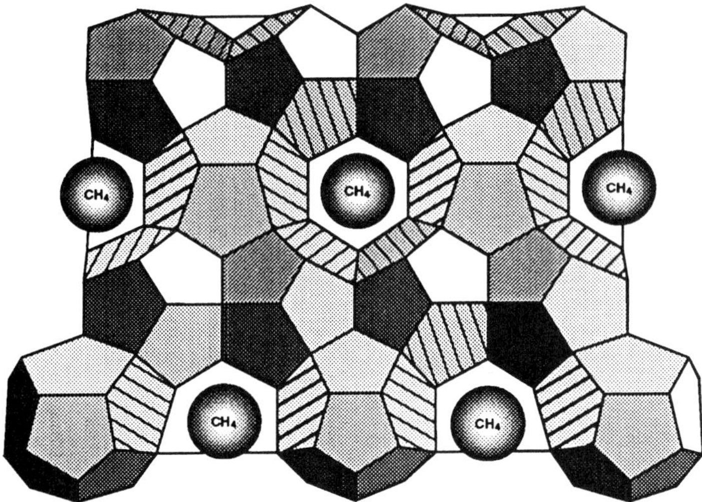

Fig. 1. Gas hydrate structure. In this Structure I methane hydrate, the rigid cages are composed of hydrogen-bonded water molecules, and each cage, both exposed and covered in this figure, contains a methane molecule. Modified from Hitchon (1974).

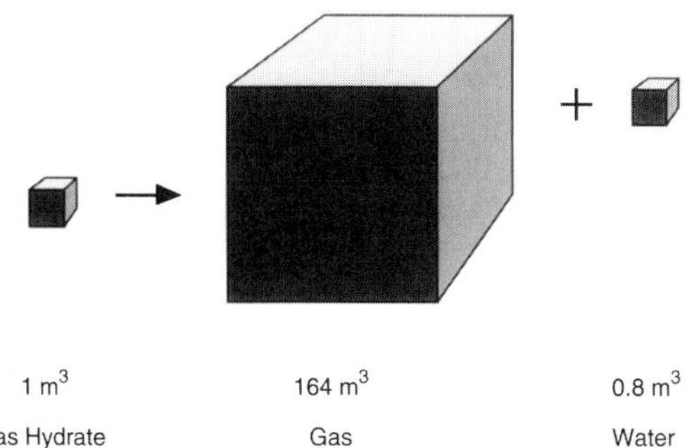

Fig. 2. One m^3 of gas hydrate yields 164 m^3 of gas and 0.8 m^3 of water at standard temperature and pressure.

temperature regime where naturally occurring gas hydrates are stable (Claypool & Kaplan 1974).

The phase-boundary information (Fig. 3) suggests that the upper depth limit for methane hydrates is about 150 m in continental polar regions, where surface temperatures are below 0°C. In oceanic sediment, gas hydrates occur where the bottom-water temperatures approach 0°C, and water depths exceed about 300 m. The lower limit of methane hydrate occurrence is determined by the geothermal gradient; the maximum lower limit is about 2000 m below the solid surface, although the lower limit is typically much less depending on local conditions. Thus, the occurrence of gas hydrates is restricted to the shallow geosphere.

One feature of gas hydrate occurrence that is not apparent from the phase diagram (Fig. 3) is the amount of gas (methane) that is necessary for gas hydrate formation. Gas hydrates will form with about 90% of the cages of the clathrate filled, that is with about 150 volumes of methane at standard conditions per volume of water (Sloan 1990). But methane solubility in sea water is very low, about 0.045 volumes of methane at standard conditions per volume of water (Yamamoto *et al.* 1976). Thus, the amount of methane required for gas hydrates greatly exceeds the solubility of methane in water. This requirement for a source of enormous amounts of methane for gas hydrate formation limits the regions on Earth where gas hydrates can be expected and are found.

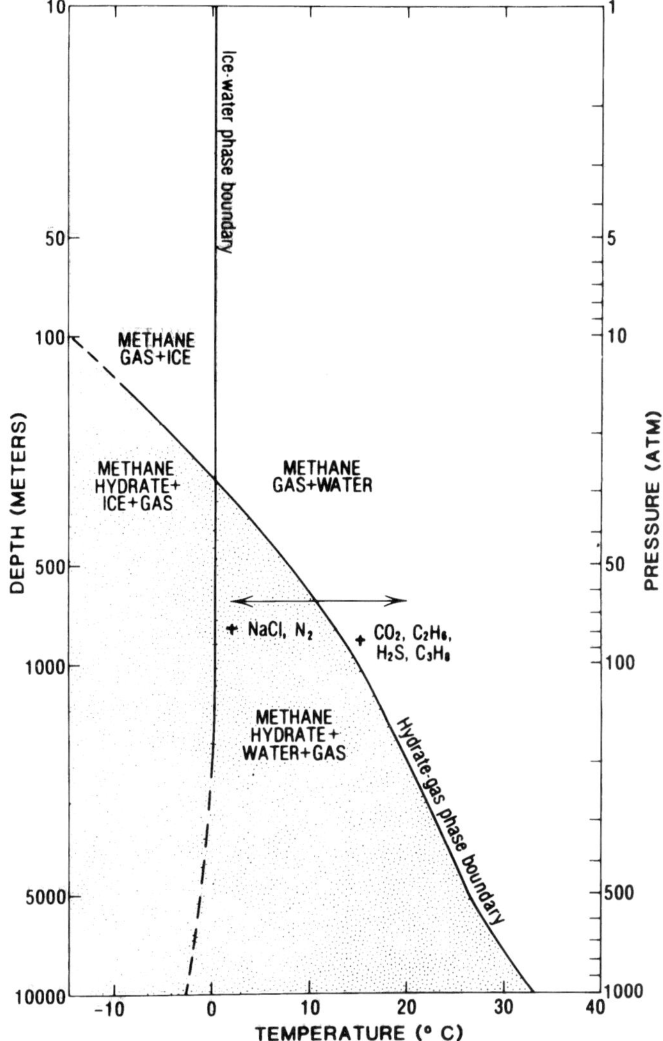

Fig. 3. Phase diagram showing the boundary between free methane gas (no pattern) and methane hydrate (pattern) for a pure water and pure methane system. Addition of NaCl to water shifts the curve to the left. Adding CO_2, H_2S, C_2H_6, C_3H_8 to methane shifts the boundary to the right and thus increases the area of the hydrate stability field. Depth scale assumes lithostatic and hydrostatic pressure gradients of 10.1 kPa m^{-1}. Redrawn after Katz *et al.* (1959).

Locations

Gas hydrates occur world-wide, but, because of the pressure–temperature and gas-volume requirements, they are restricted to two regions, polar and deep oceanic. In polar regions, gas hydrates are usually associated with permafrost both onshore in continental sediment and offshore in sediment of the continental shelves. In deep oceanic regions, gas hydrates are found in outer continental margins in sediment of slopes and rises where cold bottom water is present. The world-wide occurrence of known and inferred gas hydrates is shown in Fig. 4. Samples of gas hydrates have been recovered on land in the West Prudhoe Bay oil field in Alaska (reviewed by Kvenvolden & McMenamin 1980) and at 14 oceanic locations, providing irrefutable evidence of natural gas hydrate occurrence (Kvenvolden *et al.* 1993). Deep ocean drilling

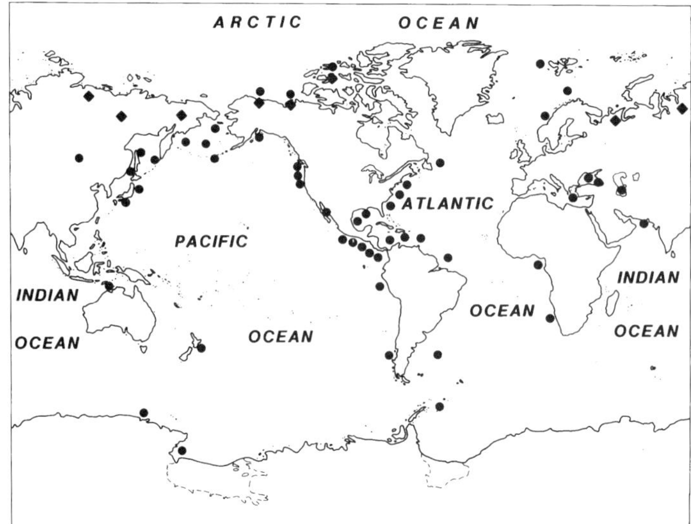

Fig. 4. Map showing world-wide locations of known and inferred gas hydrates in oceanic (aquatic) sediments (●), and in continental (permafrost) regions (□) . Modified from Kvenvolden (1988*a*).

has recovered gas hydrates at nine locations (offshore Peru, Costa Rica, Guatemala, Mexico, south-eastern United States, western United States, two locations offshore from Japan and in the Gulf of Mexico). Shallow sediment coring (piston and gravity cores) has also recovered gas hydrates at six locations (Black Sea, Caspian Sea, offshore from northern California, northern Gulf of Mexico and two locations in the Okhotsk Sea). Examples of gas hydrate samples recovered from offshore from Guatemala are shown in Fig. 5.

Geophysics

Most other oceanic occurrences of gas hydrates are inferred, based mainly on the appearance on marine seismic reflection profiles of an anomalous bottom simulating reflection (BSR). This reflection coincides with the depth predicted from the phase diagram (Fig. 3) as the base of the gas hydrate stability zone (Fig. 6). BSRs mark the interface between higher sonic velocity, hydrate-cemented, sediment above and lower sonic velocity, uncemented, sediment below. The seismic reflection from the base of gas hydrate zone is generally characterized by reflection polarity reversals (reflections opposite to those from the sea floor) and large vertical reflection coefficients, for example -0.12 ± 0.04 for BSRs of the Blake Outer Ridge offshore from the south-eastern United States (Shipley *et al.* 1979). Such negative reflection coefficients are indicative of a reflective interface between higher velocity strata overlying lower velocity strata.

Geothermal gradients

The depths to gas hydrate BSRs and bottom-water temperatures have been used in conjunction with gas hydrate phase-boundary information (Fig. 3) to estimate geothermal gradients and heat flow in oceanic sediment (Shipley *et al.* 1979; Yamano *et al.* 1982). This method has been applied to seismic data from the Blake Outer Ridge, around Central America, and in the Nankai Trough offshore from Japan. The estimated geothermal gradients and heat-flow values are consistent with results obtained by conventional means using downhole temperature and surface heat-flow measurements. An example of the use of the method comes from offshore Peru where geothermal gradients of 43 and 49.5°C km^{-1} were estimated in two areas from BSRs (Kvenvolden & Kastner 1990), and these results are similar to those obtained by direct measurements (Yamano & Uyeda 1990).

Quantitative assessments

Multi-channel seismic reflection data have been used to analyse several aspects of the zone of

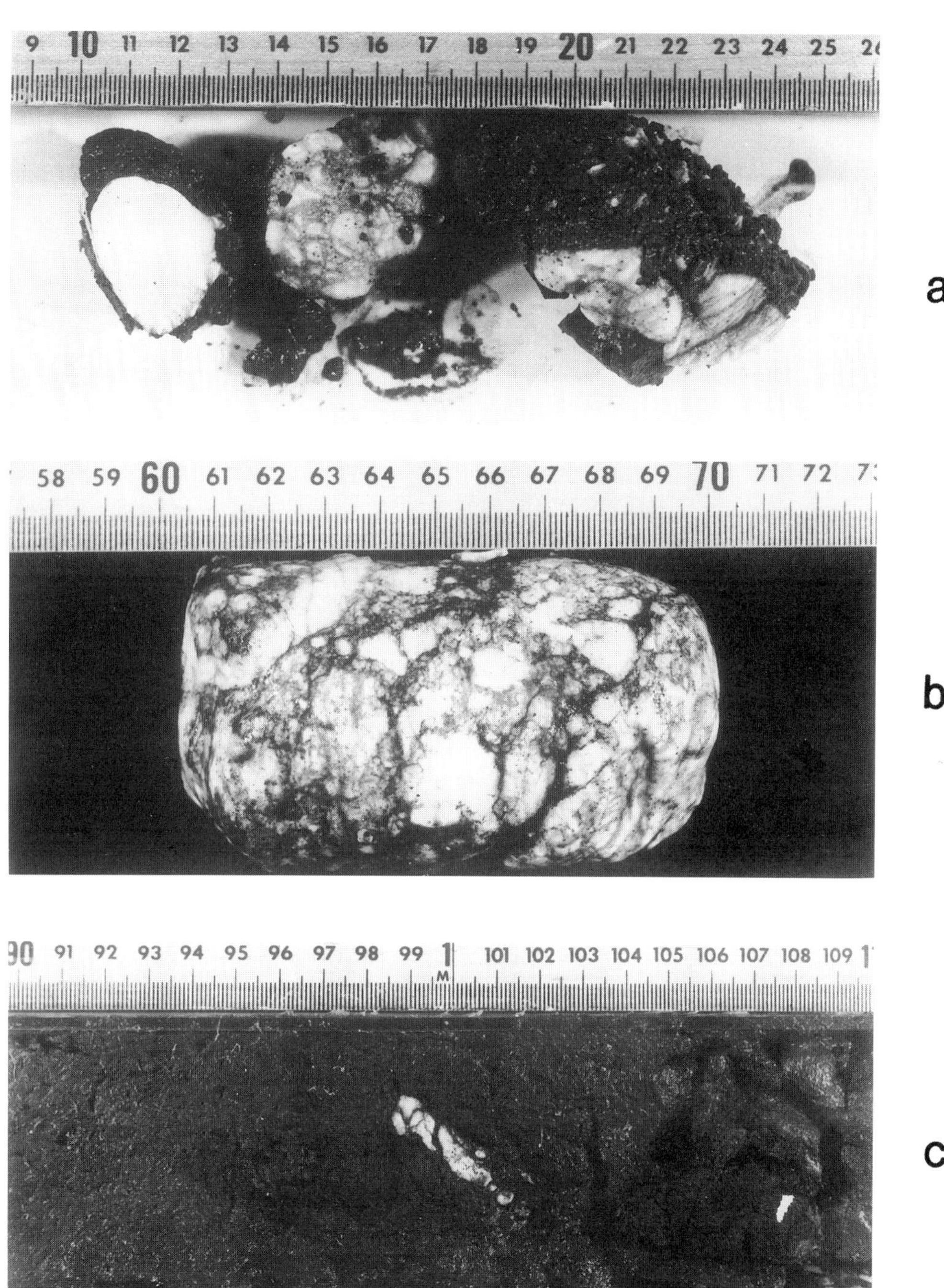

Fig. 5. Photographs of gas hydrate samples from DSDP Leg 84 offshore from Guatemala: (**a**) large pieces of gas hydrate in fractured mudstone, Site 568, 404 m sub-bottom; (**b**) gas hydrate in mudstone fracture, Site 570, 246 mbsf; (**c**) massive gas hydrate, Site 570, 249 mbsf. Scale is in cm. Modified from Kvenvolden & McDonald (1985).

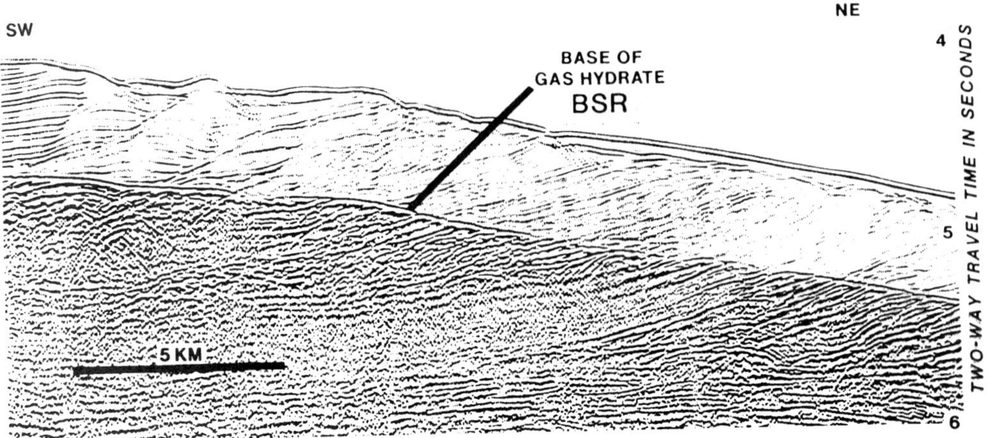

Fig. 6. A 12-fold multi-channel reflection profile from the crest and eastern flank of the Blake Outer Ridge. The strong bottom simulating reflection (BSR) is inferred to represent the base of the gas hydrate stability zone. Blanking of cross-cutting strata above the BSR is probably due to the presence of gas hydrates. The well-defined, continuous, BSR suggests that the gas hydrate here forms a seal perhaps trapping free gas beneath the BSR. Modified from Shipley *et al.* (1979).

gas hydrate stability. For example, quantitative analyses of BSRs have been undertaken with reprocessed seismic records and synthetic seismograms to estimate the amount of gas hydrate at the BSR and the thickness of the underlying free gas zone for a site offshore from Peru (Miller *et al.* 1991). The results indicate that the BSR is laterally discontinuous. A free gas zone 5.5–17 m thick beneath a zone where gas hydrate fills 10% of the porosity is estimated from BSRs of high amplitude. Where the BSR amplitude is low, the free gas zone is much thinner than 5.5 m or is absent. In a seismic study of gas hydrate BSRs, Hyndman & Spence (1992) studied BSRs from offshore Vancouver Island of Canada using amplitude vs offset (AVO) and high-resolution velocity analyses, as well as modelling of vertical incidence data. The results suggest that above the BSR is a 10–30 m thick high-velocity layer with about 30% of its pore space filled with gas hydrate. Beneath the BSR there is no seismically detectable free gas.

Nature of BSRs

The presence or absence of free gas beneath BSRs is a controversial and, as yet, unresolved issue. Two models have been proposed to account for gas hydrate formation and the development of BSRs. In the first, methane is assumed to be generated microbially from organic matter within the zone of gas hydrate stability (Claypool & Kaplan 1974). Gas hydrate formation takes place concurrent with sedimentation. As the zone of methane hydrate thickens and deepens its base eventually subsides into a temperature region where the gas hydrate is unstable. In this region free gas can occur, but this gas can migrate back into the overlying zone of gas hydrate stability if suitable migration pathways are available (Kvenvolden & Barnard 1983*a*). A consequence of this model is that gas hydrates should occur throughout the zone of gas hydrate stability and free gas may or may not be present beneath the BSR.

In the second model, gas hydrates are formed through the removal of methane from rising pore fluids as they pass from below into the gas hydrate stability zone (Hyndman & Davis 1992). In this model, most of the methane is generated microbially at depths below the stability zone but not at depths sufficient for the formation of thermogenic methane. A consequence of this model is that gas hydrates should be concentrated at the base of the stability zone, i.e. at the BSR, and free gas is not to be expected beneath the BSR.

Tests of these models have been made on Ocean Drilling Program (ODP) Legs 141 (Bangs *et al.* 1993), 146 (MacKay *et al.* 1994) and 164 (ODP Leg 164 Shipboard Scientific Party 1996) where seismic reflectors indicated by BSRs were purposely cored. Previously, the coring of these kinds of reflectors during deep ocean drilling had been avoided because of presumed safety hazards (Ocean Drilling Program 1986). The results of this drilling indicate that free gas is likely to be present beneath the BSR.

Wireline logs

In addition to seismic data, geophysical information from wireline well logs, used in combination with each other, can be valuable in the detection and evaluation of gas hydrate intervals (Kvenvolden & Grantz 1990). Well logs for gas hydrate studies include caliper, gamma ray, spontaneous potential, resistivity, sonic velocity and neutron porosity (Goodman 1980). Figure 7 illustrates the wireline log responses, including a mud-gas log, in one gas hydrate-bearing interval drilled at the NW Eileen State No. 2 well on the North Slope of Alaska. The resistivity and sonic velocity responses have proved most useful in a comprehensive study of well-log data from 445 wells on the North Slope where six gas hydrate-bearing intervals have been identified and mapped in 50 of these wells (Collett *et al.* 1988). A comparison between logs from the NW Eileen State well and logs from Deep Sea Drilling Project (DSDP) Hole 570 located offshore from Guatemala showed similar characteristics and clearly defined the gas hydrate zones (Mathews 1986). Well logs provide a basis for estimation of gas quantity, and, in conjunction with seismic data, hold the key to future worldwide gas hydrate assessment.

Geochemistry

Sources of methane

In both models for gas hydrate formation discussed previously, the methane is considered to be mainly microbial in origin. This conclusion

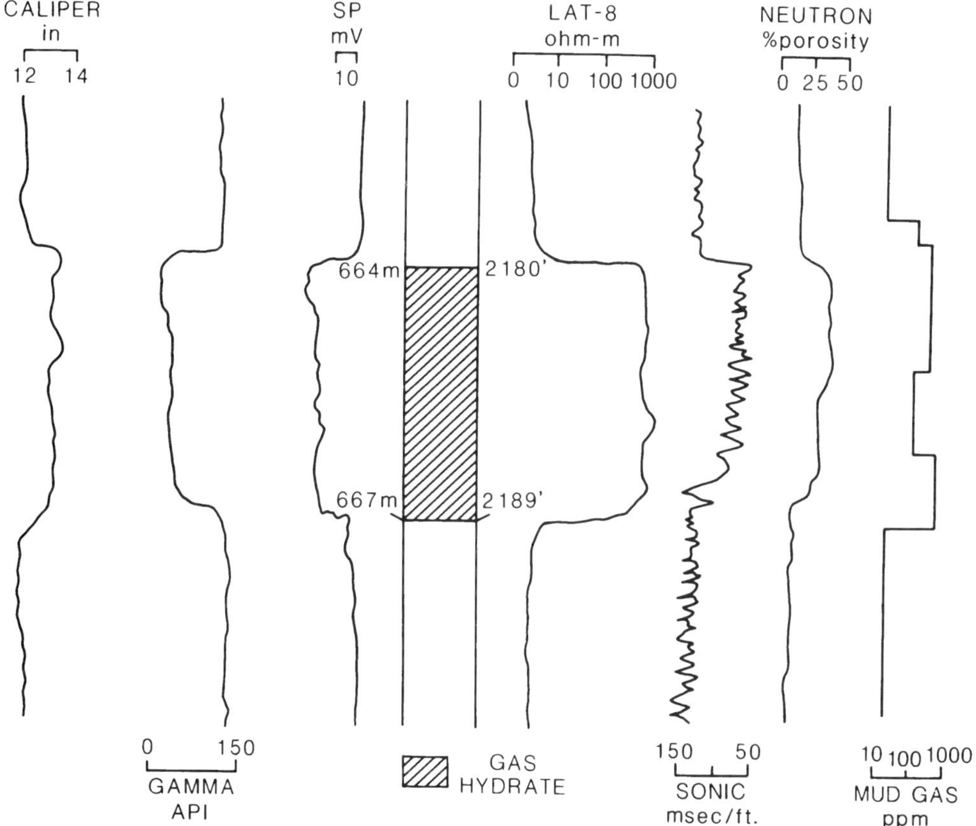

Fig. 7. Comparison of wireline logs (caliper, gamma-ray, spontaneous potential (SP), resistivity (LAT-8), sonic velocity and neutron porosity) and total mud-gas log for a gas hydrate-containing interval (664–677 m) in the NW Eileen State No. 2 Well on the North Slope of Alaska. From Kvenvolden & Grantz (1990).

Table 1. *Carbon isotopic compositions and concentrations of methane in natural gas hydrates and hydrate-containing sediment*

Regions	Sample type	CH_4 (%)	$\delta^{13}C$ (‰)	Reference
Offshore SE U.S.A.				
Blake Outer Ridge				
DSDP Leg 11	Sediment	>99	−88 to −70	Claypool *et al.* (1973)
DSDP Leg 76	Sediment	>99	−93.8 to −65.4	Kvenvolden & Barnard (1983*b*)
				Galimov & Kvenvolden (1983)
	Gas hydrate	>99	−68.0	Brooks *et al.* (1983)
Offshore Peru				
Peru–Chile Trench				
ODP Leg 112	Sediment	>99	−79 to −55	Kvenvolden & Kastner (1990)
ODP Leg 112	Gas hydrate	>99	−65.0, −59.6	Kvenvolden & Kastner (1990)
Offshore northern California				
Eel River Basin	Gas hydrate	>99	−69.1 to −57.6	Brooks *et al.* (1991)
Black Sea	Gas hydrate	>99	−63.3, −61.8	Ginsburg *et al.* (1990)
Gulf of Mexico				
DSDP Leg 96	Sediment	>99	−73.7 to −70.1	Pflaum *et al.* (1986)
DSDP Leg 96	Gas hydrate	>99	−71.3	Pflaum *et al.* (1986)
Garden Banks	Gas hydrate	>99	−70.4	Brooks *et al.* (1986)
Green Canyon	Gas hydrate	>99	−69.2, −66.5	Brooks *et al.* (1986)
Green Canyon	Gas hydrate	62,74,78	−44.6, −56.5, −43.2	Brooks *et al.* (1986)
Mississippi Canyon	Gas hydrate	97	−48.2	Brooks *et al.* (1986)
Caspian Sea	Gas hydrate	59–96	−44.8 to −55.7	Ginsburg *et al.* (1992)
Offshore Guatemala, Middle America Trench				
DSDP Leg 84	Sediment	>99	−71.4 to −39.5	Kvenvolden & McDonald (1985)
				Jeffrey *et al.* (1985)
DSDP Leg 84	Gas hydrate	>99	−43.6 to −36.1	Kvenvolden *et al.* (1984)
DSDP Leg 84	Gas hydrate	>99	−46.2 to −40.7	Brooks *et al.* (1985)

is based on geochemical investigations of recovered gas hydrates and of hydrocarbon gases from sediment sections known to contain gas hydrates, such as occur offshore from the south-eastern United States, northern California and Peru, and in the Black Sea and at some sites in the Gulf of Mexico (Table 1). These investigations have shown that the molecular composition of the hydrocarbon gases and the isotopic composition of methane are consistent with results expected from microbial gas generation processes. Methane in all cases constitutes more than 99% of the hydrocarbon gas mixtures, and the isotopic composition of methane ($\delta^{13}C$) is usually lighter than −60‰ relative to the Peedee Belemnite (PDB) standard. The molecular and isotopic ranges are diagnostic of gas of microbial origin (Bernard *et al.* 1976).

In a sedimentary section where methane is being microbially generated, there is a carbon-isotopic consistency between methane, the product and carbon dioxide, the immediate precursor, as illustrated in Fig. 8. In this example from the Blake Outer Ridge (Galimov & Kvenvolden 1983) the carbon isotopic composition of methane increases (becomes heavier) with depth from about −94‰ in the uppermost sediment to about −66‰ in the deepest sediment, reflecting a systematic but non-linear depletion of ^{12}C with depth. The carbon isotopic composition of carbon dioxide also increases with depth of sediment from about −25‰ to about −4‰, showing a depletion of ^{12}C that closely parallels the trend of isotopic composition of methane. The magnitude and parallel distribution of carbon isotopic values for both methane and carbon dioxide are consistent with the concept that the formation of methane resulted from the microbial reduction of carbon dioxide derived from organic matter. The results strongly suggest that the methane in gas hydrates at this site is microbial in origin. Other localities with similar depth trends in carbon isotopic compositions of methane and/or carbon dioxide are offshore from Guatemala (Jeffrey *et al.* 1985) and offshore from Peru (Kvenvolden & Kastner 1990) where gas hydrates have also been recovered in DSDP and ODP drilling, respectively.

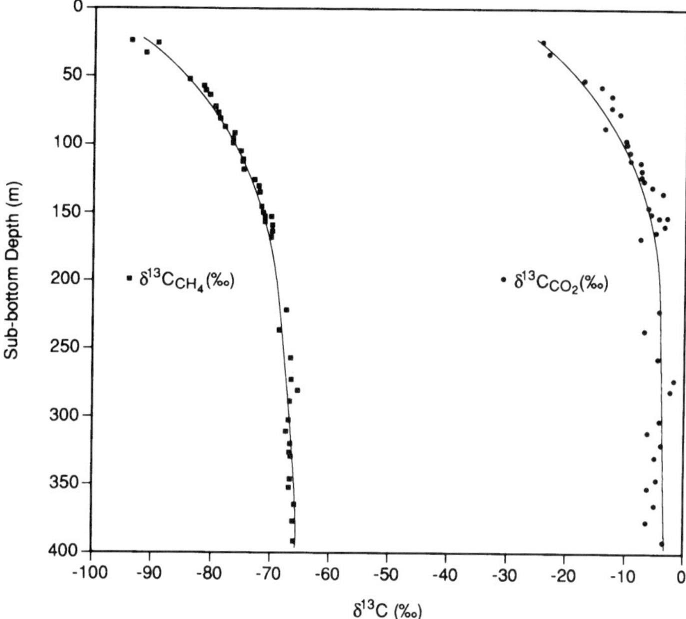

Fig. 8. Distribution with depth of carbon isotopic compositions of methane and carbon dioxide from Site 533, DSDP Leg 75, Blake Outer Ridge. Modified from Galimov & Kvenvolden (1983).

Microbial vs thermogenic methane

Methane in gas hydrates of the Gulf of Mexico (Pflaum *et al.* 1986; Brooks *et al.* 1986) is microbial at some sites and thermogenic at others (Table 1). Samples of hydrate hydrocarbon gas from Garden Banks and Green Canyon contain more than 99% methane with carbon isotopic compositions lighter than −60‰. In a different sample from Green Canyon and one from Mississippi Canyon the hydrate hydrocarbon gas is 62–78% methane, and this methane has carbon isotopic compositions ranging from −43.2 to 56.5‰. Gas hydrates associated with mud volcanoes in the Caspian Sea contain hydrocarbon gases with 59–96% methane having carbon isotopic compositions of −44.8 to −57.3‰. These samples fall within the molecular and isotopic compositional field diagnostic from thermogenic methane (Bernard *et al.* 1976).

Molecular and isotopic compositions of gases associated with gas hydrates found on DSDP Leg 84 offshore from Guatemala support either a microbial or thermal source for the methane (Kvenvolden *et al.* 1984). Hydrocarbon gases from cores of the deepest sediment samples and from dissociated gas hydrates all contain less than 99% methane (Kvenvolden & McDonald 1985). This observation suggests that the methane is microbial in origin. Although many methane samples had carbon isotopic compositions lighter than −60‰, suggesting a microbial source, at two sites, methane in sediments at depths below 210 m increased in carbon isotopic value to as heavy as about −40‰; methane from a massive gas hydrate sample had a carbon isotopic composition of −44‰ (Jeffrey *et al.* 1985). A detailed study of the dissociation of this massive gas hydrate under controlled laboratory conditions showed carbon isotopic compositions of methane ranging from −36.1 to −43.6‰ (Kvenvolden *et al.* 1984). These heavy carbon isotopic values are usually considered to reflect thermogenic gas. However, this isotopically heavy methane was accompanied by isotopically very heavy carbon dioxide, +16.3‰ for gaseous carbon dioxide (Jeffrey *et al.* 1985) and +37‰ for total dissolved carbon dioxide (Claypool *et al.* 1985).

The isotopically heavy methane could have been derived from the isotopically heavy carbon dioxide by microbial processes similar to those described previously for the Blake Outer Ridge. Alternatively, the isotopically heavy methane could have migrated up from deeper sediments where it was thermally derived from deeply buried organic matter. The association of isotopically heavy methane with isotopically heavy carbon dioxide could be coincidental or controlled by some unknown isotopic interaction between methane and carbon dioxide. The microbial explanation is preferred, but a thermal origin of the methane cannot be ruled out (Kvenvolden *et al.* 1984).

Implicit in the preceding discussion is that the ultimate source of hydrate methane is buried organic matter which can be altered microbially producing carbon dioxide as an intermediate product in the process of methane formation. Deeply buried organic matter, experiencing temperatures from about 80 to 150°C, can be altered thermally to produce methane directly (Claypool & Kvenvolden 1983). The amount and quality of organic matter present is critical to the methane generation process. Because gas hydrate formation requires amounts of methane that greatly exceed the water solubility of methane, the amount of organic matter needed for gas hydrate formation is very large. Observations on the present organic matter content of gas hydrate containing sediments offshore from Guatemala suggest that 2.0–3.5% organic carbon is sufficient only if *in situ* and not migrated methane is involved (Hesse & Harrison 1981); on the Blake Outer Ridge, the present organic matter content of sediment associated with gas hydrate occurrence averages about 1% organic carbon (Kvenvolden & Barnard 1983*b*). At the time of gas hydrate formation, the organic carbon content in the gas hydrate zone was undoubtedly higher, perhaps double the present amounts.

Inorganic geochemistry

During gas hydrate formation, water molecules crystallize into a cubic lattice structure, and, as in the formation of normal hexagonal ice, the hydrate crystals exclude salt ions from the crystal structure. Measurements of chlorinity of samples of gas hydrate water recovered offshore from Guatemala ranged from 0.51 to 3.2‰ (Kvenvolden & McDonald 1985) and from 1.8 to 8.2‰ for samples from offshore Peru (Kvenvolden & Kastner 1990). These samples, all of which were probably contaminated to varying degrees with sea water, had chlorinity values much less than average sea-water chlorinity of 19.8‰. During DSDP Leg 67 drilling offshore from Guatemala, Hesse & Harrison (1981) first noticed that the chlorinities of pore-water samples in the gas hydrate-bearing zones decreased significantly with depth from about 19 to 9‰. Concurrently the oxygen isotopic composition ($\delta^{18}O$ relative to Standard Mean Ocean Water (SMOW)) of pore water increased with depth from near zero close to the sediment surface to values as large as +2.6‰ at depth. Oxygen isotopic fractionation of pore water apparently takes place during gas hydrate formation, leading to ^{18}O enrichment with depth of pore fluids recovered after gas hydrate dissociation. This same trend of oxygen isotopic compositions was observed in pore waters of gas hydrate-bearing sediments recovered on ODP Leg 112 offshore from Peru (Kvenvolden & Kastner 1990).

The decreasing chlorinity trends with depth in gas hydrate-bearing sediments as noticed on DSDP Leg 67 have been observed in the Blake Outer Ridge (DSDP Leg 76) (Kvenvolden & Barnard 1983*b*; Jenden & Gieskes 1983), offshore from Guatemala (DSDP Leg 84) (Kvenvolden & McDonald 1985; Hesse *et al.* 1985) and offshore from Peru (ODP Leg 112) (Kvenvolden & Kastner 1990). These chlorinity trends are derived to a large extent from the dissociation of gas hydrates during core recovery. The chloride concentrations are measured on mixtures composed of *in situ* pore water and dissociated gas hydrate water. Thus, a progressive dilution artifact is produced in the pore water that can be squeezed from the sediment with increasing depths. Pore-water freshening is a useful signal for the presence of gas hydrates although it is not a unique signal. Freshwater migration in sediments near continents and clay mineral dewatering are also processes that can cause pore waters to have decreased chlorinities. However, where pore-water freshening is observed in sediments having large amounts of methane within the depth range of the gas hydrate stability zone, gas hydrates are likely to be present.

Estimates of methane in gas hydrates

Estimates of the amount of methane in gas hydrates are highly speculative and highly varied (Kvenvolden 1988*a*). Although it is generally known that gas hydrates occur worldwide (Fig. 4), knowledge of their occurrence is very incomplete, resulting in a wide range of estimates of the amount of methane that is present in them. The Potential Gas Committee (1981) summarized such estimates which show ranges of methane carbon values from a low of 7.6×10^{15} g (1.4×10^{13} m^3 of methane gas) in Arctic permafrost regions to a high of 4.1×10^{21} g (7.6×10^{18} m^3 of methane gas) in oceanic sediment.

Resource estimates before 1985

Estimates made before 1980 of the amount of methane in natural gas hydrates are highly speculative due to incomplete knowledge of gas hydrate occurrences and poor methods of estimating reserves in known occurrences. The following discussion is taken from a detailed analysis by the Potential Gas Committee (1981) of the potential supply of natural gas as of December 1980. Conventional methods of estimating gas resources are not applicable to gas hydrate deposits because: (1) on conventional well logs, gas hydrates have the same characteristics as permafrost; (2) flow tests require a means to disrupt equilibrium conditions in order to dissociate the gas hydrates to verify their presence; (3) reservoir permeability has thus far been found to be very low; and (4) the hydrate gas/free gas ratio cannot be fully determined from core samples or porosity measurements. Nevertheless regional and world estimates of gas hydrate resources have been calculated (Table 2). Trofimuk *et al.* (1977) attempted to incorporate many complex factors into their calculations, including regional and world coefficients of past and present temperatures, sediment thicknesses, per cent of organic material and per cent of methane production and retention.

World estimates for the amount of methane carbon in gas hydrate deposits range from 7.6×10^{15} to 1.8×10^{19} g for permafrost areas and from 1.7×10^{18} to 4.1×10^{21} g for oceanic sediments. The estimates in Table 2 show considerable variation, but oceanic sediments seem to be a much greater resource of hydrate methane than continental sediments because in oceanic sediments more extensive production of microbial methane occurs and the gas hydrate stability zone is more extensive. Estimates by Dobrynin *et al.* (1981) are considerably higher than the others. Their calculations are 'rough estimates' based on permafrost coverage and zones of gas hydrate stability in oceanic sediments without apparent regard for distributions of sedimentary basins or sources of methane. Their estimates provide an overly optimistic upper limit.

None of these estimates (Table 2) seem to account for reservoir quality. Proved and currently recoverable world resources of conventional methane carbon are 3.5×10^{16}–4.1×10^{16} g, and the estimated total remaining recoverable methane carbon is 1.2×10^{17}–1.4×10^{17} g (Parent 1980). If only 1% of gas hydrate methane is within porous and permeable strata, then the most conservative estimate for recoverable hydrate methane is still more than 1.7×10^{16} g of methane carbon, or half of the proved and currently recoverable conventional gas resources. Also the above estimates do not consider possible accumulations of free gas trapped beneath the gas hydrate layers.

Estimates of methane in gas hydrate deposits of the United States (Table 3) were extrapolated from the world estimates (Table 2). Based on Russian estimates for methane carbon in gas hydrates of Siberia, McIver (1981) calculated that permafrost areas in Alaska have $\sim 3.5 \times 10^{14}$ g of hydrate methane carbon. Gas hydrate resources in permafrost areas of the United States may contain from 1.7×10^{14} to 3.8×10^{17} g of methane carbon. Gas hydrate resources in continental shelf and slope areas of the United States represent ~7% of the total world resource of hydrated methane if it is assumed that 35% of oceanic gas hydrates occur in these areas (Potential Gas Committee 1981). Thus, estimates of methane carbon in gas hydrates in oceanic sediments along continental margins of the United States range from 4.1×10^{16} to 1.0×10^{20} g.

Resource estimates after 1985

As discussed above, estimates made during the 1970s and early 1980s of the amount of methane carbon in gas hydrates produced a wide range of values, with the upper limit of 4.1×10^{21} g being unrealistically large. An increased understanding

Table 2. *World estimates of the amount of methane in gas hydrates in continental and oceanic settings (from the Potential Gas Committee 1981)*

Methane carbon	Methane gas	Reference
Continental		
7.6×10^{15}	1.4×10^{13}	Meyer (1981)
1.7×10^{16}	3.1×10^{13}	McIver (1981)
3.0×10^{16}	5.7×10^{13}	Trofimuk *et al.* (1977)
1.8×10^{19}	3.4×10^{16}	Dobrynin *et al.* (1981)
Oceanic		
1.7×10^{18}	3.1×10^{15}	McIver (1981)
$(2.7–13.4) \times 10^{18}$	$(5–25) \times 10^{15}$	Trofimuk *et al.* (1977)
4.1×10^{21}	7.6×10^{18}	Dobrynin *et al.* (1981)

Table 3. *Estimates of the amount of methane in gas hydrates of the United States (from the Potential Gas Committee 1981)*

Methane carbon	Methane gas	Reference
Continental (Alaska)		
1.7×10^{14}	3.1×10^{11}	Meyer (1981)
3.5×10^{14}	6.5×10^{11}	McIver (1981)
6.4×10^{14}	1.2×10^{12}	Trofimuk *et al.* (1977)
3.8×10^{17}	7.1×10^{14}	Dobrynin *et al.* (1981)
Oceanic (shelf and slope)		
4.1×10^{16}	7.6×10^{13}	McIver (1981)
$(0.6–3.2) \times 10^{19}$	$(1.2–6.1) \times 10^{14}$	Trofimuk *et al.* (1977)
1.0×10^{20}	1.9×10^{17}	Dobrynin *et al.* (1981)

of gas hydrate occurrence since then has resulted in estimates that are in the lower range of previous values. For example, Claypool (Kvenvolden & Claypool 1988) estimated the amount of methane carbon in oceanic sediments by determining the area of the world ocean in which sediment with 1% or more organic carbon is accumulating (10×10^6 km^2). Assuming 0.5 km as the average thickness of the gas hydrate stability zone, an average porosity of 50% with gas hydrates occupying 10% of the available porosity and a gas hydrate yield of 160 standard volumes of gas per volume of hydrate, the amount of methane carbon present in oceanic gas hydrates is calculated to be $\sim 2.1 \times 10^{19}$ g.

This estimate of 2.1×10^{19} g could be refined downward by taking into consideration the fact that most of the sediment in the estimated area probably contained insufficient organic carbon to generate the volumes of methane required to occupy 10% of the available porosity with gas hydrates. The estimate could also be refined upward by taking into account that sediment accumulation in upwelling areas would have significantly more methane than is allowed for in the average values used. Such refinements are not justified, however, based on the present state of understanding.

Kvenvolden & Grantz (1990) attempted to estimate the extent of gas hydrate in the outer continental margin of the Arctic Basin by extrapolating information from the amount of gas hydrate offshore from northern Alaska as inferred from marine seismic records. Taking into account areal extent (5.25×10^5 km^2), thickness of the gas hydrate zone (40 m), average porosity (30%) and gas hydrate yield (140 standard volumes of gas per volume of hydrate) they calculated a total of $\sim 5.3 \times 10^{17}$ g of methane carbon in sediment of the outer margin of the Arctic Basin. The length of this margin is ~5% of the total length of continental margins world-wide. If oceanic gas hydrates in general are distributed as they are inferred to be in the Arctic, then the total amount of methane carbon in oceanic gas hydrates would be $\sim 1 \times 10^{19}$ g (Kvenvolden 1988*a*). This estimate is a factor of 2 less than that made by Claypool but falls within the range of values (2.7×10^{18}–13.4×10^{18} g) determined by Trofimuk *et al.* (1977). These recent estimates, however, are about two orders of magnitude smaller than that determined by Dobrynin *et al.* (1981) of 4.1×10^{21} g.

MacDonald (1990*a*) used data from Cherskiy *et al.* (1985) to estimate roughly the areal extent, thickness and volume of the gas hydrate stability zone for four major regions in northern Russia. A porosity of either 2 or 4% was assumed, depending on the age of the sediments containing gas hydrates. The volume actually occupied by gas hydrate was assumed to be 1%. Based on these assumptions, MacDonald (1990*a*) calculated that 3.5×10^{17} g of methane carbon are stored in gas hydrates in the permafrost regions of Siberia and 4.6×10^{16} g of methane carbon are present in continental gas hydrates of the North American Arctic, giving a total of 3.9×10^{17} g of methane carbon in gas hydrates of the world's permafrost regions. A higher value may actually be appropriate if Antarctica were included in the estimate.

For oceanic gas hydrates, MacDonald (1990*a*) assumed that the average thickness of the gas hydrate stability zone is 0.5 km and that the total volume of sediment in this zone between water depths of 200 and 3000 m is 31.3×10^6 km^3, of which 12.5×10^6 km^3 are available for gas hydrate formation. If only 1% of the total sedimentary pore space contains gas hydrates, then the total volume of gas hydrate is 1.25×10^6 km^3. This total volume corresponds to 1.1×10^{19} g of methane carbon if the gas hydrate structure is 90% filled. This estimate does not include gas hydrates in sediments at water depths greater than 3000 m. The total resource base of methane carbon in continental and oceanic gas hydrates would therefore be 1.1×10^{19} g.

Convergence of gas hydrate resource estimates

Estimates of the amount of methane carbon in gas hydrates are still speculative and uncertain. Although it is generally appreciated that gas hydrates occur world-wide (Fig. 4), detailed geological knowledge of their occurrence is incomplete, contributing to the wide range of estimates. Nevertheless, since 1988 there has been a convergence of ideas regarding the amount of methane in the gas hydrate deposits of the world. Current estimates of the amount of methane carbon in gas hydrates are in rough accord at 1.1×10^{19} g or 2.1×10^{16} m^3 of methane gas (MacDonald 1990*a*), and 1×10^{19} g or 2×10^{16} m^3 of methane gas (Kvenvolden 1988*a*), with 2.1×10^{19} g or 4×10^{16} m^3 (Kvenvolden & Claypool 1988) being only a factor of 2 larger. The relative magnitude of gas hydrates as a reservoir of organic carbon on the Earth is illustrated in Fig. 9. If these estimates are valid, then the amount of methane in gas hydrates is almost two orders of magnitude larger than the estimated total remaining recoverable conventional methane ($\sim 1.4 \times 10^{17}$ g or 2.5×10^{14} m^3) and about a factor of 2 larger than the methane equivalent of all known fossil fuel deposits (coal, oil and natural gas).

Aspects of gas hydrates

Chemists have known about gas hydrates since the early part of the 19th century (reviewed by Sloan 1990). The petroleum industry became aware of these substances in the 1930s when gas hydrate formation was discovered to be the cause of pipeline blockage during transmission of natural gas (Hammerschmidt 1934). In the 1960s naturally occurring gas hydrates were found in the Siberian Messoyakha gas field (reviewed by Makogon 1981), and in the 1970s it was recognized that gas hydrates occur naturally not only in polar continental regions but also in shallow, deep-water sediment of oceanic regions at outer continental margins (Claypool & Kaplan 1974). Kvenvolden & McMenamin (1980) reviewed the geological occurrences of natural gas hydrates. Since the time of this review it has become increasingly evident that naturally occurring gas hydrates are significant components of the shallow geosphere and are of societal concern in at least three major ways: resource, hazard and climate.

Potential energy resource

Two factors make gas hydrates attractive as a potential energy resource. First is the enormous amount of methane that is apparently seques-

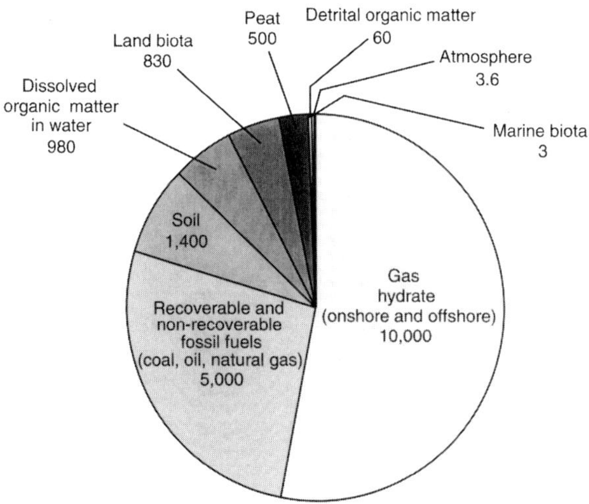

Fig. 9. Distribution of organic carbon in the Earth (excluding dispersed organic carbon such as kerogen and bitumen, which equals nearly 1000 times the total amount shown in the diagram). Numbers are in units of 10^{15} g of carbon. Adapted from Kvenvolden (1988*a*).

tered within clathrate structures at shallow sediment depths within 2000 m of the Earth's surface. Second is the wide geographical distribution of the gas hydrates (Fig. 4). The energy density (volume of methane per volume of rock) of methane hydrates is 10 times greater than the energy density of other unconventional sources of gas, such as coal beds, tight sands, black shales and deep aquifers, and 2–5 times greater than the energy density of conventional natural gas (MacDonald 1990*a*).

Production methods

Although naturally occurring gas hydrates were recognized in the 1960s, the gas industry has been slow to develop methodologies to recover methane from these substances. Three principal methods are being considered: thermal stimulation, depressurization and inhibitor injection (Holder *et al.* 1984). In thermal stimulation, thermal energy is released into the gas hydrate-bearing strata in order to increase the local temperature enough to cause gas hydrates to dissociate. In depressurization, the pressure in the gas hydrate deposit is lowered sufficiently to cause gas hydrate dissociation; heat energy for the process comes from the Earth's interior (geothermal heat flow). Injection of inhibitors such as methanol shifts the pressure–temperature equilibrium so that the gas hydrates are no longer stable at *in situ* pressure–temperature conditions. Of these three production methodologies, depressurization combined with hot-water injection appears to be the most practical where free gas is trapped beneath the gas hydrate (Holder *et al.* 1984). That free gas is trapped beneath the gas hydrate has yet to be established with certainty, as discussed previously. Gas hydrates will become a potential energy resource when it can be shown that the energy required to release methane from the hydrate is significantly less than the thermal energy of the methane that can be recovered from the dissociated gas hydrate deposits. Circulation of warm surface water into gas hydrate deposits and horizontal drilling techniques provide possible future approaches to the exploitation of hydrated methane.

Recovery of methane

The production of methane from gas hydrates requires that the deposit occurs in geological contexts wherein the elements of a conventional gas deposit are present; that is, in association with a reservoir of adequate porosity and permeability, with a methane source and with a seal to form a trap. Dissociated gas hydrate can serve as one source of methane; migrated methane from deep below the gas hydrate zone is another possible source. The ability of the gas hydrate to act as a seal, however, has not been well established. Miller *et al.* (1991) present geophysical evidence from the Blake Outer Ridge that suggests gas hydrates are trapping gas beneath the BSR (Fig. 6). Examples of possible traps caused by gas hydrates are shown in Fig. 10. These examples assume that the gas hydrate-containing interval is impermeable and that free gas can be trapped The free gas below the gas hydrate has migrated up from depth and is probably thermogenic. As methane is removed by production, however, the gas hydrate seal will begin to decompose, providing more producible methane. Thus, the seal becomes a source. Two of the trapping situations in Fig. 10 are in effect structural, caused by the domal or anticlinal shapes of the gas hydrate layer. The other trap is stratigraphic where gas-bearing strata are sealed at their up-dip ends by the presence of gas hydrates (Dillon *et al.* 1980).

Development of the Messoyakha gas field in western Siberia during the past 25 years has proven that methane can be recovered from naturally occurring gas hydrates (Makogon 1981), but the methanol injection methods used for production proved to be prohibitively expensive. Recovery of gas hydrates in pressurized core barrels at the Arco-Exxon NW Eileen State No. 2 wildcat well on the North Slope of Alaska confirmed the presence of gas hydrates in the region of the Prudhoe Bay and Kuparuk River oil fields (Kvenvolden & McMenamin 1980). It has been estimated that about $1.1 \times 10^{12}\,m^3$ of methane (440×10^{12} g methane carbon) are present in gas hydrates of this region (Collett 1992). This amount of methane is about 1.4 times the conventionally reservoired methane of the same region, but is only 13% of the conventional US gas reserves (Fig. 11). If methane from gas hydrates is ever to be recovered commercially in the United States, larger accumulations will have to be identified and exploited. Initial production, however, is likely to come from the North Slope of Alaska where the industrial infrastructure for gas handling is already in place. Methane first produced from gas hydrates on the North Slope will probably be used for repressurization of waning oil fields rather than as a source of fuel. Although methane from gas hydrates should be considered a potential energy resource, wide-scale recovery of methane from gas hydrates will probably not be accomplished until the 21st century (Sloan 1990).

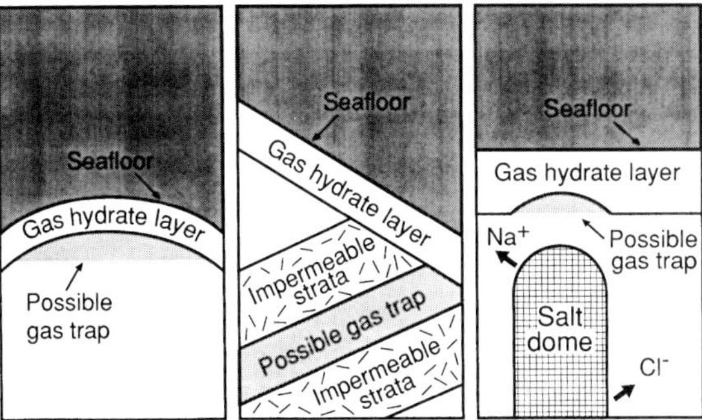

Fig. 10. Schematic diagrams of geological situations in which gas hydrates serve as seals forming traps for methane. Developed from Dillon *et al.* (1980).

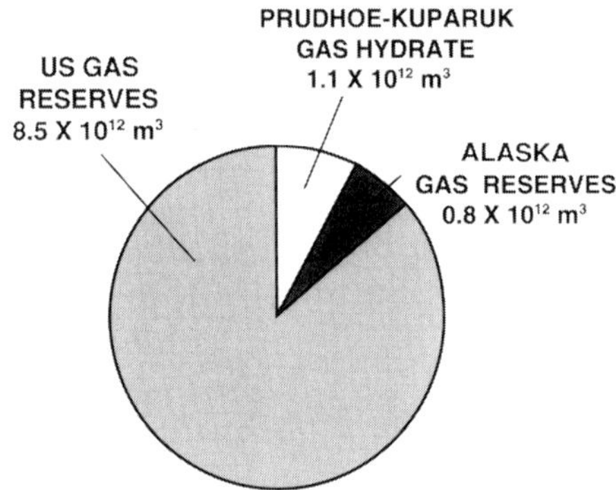

Fig. 11. Comparison of the amount of methane estimated to be present in the region of the Prudhoe Bay and Kuparuk River oil fields with the amount of conventionally reservoired methane of the same region and amount of methane in U.S. conventional gas reserves.

Geological hazard

Before gas hydrates form in usual geological settings, vast quantities of methane and water are free to migrate within the interstitial pore spaces of consolidating sediment. During gas hydrate formation, methane and water become immobilized as a solid, restricting pore space and retarding the migration of fluids. Solid water rather than liquid water occupies the pore spaces, and the sedimentological processes of consolidation and mineral cementation are greatly inhibited, although gas hydrates themselves can act as metastable cementation (bonding) agents. The permeability of the sediment to gases and liquids decreases as more gas hydrate forms. Eventually, gas hydrates may occupy much of the sedimentary section within the zone of gas hydrate stability. Continued sedimentation leads to deeper burial of the gas hydrate. Finally, the gas hydrate will be buried so deeply that temperatures at the base of the stability zone will be reached at which the gas hydrates are no longer stable. The solid gas–water mixture (i.e. the gas hydrate) will become a liquid gas–water mixture. Thus, the basal zone of the gas hydrate becomes underconsolidated, possibly over pressured due to the newly released gas, leading to a zone of weakness (low shear strength) where failure could be triggered by gravitational loading or seismic disturbances, and submarine landslides result (McIver 1982).

The same conditions that cause gas hydrate dissociation during continued sedimentation can also be brought about by the lowering of sea level or by an increase in bottom-water temperatures. These processes change the *in situ* pressure or temperature regime. In adjusting to the new pressure–temperature conditions, the gas hydrates dissociate producing an enhanced fluidized layer at the base of the gas hydrate zone. Submarine slope failure can follow giving rise to debris flows, slumps and slides, accompanied by the release of methane gas into the water column. A scenario illustrating submarine slope failure is shown in Fig. 12.

Examples. The possible connection between gas hydrate boundaries and submarine slide and slump surfaces was first recognized by McIver (1977), and several possible examples have been described later. These examples include surficial slides and slumps on the continental slope and rise of SW Africa (Summerhayes *et al.* 1979), slumps on the U.S. Atlantic continental slope (Carpenter 1981), large submarine slides on the Norwegian continental margin (Bugge *et al.* 1987; Jansen *et al.* 1987),

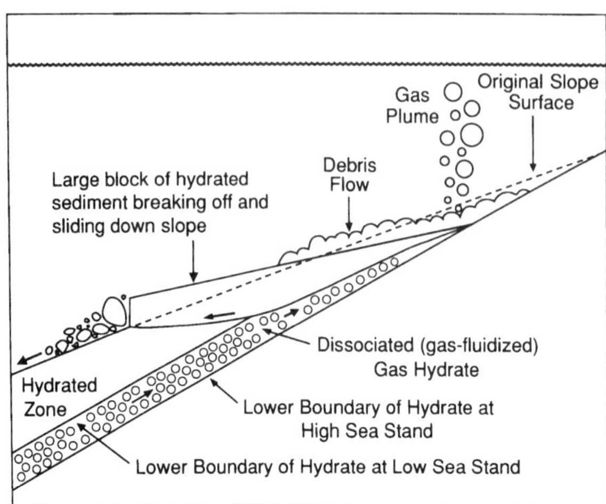

Fig. 12. Diagram showing the effects of changes in pressure and temperature on submarine gas hydrates and the resulting sea-floor failures and gas release. Adapted from McIver (1982).

sediment blocks on the sea floor in British Columbia fjords (Bornhold & Prior 1989), and massive bedding-plane slides and rotational slumps on the Alaskan Beaufort Sea continental margin (Kayen & Lee 1991). In the Caspian Sea (Ginsburg *et al.* 1992) and offshore from North Panama (Reed *et al.* 1990), submarine mud volcanoes, another kind of geohazard, have also been attributed to the release of gas from gas hydrates.

Periodic Pleistocene eustatic sea-level transgressions and regressions provide a mechanism to account for the waxing and waning of submarine gas hydrates. During the last Pleistocene regression, sea level lowered approximately 100 m between about 28 and 17 ka, resulting in a reduction of total stress acting on the sea floor of about 1000 kPa (Kayen & Lee 1991). The reduction in the total pressure initiates dissociation at the base of the gas hydrates, releasing excess methane and water. Failure follows on moderate slopes unless the increased fluid pressures can be adequately vented. On the Beaufort Sea continental slope is a zone of massive slides and slumps that coincides with a region of sediment inferred from seismic reflection studies to contain gas hydrates (Fig. 13). Fluctuations in global climate, reflected in Pleistocene sea-level lowerings, probably caused these submarine landslides and, perhaps, caused other slides on other continental margins where gas hydrates are present (Kayen & Lee 1991).

Local geohazards. Besides the global-scale geological hazards created by destabilized gas hydrates, local-scale hazards also result when the pressure–temperature conditions of gas hydrates are perturbed anthropogenically, for example, during drilling and production operations. These operations can result in hazardous situations, such as uncontrolled gas releases, casing failures and well-site subsidence, if appropriate precautions are not taken to minimize the disturbance to the gas hydrate deposit (Yakushev & Collett 1992).

Global climate change

Methane is an important trace components of the atmosphere, having a current concentration of about 4.9×10^{15} g (3.7×10^{15} g methane carbon), approximately half of the minimum amount of methane estimated by the Potential Gas Committee (1981) to occur in gas hydrates of Arctic permafrost regions. The concentration of atmospheric methane is increasing at a rate of almost 1.0% year^{-1} (Watson *et al.* 1990). Because methane is radiatively active, it is a 'greenhouse' gas that has a global warming potential 20 times larger than an equivalent weight of carbon dioxide when integrated over a 100 year time span (Shine *et al.* 1990). The Earth's atmosphere has a wide variety of sources and sinks for methane (Cicerone & Oremland 1988), including gas hydrates, which exist in metastable equilibrium with their environment and are affected by changes in pressure and temperature. The amount of methane that is trapped in gas hydrates onshore and offshore is perhaps 3000 times the amount in the atmosphere; a large release of methane from this source could

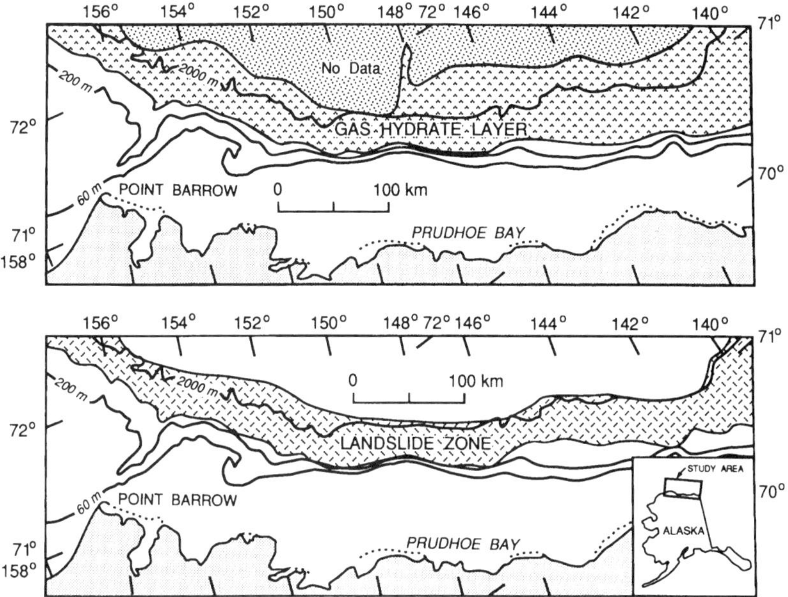

Fig. 13. Map of the continental margin of the Beaufort Sea offshore from Alaska showing the coincident regions of large landslides and gas hydrates. Adapted from Kayen & Lee (1991).

have a significant impact on atmospheric composition and thus on the radiative properties of the atmosphere that affect global climate (MacDonald 1990*b*).

Warming and cooling. Pleistocene global climate changes probably caused methane release from gas hydrate deposits, as suggested previously in Fig. 12. However, the opposite may also be true. Methane released from gas hydrates may in turn have caused changes in global climate. The arguments go as follows: during global warming, glaciers and ice caps melt, contributing water to the oceans; oceans also thermally expand, and these factors cause a rise in sea level. This sea-level rise causes an increase in subsurface hydrostatic pressure that stabilizes submarine gas hydrate deposits on outer continental margins and on polar continental shelves. Meanwhile, onshore, the increasing air temperatures eventually destabilize continental gas hydrates at time scales of hundreds to thousands of years, releasing methane that may reach the atmosphere. Water temperatures also increase during global warming, although deep-water temperatures on the outer continental margins probably do not change significantly because of the heat capacity of the large mass of water involved. Thus, for deep-water gas hydrates on outer continental margins (water depths greater than 300–500 m), the effects on hydrate stability caused by increasing sea level outweigh the destabilizing effects of any possible increase in bottom-water temperature.

The same results are not true for the gas hydrate deposits underlying polar continental shelves. Increasing air temperatures increase shallow shelfal water temperatures. More important is the increase in submarine bottom-surface temperatures caused by the transgression of the polar ocean over the exposed, colder continental shelf surface as sea level rises. Flooding of the shelf by relatively warm water offsets the effects of increasing pressure due to the sea-level rise. Gas hydrates of the polar continental shelves are thus destabilized, releasing methane. Therefore, during global warming, deep-sea gas hydrates become more stable, but gas hydrates of polar continents and continental shelves are destabilized, leading to methane release over long time scales (Kvenvolden 1988*b*). Methane reaching the atmosphere from these sources contributes to the global warming trend.

During a global cooling cycle the whole system reverses. Glaciers and ice caps grow, removing water from the oceans, and oceans thermally contract. The result is a eustatic fall in sea level and regression of the oceans from continental shelves. During regression the pressure on gas hydrates decreases, and the gas hydrates become unstable. As polar continental and exposed continental shelves become colder, the cold temperatures eventually stabilize the buried gas hydrate deposits, offsetting the effects of decreasing pressures on the gas hydrates. In polar regions, continental gas hydrate deposits buried under an increasing ice load of advancing glaciers experience increased pressure and further stabilization. As sea level falls, only the deep-water gas hydrates of continental slopes and rises become unstable, releasing methane that may reach the atmosphere. To get to the atmosphere, however, the methane must escape from the sediment and traverse a long water column without being dissolved and oxidized. Thus, during global warming and cooling, i.e. during interglacial and glacial climates, gas hydrate deposits should respond to surface pressure and temperature changes; however, the extent of influence that methane from gas hydrates has on global climate is still very uncertain.

Past climate change. Nevertheless, interesting ideas have been proposed to explain the role of gas hydrates in global climate change. For example, Nisbet (1990) suggests that methane from continental gas hydrates contributed to the rapid rise in atmospheric methane, carbon dioxide and global temperature at the end of the last major glaciation about 13.5 ka ago. The basic elements of his idea are shown in Fig. 14a, where polar continental gas hydrate deposits are destabilized by pressure reduction of melting ice sheets and temperature increases. The resulting warming provides a strong positive feedback that amplifies methane emissions and ultimately helps to end the ice age. Taken to extremes, however, this scenario would lead to uncontrolled methane release from gas hydrates and consequent enhanced global warming, but evidence from the Vostok ice core shows that no uncontrolled release of methane to the atmosphere has taken place at least during the last 160 ka (Chappellaz *et al.* 1990)

A different scenario has been proposed by Paull *et al.* (1991). They suggest that outer continental margin gas hydrate deposits release methane during a falling sea level, i.e. during global cooling. The resulting decrease in pressure causes these gas hydrates to dissociate. The released methane enhances global warming and triggers deglaciation (Fig. 14b). Thus, methane derived from outer continental margin gas hydrate deposits may be a important factor in limiting the extent of glaciation during a glacial cycle.

Present climate change. As rich as these ideas are, they are all speculative. How gas hydrates behave in the present climate regime is not known. Greater understanding of the present is needed to unravel the past and to predict the future. I have suggested that gas hydrate deposits of the polar continental shelves are presently most vulnerable to climate change (Kvenvolden 1988b). These areally extensive shelves, formerly exposed to very cold surface temperatures (-10 to −20°C) have been, and are being, transgressed by a much warmer polar ocean (~0°C). The polar shelf surface, therefore, has experienced a +10°C or more change in temperature over at least the past 10 000 years. Although pressure on shelfal gas hydrates has increased owing to a rise in sea level of about 100 m, this pressure increase that tends to stabilize the gas hydrates is more than offset by the large temperature increase that destabilizes the gas hydrates. The amount of methane released by this process has been estimated to be about 3×10^{12} g $year^{-1}$ of methane carbon (Kvenvolden 1991) or about 1% of all current sources of atmospheric methane (Fig. 15). If this suggestion and estimate are correct, then escape of methane from gas hydrate deposits of polar continental shelves should be observable.

A test of this idea was conducted on the continental shelf north from Oliktok Point, Alaska (Fig. 16) (Kvenvolden *et al.* 1992). Kvenvolden *et al.* discovered that methane concentrations in the water under the ice (Fig. 16) are 6–28 times

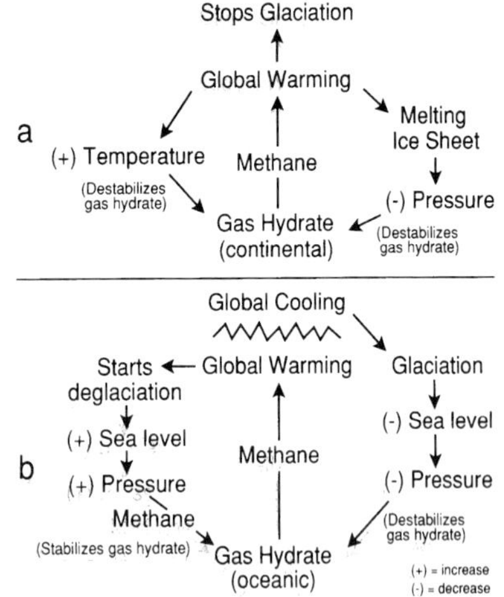

Fig. 14. Diagrams illustrating how gas hydrate decomposition may affect glacial cycles. (**a**) continental gas hydrates (positive environmental feedback loop); and (**b**) oceanic gas hydrates (negative environmental feedback loop).

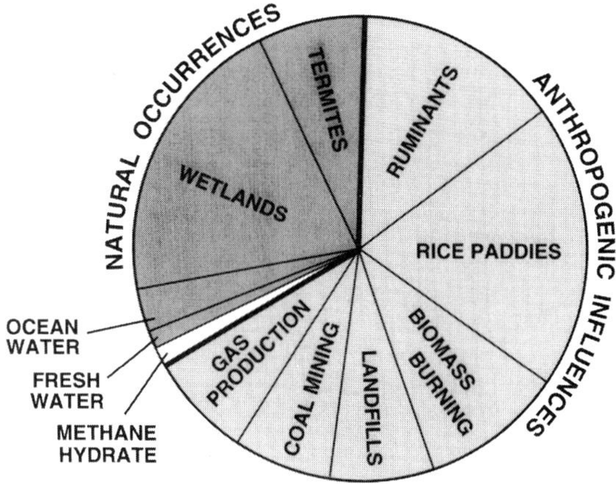

Fig. 15. Distribution of annual methane release rates for identified sources. Adapted from Cicerone & Oremland (1988).

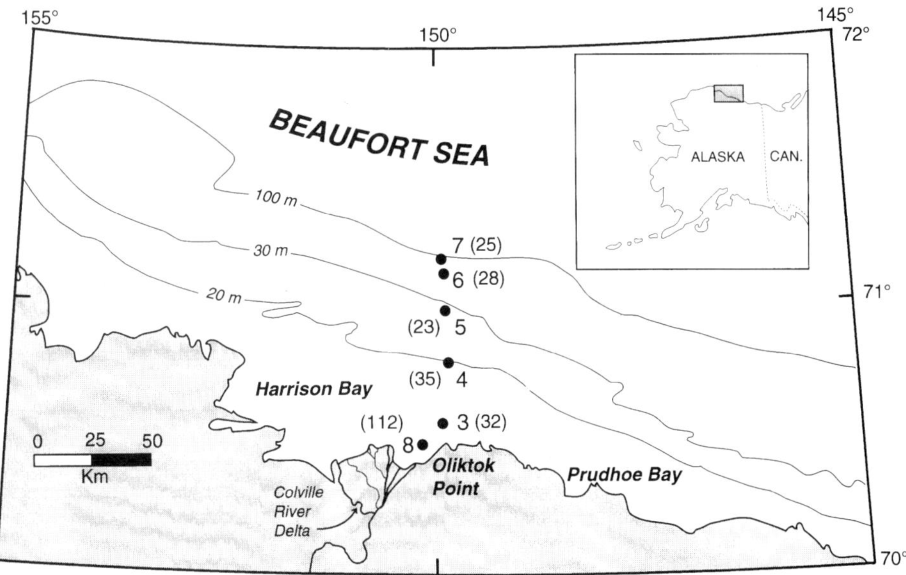

Fig. 16. Map showing stations on transect north of Oliktok Point, Alaska in the Beaufort Sea. Methane was determined in water samples recovered from under the ice. Methane concentrations in parentheses for samples taken nearest the water–ice interface are given in nM.

greater than the atmospheric equilibrium concentration of about 4 nM. Later, the methane content of water in the Beaufort Sea offshore from Alaska during ice-covered and ice-free conditions in 1993, 1994 and 1995 were compared. With very few exceptions, the average methane concentrations are higher when ice is present, with seasonal differences ranging from about 3 to 50 nM. Coastal methane concentrations can be very high with a maximum of 275 nM measured when ice was present and 11 nM when ice was absent at the same location. An intriguing ephemeral bottom-water anomaly (maximum 149 nM) at 20 m water depth north of Oliktok Point suggests seepage of methane from sediments. A carbon isotopic composition of −46.4‰ indicates that dissociating gas hydrate may be the methane source. The amount of methane that apparently enters the atmosphere each year from the Arctic Ocean, as determined from the seasonal differences in concentrations, is less that 0.1 Tg year^{-1},

(1 Tg = 1×10^9 kg), with minimal impact on the global methane budget. However, if gas hydrate is the source of the anomalous methane, then a reservoir of methane of unknown size is poised to be released during global change.

Conclusions

From a geological perspective, gas hydrates are an important feature of the shallow geosphere. If current estimates are correct, gas hydrates contain more potential fossil fuel energy than is present in conventional oil, gas and coal deposits. Uncertain, however, is the portion of this potential energy source that can actually be exploited. Because of unsolved technological problems in producing methane from gas hydrates, wide-scale recovery of methane from these substances probably will not take place until some time in the 21st century.

Besides being an unconventional potential source of methane, gas hydrates are also geological agents that affect the properties of sediments. For example, gas hydrates alter: (1) physical properties, such as shear strength, porosity and permeability; (2) geophysical properties, such as acoustic velocity and resistivity; and (3) geochemical properties, such as fluid composition and movement. The change in physical properties and flow regimes caused by gas hydrates leads to geological hazards in the form of sediment failures particularly on the sea floor. Many submarine slumps and slides are probably caused by gas hydrate decomposition.

Because methane is both sequestered and released from gas hydrates, depending on the pressure–temperature regime, the possibility exists that methane from gas hydrates can reach the atmosphere. Because methane is a strong 'greenhouse' gas, the methane from gas hydrates may influence global climate, past, present, and future, but the extent of this influence has not yet been determined even for the present climatic cycle.

References

BANGS, N., SAWYER, D. & GOLOVCHENKO, X. 1993. Free gas at the base of the gas hydrate zone in the vicinity of the Chile triple junction. *Geology*, **21**, 905–908.

BERNARD, B., BROOKS, J. & SACKETT, W. 1976. Natural gas seepage in the Gulf of Mexico. *Earth and Planetary Science Letters*, **31**, 48–54.

BROOKS, J., BARNARD, L., WIESENBURG, D., KENNICUTT, M., II & KVENVOLDEN, K. 1983. Molecular and isotopic compositions of hydrocarbons at site 533, Deep Sea Drilling Project Leg 76. *In:* SHERIDAN, R., GRADSTEIN, F. *et al.* (eds) *Initial Reports of the Deep Sea Drilling Project*, Volume 67. US Government Printing Office, Washington, D.C. 377–384.

——, COX, H., BRYANT, W., KENNICUTT, M., II, MANN, R. & MCDONALD, T. 1986. Association of gas hydrates and oil seepage in the Gulf of Mexico. *Organic Geochemistry*, **10**, 221–234.

——, FIELD, M. & KENNICUTT, M., II. 1991. Observations of gas hydrates in marine sediments, offshore northern California. *Marine Geology*, **96**, 103–108.

——, JEFFREY, A., MCDONALD, T., PFLAUM, R. & KVENVOLDEN, K. 1985. Geochemistry of hydrate gas and water from site 570, Deep Sea Drilling Project Leg 84. *In*: VON HUENE, R., AUBOUIN, J. *et al.* (eds) *Initial Reports of the Deep Sea Drilling Project*, Volume 84. US Government Printing Office, Washington, DC, 699–703.

BORNHOLD, B. & PRIOR, D. 1989. Sediment blocks on the sea floor in British Columbia fjords. *Geo-Marine Letters*, **9**, 135–144.

BUGGE, T., BEFRING, S., BELDERSON, R. *et al.* 1987. A giant three-stage submarine slide off Norway. *Geo-Marine Letters*, **7**, 191–198.

CARPENTER, G. 1981. Coincident sediment slump/clathrate complexes on the U.S. Atlantic continental slope. *Geo-Marine Letters*, **1**, 29–32.

CHAPPELLAZ, J., BARNOLA, J., RAYNAUD, D., KOROTKEVICH, Y. & LORIUS, C. 1990. Ice-core record of atmospheric methane over the past 160,000 years. *Nature*, **345**, 127–131.

CHERSKIY, N., TSAREV, V. & NIKITIN, S. 1985. Investigation and prediction of conditions of accumulation of gas resources in gas-hydrate pools. *Petroleum Geology*, **21**, 65–89.

CICERONE, R. & OREMLAND, R. 1988. Biogeochemical aspects of atmospheric methane. *Global Biogeochemical Cycles*, **2**, 299–327.

CLAYPOOL, G. & KAPLAN, I. 1974. The origin and distribution of methane in marine sediments. *In*: KAPLAN, I. (ed.) *Natural Gases in Marine Sediments*. Plenum, New York, 99–139.

—— & KVENVOLDEN, K. 1983. Methane and other hydrocarbon gases in marine sediment. *Annual Review of Earth and Planetary Sciences*, **11**, 299–327.

——, PRESLEY, B. & KAPLAN, I. 1973. Gas analyses in sediment samples from Legs 10, 11, 13, 14, 15, 18, and 19. *In*: SCHOLL, D. *et al.* (eds) *Initial Reports of the Deep Sea Drilling Project*. Volume 19. US Government Printing Office, Washington, DC, 879–884.

——, THRELKELD, C., MANKEWICZ, P. *et al.* 1985. Isotopic composition of interstitial fluids and origin of gases in sediments of the Middle America Trench, DSDP Leg 84. *In*: VON HUENE, R., AUBOUIN, J. *et al.* (eds) *Initial Reports of the Deep Sea Drilling Project*, Volume 84. US Government Printing Office, Washington, DC, 683–691.

COLLETT, T. 1992. Potential of gas hydrates outlined. *Oil & Gas Journal*, **90**(25), 84–87.

——, BIRD, K., KVENVOLDEN, K. & MAGOON, L. 1988. Geologic Interrelations Relative to Gas Hydrates within the North Slope of Alaska. *US Geological Survey Open File* Report, **88–389**.

DILLON, W., GROW, J. & PAULL, C. 1980. Unconventional gas hydrate seals may trap gas off southeast U.S. *Oil & Gas Journal*, **78**(1), 124–130.

DOBRYNIN, V., KOROTAJEV, Y. & PLYUSCHEV, D. 1981. Gas hydrates: a possible energy resource. *In*: MEYER, R. & OLSON, J. (eds) *Long-term Energy Resources*. Pitman, Boston, MA, 727–729.

GALIMOV, E. & KVENVOLDEN, K. 1983. Concentrations and carbon isotopic compositions of CH_4 and CO_2 in gas from sediments of the Blake Outer Ridge. Deep Sea Drilling Project Leg 76. *In*: SHERIDAN, R., GRADSTEIN, F. *et al.* (eds) *Initial Reports of the Deep Sea Drilling Project*, Volume 76. US Government Printing Office, Washington, DC, 403–407.

GINSBURG, G., GUSEYNOV, R., DADASHEV, A. *et al.* 1992. Gas hydrates of the southern Caspian. *International Geology Review*, **34**, 765–782.

——, KREMLEV, A., GRIGOR'EV, M. *et al.* 1990. Filtrogenic gas hydrates in the Black Sea. *Soviet Geology and Geophysics*, **31**(3), 8–16.

GOODMAN, M. 1980. In situ gas hydrates: past experience and exploitation concepts. *In: Proceedings 1st International Gas Research Conference, Chicago*. Government Institute, Rockville, MD, IL, 376–391.

HAMMERSCHMIDT, E. 1934. Formation of gas hydrates in natural gas transmission lines. *Industrial Engineering Chemistry*, **26**, 851–855.

HESSE, R. & HARRISON, W. 1981. Gas hydrates (clathrates) causing pore-water freshening and oxygen isotope fractionation in deep-water sedimentary sections of terrigenous continental margins. *Earth and Planetary Science Letters*, **55**, 453–562.

——, LEBEL, J. & GIESKES, J. 1985. Interstitial water chemistry of gas-hydrate-bearing sections on the Middle America Trench slope, Deep-Sea Drilling Project Leg 84. *In*: VON HUENE, R., AUBOUIN, J. *et al.* (eds) *Initial Reports of the Deep Sea Drilling Project*, Volume 84. US Government Printing Office, Washington, DC, 727–736.

HITCHON, B. 1974. Occurrence of natural gas hydrates in sedimentary basins. In: KAPLAN, I. R. (ed.) *Natural Gases in Marine Sediments*. Plenum, New York, 195–225.

HOLDER, G., KAMATH, V. & GODBOLE, S. 1984. The potential of natural gas hydrates as an energy resource. *Annual Review of Energy*, **9**, 427–445.

HUNT, J. 1979. *Petroleum Geochemistry and Geology*. Freeman, San Francisco, CA.

HYNDMAN, R. & DAVIS, E. 1992. A mechanism for the formation of methane hydrate and sea-floor bottom-simulating reflectors by vertical fluid expulsion. *Journal of Geophysical Research*, **97**(5), 7025–7041.

—— & SPENCE, G. 1992. A seismic study of methane hydrate marine bottom simulating reflectors. *Journal of Geophysical Research*, **97**(B5), 6683–6698.

JANSEN, E., BEFRING, S., BUGGE, T. *et al.* 1987. Large submarine slides on the Norwegian continental margin: sediments, transport and timing. *Marine Geology*, **78**, 77–107.

JEFFREY, A., PFLAUM, R., MCDONALD, T. *et al.* 1985. Isotopic analysis of core gases at sites 565–570, Deep Sea Drilling Project Leg 84. *In*: VON HUENE, R., AUBOUIN, J. *et al.* (eds) *Initial Reports of the Deep Sea Drilling Project*, Volume 84. US Government Printing Office, Washington, DC, 719–726.

JENDEN, P. & GIESKES, J. 1983. Chemical and isotopic compositions of interstitial water from Deep Sea Drilling Project sites 533 and 534. *In*: SHERIDAN, R., GRADSTEIN, F. *et al.* (eds) *Initial Reports of the Deep Sea Drilling Project*. Volume 76. US Government Printing Office, Washington, DC, 453–461.

KATZ, D., CORNELL, D. *et al.* 1959. *Handbook of Natural Gas Engineering*. McGraw-Hill, New York.

KAYEN, R. & LEE, H. 1991. Pleistocene slope instability of gas hydrate-laden sediment on the Beaufort Sea margin. *Marine Geotechnology*, **10**, 125–141.

KVENVOLDEN, K. 1988a. Methane hydrate – a major reservoir of carbon in the shallow geosphere? *Chemical Geology*, **71**, 41–51.

—— 1988*b*. Methane hydrates and global climate. *Global Biogeochemical Cycles*, **2**, 221–229.

—— 1991. A review of Arctic gas hydrates as a source of methane in global change. *In*: WELLER, G., WILSON, C. & SEVERIN, B. (eds) *International Conference on the Role of the Polar Regions in Global Change, II*. Geophysical Institute and Center for Global Change and Arctic System Research, University of Alaska Fairbanks, Fairbanks, AK. 696–701.

—— & BARNARD, L. 1983*a*. Hydrates of natural gas in continental margins. *In*: WATKINS, J. & DRAKE, C. (eds.) *Studies in Continental Margin Geology*. American Association of Petroleum Geologists, Memoir, **34**, 631–640.

—— & Barnard, L. 1983*b*. Gas hydrates of the Blake Outer Ridge, Site 533, Deep Sea Drilling Project Leg 76. *In*: SHERIDAN, R., GRADSTEIN, F. *et al.* (eds) *Initial Reports of the Deep Sea Drilling Project*, Volume 76. US Government Printing Office, Washington, DC, 353–365.

—— & CLAYPOOL, G. 1988. Gas Hydrates in Oceanic Sediment. *US Geological Survey Open File Report*, **88–216**.

—— & A. GRANTZ. 1990. Gas hydrates of the Arctic Ocean region. *In*: GRANTZ, A., JOHNSON, L. & SWEENEY, J. (eds) *The Geology of North America. Vol. L – The Arctic Ocean Region*. Geological Society of America, Boulder, CO, 539–549.

—— & KASTNER, M. 1990. Gas hydrates of the Peruvian outer continental margin. *In*: SUESS, E., VON HUENE, R. *et al.* (eds) *Proceedings of the Ocean Drilling Program, Scientific Results*. College Station, TX. Ocean Drilling Program, **112**, 517–526.

—— & MCDONALD, T. 1985. Gas hydrates of the Middle America Trench – Deep Sea Drilling Project Leg 84. *In*: VON HUENE, R., AUBOUIN, J. *et al.* (eds) *Initial Reports of the Deep Sea Drilling Project*, Volume 84. US Government Printing Office, Washington, DC, 66–682.

—— & McMENAMIN, M. 1980. *Hydrates of Natural Gas: A Review of Their Geologic Occurrence. US Geological Survey Circular*, **825**.

——, CLAYPOOL, G., THRELKELD, C. & SLOAN, E. 1984. Geochemistry of a naturally occurring massive marine gas hydrate. *Organic Geochemistry*, **6**, 703–713.

——, GINSBURG, G. & SOLOVYEV, V. 1993. Worldwide distribution of subaquatic gas hydrates. *Geo-Marine Letters*, **13**, 32–40.

——, LORENSON, T & LILLEY, M. 1992. Methane in the Beaufort Sea on the continental shelf of Alaska. (Abstract). *EOS, Transactions, American Geophysical Union*, **73**(43), 309.

LEGGETT, J. 1990. The nature of the greenhouse threat. *In*: LEGGETT, J. (ed.) *Global Warming, The Greenpeace Report*. Oxford University Press, Oxford, 14–43.

MACDONALD, G. 1990*a*. The future of methane as an energy resource. *Annual Review of Energy*, **15**, 53–83.

——. 1990*b*. Role of methane clathrates in past and future climates. *Climatic Change*, **16**, 247–281.

MACKAY, M., JARRARD, R., WESTBROOK, G., HYNDMAN, R. & SHIPBOARD SCIENTIFIC PARTY. 1994. Origin of bottom-simulating reflectors: Geophysical evidence from the Cascadia accretionary prism. *Geology*, **22**, 459–462.

MATHEWS, M. 1986. Logging characteristics of methane hydrate. *The Log Analyst*, **27(3)**, 26–63.

MAKOGON, Y. F. 1981. *Hydrates of Natural Gas*. Pennwell, Tulsa, OK.

McIVER, R. D. 1977. Hydrates of natural gas – important agent in geological processes. (Abstract.). *Geological Society of America*, **9**, 1089–1090.

—— 1981. Gas hydrates. *In*: MEYER, R. & OLSON, J. (eds) *Long-term Energy Resources*. Pitman, Boston, MA, 713–726.

—— 1982. Role of naturally occurring gas hydrates in sediment transport. *AAPG Bulletin*, **66**, 789–792.

MEYER, R. 1981. Speculations on oil and gas resources in small fields and unconventional deposits. *In*: MEYER, R. & OLSON, J. (eds) *Long-term Energy Resources*. Pitman, Boston, MA, 49–72.

MILLER, J., LEE, M. & VON HUENE, R. 1991. An analysis of a seismic reflection from the base of a gas hydrate zone, offshore Peru. *AAPG Bulletin*, **75**, 910–924.

NISBET, E. 1990. The end of the ice age. *Canadian Journal of Earth Science*, **27**, 148–157.

OCEAN DRILLING PROGRAM. 1986. Guidelines for pollution prevention and safety. *JOIDES Journal*, **12**(5).

ODP LEG 164 SHIPBOARD SCIENTIFIC PARTY. 1996. Methane gas hydrate drilled at Blake Ridge. *EOS, Transactions, American Geophysical Union*, **77**, 219.

PARENT, J. 1980. *A Survey of the United States and Total World Pproduction, Proved Reserves, and Remaining Recoverable Resources of Fossil Fuels and Uranium*. The Institute of Gas Technology, Chicago, IL.

PAULL, C., USSLER, W., III & DILLON, W. 1991. Is the extent of glaciation limited by marine gas-hydrates? *Geophysical Research Letters*, **18**, 432–434.

PFLAUM, R., BROOKS, J., COX, H., KENNICUTT, M., II & SHEU, D.-D. 1986. Molecular and isotopic analysis of core gases and gas hydrates, Deep Sea Drilling Project Leg 96. *In*: BOUMA, A., COLEMAN, J., MEYER, A. *et al.* (eds) *Initial Reports of the Deep Sea Drilling Project*. Volume 96. U.S. Government Printing Office, Washington, DC, 781–784.

POTENTIAL GAS COMMITTEE. 1981. *Potential Supply of Natural Gas in the United States*. Potential Gas Agency, Colorado School of Mines, Golden, CO.

REED, D., SILVER, E., TAGUDIN, J., SHIPLEY, T. & VROLIJK, P. 1990. Relations between mud volcanoes, thrust deformation, slope sedimentation, and gas hydrate, offshore north Panama. *Marine and Petroleum Geology*, **7**, 44–54.

SHINE, K., DERWENT, R., WUEBBLES, D. & MORCRETTE, J.-J. 1990. Radiative forcing of climate. *In*: HOUGHTON, J., JENKINS, G. & EPHRAUMS, J. (eds) *Climate Change, The IPCC Scientific Assessment*. Cambridge University Press, Cambridge, 41–68.

SHIPLEY, T., HOUSTON, M., BUFFLER, R. *et al.* 1979. Seismic reflection evidence for the widespread occurrence of possible gas-hydrate horizons on continental slopes and rises. *AAPG Bulletin*, **63**, 2204–2213.

SLOAN, E. D. 1990. *Clathrate Hydrates of Natural Gas*. Marcel Dekker, New York.

SUMMERHAYES, C., BORNHOLD, B. & EMBLEY, R. 1979. Surficial slides and slumps on the continental slope and rise of South West Africa: a reconnaissance study. *Marine Geology*, **31**, 265–277.

TROFIMUK, A., CHERSKIY, N. & TSAREV, V. 1977. The role of continental glaciation and hydrate formation on petroleum occurrences. *In*: MEYER, R. (ed.) *Future Supply of Nature-made Petroleum and Gas*. Pergamon, New York, 919–926.

WATSON, R., RODHE, H., OESCHGER, H. & SIEGENTHALER, U. 1990. Greenhouse gases and aerosols. *In*: HOUGHTON, J. JENKINS, G. & EPHRAUMS, J. (eds) *Climate Change, The IPCC Scientific Assessment*. Cambridge University Press, Cambridge, 1–40.

YAKUSHEV, V. & COLLETT, T. 1992. Gas hydrates in Arctic regions – risks to drilling and production. *In*: *International Offshore and Polar Engineering Conference*, Volume 1. ISOPE, Golden, CO, **1**, 669–673.

YAMAMOTO, S., ALCAUSKAS, J. & CROZIER, T. 1976. Solubility of methane in distilled water and sea water. *Journal of Chemical and Engineering Data*, **21**, 78–80.

YAMANO, M. & UYEDA, S. 1990. Heat-flow studies in the Peru Trench subduction zone. *In*: SUESS, E., VON HUENE, R. *et al.* (eds) *Proceedings of the Ocean Drilling Program, Scientific Results*, College Station, TX. Ocean Drilling Program, **112**, 653–661.

YAMANO, M., UYEDA, S., AOKI, Y. & SHIPLEY, T. 1982. Estimates of heat flow derived from gas hydrates. *Geology*, **10**, 339–343.

Physical/chemical properties of gas hydrates and application to world margin stability and climatic change

E. D. SLOAN, JR

Center for Hydrate Research, Colorado School of Mines, Golden, CO 80401, USA

Abstract: The major points in this paper concern: (a) physical and chemical properties and (b) applications of those properties. Three questions are addressed: What are hydrates? What is our knowledge about their thermodynamic and kinetic properties? What are the applications to the environment and climate stability?

The physical and chemical characteristics of hydrates are approximated by three heuristics: (1) physical properties approximate those of ice, (2) phase equilibrium characteristics are set by the size ratio of guest within host cages, and (3) thermal properties are set by hydrogen-bonded crystals with cavity size ratios. Knowledge of hydrate kinetics is substantially lacking, but it appears that formation kinetics derive from aggregation of water clusters at interfaces. A significant future challenge is to characterize hydrates directly (through NMR, Raman, diffraction, etc.) for both thermodynamics and kinetics.

Hydrocarbons in natural hydrates represent twice the amount of all combined fossil fuels. Most recovered samples have been small, dispersed (even dissociated) particles with isolated examples of massive hydrates. Hydrates probably will not contribute significant methane to the atmosphere in the near future. Ocean hydrates and air hydrates from Antarctic ice are indicators of ancient climatic changes.

Gas clathrates (commonly called hydrates) are crystalline compounds which occur when water forms a cage-like structure around smaller guest molecules. The proper name 'clathrate' was given to the class by Powell (1948) from the Latin 'clathratus' meaning to encage. While they are more commonly called hydrates, a careful distinction should be made between these non-stoichiometric clathrate hydrates and other stoichiometric hydrate compounds which occur when water combines with various salts via coulombic forces, but without cages.

Gas hydrates of current interest are composed of water and the following eight molecules: methane, ethane, propane, isobutane, normal butane, nitrogen, carbon dioxide and hydrogen sulphide. Yet, other apolar components between the sizes of argon (0.35 nm) and ethylcyclohexane (0.9 nm) can form hydrates. Hydrate formation is a possibility anywhere water exists in the vicinity of such molecules, both naturally and artificially, at temperatures above and below 273 K when the pressure is elevated.

Since the time of their discovery in 1810 by Sir Humphrey Davy, hydrates have been a laboratory curiosity, displaying many unusual properties. However, it is primarily due to their crystalline, non-flowing nature that hydrates became of interest to the hydrocarbon industry at the time of their first observance in pipelines (Hammerschmidt 1934). For the last 60 years hydrates have been considered a nuisance because they block hydrocarbon flow channels, jeopardize the foundations of deep-water platforms and pipelines, collapse drilling tubing, and foul process heat exchangers and expanders.

Another application of hydrates is as a potential future energy resource. Hydrates act to concentrate hydrocarbons; 1 m^3 of hydrates may contain as much as 180 SCM of gas. Three decades ago (Makogon 1965) it was recognized that large natural reserves of hydrocarbons exist in hydrated form, both in deep oceans and in the permafrost. Evaluation of these reserves is highly uncertain, yet even the most conservative estimates concur that there is twice as much energy in hydrated form as in all other hydrocarbon sources combined. While one commercial example exists of gas recovery from hydrates (Sloan 1998, p. 525 ff), the economics of *in situ* hydrate dissemination in deep-water–permafrost environments will prevent their recovery until the next millennium. There is a national project to drill hydrates in 1999 in offshore Japan.

Questions relating hydrate stability to atmospheric methane were first raised by Kvenvolden & McMenamin (1980), but degrees of ocean methane hydrate release scenarios have been considered by Nisbet (1989, 1992), MacDonald (1990), Legett (1990) and others (Fei 1991; Englezos 1992; Hatzikiriakos & Englezos 1993;

SLOAN, E. D. JR. 1998. Physical/chemical properties of gas hydrates and application to world margin stability and climatic change. *In:* HENRIET, J.-P. & MIENERT, J. (eds) *Gas Hydrates: Relevance to World Margin Stability and Climate Change*. Geological Society, London, Special Publications, **137**, 31–50.

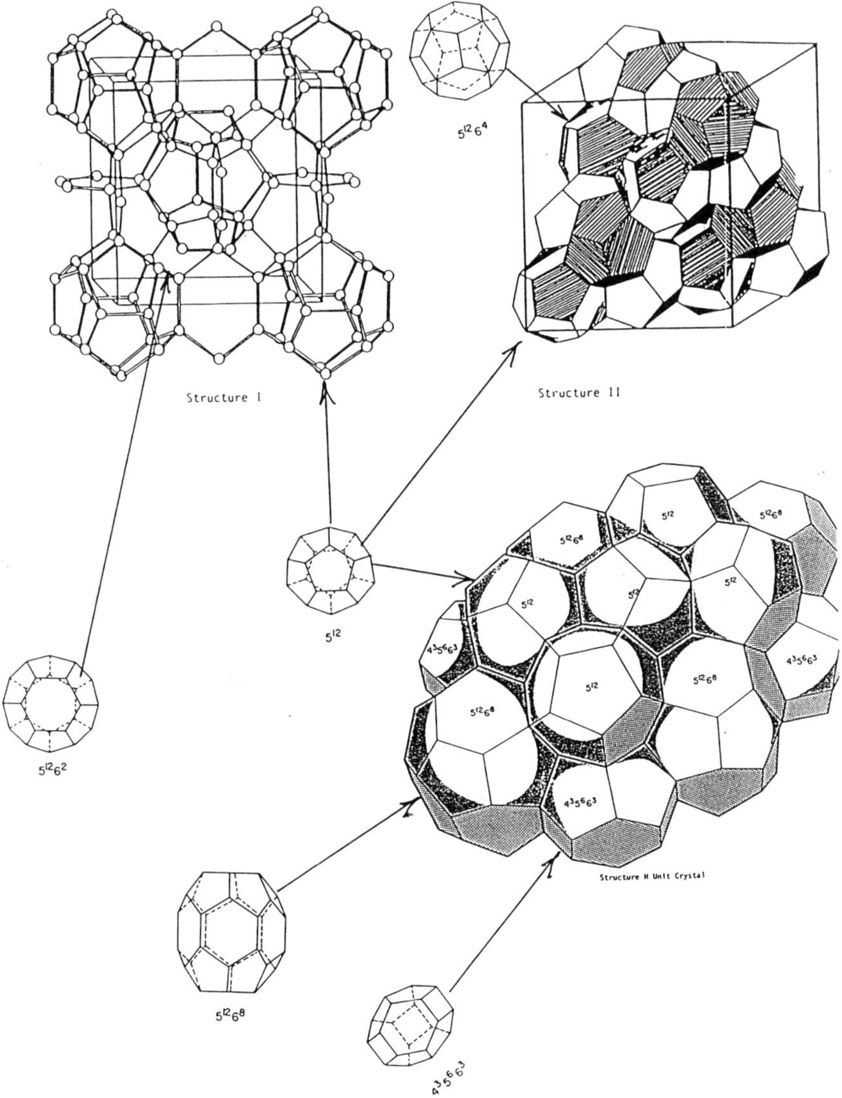

Fig. 1. Three unit crystals and their component cavities.

Kvenvolden 1993; Cranston 1994; Yakushev 1994; Harvey & Huang 1995). The purpose of this paper is to review physico-chemical properties of gas hydrates as applied to world margin stability and climatic changes.

Following this introduction, the second section addresses the question 'What are hydrates and how do they form?' In parallel with this foundation, the second section also considers the question, 'What are the physico-chemical properties of hydrates?' The third section deals with applications of physico-chemical properties to questions of margin stability and climatic change. The third section also provides a brief literature review of methane dissociation from hydrates. The fourth and final section deals with some basic research needs. The reader may wish to investigate these details further via references contained in several monographs (Makogon 1997; Sloan 1998).

What are hydrates and how do they form?

Hydrates normally form in one of three repeating crystal structures shown in Fig. 1. Structure I (sI), a body-centred cubic structure forms with small natural gas molecules found *in situ* in deep oceans. Structure II (sII), a diamond lattice within a cubic framework, forms when natural gases or oils contain molecules larger than ethane but smaller than pentane; sII represents hydrates which commonly occur in production and processing conditions, as well as in many cases of gas seeps from faults in ocean environments.

Physical properties of the newest hydrate structure H (Ripmeester *et al.* 1987; Mehta & Sloan 1993, 1994*a,b*, 1996) are in the initial stages of description. The hexagonal structure H (sH) has been shown by Ripmeester (1991) to have cavities large enough to contain molecules the size of common components of naphtha and gasoline. In addition, Sassen & MacDonald (1994) have found one instance of *in situ* sH in the Gulf of Mexico.

Hydrate crystal structures

Table 1 provides a hydrate structure summary for the three hydrate unit crystals (sI, sII and sH) shown in Fig. 1. The crystals structures are given with reference to the water molecule skeleton, composed of a basic 'building block' cavity which has 12 faces with five sides per face (abbreviated as 5^{12}). By linking the vertices of 5^{12} cavities one obtains sI. Linking the faces of 5^{12} cavities results in sII. In sH a layer of linked 5^{12} cavities connects layers of other cavities.

Interstices between the 5^{12} cavities are larger cavities which contain 12 pentagonal faces and either two, four or eight hexagonal faces (denoted as $5^{12}6^2$ in sI, $5^{12}6^4$ in sII or $5^{12}6^8$ in sH). In addition sH has a cavity with square, pentagonal and hexagonal faces ($4^35^66^3$). Figure 1 depicts the four cavities of sI, sII and sH. In Fig. 1 an oxygen atom is located at the vertex of each cavity angle; the lines represent hydrogen bonds by which one chemically bonded hydrogen connects to lone pair electrons on a neighbouring oxygen atom.

Inside each cavity resides a maximum of one guest molecule, typified by the eight guests associated with 46 water molecules in sI ($2[5^{12}]$ $6[5^{12}6^2]$ $46H_2O$), indicating two 5^{12} cavities and six $5^{12}6^2$ cavities in sI. Similar formulae for sII and sH are ($16[5^{12}]$ $8[5^{12}6^4]$ $136H_2O$) and ($3[5^{12}]$ $2[4^35^66^3]$ $1[5^{12}6^8]$ $34H_2O$), respectively.

Structure I, a body-centred cubic structure, forms with natural gases containing molecules smaller than propane; consequently sI hydrates are found *in situ* in deep oceans with biogenic gases containing mostly methane, carbon dioxide and hydrogen sulphide. Structure II, a diamond lattice within a cubic framework, forms when natural gases or oils contain molecules larger than ethane but smaller than pentane; sII represents hydrates from thermogenic gases. Formation of Structure H hydrate requires a small occupant (like methane, nitrogen or carbon dioxide) for the 5^{12} and $4^35^66^3$ cages, but the molecules in the $5^{12}6^8$ cage should be larger than 0.7 nm but smaller than 0.9 nm (e.g. methylcyclohexane).

From this point onward the review will emphasize sI hydrates which form with biogenic gases. As will be shown later, most oceanic hydrates are believed to be of biogenic gas origin, with only anecdotal evidence for thermogenic gas hydrates.

However, sII will also be briefly discussed in case thermogenic hydrates are found in substantial quantities in the future. Booth *et al.* (1996) suggests that most *in situ* hydrates have been found near faults, so that gas migration pathways might be available for both biogenic and thermogenic gases.

Table 1. *Geometry of cages in three hydrate crystal structures*

Hydrate crystal structure	I		II		H		
Cavity	Small	Large	Small	Large	Small	Medium	Large
Description	5^{12}	$5^{12}6^2$	5^{12}	$5^{12}6^4$	5^{12}	$4^35^66^3$	$5^{12}6^8$
Number of cavities/unit cell	2	6	16	8	3	2	1
Average cavity radius (Å)	3.95	4.33	3.91	4.73	3.91	4.06	5.71
Variation in radius* (%)	3.4	14.4	5.5	1.73		Not available	
Coordination number†	20	24	20	28	20	20	36
Number of waters molecules/unit cell		46		136		34	

* Variation in distance of oxygen atoms from the centre of the cage.
† Number of oxygen atoms at the periphery of each cavity.

Time-independent properties resulting from hydrate crystal structures

In this section three types of properties related to the foregoing crystal structures are discussed: mechanical properties; the guest/host size ratio concept; and how to use the size ratio to qualitatively explain phase equilibrium conditions. In this section phase equilibria and the size ratio are qualitatively shown to explain several thermal properties, and a property summary is provided for methane hydrates for those who wish to assume that methane alone is the gas present in ocean hydrates.

Mechanical properties of hydrates. As may be calculated via Table 1, if all the cages of each structure are filled, all three hydrates have the amazing property of being approximately 85% (mol) water and 15% gas. The fact that the water content is so high suggests that the mechanical properties of the three hydrate structures should be similar to that of ice. A comparison of ice with sI and sII hydrate mechanical properties is shown in Table 2. Many mechanical properties of sH have not been measured to date.

Guest/cavity size ratio: a basis for property understanding. The occupied hydrate cavity is a function of the size ratio of the guest molecule within the host cavity. To a first approximation, the concept of 'a ball fitting within a ball' is a key to understanding many hydrate properties. After an introduction, the concept is related to phase equilibrium conditions before relating the same concept to thermal properties later in this section.

Figure 2 (corrected from von Stackelberg 1949) may be used to illustrate five points regarding the guest/cavity size ratio for hydrates formed of a *single* guest component in either sI or sII.

- The sizes of stabilizing guest molecules range between 0.35 and 0.75 nm. Below 0.35 nm molecules will not stabilize sI, and above 0.75 nm molecules are too large for sII cavities.
- Some molecules are too large to fit the smaller cavities of each structure (e.g. C_2H_6 fits in the $5^{12}6^2$ of sI; or C_3H_8 and i-C_4H_{10} each fit the $5^{12}6^4$ of sII).
- Other molecules such as CH_4 and N_2 are small enough to enter both cavities (denoted as either $5^{12}+5^{12}6^2$ in sI or $5^{12}+5^{12}6^4$ in sII) when hydrate forms with those single components. Kuhs *et al.* (1996) have recently shown that two N_2 molecules can be accommodated in the $5^{12}6^4$ cavity at pressures greater than 0.5 kbar.
- The largest molecules of a gas mixture

Table 2. *Comparison of properties of ice and sI and sII hydrates*

Property	Ice	Structure I	Structure II
Spectroscopic			
Crystallographic unit cell			
Space group	P6$_3$/mmc	Pm3n	Fd3m
No. of H_2O molecules	4	46	136
Lattice Parameters at 273 K	$a=4.52$ $c=7.36$	12.0	17.3
Dielectric constant at 273 K	94	~58	58
Far infrared spectrum	Peak at 229 cm^{-1}	Peak at 229 cm^{-1} with others	
H_2O diffusion correlation time, (μsec)	220	240	25
H_2O diffusion activation energy (kJ m^{-1})	58.1	50	50
Mechanical property			
Isothermal Young's modulus at 268 K (10^9 Pa)	9.5	8.4^{est}	8.2^{est}
Poisson's ratio	0.33	~0.33	~0.33
Bulk modulus (272 K)	8.8	5.6	NA
Shear modulus (272 K)	3.9	2.4	NA
Velocity ratio (Comp/shear): 272 K	1.88	1.95	NA
Thermodynamic property			
Linear thermal expansion 200 K (K^{-1})	56×10^{-6}	77×10^{-6}	52×10^{-6}
Adiabatic bulk compression: 273 K (10^{-11} Pa)	12	14^{est}	14^{est}
Speed long sound: 273 K (km^{-1})	3.8	3.3	3.6
Transport			
Thermal conductivity: 263 K (W m^{-1} K^{-1})	2.23	0.4 ± 9.02	0.51 ± 0.02

Modified after Davidson (1983) and Ripmeester *et al.* (1994).

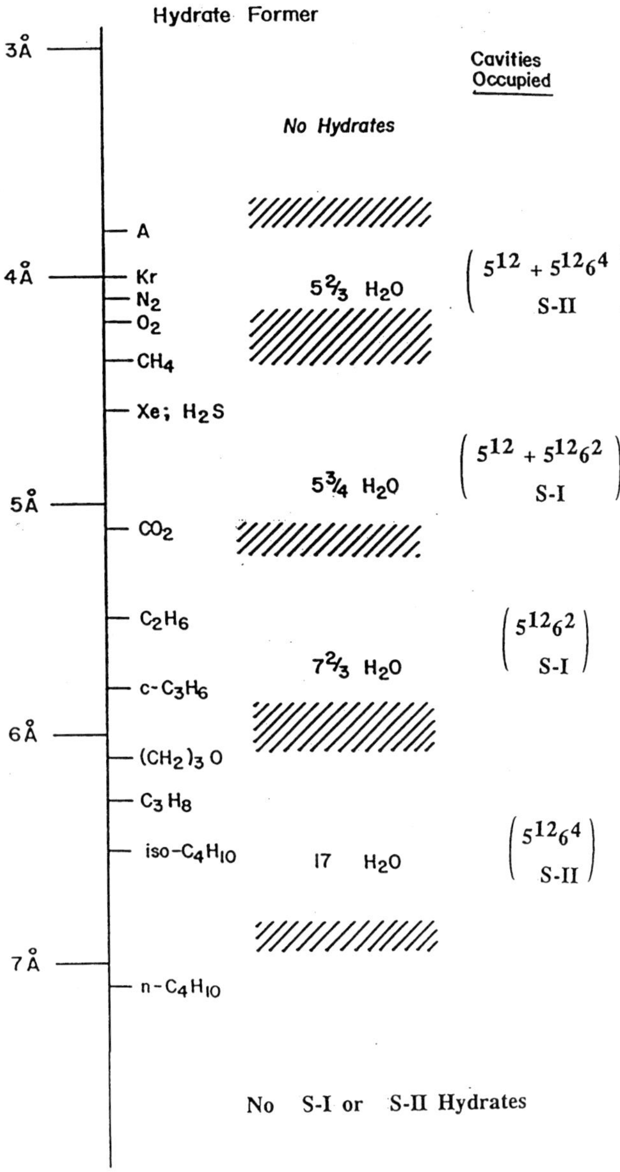

Fig. 2. Sizes of hydrate cavities and guest molecules.

usually determines the structure formed. For example, because propane and i-butane are present in many thermogenic natural gases, they will cause sII, to form. In such cases, methane will distribute in both cavities of sII and ethane will enter only the $5^{12}6^4$ cavity of sII.

- Molecules which are very close to the hatched lines separating the cavity sizes appear to exhibit the most non-stoichiometry due to their inability to fit securely within the cavity.

Table 3 shows the size ratio of several common gas molecules within each of the four cavities of sI and sII. Note that a size ratio (guest molecule/cavity) of approximately 0.9 is necessary for stability of a simple hydrate, indicated by ‡. When the size ratio exceeds unity, the molecule will not fit within the cavity and the

Table 3. *Ratios of molecular diameters* to cavity diameters† for some molecules including natural gas hydrate formers*

Molecule	Cavity type = > Guest diameter (Å)	(Molecular diameter) / (cavity diameter)			
		Structure I		Structure II	
		5^{12}	$5^{12}6^2$	5^{12}	$5^{12}6^4$
N_2	4.1	0.804	0.700	0.817‡	0.616‡
CH_4	4.36	0.855‡	0.744‡	0.868	0.655
H_2S	4.58	0.898‡	0.782‡	0.912	0.687
CO_2	5.12	1.00	0.834‡	1.02	0.769
C_2H_6	5.5	1.08	0.939‡	1.10	0.826
C_3H_8	6.28	1.23	1.07	1.25	0.943‡
i-C_4H_{10}	6.5	1.27	1.11	1.29	0.976‡
n-C_4H_{10}	7.1	1.39	1.21	1.41	1.07

* Molecular diameters obtained from von Stackelberg & Müller (1954).
† Cavity radii from Table 1 minus 1.4 Å water radii.
‡ Indicates the cavity occupied by the single hydrate guest.

structure will not form. When the ratio is significantly less than 0.9 the molecule cannot lend significant stability to the cavity.

Consider the single guest ethane, which forms in the $5^{12}6^2$ cavity in sI because ethane is too large for the small 5^{12} cavities in either structure and too small to give much stability to the large $5^{12}6^4$ cavity in sII. Similarly, propane is too large to fit any cavity except the $5^{12}6^4$ cavity in sII, so that gases of pure propane form sII hydrates from free water. On the other hand, methane's size is sufficient to lend stability to the 5^{12} cavity in either sI or sII, with a preference for sI because CH_4 lends slightly higher stability to the $5^{12}6^2$ cavity in sI than the $5^{12}6^4$ cavity in sII.

Phase equilibrium conditions. In Fig. 3 pressure is plotted against temperature, with gas composition as a parameter, for methane + propane mixtures. Consider a gas of any given composition (marked 0–100% propane) on a line in Fig. 3. At conditions to the right of the line, a gas of that composition will exist in equilibrium with liquid water. The mutual solubility of the aqueous and hydrocarbon phases is only a few parts per thousand. The interface is the only point where the two ingredients are in sufficient concentrations (85% water, 15% hydrocarbon) to form hydrates.

As the temperature is reduced (or as the pressure is increased) hydrates form from gas and liquid water at the Fig. 3 line for the given gas composition. At that condition three phases (liquid water + hydrates + gas (L_W + H + V)) will be in equilibrium. With further reduction of temperature (or increase in pressure) the fluid phase which is not in excess (gas in ocean environments) will be exhausted. To the left of the line hydrate will exist in two-phase equilibrium with excess water.

All of the conditions given in Fig. 3 are for temperatures above 273 K and pressures along the lines vary exponentially with temperature. Put explicitly, hydrate stability at the three-phase (L_W–H–V) condition is always much more sensitive to temperature than to pressure.

Figure 3 also illustrates the dramatic effect of gas composition on hydrate stability; as any amount of propane is added to methane the structure changes (sI → sII) to a hydrate with much wider stability conditions. Note that at 275 K a 50% decrease in pressure is needed to form sII hydrates, when only 1% propane is in the gas phase. sII forms at higher temperatures than sI.

Figure 3 provides a convenient illustration of two common ways to dissociate hydrates. By increasing the temperature at constant pressure, the system is moved first to the three-phase line, where dissociation occurs at constant temperature and pressure with input of the heat of dissociation. Alternatively by decreasing the pressure the system is moved to the three-phase line, so that the temperature is lower than ambient and heat flows to dissociate the hydrate.

When the hydrate is massive and the initial temperature is close to the ice-point, removal of the heat of formation will cause the temperature to drop below the 273 K so that any residual water may be converted to ice. Yakushev & Istomin (1991) and Gudmundssen *et al.* (1994) both

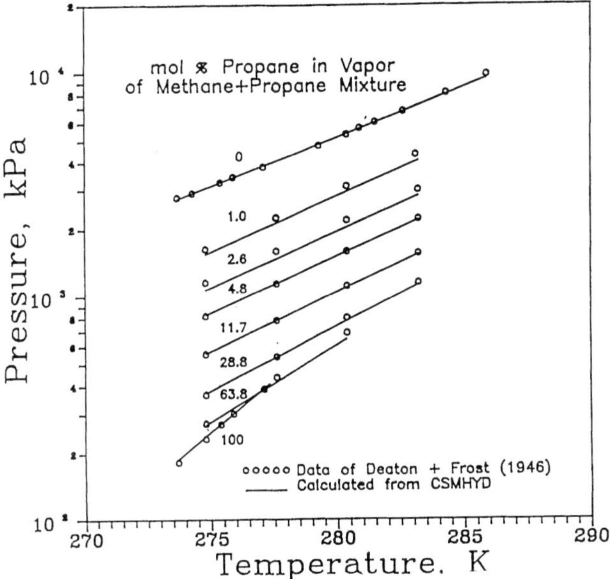

Fig. 3. Hydrate formation conditions for mixtures of methane and propane from water and gas.

suggest that ice cladding can inhibit further hydrate dissociation.

Any discussion of hydrate phase equilibrium would be incomplete without remarking that hydrates provide the most-used case of statistical thermodynamics to predict phase equilibria by industry. The van der Waals & Platteeuw (1959) model was formulated after the determination of sI and sII crystal structures shown in Fig. 1. With this elegant mathematical model, one may predict the three-phase pressure or temperature of hydrate formation, knowing the gas composition. For further discussion see Sloan (1998, Chap. 5).

Thermal properties. In this subsection three properties (heat of dissociation, thermal conductivity and thermal expansivity) are discussed in relation to the above size ratio of guest : host and phase equilibrium conditions.

Heat of dissociation. The heat of dissociation (ΔH_d) is defined as the enthalpy change to dissociate the hydrate phase to a vapour and aqueous liquid, with values given at temperatures just above the ice-point. The heat effect due to this phase change is generally much larger than the sensible heat effect (which uses heat capacities, C_p) without a phase change.

Thermodynamics provide a convenient result of being able to obtain properties like ΔH_d, which are difficult to measure, using the easily measured three-phase lines like those shown in Fig. 3, along with the Clausius–Clapeyron equation:

$$\Delta H_d = -\frac{zR\mathrm{d}(\ln P)}{d\frac{1}{T}} \tag{1}$$

where z and R represent the compressibility and the universal gas constant, respectively. Equation (1) provides the surprising facility of being able to estimate values from ΔH_d from the slope of the ln P vs $(1/T)$ lines.

Sloan & Fleyfel (1991) show that to a fair engineering approximation ($\pm 10\%$) ΔH_d is: (1) a function not only of the hydrogen bonds in the crystal but also of cavity occupation; (2) independent of guest components; and mixtures of similarly-size components, and (3) without an occupant, cavity dissociation is more difficult, resulting in a higher ΔH_d.

As one illustration, simple hydrates of C_3H_8 or i-C_4H_{10} have a similar ΔH_d of 129 and 133 kJ mol^{-1} (Handa 1986) because they both occupy the $5^{12}6^4$ cavity, although their size ratios are somewhat different (0.943 and 0.976). This similarity of ΔH_d is remarkable, but it is due to both the stabilization of the $5^{12}6^4$ cavity and the similar hydrogen-bonded water unit cell skeleton.

Similar statements could be made about the ΔH_d values for other simple hydrate formers which occupy similar size cavities, such as C_2H_6

($\Delta H_d = 72\,kJ\,mol^{-1}$, Handa 1986) and CO_2 ($\Delta H_d = 73\,kJ\,mol$, Long 1994) in the $5^{12}6^2$ cavity, or CH_4 and H_2S (ΔH_d within 3% of each other, Long 1994) which occupy both 5^{12} and $5^{12}6^2$ as simple hydrates.

As a second illustration, mixtures of C_3H_8 + CH_4 shown in Fig. 3 have a value of $\Delta H_d = 79\,kJ\,mol^{-1}$ over a wide range of composition. In such mixtures C_3H_8 occupies most of the $5^{12}6^4$ cavities, while CH_4 occupies a small number of $5^{12}6^4$ and many 5^{12} cavities. In fact, most natural gases (which form sII) have similar values of ΔH_d.

The reader should note that Skovborg & Rasmussens (1994) concerns about the above approximation were addressed by Sloan & Fleyfel (1992), and that the approximations were later confirmed by Long (1994) and shown to apply to Structure H by Mehta (1996).

Thermal conductivity. Stoll & Bryan (1979) first measured thermal conductivity of propane hydrates to be a factor of 5 less than that of ice ($2.23\,W\,m^{-1}\,K^{-1}$). Cook & Laubitz (1981), Ross & Andersson (1982), Cook & Leaist (1983) and Asher (1987) confirmed the low thermal conductivity of hydrates, as well as similarities of the values for each structure. The thermal conductivity of the solid hydrate ($0.49\,W\,m^{-1}\,K^{-1}$) more closely resembles that of liquid water ($0.605\,W\,m^{-1}\,K^{-1}$).

Ross *et al.* (1981) also determined that tetrahydrofuran hydrate thermal conductivity was proportionally dependent upon temperature, but had no pressure dependence. Ross & Andersson (1982) suggested that this behaviour, which had never before been reported for crystalline organic materials, was associated with the properties of glassy solids.

In the hydrate lattice structure, the water molecules are largely restricted from translation or rotation, but they do vibrate anharmonically about a fixed position. This anharmonicity provides a mechanism for the scattering of phonons (which normally transmit energy) providing a lower thermal conductivity. Tse *et al.* (1983*b*, 1984) and Tse & Klein (1987*a*) used molecular dynamics to show that frequencies of the guest molecule translational and rotational energies are similar to those of the low frequency lattice (acoustic) modes.

Tse (1994) notes that weak coupling between the guest and host lattice does not noticeably affect most structural thermodynamic and mechanical properties, but such coupling has a marked effect on the transport of heat. As defects normally serve to decrease any crystal thermal conductivity, hydrate cavities might be considered as severe defects which result in anomalously low values of thermal conductivity.

Thermal expansion of hydrates. Linear thermal expansion coefficients of hydrate ($dl/l dT$) for structures I, II and ice have recently been determined through dilatometry by Roberts *et al.* (1984) and through X-ray powder diffraction by Tse *et al.* (1987). The values for sH hydrate at 200 K have been measured for hexamethylethane (HME) and 2,2-dimethylbutane (DMB) at 150 and 200 K by Tse (1990) who noted that cubic expansion values are similar to those of sI and sII, but that there is a difference in the direction of linear expansion for sH. At 200 K linear thermal expansions are: sI ($77 \times 10^{-1}\,K^{-1}$); sII ($52 \times 10^{-6}\,K^{-1}$); sH ($a = 67 \times 10^{-6}$, $c = 59 \times 10^{-6}\,K^{-1}$ for DMB) and ice ($a = 56 \times 10^{-6}$, $c = 57 \times 10^{-6}\,K^{-1}$).

Through constant pressure molecular dynamic calculations for thermal expansion of ice and of empty structure I, Tse & Klein (1987) determined that the high hydrate thermal expansivity is due to anharmonic behaviour in the water lattice. Tse (1994) suggests that this results from collisions of the guest molecule with the cage wall, which exerts an internal pressure to weaken the interaction between the water hydrogen bonds.

Summary of physico-chemical properties of methane hydrates. In the next section, on 'Applications to margin stability and climatic changes, it is argued that most oceanic hydrates are currently assumed to be sI of biogenic methane, due to the anecdotal instances of thermogenic hydrates with significant amounts of propane and higher hydrocarbons. While there seems to be concurrence on this point in the literature, there are several exceptions cited in the Gulf of Mexico and the Caspian Sea.

As a summary of the physico-chemical properties, Table 4 provides a listing of the methane hydrates properties which will be of interest in quantifying any exploration, formation or dissociation modelling.

The modeller will, of course, wish to account for the system fraction which is hydrate, relative to free gas, water and sediment. In the absence of measurements or theory, a linear combination on a mole fraction basis is usually assumed. Handa & Stupin (1992), Zakrzewski & Handa (1993) and, recently, Bondarev *et al.* (1996) have indicated that the linear approximation is flawed.

Kinetics of formation as related to hydrate crystal structures

The answer to the questions, 'What are hydrates?' and 'Under what condition do hydrates form?' in the previous sections is much

Table 4. *Selected properties for methane hydrate*

Property (unit)	Value or correlation
Dissociation pressure (KPa)	exp [38.98 − 8533.8/T(K)]
Heat of dissociation (kJ mol^{-1})	54.2 at 273 K to water and vapour
Heat capacity (J K^{-1} mol^{-1})	260.6 at 273 K
Thermal conductivity (W m^{-1}K^{-1})	0.49
Density (g cm^{-3})	0.90
Poisson's ratio	0.33
Velocity of sound (km s^{-1})	3.3

more certain than answers to 'How do hydrates form?' The question of 'Why do hydrates form?' is not considered. The mechanism and rate (i.e. the kinetics) of hydrate formation are controversial topics at the forefront of current research.

The kinetics of hydrate formation are clearly divided into three parts: (a) nucleation of a critical crystal radius, (b) growth of the solid crystal, and (c) the transport of components to the growing solid–liquid interface. All three kinetic components are currently under study, but a satisfactory answer has not been found to any of them. Skovborg (1993) proposed the most recent quantitative model, based upon mass transport as a limiting factor. Skovborgs model was based in part on a re-analysis of the data measured by Englezos *et al.* (1987*a*,*b*). The latter researchers proposed the best extant crystal growth model. In the current work an hypothesis is summarize for a nucleation mechanism of hydrates, based upon the above overview of the crystal structures.

In a series of successively revised mechanisms for the nucleation hypothesis (Sloan 1990; Sloan & Fleyfel 1991; Müller-Bongartz *et al.* 1992) it has been proposed that clusters at the interface may grow to achieve a critical radius as shown schematically in Fig. 4. Christiansen & Sloan (1994) extended the hypothesis model, with the following elements.

- Pure water exists with many transient, labile ring structures of pentamers and hexamers.
- Water molecules form labile clusters around dissolved guest molecules. These clusters are quantified in units of four water molecules as a function of dissolved guest size range. The number of water molecules in each cluster shell (i.e. the coordination number) for natural gas components are: methane (20), ethane (24), propane (28), iso-butane (28), nitrogen (20), hydrogen sulphide (20) and carbon dioxide (24)
- Clusters of dissolved species combine to form unit cells. To form sI, coordination numbers of 20 and 24 are needed for the 5^{12} and the $5^{12}6^2$ cavities, respectively; sII requires coordination numbers of 20 and 28 for the 5^{12} and $5^{12}6^4$ cavities. Nucleation is facilitated if labile clusters are available with both types of coordination numbers for either sI (e.g. $CH_4+C_2H_6$ mixtures)or sII (e.g. $CH_4+C_3H_8$ or most unprocessed natural gases). If the liquid phase has clusters of only one coordination number, nucleation is inhibited until the clusters can transform to the other size, by making and breaking hydrogen bonds.
- An activation barrier is associated with the cluster transformation. If the dissolved gas is methane, the barrier for transforming the cluster coordination number from 20 (for the 5^{12}) to 24 (for the $5^{12}6^2$) is high, both because the guest cannot lend much stability to the larger cavity and because the $5^{12}6^2$ cavities outnumber the 5^{12} in sI by a factor of 3. Energy for transformation of methane–water clusters from 20 to 28 is higher than that from 20 to 24, because methane is not large enough to stabilize the $5^{12}6^4$ cavity.
- If the dissolved gas is ethane with a water coordination number of 24, the transformation of empty cavities with coordination numbers is likely to be rapid, due to the high ratio (3:1) of $5^{12}6^2$ to 5^{12} cavities in sI. If the dissolved gas is propane with a coordination number of 28, transformation to sII is likely to be slow because $5^{12}6^4$ cavities are outnumbered by the 5^{12} cavities by a factor of 2.
- Table 5 shows limited experimental confirmation of nucleation rate as a function of available labile cavities. Data in the table were measured at constant pressure difference ($P^{\exp} - P^{\mathrm{eq}}$) at 0.5°C and shows that methane and propane have long induction times, while short induction times were obtained for ethane, $CH_4+C_2H_6$, $CH_4+C_3H_8$, and natural gas mixtures.

Table 5. *Experimental and cluster hypothesis predictions of induction times*

Crystal structure	Gas component	Pressure at 0.5°C		Induction time	
		Experimental (MPa)	Equilibrium (MPa)	Hypothesis	Measured (h)
sI	CH_4	5.51	2.85	Long	>24
sI	CO_2	3.38	1.43	Short	2.0
sI	N_2O	3.03	1.20	Short	0.3
sI	Xe	0.51	0.23	Short	2.6
sI	C_2H_4	2.34	0.66	Short	1.5
sI	C_2H_6	2.14	0.56	Short	0.8
sI	$CH_4 + C_2H_6$ blends			Short	
sII	C_3H_8	0.31	0.19	Long	>24
sII	i-C_4H_{10}			Long	
sII	95/5 blend of CH_4/C_3H_8	2.89	0.70	Short	3.9
sII	CH_4/i-C_4H_{10} blends			Short	
sII	Natural gas[1]	3.33	0.76	Short	4.0

[1] 0.4% N_2, 87.2% CH_4, 7.6% C_2H_6, 3.1% C_3H_8, 0.5% i-C_4H_{10}, 0.8% n-C_4H_{10}, 0.2% ni-C_5 H_{12}, 0.2% n-C_5H_{12}.

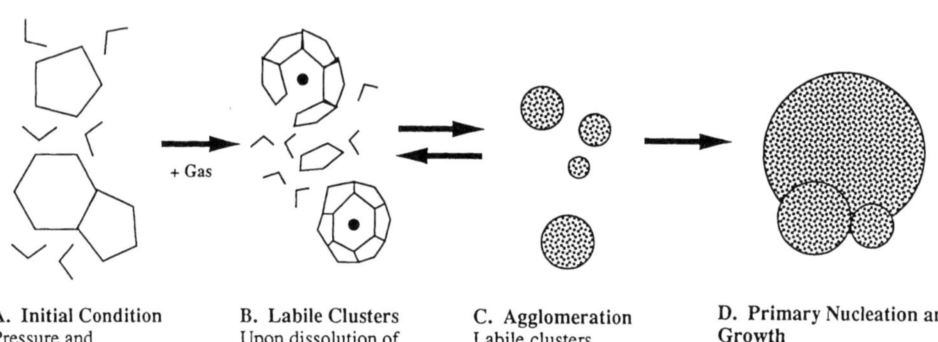

A. Initial Condition
Pressure and temperature in hydrate forming region, but no gas molecules dissolved in water

B. Labile Clusters
Upon dissolution of gas in water, labile clusters form immediately.

C. Agglomeration
Labile clusters agglomerate by sharing faces, thus increasing disorder.

D. Primary Nucleation and Growth
When the size of cluster agglomerates reaches a critical value, growth begins.

Fig. 4. Mechanism hypothesis for kinetics of hydrate formation.

- Alternating structures arise which provide parallel formation pathways and consequently slow nucleation kinetics. Consider, connections of hexagonal faces in large cavities of $5^{12}6^2$ and $5^{12}6^4$ to make sI and sII, for example, from pure components of ethane and propane, respectively. In building sI from ethane, a hexagonal face of one $5^{12}6^2$ cavity is joined to a hexagonal face of another $5^{12}6^2$, but all orientations appear to give similar crystal structures, so there is only one formation path to connect $5^{12}6^2$ at hexagonal faces. However, the $5^{12}6^4$ has two alternative connections, leading to two types of sII (cubic and hexagonal) and consequently slower formation of the normal hexagonal sII.

The final point (the alternating structures component in the hypothesis) has come under criticism, first by Skovborg (1993) and then by Natarajan (1993). However, Skovborg noted that alternating structures may account for some of his nucleation data.

The above cluster model hypothesis is not restricted to the bulk solution, but can occur at the interface, either in the liquid or the vapour side. Such models were recently proposed by Long & Sloan (1994) and by Kvamme (1996). The reader should note that the above is a largely unproven hypothesis, whose only justification is to serve as a mental picture for qualitative predictions and future corrections.

It should be emphasized that in direct contrast to well-determined thermodynamic properties, kinetic characterization of hydrate formation/dissociation is very illdetermined. One has only to turn to the recent review of hydrate kinetics by Englezos (1995) or to the authors 1997 monograph (Sloan 1997) to determine the following unsettling facts which act as a state-of-the-art summary.

- Hydrate nucleation is both heterogeneous and stochastic, and therefore is only approachable by very approximate models. Most hydrate nucleation models assume homogeneous nucleation and typically cannot fit *c*. 20% of the data generated in the laboratory of the modeller.
- Hydrate kinetics are apparatus dependent, i.e. the results from one laboratory are not transferable to another laboratory or field situation.
- As in thermodynamic studies, the hydrate phase is almost never measured in kinetic studies.
- While the best hydrate dissociation models are derived from solid moving-boundary differential equations (e.g. Yousif & Sloan 1991), it is clear those models do not account for the porous, surface formation and occlusion nature of hydrates on a macroscopic scale.
- No satisfactory kinetic model currently exists for formation or dissociation.

Applications to margin stability and climatic change

Amount, source and phase equilibria of naturally occurring hydrates

Table 6 provides a decade of estimates of natural gas in hydrates in the geosphere. These estimates range from the maximum value of Dobrynin *et al.* (1981), who apparently assumed that hydrates could occur wherever satisfactory thermodynamic conditions exist, to the minimum values of McIver (1981) and Meyer (1981) who considered more limiting factors, such as availability of methane, limited porosity and percentages of organic matter, thermal history of various regions, etc. All of the estimates of natural gas hydrates are not well defined, and therefore somewhat speculative. However, the most recent estimates, made by independent investigators through different methods, converge on very large values of gas reserves in hydrated form.

One appraisal of the amount of *in situ* hydrates (Kvenvolden 1988*a*) was obtained by scaling a gross estimate of the amount of hydrates in the continental margin of northern Alaska to the total length of continental margins world-wide. In Table 6, note that each investigator determined the hydrate reserve in the ocean to be at least 2 orders of magnitude greater than that in the permafrost.

Estimates of the oceanic hydrate reserves are so large that any error in those approximations would encompass the entire permafrost hydrate reserves. Even the most conservative estimates of gas in hydrates in Table 6 indicate their enormous energy potential. Kvenvolden (1988*a*) indicated that the 10 000 Gigatons (1 Gt $= 10^{15}$ g) or 1.8×10^{16} m^3 of carbon in hydrates may surpass the available carbon in the global cycle by a factor of 2.

The most recent gas composition discussions of ocean hydrates by Kvenvolden (1993) and by Collett (1995) indicate that hydrates can usually be represented by methane gas without other gas components. Biogenic gases are thought to be pervasive. Thermogenic gases are much less usual and result from migration from depths along faults.

Several notable exceptions exist to the above generalization. Brooks *et al.* (1986) noted that in the Gulf of Mexico approximately equal numbers of biogenic (sI) and thermogenic (sII) hydrates have been found. Ginsburg (1994) notes examples of thermogenic hydrates from mud volcanoes in the Caspian Sea, and Sassen & MacDonald (1994) provided the initial finding of sH hydrates, also in the Gulf of Mexico.

Table 6. *Estimates of methane in* in situ *hydrates*

Permafrost hydrates (m^3)	Oceanic hydrates (m^3)	Reference
5.7×10^{13}	$5 \times 25 \times 10^{15}$	Trofimuk *et al.* (1977)
3.1×10^{13}	3.1×10^{15}	McIver (1981)
3.4×10^{16}	7.6×10^{18}	Dobrynin *et al.* (1981)
1.4×10^{13}		Meyer (1981)
1.0×10^{14}	1.0×10^{16}	Makogon (1988)
	1.8×10^{16}	Kvenvolden (1988*a*)

In a recent database summarizing the *in situ* recovered hydrate sample database, Booth *et al.* (1996) gives three important generalizations.

- Seventy per cent of hydrates are located at higher pressures or lower temperatures than the three-phase (L_W–H–V) boundary prediction. Evidence for this generalization is shown in a phase diagram in Fig. 5. While it is not explicitly stated, this result suggests that the system is at two-phase (L_W–H) equilibrium, wherein the gas phase is exhausted and the liquid phase (with dissolved methane) exists in equilibrium solely with hydrates.
- The extreme of any hydrate equilibria condition is coincident with the three-phase condition predicted using salt water.
- Most recovered samples have been small, dispersed (even dissociated) particles with isolated examples of massive hydrates

The importance of the above generalizations can be related to the above phase equilibrium principles by an example. Consider the case for the dissociation of a hydrate sample such as MAT Guatemala 2 (the uppermost point on Fig. 5) which is over 15°C cooler than the three-phase boundary. Before the hydrate melts, it must be heated to the three-phase boundary at constant pressure. The hydrate and the surrounding media must have a considerable sensible heat input term ($\Delta H = m\ C_p \Delta T$, where H is enthalpy, m is mass, Cp is heat capacity and ΔT is the temperature difference require to move the system to the three-phase boundary). This heat input is more than the few degrees usually cited as a consequence of, for example,

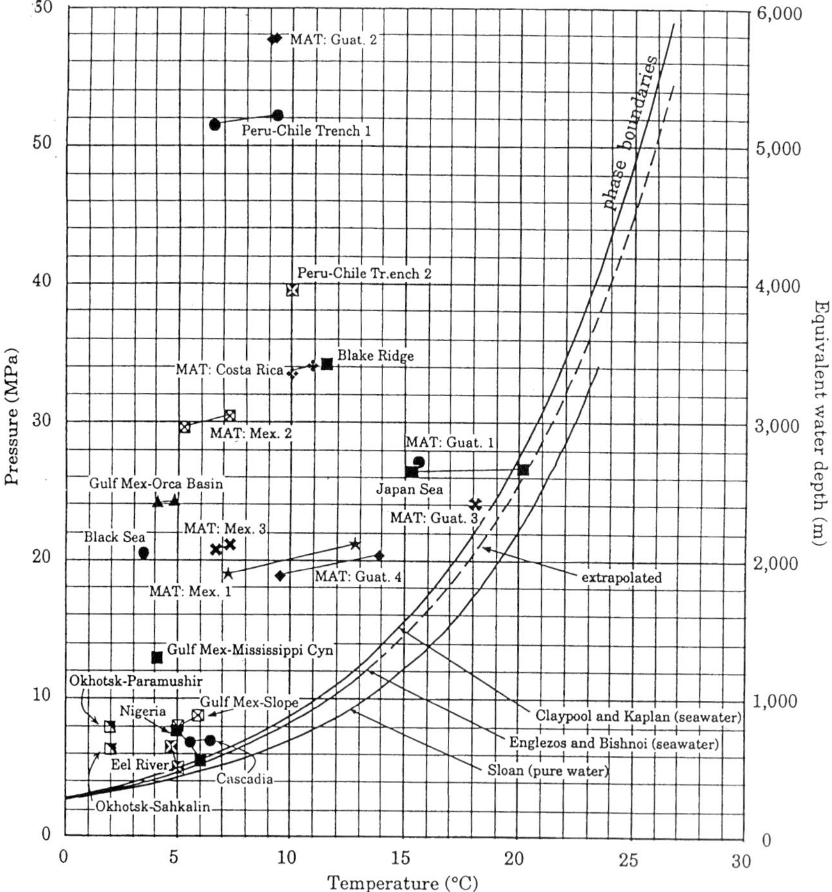

Fig. 5. Relation of *in situ* temperatures and pressures of hydrate samples to the three-phase (L_w–H–V) boundary (Booth *et al.* 1996).

global warming or warming by an eddy from the Gulf Stream. At the three-phase boundary, hydrates melt at constant temperature and pressure as the heat of dissociation ($\Delta H = m\ \Delta H_d$) is input.

The above example stands in contrast to those cases by MacDonald *et al.* (1994) who have reported incidences of Gulf of Mexico hydrate dissociation over a short period. In these cases hydrates formed from seeping gas at a fault; these hydrates were already at the three-phase boundary, so they could be dissociated by a small heat input from the surrounding water.

On the other hand, the hydrate samples found by MacDonald *et al.* (1994) at faults indicate that methane did not escape as rapidly as it would have if hydrates had not formed. In such cases it is clear that hydrates act as a trap/storage for venting methane, as well as a release of methane. As mentioned earlier, Yakushev & Istomin (1991) suggest that *in situ* permafrost hydrates self-preserve when they melt by using the heat of dissociation to cool the residual water below the ice-point, so that a cladding of ice prevents further dissociation.

Review or literature regarding hydrate effects on climate

Because the author has not done research on the question relating hydrates to climatic change, he must claim ignorance of the area, other than the below cursory literature review. In fact, every other author in the present volume is more qualified than the author on this topic, so this brief overview should be regarded with some skepticism, serving only as a personal perspective and without intent to pre-empt other speakers.

The original concern for hydrates as a input to the climate arose as a result of the review by Kvenvolden & McMenamin (1980), shortly thereafter Bell (1982), Kvenvolden (1988*b*), MacDonald (1990) and Nisbet (1989) expressed concern about methane release by *in situ* hydrates. In an alarmist perspective, Leggett (1990) notes that, "the uncertainties are enormous and the stakes are probably higher than with any other potential feedback (mechanism)".

The original concerns for current release of methane from hydrates were highly speculative and have since been mitigated by many researchers. Makogon (pers. comm. 1997) notes that if the sea temperature were 1°C higher and if the geothermal gradient were increased by 1°C per 100 m, then up to $3 \times 10^9\ m^3$ of free methane can be released per km^2 of hydrate deposits. However, currently a sanguine view appears to hold consensus in the literature by researchers such as Fei (1991), Cranston (1994), Yakushev (1994) and Kvenvolden *et al.* (1993). However, there is general agreement that the most jeopardy exists for hydrates at the three-phase boundary which can be affected by small warming trends.

The most detailed mathematical models (with the fewest assumptions) have been done by Englezos (1992) and Hatzikiriaos & Englezos (1993), and recent estimates by Harvey & Huang (1995). Both sets of modelers assume 'worst case' conditions for methane release. It appears that Englezos *et al.* would agree with the final conclusion from Harvey & Huang (1995):

> Uncertainty in future global warming due to potential methane clathrate destabilization is thus smaller than the uncertainty due to future fossil fuel use or climate sensitivity.

Ancient climates and CO_2 sequestering. There are two other situation relating hydrates to environmental concerns which deserve attention: (1) ancient age methane releases and ice hydrates of air from Antarctica; and (2) carbon dioxide storage as hydrates in the ocean.

Nisbet (1989) and MacDonald (1990) argue that destabilization of methane hydrates was the principal mechanism that caused the warming trend to end the ice age. More recently, Dickens *et al.* (1995) suggest that during the latent Paleocene (~55.6 Ma) thermal dissociation of methane hydrate accounted for the −2 to −2‰ excursion observed in the $\delta^{13}C$ isotopic record.

Ancient climate revelations may also be determined by recovery of air hydrates from Antarctica, as reported by Hondoh (1996). In these conditions, Hondoh has observed that two air bubbles were buried at equivalent depths (*c.* 200 m) for about 60 000 years. One bubble completely transformed to a single hydrate crystal, while the second remained an air bubble with no hydrate formation. The gas concentrations in these bubbles seem to follow the Earths ancient temperature oscillations.

The case of CO_2 storage in ocean hydrate was proposed as a means of power plant stack gas disposal by many workers in Japan, as illustrated by the work of Aya *et al.* (1992). Later Aya *et al.* (1995) showed that hydrate would dissolve in sea water (which was unsaturated with CO_2) and thus become unstable. A Japanese summary perspective on this question is in a book representing a two-conference collection of articles edited by Handa & Ohsumi (1995). Harrison *et al.* (1995) reached the conclusion of frequent hydrate instability, but suggested that the sequestered CO_2 might become rock, or it might raise the pH of the ocean to a level which is intolerant of life, dependent upon the surrounding geology and fluids. The most recent model on CO_2 hydrate formation and dissolution is by Mori & Mochizuki (1997).

Challenges for future fundamental research on hydrate properties

It is clear that there are three sources of concern for *in situ* oceanic hydrates: (1) as a resource; (2) a geohazard; or (3) a factor in current or ancient climate changes. All three perspectives are long range but, in the opinion of the author, addressing the first concern may be the most fruitful for the future. However, the latter two concerns may provide vehicles to do fundamental research which can be applied to many situations. Similarly, the funding of research to prevent hydrate pipeline pluggage has provided fundamental results with multiple applications.

There are several areas which require definition if we are to explore hydrates in the ocean. First we must characterize the solid phase of laboratory specimens. Secondly, we must characterize the hydrates *in situ*. Finally, we must compose comprehensive models which describe the formation and (more importantly) the dissociation of hydrates in the ocean. Because other tutorials likely to address the last two challenges, here I concentrate on the first challenge, both by measurements and by calculation.

Characterizing the solid hydrate phase in the laboratory

Importance of sample characterization. The essence of all laboratory work is repeatability. That is, the same results should be obtained for similar hydrate samples in laboratories in Europe, Asia or the Americas. To ensure repeatability, the sample itself should be well characterized so that various laboratories can determine reproducibility of measurements. For example, it is mandatory that one should know not only how much of a sample is hydrated, but also whether the hydrate amount is evenly or locally distributed.

Unfortunately in most laboratory experiments, the hydrate phase has not been characterized directly. Normally measurement(s) of the gas or liquid phase is made over the course of the experiment; and the difference is attributed to hydrate phase formation. For example, one typically measures the change in gas pressure and composition in a constant volume experiment, and uses a mass balance to determine that the pressure reduction and composition change is due to hydrate formation. These indirect measurements are often aided by sophisticated models (e.g. van der Waals & Platteeuw 1959) which predict properties of the hydrate phase.

However, it is unsatisfactory only to make indirect measurements of the hydrate phase, with the current state-of-the-art experimental techniques. The indirect measurement of the hydrate phase has arisen from three causes: (1) it is difficult to make a reproducible hydrate sample, (2) there are few means to characterize the solid hydrate phase; and (3) the accurate method of van der Waals & Platteeuw (1959) allowed sophisticated prediction of the hydrate phase. The latter reason may have caused researchers to become somewhat complacent, satisfied with the *status quo* of prediction techniques, when new measurement techniques were at hand.

Producing reproducible samples. When hydrates form with hydrocarbons and water, the solid hydrate phase at the interface prevents further contact between the vapour and liquid phase. The interface is the site of formation because the water content of hydrates (85 mol%) and the hydrocarbon content (15%) is higher than either the solubility of water in the hydrocarbon or the hydrocarbon solubility in water. The result is that hydrates form an open, porous structure (even when sediment is not present) with significant amounts of occluded water between solid boundaries. As a consequence it is very difficult to convert a constant amount of water to hydrate during repeated experiments, so that the hydrate sample amount and composition may vary between experiments.

Experimenters have often been able to circumvent the above difficulty by using a miscible hydrate former such as ethylene oxide (a sI former) or tetrahydrofuran (sII). With a miscible former the liquid solution is formulated at the hydrate composition (e.g. just above the termination of the vertical line in the tetrahydrofuran phase diagram of Fig. 6) so that cooling produces hydrates with no change in composition, either at 4.4°C (sII) or at 12.7°C (for ethylene oxide sI). Miscible hydrate former samples are easily made, with acceptably small compositional variations.

As indicated in the physical properties section, many mechanical and thermal properties are determined by the water crystal skeleton which composes 85% of the structure. A miscible

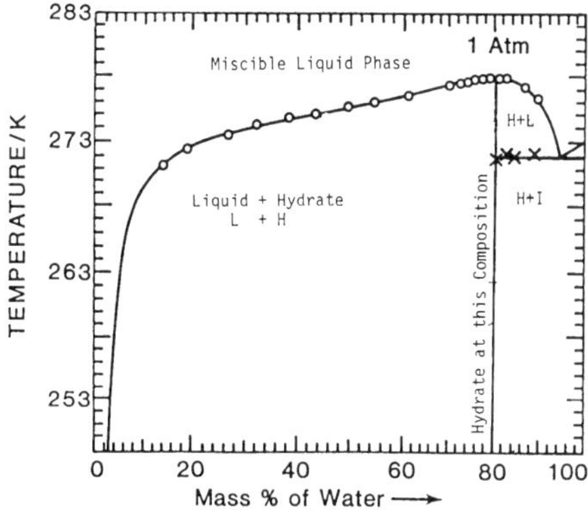

Fig. 6. Phase diagram for the binary system water + tetrahydrofuran (THF).

hydrate former provides a reproducible hydrate structure and composition for such measurements. As indicated in the section on 'What are hydrates and how do they form?' the hydrate guest will affect such measurements as phase equilibria conditions, but mechanical and thermal properties are relatively independent of guest.

Laboratory/pipeline hydrates differ from in situ *ocean*. hydrates Kinetics cause hydrate samples in the laboratory and the pipeline to be very unlike ocean floor hydrates. Hydrates in laboratories and pipelines are porous plugs which have the unusual property of transmitting pressure while inhibiting flow, as is clearly shown from the laboratory experiments of Lysne (1995), as well as results from Norsk Hydro (Lingelem *et al.* 1994) and Statoil (Austvik *et al.* 1995). Hydrate plug porosity is caused by the fact that hydrates form at the interface and the solid prevents further contact of the gas and liquid, inhibiting growth. Thus, laboratory and pipeline hydrate plugs tend to be open, porous masses. Growth to a solid, ice-like mass will only occur after several weeks or longer, during which time diffusion and flow must occur through the hydrate solid shell fissures.

In contrast, hydrate samples recovered form the ocean floor are solid, ice-like masses, considerably different to the laboratory samples. However, a new technique for generating ocean floor-like hydrates has recently been proven. Using X-ray diffraction Stern *et al.* (1996) have been remarkably successful at converting 97% of ice to water by raising ice grain temperatures above the solidus while under high pressure in an annealing-like procedure. Workers who wish to simulate ocean hydrates may wish to consider these results carefully.

This dissimilarity of normal laboratory samples, and natural samples, reinforces the conclusion that solid-phase characterization is one of the most important, yet most often neglected, steps in hydrate experiments.

Thermodynamic measurements of the solid hydrate phase. For hydrate phase measurements there are two sorts of devices, diffraction and spectroscopic measurements. The most recent diffraction measurements are of the neutron-type, as neutron diffraction can detect both hydrogens and guest molecules, while X-ray diffraction detects the oxygen positions. Recent neutron diffraction measurements are typified by Tse (1994) and Kuhs *et al.* (1996).

Three types of spectroscopy have been used for hydrates: (1) nuclear magnetic resonance (NMR) with cross polarization (CP) and magic-angle spinning (MAS); (2) Fourier transform infrared (FT-IR) spectroscopy; and (3) Raman spectroscopy.

The first comprehensive review of NMR studies of clathrates was written by Davidson & Ripmeester (1984), and a thorough update has been written by Ripmeester & Ratcliffe (1990). Most of this overview is taken from the latter reference. Of NMR hydrate compounds ^{129}Xe has obtained prominence due to its large (*c.* 100 ppm) chemical shift. Figure 7 shows cross-polarization spectra for ^{129}Xe in hydrates of sI, sII and sH cavities. Ripmeester & Ratcliffe (1990) state that the figure illustrates the following useful points from the NMR spectra. Each type of hydrate has a characteristic spectrum. Lines for ^{129}Xe in the different cages can be distinguished. The line shape of ^{129}Xe in the two small cages of sH overlap but have been resolved in a MAS experiment.

Using Fourier transform infra-red (FTIR) spectroscopy, Woolridge *et al.* (1987) determined Bjerrum L-defect activity was necessary for epitaxial EO hydrate growth. Fleyfel & Devlin (1988, 1989, 1991) studied simple and mixed hydrates of carbon dioxide (CO_2) with EO and THF using epitaxial deposition. Using defect theory in hydrates, these workers presented mechanisms for the kinetics of CO_2 hydrate growth. However, owing to signal interference between guests and water, not many FTIR experiments recently been carried out on hydrates.

Recently Sum *et al.* (1996) have shown that Raman spectroscopy can be used to determine the fraction of filled cages in hydrates, and the fraction of various components in the cages. Uchida *et al.* (1996) have also shown this type of spectroscopy to be useful in determining the relative cage occupation. As Raman appears to

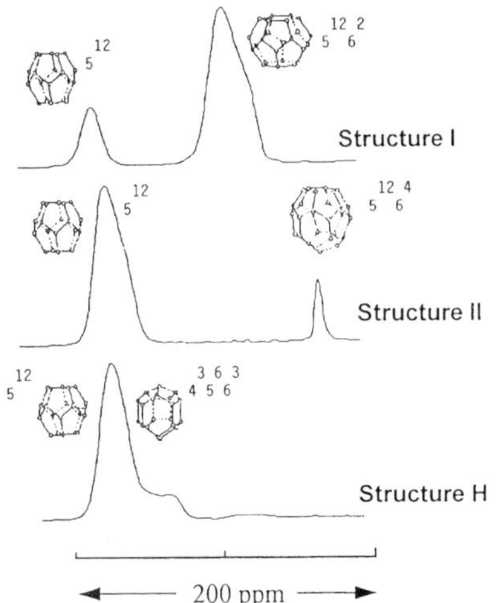

Fig. 7. Typical ^{129}Xe NMR spectra of static samples at 77 K. Distinct line positions and shapes are observed for each cage of sI, sII and sH hydrate (after Ripmeester *et al.* 1987).

be both more versatile and less resource intensive, it may predominate in the future.

Kinetics of formation and dissociation. This area constitutes the largest challenge to fundamental hydrate research. It will probably be best approached through spectroscopic measurements of hydrate phase kinetics. As indicated in the above section, time-independent (thermodynamic) spectroscopy is just coming into usage for hydrates. Time-dependent (kinetic) spectroscopy is sure to be more challenging.

Because the above means of measurement are made on a molecular scale they require some interpretation, or statistical thermodynamics, to achieve translation to a macroscopic scale. An alternative, but as yet qualitative, means of property determination may be obtained by computer simulation.

Characterizing hydrates through calculations

In computer simulation, an assembly (or ensemble) of molecules are simulated to predict macroscopic properties. Two simulation techniques have been commonly used: (1) molecular dynamics (MD); and (2) Monte Carlo (MC) analysis. Recently, lattice dynamics (LD), a new technique, has been used for the hydrate phase (Sparks & Tester 1992; Belosludov *et al.* 1996) at considerable savings in computation. A very significant long-range lattice dynamics effort is due to Tanaka (1993, 1994*a,b*, 1995) and Tanaka & Kiyohara (1993*a,b*) pointing to flaws in the van der Waals & Platteeuw (1959) model.

Molecular dynamics studies of hydrates. Work in three laboratories comprises most of the molecular dynamics hydrate studies. The pioneering work of Tse *et al.* (1983*a,b*, 1984, 1987) and Tse & Klein (1987) are exemplary in comparing simulation calculations to measurements, principally through macroscopic or spectroscopic techniques.

The second major study in molecular dynamics was made by Rodger (1989, 1990*a,b*, 1991*a,b*, 1994) who considered structural stability. A third significant effort (including the aforementioned lattice dynamics work) comes from Tanaka (1993, 1994*a,b*, 1995) and Tanaka & Kiyohara (1993*a,b*) who considered a revised molecular model, which might be applied on a microscopic scale.

Molecular dynamic studies in Holders laboratory (Hwang 1989; Hwang *et al.* 1993; Zele 1994) have calculated Langmuir coefficients, in the van der Waals & Platteeuw (1959) model and considered the effect of guests that stretch the host lattice. Work in this laboratory has concentrated on the clustering of water around water molecules (Long & Sloan 1993) and system behaviour at the hydrate–water interface (Pratt & Sloan 1995). Itoh *et al.* (1996) used MD to explain the CO_2 bending and stretching peaks in Raman spectra.

Monte Carlo studies of hydrates. There are substantially fewer Monte Carlo studies of hydrates than there are molecular dynamics studies. The initial Monte Carlo study of hydrates was by Tester *et al.* (1972), followed a decade later by Tse & Davidson (1982), who checked the Lennard–Jones–Devonshire spherical cell approximation for interaction of guest with the cavity. Lund (1990) studied guest–guest interactions within the lattice. More recently Natarajan (1993) studied the technique for calculation of the Langmuir coefficients, who (somewhat surprisingly) suggested that the technique provided unacceptable results. However, the many past successes of the Monte Carlo technique will promote future studies of equilibrium properties.

Conclusions

The purpose of this paper was to specify physical and chemical properties which might by applied to questions related to ocean and climate stability. The major points of this article are given in the abstract. While time-independent properties of sI and sII hydrates are determined, those for sH are just beginning to be explored. The time-dependent characteristics of all three hydrate structures are largely unspecified and kinetic models to date are all unsatisfactory.

The above properties have direct application to ocean and climate stability. In brief, only biogenic methane is considered to be a wide-spread hydrate former in the ocean, and consequently the problem is bounded. Through a study of hydrate samples and phase equilibria one may realize that several degrees of sediment warming will be required to dissociate most (>70%) of the *in situ* hydrates. Therefore, global warming is unlikely to be caused by methane release from hydrates. Ocean CO_2 hydrates are discussed as well as hydrates indicators of ancient climate changes.

The final portion of the work provides some future challenges for researchers. Among the greatest challenge is the measurement of the hydrate phase itself, which has been neglected.

References

ASHER, G. B. 1987. *Development of a Computerized Thermal Conductivity Measurement System Utilizing the Transient Needle Probe Technique: An Application to Hydrates in Porous Media*. Dissertation, Colorado School of Mines, Golden, CO.

AUSTVIK, T., HUSTVEDT, E., MELAND, B., BERGE, L. & LYSNE, D. 1995. Tommeliten gamma field hydrate experiments. *In*: *7th International Conference on Multiphase Production*. BHRA Group Conference Series, Publication **14.**

AYA, I., YAMANE, K. & YAMADA, N. 1992. Stability of clathrate-hydrate of carbon dioxide in highly pressurized water. *Fundamentals of Phase Change*, **215**, 17.

——, —— & ——. 1995. Dissolution test of a CO_2 droplet through clathrate film at high pressure. *In*: HANDA, Y. P. & OHSUMI, T. (eds) *Direct Ocean Disposal of Carbon Dioxide*. Terra Publications, Japan.

BELL, P. R. 1982. Methane hydrate and the carbon dioxide question. *In*: CLARK, W. (ed.) *Carbon Dioxide Review*. Oxford Univiversity Press.

BELOSLUDOV, R., GRACHEV, E., DYADIN, Y. & BELOSLUDOV, V. 1996. Dynamic properties and stability of helium gas hydrates based on ice framework from lattice dynamics computer simulations. *In*: MONFORT, J. P. (ed.) *Second International Conference on Natural Gas Hydrates*, Toulouse, 303.

BONDAREV, E., GROISMAN, A. & SAVVIN, A. 1996. Porous medium effect of phase equilibrium of tetrahydrofuran hydrate. *In*: MONFORT, J. P. (ed.) *Second International Conference on Natural Gas Hydrates*, Toulouse, p. 89.

BOOTH, J., ROWE, M. & FISCHER, K. 1996. *Offshore Gas Hydrate Sample Database*. U.S. Geological Survey Open File Report **96** (draft copy).

BROOKS, J., COX, H., BRYANT, W., KENICCUT, M., MANN T. & MCDONALD, T. 1986. Association of gas hydrates and oil seepage in the Gulf of Mexico. *Organic Geochemistry*, **10**, 221.

CHRISTIANSEN, R. & SLOAN, E. D., JR. 1994. Mechanism and kinetics of hydrate formation. *In:* SLOAN, E., HAPPEL, J. & HNATOW, M. (eds) *First International Conference on Natural Gas Hydrates*. Anals of the New York Academy of Sciences, **715**, 283.

COLLETT, T. S. 1995. Gas hydrate resources of the United States. *In*: GAUTIER, D. & DOLTON, G. (eds) *National Assessment of US Oil and Gas Resources*. USGS Series, **30**, 78.

COOK, J. & LAUBITZ, M. 1981. The thermal conductivity of two clathrate hydrates. *In: Proceedings of the 17th International Thermal Conductivity Conference*, Gaithersburg, MD, 13.

—— & LEAIST, D. 1983. An exploratory study of the thermal conductivity of methane hydrates. *Geophysical Research Letters*, **10**, 397.

CRANSTON, R. 1994. Marine sediments as a source of atmospheric methane. *Bulletin of the Geological Society of Denmark*, **41**, 101.

DAVIDSON, D. 1983. Gas hydrates as clathrate ices. *In:* COX J. L. (ed.) *Natural Gas Hydrates: Properties, Occurrence, and Recovery*. Butterworths, Boston, MA, 1.

—— & RIPMEESTER, J. 1984. NMR, NQR, and dielectric Properties of clathrates. *In*: ATWOOD, J.L., DAVIES, J. E. D. & MACNICHOL, D. D. (eds) *Inclusion Compounds*, Volume 3. Academic, New York, 69.

DEATON, W. M. & FROST, E. M. 1946. *Gas Hydrates and Their Relation to the Operation of Natural Gas Pipelines*. U.S. Bureau of Mines, Monographs.

DICKENS, G., O'NEIL, J., REA, D. & OWEN, R. 1995. Dissociation of oceanic methane hydrate as a cause of the carbon isotope excursion at the end of the Paelocene. *Paleoceanograpy*, **10**(6), 965.

DOBRYNIN, V., KOROTAJEV, Y. & PLYUSCHEV, D. 1981. Gas hydrates – one of the possible energy sources. *In*: MEYER, R. (ed.) *Long Term Energy Resources*. Pitman, Boston, MA, 727.

ENGLEZOS, P. 1992. Atmospheric climate changes and the stability of the *in-situ* methane hydrates in the Arctic. *In*: *Procedings of the Second International Offshore & Polar Engineering Conference*, San Francisco, CA, 644.

—— 1995. Kinetics of gas hydrate formation and kinetic inhibition in offshore oil and gas operations. *In: Proceedings of the 5th International Offshore & Polar Engineering Conference*, the Hague, The Netherlands, 289.

——, KALOGERAKIS, N., DHOLABHAI, P. & BISHNOI, P. 1987*a*. Kinetics of formation of methane and ethane gas hydrates. *Chemical Engineering Science*, **42**(11), 2647.

——, KALOGERAKIS, N., DHOLABHAI, P. D. & BISHNOI, P.R. 1987*b*. Kinetics of gas hydrate formation from mixtures of methane and ethane. *Chemical Engineering Science*, **42**(11), 2659.

FEI, T. 1991. *A Theoretical Study of the Effects of Sea*

Level and Climatic Change on Permafrost Temperatures and Gas Hydrates. MS. thesis, University of Alaska, Fairbanks.

FLEYFEL, F. & DEVLIN, J. 1988. FTIR Spectra of 90 K films of simple, mixed and double clathrate hydrates of TMO, CH_3Cl, CO_2, THF, and EO containing decoupled D_2O. *Journal of Physical Chemistry*, **92**, 631.

—— & —— 1989. Infrared spectra of large clusters containing small ether molecules: liquid crystalline and clathrate–hydrate cluster spectra. *Journal of Physical Chemistry*, **93**, 7292.

—— & —— 1991. FTIR spectra of CO_2 clusters. *Journal of Physical Chemistry*, **95**, 3811.

GINSBURG, G. D. 1994. Mud volcano gas hydrates in the Caspian Sea. *Bulletin of the Geological Society of Denmark*, **41**, 95.

GUDMUNDSSON, J. S., PARLAKTUNA, M. & KHOKHAR, A. 1994. Storage of natural gas as frozen hydrate. *SPE Prod. Facil.*, 69.

HAMMERSCHMIDT, E. G. 1934. Formation of gas hydrates in natural gas transmission lines. *Industrial Engineering Chemistry*, **26**, 851.

HANDA, Y. P. 1986. Composition, enthalpies of dissociation and heat capacities in the range 85 to 270 K for clathrate hydrates of methane, ethane, propane and enthalpies. *Journal of Chemical Thermophysics*, **18**, 915.

—— & OHSUMI, T. 1995. *Direct Ocean Disposal of Carbon Dioxide*. Terra Publications, Japan.

—— & STUPIN, D. 1992. Thermodynamic properties and dissociation characteristics of methane and propane hydrates in 70 Å – radium silica gel pores. *Journal of Physical Chemistry*, **96**, 8599.

HARRISON, W. J., WENDLANDT, R. F. & SLOAN, E. D., JR. 1995. Geochemical interactions resulting from carbon dioxide disposal on the seafloor. *Applied Geochemistry*, **10**, 461.

HARVEY, L. D. & HUANG, Z. 1995. Evaluation of the potential impact of methane clathrate destabilization on future global warming. *Journal of Geophysical Research*, **100**(D2), 2905.

HATZIKIRIAKOS, S. G. & ENGLEZOS, P. 1993. The relationship between global warming and methane gas hydrates in the Earth. *Chemical Engineering Science*, **48**, 3963.

HONDOH, T. 1996. Clathrate hydrates in polar ice sheets. *In: Second International Conference on Natural Gas Hydrates*, Toulouse, 131.

HWANG, M.-J. 1989. *A Molecular Dynamics Simulation Study of Gas Hydrates*. Ph.D. thesis, University of Pittsburgh.

——, HOLDER, G. D. & ZELE, S. R. 1993. Lattice distortion by guest molecules in gas-hydrates. *Fluid Phase Equilibria*, **83**, 437.

ITOH, H., HORIKAWA, S., KAWAMURA, K., UCHIDA, T., HONDOH, T. & MAE, S. 1996. Molecular dynamics of Structure I clathrate hydrates of carbon. *In: Second International Conference on Natural Gas Hydrates*, Toulouse, 341.

KUHS, W. F., CHAZALLON, B., RADELLI, P., PAUER, F. & KIPFSTUHL, J. 1996. Raman spectroscopic and neutron diffraction studies on natural and synthetic clathrates of air and nitrogen. *In: Second International Conference on Natural Gas Hydrates*, Toulouse, 341.

KVAMME, B. 1996. New theory for the kinetics of hydrate formation. *In: Second International Conference on Natural Gas Hydrates*, Toulouse, 131.

KVENVOLDEN, K. 1988*a*. Methane hydrates – a major reservoir of carbon in the shallow geosphere. *Chemical Geology*, **71**, 41.

——. 1988*b*. Methane hydrates and global climate. *Global Biogeochemical Cycles*, **2**(3), 221.

——. 1993. Gas hydrates – geological perspective and global change. *Reviews of Geophysics*, **31**, 173.

—— & MCMENAMIN, M. 1980. Hydrates of natural gas: a review of their geologic occurrence. US Geological Survey Circular, **825**.

——, COLLETT, T. & LORENSON, T. 1993. Studies of permafrost and gas-hydrates as possible sources of atmospheric methane at high latitudes. *In:* ORENLAND, R. (ed.) *Biogeochemistry of Global Change*. Chapman & Hall, New York.

LEGGETT, J. 1990. The nature of the greenhouse threat. *In: Global Warming: The Greenpeace Report*. Oxford University Press, 30, 40 and 41.

LINGELEM, M., MAJEED, A. & STANGE, E. 1994. Industrial experience in evaluation of hydrate formation, inhibition, and dissociation in pipeline design and operation. *In:* SLOAN, E., HAPPEL, J. & HNATOW, M. (eds) *First International Conference on Natural Gas Hydrates*. Annals of the New York Academy of Sciences, **715**, 75.

LONG, J. 1994. *Gas Hydrate Formation Mechanism and Kinetic Inhibition*. Dissertation, Colorado School of Mines, Golden, CO.

—— & SLOAN, E. 1993. Quantized water clusters around apolar molecules. *Molecular Simulation*, **11**, 145.

—— & SLOAN, E., JR. 1996. Hydrates in the ocean and evidence for the location of hydrate formation. *International Journal of Thermophysics*, **17**, 1.

LUND, A. 1990. *The Influence of Gas–Gas Interactions on the Stability of Gas Hydrates*. Ph.D. thesis, University of Trondheim, Norway.

LYSNE, D. 1995. *An Experimental Study of Hydate Plug Dissocation by Pressure Reduction*. D. Ing. thesis, Norwegian Institute of Technology, University of Trondheim.

MACDONALD, G. J. 1990. Role of methane clathrates in past and future climates. *Climatic Change*, **16**, 247.

MACDONALD, I. R., GUINASSO, N., SASSEN, R. *et al.* Gas hydrate that breaches the sea floor on the continental slope of the Gulf of Mexico. *Geology*, **22**, 699.

MAKOGON, Y. F. 1965. Hydrate formation in the gas-bearing beds under permafrost conditions. *Gazov. Promst.*, 14.

—— 1988. *Natural Gas Hydrates: The State of Study in the USSR and Perspectives for its Use*. Presented at the Third Chemical Congress of North America, Toronto.

—— 1997. *Hydrates of Hydrocarbons*. Pennwell, Tulsa, OK.

MCIVER, R. D. 1981. Gas hydrates. *In:* MEYER, R. (ed.) *Long Term Energy Resources*. Pitman, Boston, MA, 713.

MEHTA, A. P. 1996. *A Thermodynamic Investigation of Structure H Clathrate Hydrates*. Dissertation, Colorado School of Mines, Golden, CO.

—— & SLOAN, E. D., JR. 1993. Structure H hydrate phase equilibria of methane + liquid hydrocarbon mixtures. *Journal of Chemical Engineering Data*, **38**, 580.

—— & SLOAN, E. D., JR. 1994*a*. Structure H hydrates phase equilibria of paraffins, naphthenes, and olefins with methane. *Journal of Chemical Engineering Data*, **39**, 887.

—— & SLOAN, E. D., JR. 1994*b*. A thermodynamic model for Structure H hydrates. *American Institute of Chemical Engineering Journal*, **40**, 312.

—— & SLOAN, E. D. JR. 1996. Structure H hydrates: the state-of-the-art. *In: Second International Conference on Natural Gas Hydrates*, Toulouse, 1.

MEYER, R. F. 1981. *In*: MEYER, R. & OLSON, J. (eds) *Long Term Energy Resources*. Pitman, Boston, MA, 49.

MORI, Y. & MOCHIZUKI, T. 1998. Dissolution of liquid CO_2 into water at high pressures: a search for the mechanism of retarding dissolution caused by hydrate film formation. *Energy Conversion and Management*, in press.

MÜLLER-BONGARTZ, B., WILDEMAN, T. & SLOAN, E., JR. 1992. A hypothesis for hydrate nucleation phenomena. *In: Proceedings of the Second International Offshore and Polar Engineering Conference*, San Francisco, CA, 638.

NATARAJAN, V. 1993. *Thermodynamic and Nucleation Kinetics of Gas Hydrates*. Ph.D. thesis, University of Calgary, Canada.

NISBET, E. G. 1989. Some northern sources of atmospheric methane: production, history, and future implications. *Canadian Journal of Earth Sciences*, **26**, 1603.

—— 1992. Sources of atmospheric CH_4 in early postglacial time. *Journal of Geophysical Research*, **97**(D12), 12,849.

POWELL, H. J. M. 1948. The structure of molecular compounds. Part IV, clathrate compounds. *Journal of the Chemical Society of London*, 61.

PRATT, R. & SLOAN, E., JR. 1995. A computer simulation and investigation of liquid–solid interfacial phenomena for ice and clathrate hydrates. *Molecular Simulation*, **15**, 247.

RIPMEESTER, J. 1991. *The Role of Heavier Hydrocarbons in Hydrate Formation*. Paper Presented at the 1991 *American Institute of Chemical Engineering Spring Meeting*, Houston, TX.

—— & RATCLIFFE, C. 1990. Solid state NMR studies of inclusion compounds. In: ATWOOD, J., DAVIES, J. & MACNICHOL, D. (eds) *Inclusion Compounds*, Volume 5. Oxford University Press, Oxford.

——, RATCLIFFE, C., KLUG, D. & TSE, J. 1994. Molecular perspectives in structure and dynamics in clathrate hydrates. *In*: SLOAN, E., HAPPEL, J. & HNATOW, M. (eds) *First International Conference on Natural Gas Hydrates*. Annals of the New York Academy of Sciences, **715**, 161.

——, TSE, J., RATCLIFFE, C. & POWELL, B. 1987. A new clathrate hydrate structure. *Nature*, **325**, 135.

ROBERTS, R. B., ANDRIKIDIS, C., TAINISH, R. J. & WHITE, G. K. 1984. Thermal expansion of an ice clathrate. *Proceedings of the 10th International Cryogenic Engineering Conference, Helsinki*, 499.

RODGER, P. 1989. Cavity potential in Type I gas hydrates. *Journal of Physical Chemistry*, **93**, 50.

——. 1990*a*. The stability of gas hydrates. *Journal of Physical Chemistry*, **94**, 6080.

——. 1990*b*. Mechanisms for stabilising water clathrates. *Molecular Simulation*, **5**, 315.

——. 1991*a*. *Computer Simulations of Gas Hydrates*. Paper presented at the 1991 American Institute of Chemical Engineering Spring Meeting, Houston, TX.

——. 1991*b*. Lattice Relaxation in Type I Gas Hydrates. *American Institute of Chemical Engineering Journal*, **37**, 1511.

——. 1994. Towards a microscopic understanding of clathrate hydrates. *In:* SLOAN, E., HAPPEL, J. & HNATOW, M. (eds) *First International Conference on Natural Gas Hydrates*. Annals of the New York Academy of Sciences, **715**, 187.

ROSS, R. & ANDERSSON, P. 1982. Clathrate and other solid phases in the THF-water system: thermal conductivity and heat capacity under pressure. *Canadian Journal of Chemistry*, **60**, 881.

——, ANDERSSON, P. & BACKSTROM, G. 1981. Unusual PT dependence of thermal conductivity for a clathrate hydrate. *Nature*, **290**, 322.

SASSEN, R. & MACDONALD, I. R. 1994. Evidence for Structure H hydrate, Gulf of Mexico continental slope. *Organic Geochemistry*, **22**(6), 1029.

SKOVBORG, P. 1993. *Gas Hydrate Kinetics*. PhD. thesis, Danmarks Tekniske Hojskole, Lyngby.

—— & RASMUSSEN, P. 1994. On the dissociation enthalpies of gas hydrates. *Fluid Phase Equilibria*, **96**, 223.

SLOAN, E. D., JR. 1990. Hydrate nucleation from ice. *Proceedings of the 69th Annual Gas Proceedings Conference*, Phoenix, AZ, 8.

—— 1998. *Clathrate Hydrates of Natural Gases* (2nd edition). Marcel Dekker, New York.

—— & FLEYFEL, F. A. 1991. Molecular mechanism for gas hydrate nucleation. *American Institute of Chemical Engineering Journal*. **37**(9), 1281.

—— & FLEYFEL, F. 1992. Hydrate dissociation enthalpy and guest size. *Fluid Phase Equilibria*, **76**, 123.

—— & FLEYFEL, F. 1992. Reply to Comments on hydrate dissociation enthalpy and guest size. *Fluid Phase Equilibria*, **76**, 123.

——, HAPPEL, J. & HNATOW, M. (eds) 1994. *First International Conference on Natural Gas Hydrates*. New York Academy of Sciences, **715**.

SPARKS, K. A. & TESTER, J. W. 1992. Intermolecular potential energy of water clathrates: The inadequacy of the nearest neighbour approximation. *Journal of Physical Chemistry*, **96**, 11022.

STERN, L., KIRBY, S. & DURHAM, W. 1996. Peculiarities of methane clathrate hydrate formation and subcooling, including possible superheating of ice. *Science*, **273**, 1843.

STOLL, R. & BRYAN, G. 1979. Physical properties of sediments containing gas hydrates. *Journal of Geophysical Research*, **84**, 1629.

SUM, A., BURRUSS, R. & SLOAN, E., JR. 1996. Measurements of clathrate hydrate properties via Raman spectroscopy. *In: Second International Conference on Natural Gas Hydrates*, Toulouse.

TANAKA, H. 1993. The stability and Xe and CF_4 clathrate hydrates. Vibrational frequency, a modulation and cage distortion. *Chemical Physics Letters*, **202**, 345.

—— 1994*a*. The thermodynamic stability of clathrate hydrate. Encaging non-spherical propane molecules. *Chemical Physics Letters*, **220**, 371.

—— 1994*b*. The thermodynamic stability of clathrate hydrate – Part III. Accommodation of nonspherical propane and ethane molecules. *Journal of Chemical Physics*, **101**, 10833.

—— 1995. A novel approach to the stability of clathrate hydrate. *Fluid Phase Equilibria*, **104**, 331.

—— & KIYOHARA, K. 1993*a*. The thermodynamic stability of clathrate hydrates – Part I. *Journal of Chemical Physics*, **98**, 4098.

—— & KIYOHARA, K. 1993*b*. The thermodynamic stability of clathrate hydrate – Part II. Simultaneous occupation of larger and smaller cages. *Journal of Chemical Physics*, **98**, 8110.

—— & NAKANISHI, K. 1994. The stability of clathrate hydrates: temperature dependence of dissociation pressure in Xe and Ar hydrate. *Molecular Simulation*, **12**(3-6), 317.

TESTER, J., BIVINS, R. & HERRICK, C. 1972. Use of Monte Carlo in calculating the thermodynamic properties of water clathrates. *American Institute of Chemical Engineering Journal*, **18**(6), 1220.

TROFIMUK, A., CHERSKIY, N. & TSARAEV, V. 1977. The role of continental glaciation and hydrate formation on petroleum occurrences. *In*: MEYER, R. (ed.) *Future Supply of Nature-made Petroleum and Gas*. Pergamon, New York, 919.

TSE, J. S. 1994. Dynamical properties and stability of clathrate hydrate. *In*: SLOAN, E., HAPPEL, J. & HNATOW, M. (eds) *First International Conference on Natural Gas Hydrates*. Annals of the New York Academy of Sciences, **715**, 187.

—— 1990. Thermal expansion of structure H clathrate hydrates. *Journal of Inclusion Phenom*, **8**, 25.

—— & DAVIDSON, D. W. 1982. Intermolecular potentials in gas hydrates. *In: Proceedings of the 4th Canadian Permafrost Conference*, Calgary Alberta, 329.

—— & KLEIN, M. 1987. Dynamical properties of the Structure II clathrate hydrate of krypton. *Journal of Chemical Physics*, **91**, 4188.

——, KLEIN, M. L. & MCDONALD, I. R. 1983*a*. Molecular dynamic studies of ice Ic and the Structure I clathrate hydrate of methane. *Journal of Physical Chemistry*, **87**, 4198.

——, KLEIN, M. L. & MCDONALD, I. R. 1983*b*. Dynamical properties of the Structure I clathrate hydrate of xenon. *Journal of Chemical Physics*, **78**, 2096.

——, KLEIN, M. & MCDONALD, I. 1984. Computer simulation studies of the Structure I clathrate hydrates of methane, tetrafluoromethane, cyclopropane, and ethylene oxide. *Journal of Chemical Physics*, **81**, 6146.

——, MCKINNON, W. & MARCHI, M. 1987. Thermal expansion of Structure I ethylene oxide hydrate. *Journal of Chemical Physics*, **91**, 4188.

UCHIDA, T. TAGAKI, A., HIRANO, T., NARITA, H. KAWABATA, J., HONDOH, T. & MAE, S. 1996. Measurements on guest-host molecular density ratio of CO_2 and CH_4 hydrates by Raman spectroscopy. *In: Second International Conference on Natural Gas Hydrates*, Toulouse.

VAN DER WAALS, J. H. & PLATTEEUW, J. C. 1959. Clathrate solutions. *Advances in Chemical Physics*, **2**, 1.

VON STACKELBERG, M. 1949. Solid gas hydrates. *Naturwissenschaft*, **36**, 327, 359.

—— & MÜLLER, H. 1954. Fete Gashydrate II. Struktur and Raumchemie. *Zeitschrift für Elektrochem*, **58**, 25.

WOOLRIDGE, P., RICHARDSON, H. & DEVLIN, J. 1987. Mobile Bjerrum defects: a criterion for ice-like crystal growth. *Journal of Chemical Physics*, **87**, 4126.

YAKUSHEV, V. 1994. Some environmental problems of gas recovery related to permafrost and natural gas hydrates. *In: Proceedings of the 19th World Gas Conference*. International Gas Union, Milan.

—— & ISTOMIN, V. 1991. Gas hydrates self preservation effect. *In: Proceedings of the IPC-91 Symposium*, Sapporo, Japan, 1.

YOUSIF, M. & SLOAN, E., JR. 1991. Experimental investigation of hydrate dissociation in consolidated porous media. *SPE Reservoir Engineering, Novemver*, 452.

ZAKRZEWSKI, M. & HANDA, Y. 1993. Thermodyanmic properties of ice and of tetrahydrofuran hydrate in confined geometries. *Journal of Chemical Thermodynamics*, **25**, 631.

ZELE, S. R. 1994. *Molecular Dynamics and Thermodynamic Modeling of Gas Hydrates* PhD thesis, University of Pittsburgh.

Gas hydrate accumulation in deep-water marine sediments

G. D. GINSBURG

Research Institute for Geology and Mineral Resources of the Ocean, 1 Angliyskiy prospekt, 190121, St. Petersburg, Russia

Abstract: The accumulation of gas hydrates largely depends on the spatial variability of gas solubility in pore waters. Within the submarine gas hydrate stability zone the solubility of methane in water significantly decreases towards the sea floor in response to temperature lowering. Gas hydrates can precipitate from methane-saturated water seeping up. They also accumulate from diffusing gas and segregated pore water within diffusion aureoles associated with the ascending fluid flows and with the zones where biochemical methane is generated at high rates. Hydrates more readily form in sediments where pore waters are relatively fresh and pores are rather large. The thermobaric gas hydrate stability zone is a geochemical barrier for hydrocarbon gases migrating from sediments into the sea water. However, the rising gas is not completely preserved in the hydrate.

During the last 15 years the author, together with V. A. Soloviev, has been studying the generation and accumulation of natural gas hydrates. The results of these investigations relative to deep-water submarine gas hydrates have been summarized in a monograph (Ginsburg & Soloviev 1994*a*) and some papers (Ginsburg 1990; Ginsburg & Soloviev 1994*a*, 1997; Ginsburg *et al.* 1990, 1992, 1993; Soloviev & Ginsburg 1994, 1997), and have been reported at several conferences (Ginsburg & Soloviev 1994*b*, 1995). The above-listed publications form the basis of this paper.

Discrete distribution of submarine gas hydrates

The published submarine gas hydrate estimates are based on the concepts of their continuous extent over large areas and depth intervals, and regionally high hydrate concentrations in sediments (Ginsburg & Soloviev 1995). However, these concepts are in conflict with the knowledge of the geological medium inhomogeneity. Actually, the analysis of the world-wide observational data suggests that submarine hydrates largely occur in local accumulations (Ginsburg & Soloviev 1994*a*).

It is convenient to separate the totality of the known submarine gas hydrate accumulations into two groups: shallow sub-bottom accumulations (occurring at a depth of several metres and less beneath the sea floor); and deep-seated accumulations (occurring at a depth of several tens of metres and deeper). Figures 1–3 provide examples of separate locations of shallow sub-bottom gas hydrates; these accumulations are associated with the mud volcano crater field, the diapir crest and the fault-complicated buried anticline. The discontinuous occurrence of deep sub-bottom gas hydrate shows in boreholes is exhibited in Figs 4 and 5; these two sections are probably the most representative of deep-seated hydrates: the vertically integrated gas contents of sediments have been estimated to be as large as 4×10^8 and $2 \times 10^9 m^3$ STP km^{-2}, respectively (Mathews & von Huene 1985; Ginsburg & Soloviev 1995) – among the greatest values known.

The discontinuous distribution of hydrate-bearing intervals in boreholes is supported by gas-geochemical data. Gas-undersaturated pore waters have been revealed close to and just between hydrate shows, in particular in DSDP holes 490 and 568 (Watkins *et al.* 1981; von Huene *et al.* 1985). Gas cracks and voids have never been observed along the entire length of cored intervals.

Gas hydrates: linkage to fluid flows

Globally, all known and inferred submarine deep-water gas hydrate localities occur below continental slopes and rises, as well as below mediterranean and marginal seas. The set of conditions favourable for gas hydrate formation is characteristic of these areas. They are precisely the areas which represent a combination of appropriate thermobaric conditions (relatively low bottom-water temperature and relatively high pressure) and availability of biochemical and/or catagenic gas (due to the high organic matter content of sediments and their great thickness). In addition, fluid transport is inherent in these regions as manifested by the widespread

GINSBURG, G. D. 1998. Gas hydrate accumulation in deep-water marine sediments. *In:* HENRIET, J.-P. & MIENERT, J. (eds) *Gas Hydrates: Relevance to World Margin Stability and Climate Change*. Geological Society, London, Special Publications, **137**, 51–62.

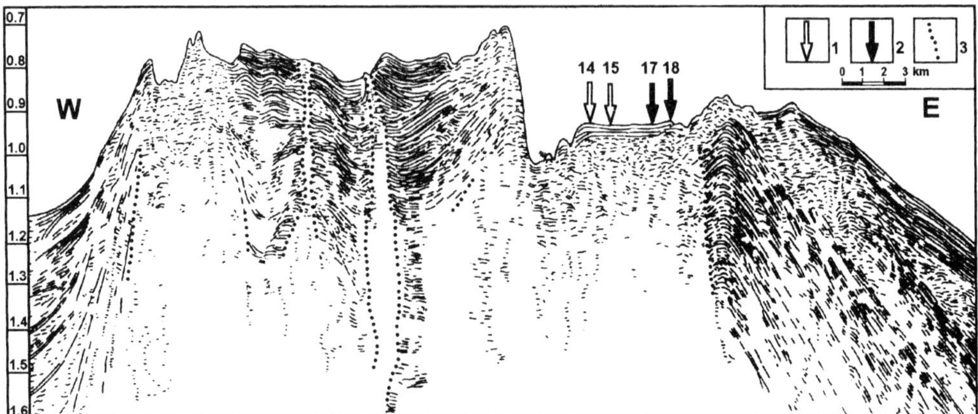

Fig. 1. Interpreted seismo-acoustic time section through the Azizbekov high and gas hydrate-bearing mud volcano Elm, southern Caspian Sea (after Ginsburg *et al.* 1992). The scale on the left is in seconds TWTT. 1, Shallow sampling sites without gas hydrate; 2, sites where gas hydrates have been recovered; 3, diapir boundaries.

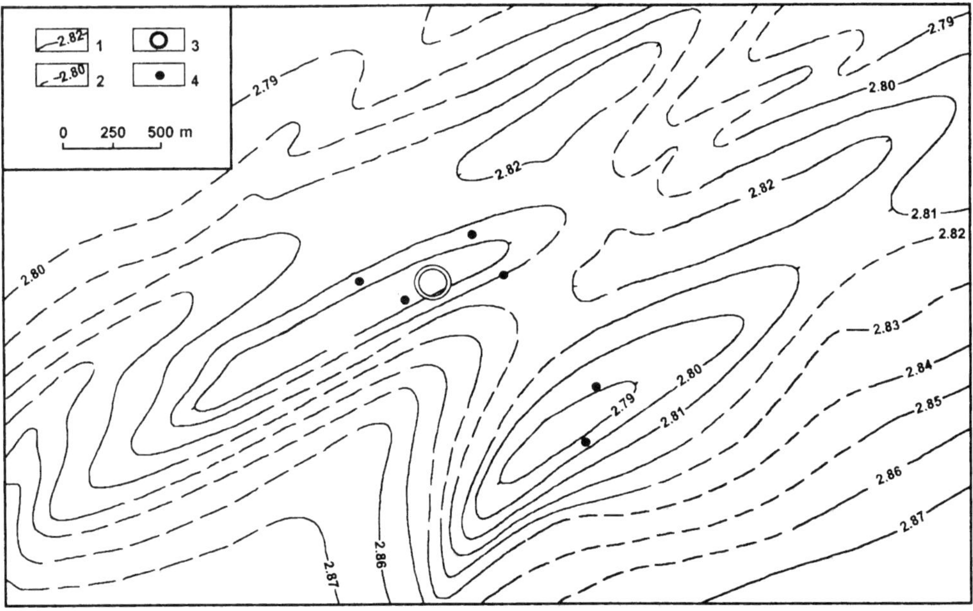

Fig. 2. Bathymetric contours of the gas hydrate area near Crimea, Black Sea (after Ginsburg *et al.* 1990). 1 and 2, Bathymetry in seconds TWTT (1, reliable, 2, supposed); 3, site of gas hydrate recovery (seven shallow cores); 4, shallow coring sites without gas hydrates.

occurrence of localized submarine seeps (Hovland & Judd 1988).

Let us next consider the linkage between gas hydrates and a fluid transport at a more detailed scale. All observed shallow sub-bottom gas hydrates are evidently associated with fluid vents. These hydrates have been observed in the Caspian, Black and Okhotsk seas, the Gulf of Mexico and in several other sites (altogether in 11 regions, Fig. 6). In these sites, hydrates are controlled by such fluid conduits as mentioned before: mud volcanoes, diapirs and faults.

The control of the deep-seated gas hydrates by fluid flow is not as apparent as for the shallow

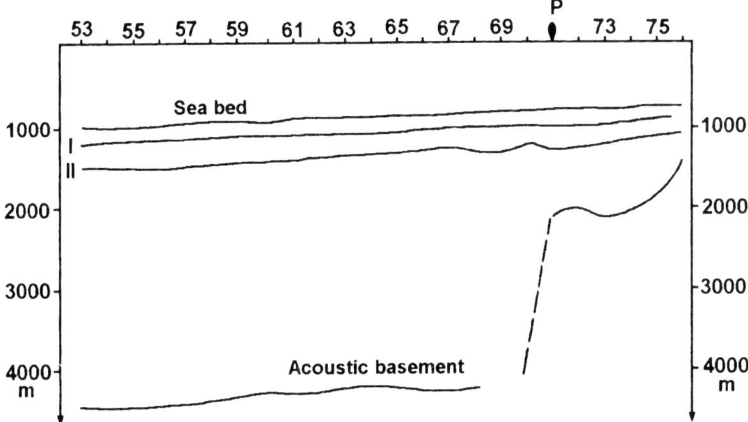

Fig. 3. Schematic cross-section through the gas hydrate accumulation near Paramushir Island, Okhotsk Sea (after Ginsburg & Soloviev 1994). Horizontal graduation line is marked in km. P is the location of gas seepage and associated gas hydrate accumulation; I and II and seismic reflectors; the dashed line designates a fault.

sub-bottom hydrates. However, the fluid flow control of deep-seated hydrates can be deduced from an association with indirect borehole indications of fluid flows, such as relatively coarse-grained sediments and anomalies of pore-water chlorinity (Figs 4, 5 and 7). As the coarse-grained sediments are, in general, relatively permeable, they are believed to be the preferable fluid conduits. Chlorinity anomalies may result from the intrusion of foreign water.

It is worthy of note that anomalies of low chlorinity were initially attributed to pore-water dilution by water released from gas hydrates decomposed in cores (Hesse & Harrison 1981; Harrison & Curiale 1982). At DSDP sites 496 and 497 chlorinity decreased with sub-bottom depth from $19\,g\,l^{-1}$, typical of sea water, to $9.5–10\,g\,l^{-1}$ at 400 m depth. Based on these data the hydrate potential of sediments near the bottom of the holes was estimated as 50% of the pore-space volume. In DSDP Leg 84, Site 568 was drilled in the vicinity of Site 496 and similar chlorinity data were obtained. At this site, along with the usual squeezing from sediment samples, pore-water samples were taken by a downhole sampler *in situ*, and water chlorinity in these samples appeared to be much the same as that of squeezed samples (von Huene *et al.* 1985). Hence, the chlorinity anomalies at Sites 496, 497 and 568 were probably not artificial (not resulted from hydrate decomposition in cores) but might have also resulted from fluid flows. These flows are supposed to induce gas hydrate formation.

A point that should be mentioned is that as early as 1979 Shipley & Didyk (1982), who had the luck to observe hydrates in DSDP cores for the first time, had noted their confinement to relatively coarse-grained and fractured sediments.

Methane solubility in pore water within the submarine gas hydrate stability zone

It is apparent that the generation, accumulation and disappearance of any water-soluble naturally occurring compound in terms of water availability is governed by solubility variations of this compound. Of course, this is also true in regard to gas hydrates.

It is extremely important for natural gas hydrate formation that the solubility of methane in water, in terms of hydrate stability, is little affected by the general (hydrostatic) pressure but is dictated essentially by the equilibrium pressure of hydrate formation. This concept was first qualitatively justified by Barkan & Voronov (1983), quantitatively described by Makogon & Davidson (1983) and later confirmed by Handa (1990). As the equilibrium pressure of hydrate stability is diminished with decreasing temperature, methane solubility in water also decreases (Fig. 8, solid line). Because of this, the solubility of methane in pore water generally decreases towards the sea floor within the submarine gas hydrate stability zone (Fig. 9). The higher the geothermal gradient, accordingly the thinner the hydrate stability zone, and the sharper the methane solubility decrease.

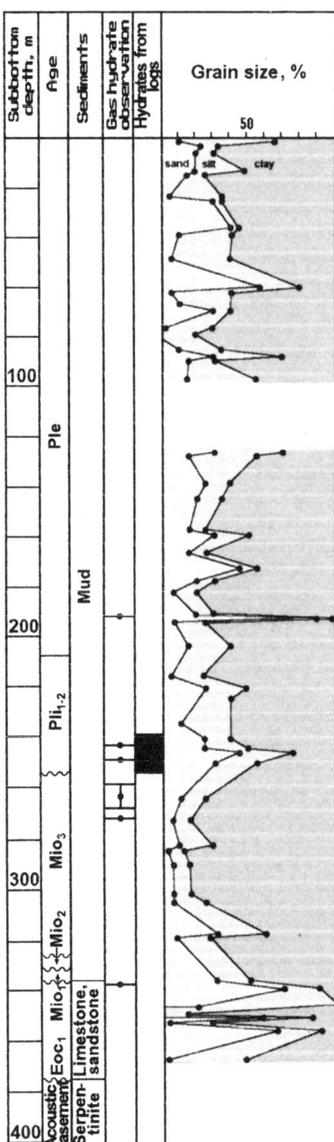

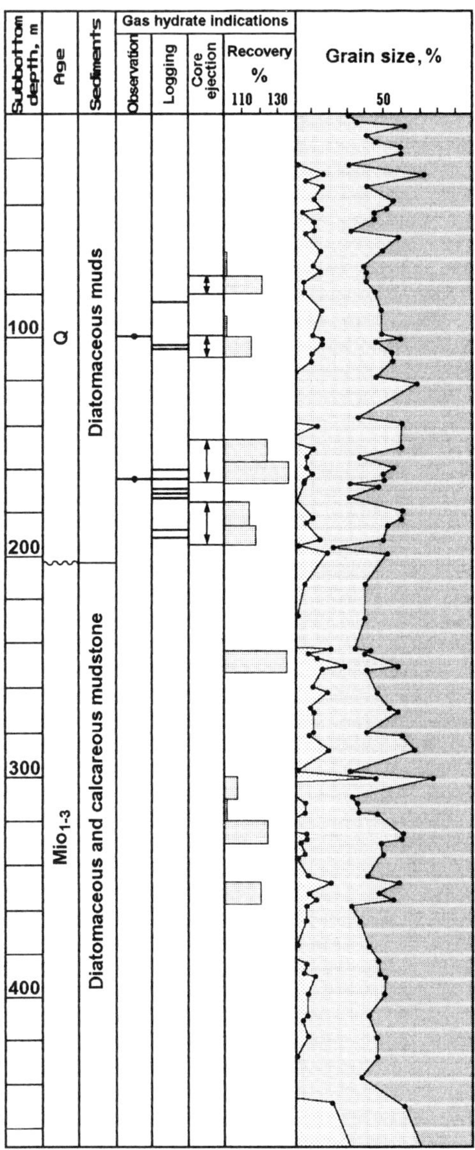

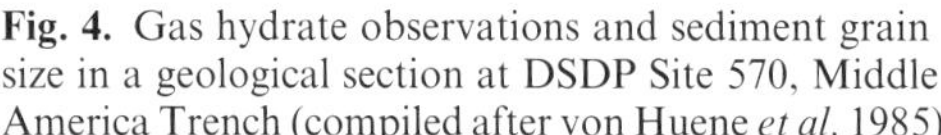

Fig. 4. Gas hydrate observations and sediment grain size in a geological section at DSDP Site 570, Middle America Trench (compiled after von Huene *et al.* 1985).

Fig. 5. Gas hydrate observations and sediment grain size in a geological section at DSDP Site 685, Peru Trench (compiled after Suess *et al.* 1988). For symbols of sediment grain size see Fig. 4.

Three modes of methane migration in sediments and two mechanisms of gas hydrate accumulation

Three major mechanisms of methane transport in sediments can be distinguished: dissolved in pore water flows; as free gas flows; and molecular diffusion.

Hydrate precipitation from ascending methane-saturated water is thought to be the most straightforward (Ginsburg 1990; Ginsburg & Soloviev 1994*a*,*b*; Soloviev & Ginsburg 1994). The hydrate zone forms a gas-geochemical barrier for methane-saturated waters which rise either from below or from within this zone: as

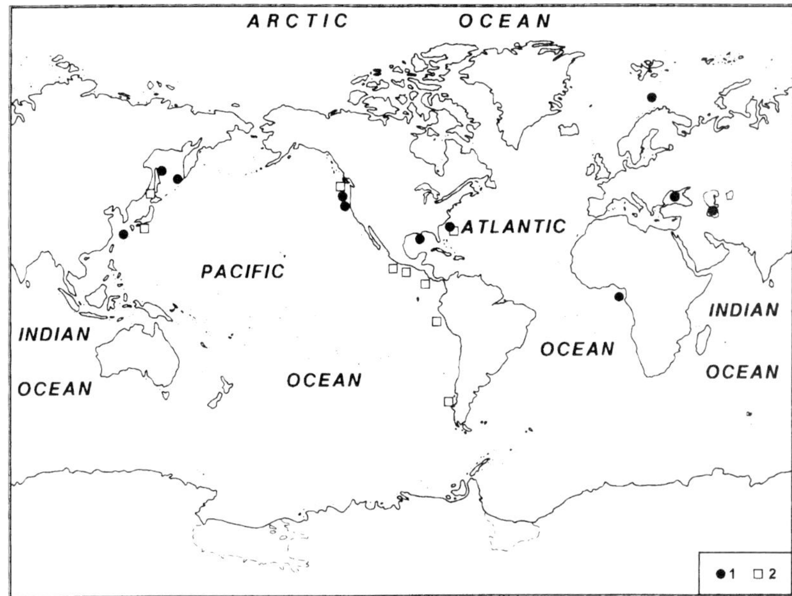

Fig. 6. World-wide locations of observed submarine gas hydrates (after Ginsburg & Soloviev 1994; additional locations from Brooks *et al.* 1994 (offshore Nigeria), Sakai *et al.* 1990 (mid-Okinawa Trough), Paull *et al.* 1995 (Carolina Continental Rise) and authors data (Barents Sea)). 1 and 2, Sea-floor seepage-associated and non-associated gas hydrates, respectively.

the water cools it should precipitate hydrate. The amount of precipitated hydrate obviously corresponds to the excess of dissolved methane (i.e. over the solubility). Clearly the effectiveness of this process depends, in particular, on the rate of water flow and the water temperature; in the case of focused flow of warm water, the thickness of the hydrate zone can decrease to zero (Fig. 10A) and a branched flow (Fig. 10B) has an advantage for hydrate accumulation over the focused one (other conditions being equal).

Gas hydrates precipitated from infiltrated waters progressively fill the sediment pore space and/or fracture porosity, and eventually cement them, producing massive and vein hydrate sediment structures. Of 23 hydrated core intervals revealed by drilling in the Middle America Trench, one of the most representative hydrate-bearing regions (Shipley & Didyk 1982; Harrison & Curiale 1982; Kvenvolden & McDonald 1985), 14 intervals show just these structures (Ginsburg & Soloviev 1994).

Gas hydrates associated with free gas flows discharging on the sea floor were observed in the Gulf of Mexico (Brooks *et al.* 1994) and in the Okhotsk Sea (Zonenshain *et al.* 1987; Ginsburg *et al.* 1993). Clearly, the gas seeping through the hydrate stability zone has no time to crystallize as a hydrate. After a hydrate film forms at the gas–water interface, each succeeding portion of free gas, prior to hydration, has to penetrate this film. Thus, the rate of hydrate formation in the vicinity of free gas flows is limited by the rate of this penetration (presumably, the rate of molecular diffusion), and hydrates are accumulated primarily from the water-dissolved gas – i.e. a solid (hydrate) phase grows at a distance from free gas.

In the Okhotsk Sea the hydrate-bearing sediments associated with free gas seeps exhibited a structure caused by sub-horizontal layers and lenses of gas hydrates (Ginsburg *et al.* 1993). This structure is considered to be the key to understanding the mechanism of gas hydrate accumulation in the vicinity of free gas flows. This sub-horizontal structure suggests that hydrates should be formed from water-dissolved gas rather than immediately from the ascending free gas. The lateral outward diffusion of methane of the ascending gas flow appears to be governed by the difference between chemical potentials of gaseous and dissolved methane at common depths. The above difference is deduced from the difference between the pressure of a free methane close to the hydrostatic pressure and the vapour pressure of dissolved methane; the latter, in terms of pore-water saturation, should be close to the equilibrium pressure of gas hydrate

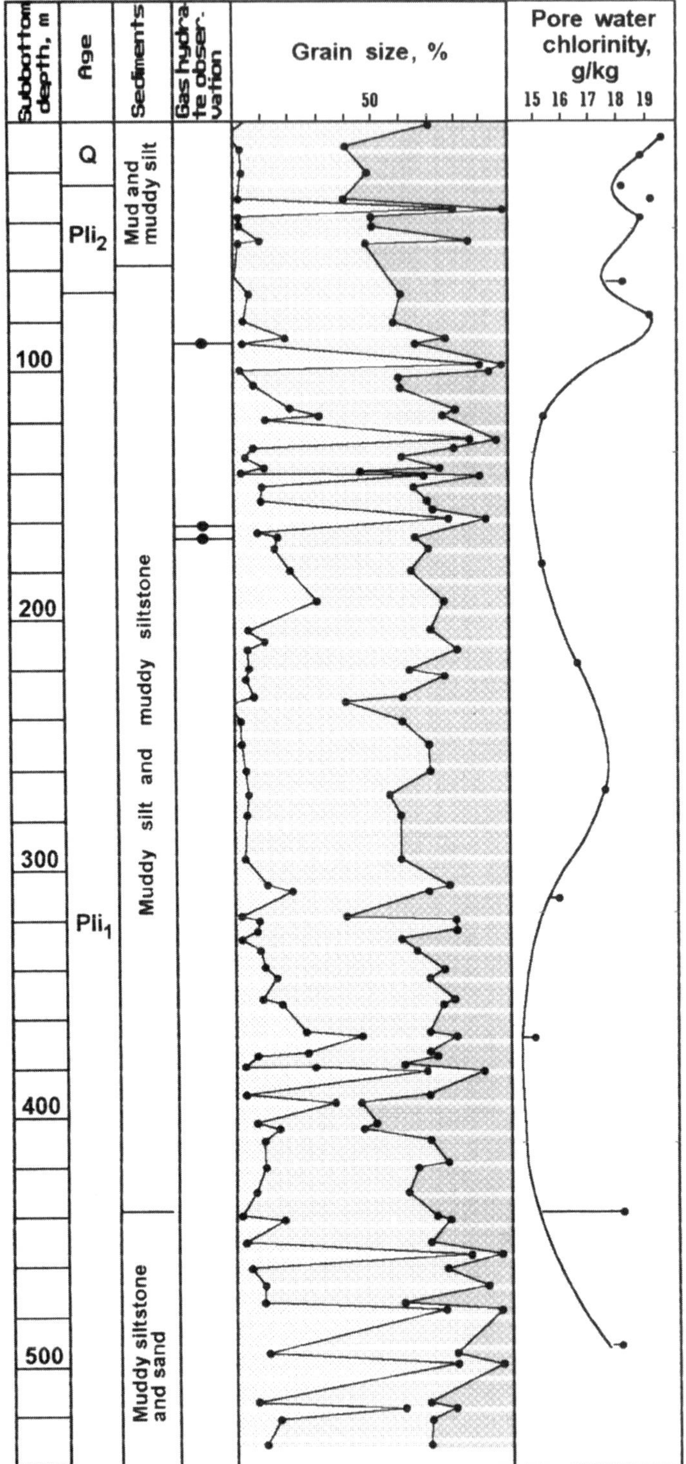

Fig. 7. Gas hydrate observations, sediment grain size and pore-water chlorinity in a geological section at DSDP Site 491, Middle America Trench (compiled from Watkins *et al.* 1981; Gieskes *et al.* 1985). For symbols of sediment grain size see Fig. 4. The chlorinity curve is drawn using sulphate as a measure of sample contamination with sea water.

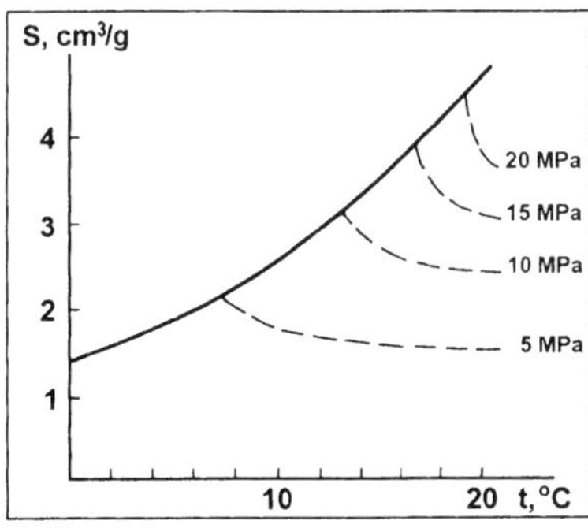

Fig. 8. Solubility of methane (S) in pure water plotted against temperature (t): isobars of solubility in terms of gas hydrate instability (set of dashed lines), and solubility in equilibrium with hydrate (solid line). Compiled using the data of Makogon & Davidson (1983) and Namiot (1991).

formation (compare P_h and P_{eq} in Fig. 11). As the difference between P_h and P_{eq} decreases with increasing sub-bottom depth, hydrate accumulations associated with ascending free gas flows are assumed to taper off downward. Accumulations of this type at great water depths should be more extensive than shallow ones (other factors being equal) because ΔP increases with deepening water. It is self-evident that this model simplifies the matter. In fact, the heat release caused by hydrate formation enhances the outward methane transport and extends the diffusion aureole around ascending gas flow.

Within this aureole the hydrates are thought to result not only from outward diffusing methane but also from upward diffusion. The intensity of the upward diffusion is controlled by high gradients of concentration and vapour pressure of water-dissolved methane in the hydrate zone (in terms of methane-saturated water); these gradients greatly exceed values outside the hydrate zone (Figs 9 and 11).

A similar pattern of methane diffusion and gas hydrate accumulation should also characterize the vicinity of ascending flows of gas-saturated water. In particular, this is possible around the water flows that are too warm for hydrate precipitation (Fig. 10A). We suppose that gas hydrates of this origin were recovered at ODP Site 892 on the Cascadia margin (Westbrook *et al.* 1994).

It is generally believed that diffusion plays only a destructive role in the history of hydrocarbon accumulations. In contrast, Egorov and his colleagues have put forward the concept of 'directional diffusion recondensation' (Egorov 1988; Geodekyan *et al.* 1984). This implies the diffusional transfer of hydrocarbons which saturate water in the presence of a temperature-controlled solubility gradient. According to this concept, the formation and accumulation of a hydrocarbon phase in the region of lower temperature results from such a transfer.

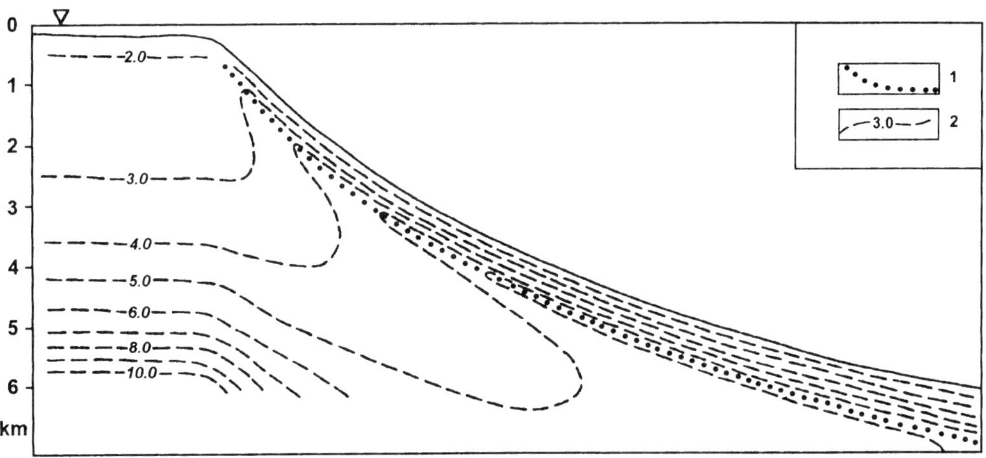

Fig. 9. Vertical cross-section demonstrating solubility of methane in water under thermobaric conditions of continental margins. 1, Base of thermobaric gas hydrate stability zone; 2, isolines of solubility numbered in STP $cm^3 g^{-1}$. Dotted line is the base of thermobaric gas hydrate stability zone. Compiled using the data of Makogon & Davidson (1983) and Namiot (1991). Accepted assumptions: water is pure; bottom-water temperature is 5°C for water depths down to 500 m, and 2°C at greater depths; geothermal gradient is 30°C km^{-1}; hydrobaric gradient is 10 MPa km^{-1}.

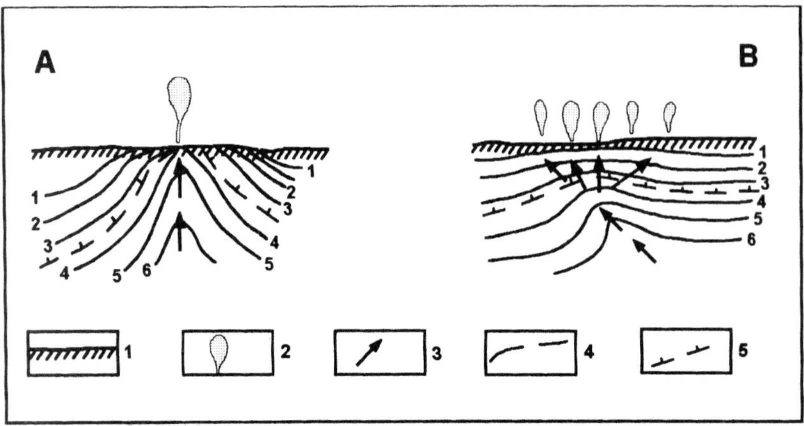

Fig. 10. Thermobaric gas hydrate stability zone in the vicinity of focused (**A**) and branched (**B**) warm fluid flows. Schematic vertical cross-section. 1, Sea bottom; 2, fluid venting; 3, direction of fluid flow; 4, isotherms (numbered in arbitrary units); 5, bottom of the hydrate zone.

It is suggested that directional diffusion recondensation is just the process that governs gas hydrate accumulation in the vicinity of free gas and gas-saturated water flows, as well as within and above the sediment sections where biochemical methane is intensively generated. Relatively impervious sediments may act as a cap in this process. DSDP-ODP data offer examples of gas hydrate ccurrences close to the boundary between relatively coarse- and fine-grained sediments (Ginsburg & Soloviev 1994).

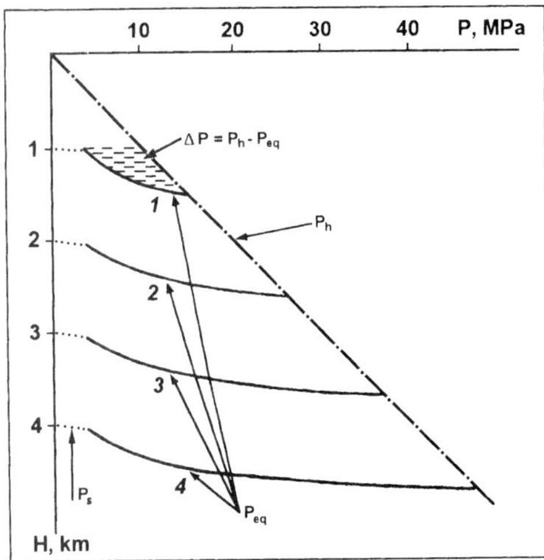

Fig. 11. Relationship between different kinds of pressure (P) affecting diffusion of methane in sub-bottom conditions. H is total depth = water depth + sub-bottom depth. P_h is conventional hydrostatic pressure. P_{eq} is the equilibrium pressure of methane hydrate; curves 1–4 relate to water depths of 1, 2, 3 and 4 km, respectively. P_s is the saturation pressure of dissolved methane within the sulphate reduction zone. Accepted assumptions: water is pure, gas is pure methane (see also Fig. 9). The P_{eq} curves are the usual PT gas hydrate equilibrium curves but the temperature axis is replaced by the depth axis based on the accepted assumptions.

Thus, gas hydrates accumulate from water solutions, no matter whether methane is delivered into the reaction zone by infiltration or diffusion. The important distinction between the two modes of hydrate accumulation in sediments (aside from the process rate) lies in the source of the hydrate water. In the case of hydrate precipitation from infiltrated gas-saturated water this source is flow itself; in the case of methane delivery by diffusion, the hydrate water is extracted from sediment pore water *in situ*. We have proposed the term segregation to designate the mechanism of hydrate accumulation from diffusing gas and from water extracted from sediments (Ginsburg & Soloviev 1994*a*, 1997; Soloviev & Ginsburg 1997). A continuous delivery of methane and the associated formation of hydrate generates a migration of pure water into the reaction zone from the adjacent sediments. This mechanism of water migration is thought to be diffusion-osmotic. Hydrate inclusions of a different shape are formed during this process due to the dewatering of surrounding sediments if the latter are compacted. In the Middle America Trench, where such inclusions were first discovered, they were called 'ice inclusions' (Shipley & Didyk 1982). The shape of inclusions is obviously caused by the factors controlling the fields of gas and water chemical potentials. In

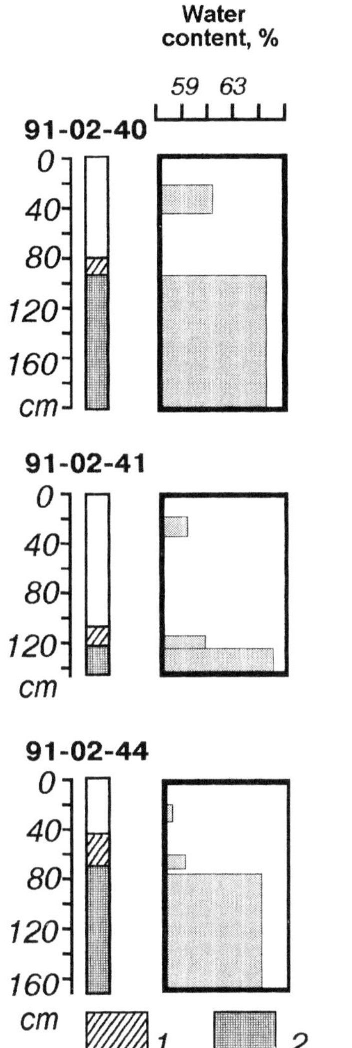

Fig. 12. Water content of sediments (wet) recovered at shallow coring stations 90-02-40, 41, 91-02-41 and 44, near Sakhalin Island, Okhotsk Sea (after Ginsburg *et al.* 1993). 1, Gassy sediments; 2, gas hydrate-bearing sediments (water content determined after hydrate decomposition).

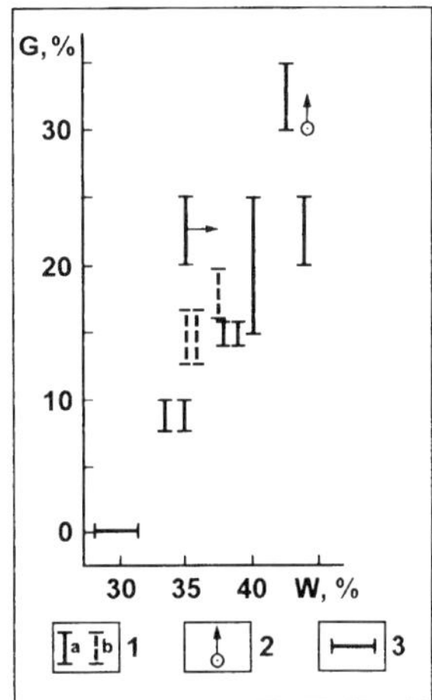

Fig. 13. Water content (W, wt%) of sediments (wet) determined after gas hydrate decomposition plotted against gas hydrate content (G); southern Caspian Sea (after Ginsburg *et al.* 1992). 1, Range of visual estimates of hydrate content of clay breccia from mud volcanoes Buzdag (a) and Elm (b); arrow pointing to the right indicates underestimated value of water content; 2, underestimated value of hydrate content; 3, range of water content of unhydrated clay breccia.

particular, the sub-horizontal lenticular-bedded hydrate sediment structure observed in association with submarine gas vents in the Okhotsk Sea (see above) may result from the sub-horizontal extention of isotherms.

As a result of water redistribution during segregational gas hydrate accumulation, the total water content of hydrate-bearing sediments may turn out to be higher than that of adjacent non-hydrated ones. A difference like this has been observed in all sampled hydrated cores in the Okhotsk Sea (Fig. 12). A water content of sediments directly proportional to their hydrate content has been demonstrated in the Caspian Sea (Fig. 13). I would like to emphasize using these examples that hydrate accumulation in sediments may imply not only a gathering of gas but also of water. Owing hydrate water abundance, a sediment may become fluidized upon decomposition of the hydrate. Hence, a core fluidity may be considered as an indication of gas hydrate decomposition before observation. In fact, the 'soupy' appearance of cores was used as just such an indication in ODP Leg 146 (Westbrook *et al.* 1994).

Diffusion is known to be a ubiquitous process in marine sediments. As a hydrate of any origin is subject to subsequent decomposition and possible directional diffusion recondensation of the hydrate methane, segregational hydrates are thought to be more common than those precipitated by infiltrated water.

Gas hydrates prefer to form from low saline pore waters and in large voids

I have mentioned two kinds of inhomogeneity of the geological medium exerting an influence on gas hydrate accumulation: permeability variations, which control fluid conduits and gas hydrate caps; and geothermal inhomogeneity (geothermal gradient), which predominantly governs gas solubility in water. Below I would like also to draw attention to two other kinds of inhomogeneity, namely hydrochemical and lithological.

It is well known that water-dissolved salts inhibit (prevent) gas hydrate formation (Sloan 1990), i.e. hydrates form more readily from fresh water. Therefore, a gradient of water salinity within the hydrate zone under gas-saturation conditions must provoke a diffusional flux of methane into less saline water, where this arriving methane should be hydrated. Such a situation may occur near boundaries of water flows. It is necessary to emphasize here that the solubility of methane in the fresh gas-saturated water is known to be higher than in saline water, whereas the corresponding methane fugacity, which actually should be considered as a driving force of diffusion, is higher in saline water (Handa 1990).

A lithological (or in more exact terms, a porometric) inhomogeneity implies, in particular, a distinction of sediment pore size (I do not consider here the shape of sediment pores and their specific properties which, of course, also affect gas hydrate accumulation). The pore medium influences the hydrate equilibrium (thermodynamic effect) and the kinetics of hydrate formation. The thermodynamic effect essentially is as follows: a pore surface is hydrophilic and therefore lowers the pore-water chemical potential. As a result, a higher thermodynamic concentration of methane is required for the formation of hydrate. In principle, this effect is similar to the influence of salts dissolved in water. This surface effect was studied by many authors (Makogon 1974, 1985; Tsarev 1976; Cha *et al.* 1988; Yousif & Sloan 1991; Handa & Stupin 1992) and had been found negligible in terms of natural sediment water content. The kinetic effect lies in the fact that a pore size may be less than a gas hydrate critical nucleus size at a given temperature. In this case, for hydrate formation to start, more significant overcooling or oversaturation is required. This effect had been studied by Chersky & Mikhailov (1990). We suggest that the essence of both effects (thermodynamic and kinetic) can be understood by examination of hydrate formation in adjacent sediments having different pore sizes. It is evident that the hydrate formation in coarse-pored sediments has an advantage over fine-pored ones – the same gas concentration in water may turn out to be sufficient to form hydrates in the former case and insufficient in the latter. What this means is that hydrate can accumulate in relatively large pores in the course of sediment compaction and/or biochemical gas generation.

Submarine gas hydrate stability zone as a geochemical barrier for methane migrating into sea water

As mentioned before, the hydrate zone forms a geochemical barrier for water-dissolved methane ascending in water flows either from below or from within this zone. However it should be stressed that methane that is dissolved in emigrating water at concentrations lower than the solubility limit (1–1.5 $cm^3 g^{-1}$, see Figs 4 and 5) is not captured in this zone.

Free gas vents have been observed on the sea floor in many regions where a hydrate stability zone occurs (in particular in the Okhotsk Sea, see Fig. 3). The mere fact that ascending gas can penetrate through a hydrate zone without being converted into hydrate seems to be surprising. That is, hydrate-bearing sediments by no means form an impermeable barrier to free gas flows. Cranston *et al.* (1994) calculated that the amount of gas being captured in one of the hydrate accumulations in the Okhotsk Sea was 5 orders of magnitude lower than the amount passing through the hydrate zone into the water column.

The diffusional runoff of dissolved methane from the submarine hydrate stability should also occur. Moreover, the runoff of methane in deep water appears to be more significant than the runoff from shelf areas and continents owing to high gradients of dissolved methane concentration (Fig. 9) and vapour pressure (Fig. 11). Hence, methane continuously escapes the hydrate zone.

Gas hydrate evaluation calls for the study of the gas balance of marine sediments

In as much as only a portion of migrating gas is trapped in submarine gas hydrates, it is evident that the theoretical evaluation of gas hydrates requires consideration of all constituents of the gas balance of marine sediments: generation, migration, consumption and discharge on the sea floor. For the time being these processes, as well as the total gas amount, cannot be quantitatively evaluated with any degree of accuracy (Ginsburg & Soloviev 1995).

Bacterial oxidation appears to fill a highly important place in the balance of submarine methane. This process is today recognized as a major governor of the methane flux: methane-oxidizing bacteria form a potent 'biofilter' for methane migrating to the atmosphere both from marine sediments and from sea water (Gal'tsenko 1995). Whether or not this sink can consume the methane flux, if submarine hydrates rapidly decompose, remains to be seen.

Conclusions

- Submarine gas hydrates mostly occur in discrete accumulations which are controlled by fluid flows.
- Hydrates accumulate from methane-saturated water in the course of pore-water infiltration and methane diffusion. The accumulation of hydrates is controlled by physical factors such as temperature gradient, pore-water salinity gradient and lithological variability. The hydrates precipitate at lower temperatures and from less saline water; relatively coarse-grained sediments make better hydrate reservoirs than fine-grained sediments.
- The submarine gas hydrate stability zone creates a geochemical barrier for upward-migrating methane. However, a quantity of dissolved methane and at least some free methane pass through this barrier. Furthermore, favourable conditions exist within the zone of hydrate stability for the diffusive runoff of methane. It is as yet unknown what proportion of ascending methane is detained within this barrier zone.
- There is no way to theoretically estimate the magnitude of the gas hydrate methane reservoir without studying methane migration. This idea has not yet been realized as clearly as the complementary idea that evaluation of the carbon exchange between sediments and sea water requires an assessment of gas hydrates.

The contribution of the Russian Ministry of Natural Resources and the Ministry of Science (the Programs 'Ocean' and 'Global Change of Natural Environment and Climate') have made it possible to carry out this work. I would like to express my gratitude to the organizers of the Gas Hydrates Workshop, in particular J.-P. Henriet and J. Mienert for offering me the possibility of presenting our results. I wish also to thank P. Vogt and the anonymous reviewer for suggesting improvements to the manuscript, and A. Egorov for a fruitful discussion of problems touched on in the paper. I am very much obliged to my regular co-author V. A. Soloviev.

References

BARKAN, E. & VORONOV, A. 1983. [Assessment of gas resources in hydrate-prone areas.] *Sovetskaya Geologia*, **8**, 26–29 (in Russian).

BROOKS, J., ANDERSON, A., SASSEN, R. *et al.* 1994. Hydrate occurrences in shallow subsurface cores from continental slope sediments. *In*: SLOAN, E., JR, HAPPEL, J. & HNATOW, M. (eds) *International Conference on Natural Gas Hydrates*. Annals of the New York Academy of Sciences, **715**, 381–391.

CHA, S., OUAZ, H. & WILDERMAN, T. 1988. A third-surface effect on hydrate formation. *Journal of Physical Chemistry*, **92**(23), 6492–6494.

CHERSKY, N. & MIKHAILOV, N. 1990. [Size of equilibrium critical nuclei of gas hydrates.] *Doklady Akademii Nauk SSSR*, **312**(4), 968–971 (in Russian).

CRANSTON, R., GINSBURG, G., SOLOVIEV, V. & LORENSON, T. 1994. Gas venting and hydrate deposits in the Okhotsk Sea. *In*: JORGENSEN, N. (ed.) *Gas in Marine Sediments, 2nd Conference*. Bulletin of the Geological Society of Denmark, Copenhagen, **41**(1), 80–85.

EGOROV, A. 1988. [*Diffusion Mechanisms of Hydrocarbons Primary Migration and Accumulation in offshore Sedimentary Basins*.] Candidate thesis, Institut Okeanologii Akademii Nauk SSSR, Moscow (in Russian).

GAL'TSENKO, V. 1995. [Bacterial cycle of methane in marine ecosystems.] *Priroda*, **6**, 35–48 (in Russian).

GEODEKYAN, A., EGOROV, A. & LOPATNIKOV S. 1984. [Mechanism of directional diffusion recondensation and primary migration of hydrocarbons.] *Doklady Akademii Nauk SSSR*, **274**(3), 691–694 (in Russian).

GIESKES, J., JOHNSTON, K. & BOEHM, M. 1985. Appendix. Interstitial water studies. Leg 66. *In*: VON HUENE, R., AUBOUIN, J. *et al.* (eds) *Initial Reports of the Deep Sea Drilling Project*, Volume 84. US Government Printing Office, Washington, DC, 961–967.

GINSBURG, G. D. 1990. [Submarine gas hydrate formation from seeping gas-saturated underground waters.] *Doklady Akademii Nauk SSSR*, **313**(2), 410–412 (in Russian).

—— & SOLOVIEV, V. A. 1994*a*. [*Submarine Gas Hydrates*.] VNII Okeangeologia, St. Petersburg (in Russian).

—— & SOLOVIEV, V. A. 1994*b*. Russian research on submarine gas hydrate geology. *In*: SLOAN, E.D., JR, HAPPEL, J. & HNATOW, M.A. (eds) *International Conference on Natural Gas Hydrates*. Annals of the New York Academy of Sciences, **715**, 484–486.

—— & SOLOVIEV, V. A. 1995. Submarine gas hydrate estimation: theoretical and empirical approaches. *In: Proceedings of the 27th Annual Offshore Technology Conference*, Houston, TX, 513–518.

—— & SOLOVIEV, V. A. 1997. Methane migration within the submarine gas hydrate stability zone under deep-water conditions. *Marine Geology*, **137**, 49–57.

——, GUSEYNOV, R., DADASHEV, G. *et al.* 1992. Gas hydrates of the Southern Caspian. *International Geology Review*, **34**(8), 765–782.

——, KREMLEV, A., GRIGOR'EV, M. *et al.* 1990. Filtrogenic gas hydrates in the Black Sea (21st voyage of the research vessel "Evpatoriya"). *Soviet Geology and Geophysics (Geologia i Geofizika)*, **31**(3), 8–16.

——, SOLOVIEV, V. A., CRANSTON, R., LORENSON, T. & KVENVOLDEN, K. 1993. Gas hydrates from continental slope offshore Sakhalin Island, Okhotsk Sea. *Geo-Marine Letters*, **13**, 41–48.

HANDA, Y. P. 1990. Effect of hydrostatic pressure and salinity on the stability of gas hydrates. *Journal of Physical Chemistry*, **94**(6), 2652–2657.

—— & STUPIN, D. 1992. Thermodynamic properties and dissociation characteristics of methane and propane hydrates in 70Å-radius silicagel pores. *Journal of Physical Chemistry*, **96**(21), 8599–8603.

HARRISON, W. & CURIALE, J. 1982. Gas hydrates in sediments of Holes 497 and 498A, Deep Sea Drilling Project Leg 67. *In:* AUBOUIN, J., VON HUENE, R. *et al.* (eds) *Initial Reports of the Deep Sea Drilling Project*. Volume 67. US Government Printing Office, Washington, DC, 591–594.

HESSE, R. & HARRISON, W. 1981. Gas hydrates (clathrates) causing pore-water freshening and oxygen isotope fractionation in deep-water sedimentary section of terrigenous continental margins. *Earth and Planetary Science Letters*, **55**, 453–462.

HOVLAND, M. & JUDD, A. G. 1988. *Seabed Pockmarks and Seepages. Impact on Geology, Biology and Marine Environment*. Graham and Trotman, London.

KVENVOLDEN, K. A. & MCDONALD, T. J. 1985. Gas hydrates of the Middle America Trench, Deep Sea Drilling Project Leg 84. *In*: VON HUENE, R., AUBOUIN, J. *et al.* (eds) *Initial Reports of the Deep Sea Drilling Project*, Volume 84.US Government Printing Office, Washington, DC, 667–682.

MAKOGON, Y. F. 1974. [*Hydrates of Natural Gases.*] Nedra, Moscow (in Russian).

—— 1985. [*Gas Hydrates, Prevention of Their Formation, and Utilization.*] Nedra, Moscow (in Russian).

—— & DAVIDSON, D. W. 1983. [Influence of excessive pressure on methane hydrate stability.] *Gazovaya promyshlennost*, **4**, 37–40 (in Russian).

MATHEWS, M. & VON HUENE, R. 1985. Site 570 methane hydrate zone. *In:* VON HUENE, R., AUBOUIN, J. *et al.* (eds) *Initial Reports of the Deep Sea Drilling Project*. Volume 84. US Government Printing Office, Washington, DC, 773–790.

NAMIOT, A. Y. 1991. [*Solubility of Gases in Water. Reference textbook.*] Nedra, Moscow (in Russian).

PAULL, C., SPIESS, F., USSLER, W., III, & BOROWSKI, W. 1995. Methane-rich plumes on the Carolina Continental Rise: association with gas hydrates. *Geology*, **23**, 89–92.

SAKAI, H., GAMO, T., KIM, E.-S., *et al.* 1990. Venting of carbon dioxide-rich fluid and hydrate formation in Mid-Okinawa Trough backarc basin. *Science*, **248**, 1093–1096.

SHIPLEY, T. & DIDYK, B. 1982. Occurrence of methane hydrates offshore southern Mexico. *In*: WATKINS, J., MOORE, J. *et al.* (eds) *Initial Reports of the Deep Sea Drilling Project*, Volume 66. US Government Printing Office, Washington, DC, 547–555.

SLOAN, E., Jr. 1990. *Clathrate Hydrates of Natural Gases*. Marcel Dekker, New York.

SOLOVIEV, V. A. & GINSBURG, G. D. 1994. Formation of submarine gas hydrates. *In*: JORGENSEN, N. (ed.) *Gas in Marine Sediments*. Bulletin of the Geological Society of Denmark, Copenhagen, **41**(1), 86–94.

—— & GINSBURG, G. D. 1997. Water segregation in the course of gas hydrate formation and accumulation in submarine gas seepage fields. *Marine Geology*, **137**, 59–68.

SUESS, E., VON HUENE, R. *et al.* 1988. Proceedings of the Ocean Drilling Project, Initial Reports. College Station, TX Ocean Drilling Program, **112**.

TSAREV, V.P. 1976. [Formation Peculiarities and Methods of Prospecting and Development of Hydrocarbon Accumulations Under Permafrost Conditions.] Yakutsk, (in Russian).

VON HUENE, R., AUBOUIN, J. *et al.* 1985. *Initial Reports of the Deep Sea Drilling Project*. Volume 84 US Government Printing Office, Washington, DC.

WESTBROOK, G., CARSON, B., MUSGRAVE, R. *et al.* 1994. *Proceeding of the Ocean Drilling Project, Initial Reports*. College Station, TX. Ocean Drilling Program, **146** (Part 1).

WATKINS, J.S., MOORE, J.C. *et al.* 1981. *Initial Reports of the Deep Sea Drilling Project*, Volume 66. US Government Printing Office, Washington, DC.

YOUSIF, M. & SLOAN, E., JR. 1991. Experimental investigation of hydrate formation and dissociation in consolidated porous media. *SPE Reservoir Engineering*, November, 452–458.

ZONENSHAIN, L., MURDMAA, I., BARANOV, B. *et al.* 1987. An underwater gas source in the Sea of Okhotsk west of Paramushir Island. *Oceanology*, **27**(5), 598–602.

Mathematical models of gas hydrate accumulation

A. W. REMPEL[1] & B. A. BUFFETT[2]

[1]*Institute of Theoretical Geophysics, Department of Applied Mathematics and Theoretical Physics, University of Cambridge, Silver Street, Cambridge, UK*

[2]*Department of Earth and Ocean Sciences, University of British Columbia, 2219 Main Mall, Vancouver, Canada*

Abstract: Gas hydrate reservoirs are widespread on the world's continental margins and in the Arctic, but little is known about the way in which they form. We use conservation principles to derive a set of equations that describe hydrate formation in uniform porous media. Using scaling arguments, we identify the physical processes that are most important to hydrate accumulation in different environments. This knowledge is used to construct models that quantitatively predict the development of hydrate layers under a variety of circumstances. These models compare favourably with recent field observations of hydrates in marine sediments.

How much gas is contained in a given hydrate reservoir and how is it distributed? These questions are of paramount importance in determining the impact of these deposits on margin stability, the global climate, future energy concerns and the methane budget. To address these issues properly, we must also know why the gas is distributed as it is and how long it takes to accumulate. Extensive exploration using bore hole (e.g. Kvenvolden & Bernard 1983; Brooks *et al.* 1985; Kastner *et al.* 1995; Dickens *et al.* 1997) and seismic techniques (e.g. Hyndman & Spence 1992; Minshull *et al.* 1994; Yuan *et al.* 1996) has provided insight into the present-day characteristics of hydrate deposits at numerous locations around the world. Researchers have extrapolated these observations to estimate the total volume of carbon contained in these reserves globally (Kvenvolden 1993; Holbrook *et al.* 1996). These ongoing studies will continue to improve our understanding of the current state of the hydrate deposits found on our continental margins and in the Arctic. Unfortunately, however, these methods provide little information regarding the reasons for the observed spatial distribution of hydrate at a given reservoir and the duration over which the hydrate has accumulated.

The accumulation rate and spatial distribution of hydrate are determined by the physical conditions at each location. We can predict the amount of hydrate we expect to encounter in a particular reservoir by modelling the ways in which the various physical processes interact. Field observations and laboratory simulations can be used to help refine these models and improve our understanding of the potential of these deposits. Alternatively, with accurate data on the distribution of hydrates within a reservoir, we can use models of hydrate accumulation to address the question of how the hydrate saturation acquired its present profile. Models of hydrate dissociation have been developed previously (Selim & Sloan 1989; Tsypkin 1991, 1992*a*,*b*), but little quantitative work on modelling hydrate formation under geological conditions has been reported in the literature.

In the current paper we present models of hydrate accumulation in different physical environments. We begin by outlining the stability requirements necessary for hydrates to be in equilibrium with their surroundings. This leads to a discussion of the governing equations that describe how hydrate deposits form in uniform porous media. Next, we use scaling arguments to delineate physical effects that are most important in various circumstances. Finally, we give a few illustrative examples of the predictions of some models of hydrate accumulation in uniform porous media and discuss their implications.

Stability requirements

The three-phase equilibrium conditions necessary for the stable coexistence of hydrate, water, and free gas are determined by the hydrate former and the pore-water chemistry (e.g. Englezos & Bishnoi 1988; Sloan 1990; Dickens & Quinby-Hunt 1997). On continental margins, the high pressures and relatively low temperatures in the shallow sediments provide a stable environment for hydrates when there is sufficient gas (see Fig. 1). Knowledge of the hydrostatic pressure, the geothermal gradient and the appropriate three-phase equilibrium curve is sufficient

Rempel, A. W. & Buffett, B. A. 1998. Mathematical models of gas hydrate accumulation. *In:* Henriet, J.-P. & Mienert, J. (eds) *Gas Hydrates: Relevance to World Margin Stability and Climate Change*. Geological Society, London, Special Publications, **137**, 63–74.

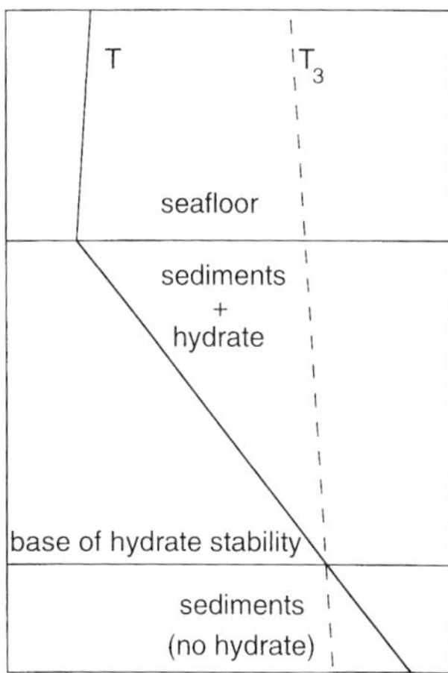

Fig. 1. A schematic representation of the thermal conditions near the sea floor. The dashed line represents the temperature T_3 for three-phase equilibrium between hydrate, pore fluid and free gas at the hydrostatic pressure. Hydrate is stable in the upper region of the sediment column where the temperature T falls below T_3; beneath this level, hydrates are no longer stable.

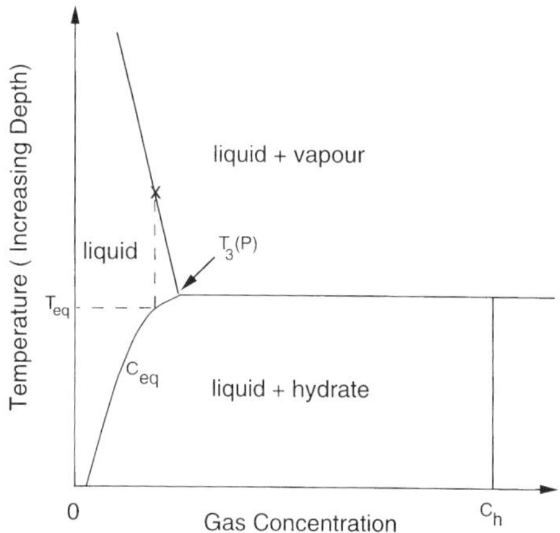

Fig. 2. A schematic representation of a portion of the phase diagram of a gas–water system at constant pressure. In the case of a closed system which is cooled from the point marked X, hydrate does not begin to form until some temperature T_{eq}, which is less than the three-phase equilibrium temperature $T_3(P)$. As the system is cooled below T_{eq} hydrate continues to form with gas mass fraction c_h as the gas concentration in solution is depleted along the curve marked c_{eq}.

to estimate where and to what depth the stability zone extends. (In general, sediment properties may alter the three-phase equilibrium through the (Gibbs–Thompson) effects of curvature for example (Clennell *et al.* 1995); these complications act to adjust the pressure–temperature conditions for three-phase equilibrium, and in principle they may be accounted for in the choice of the appropriate three-phase equilibrium curve.) The stability boundary is often marked by a bottom simulating reflector (BSR) in seismic experiments, which is caused by a jump in acoustic impedance as hydrate-bearing sediments give way to fluid-saturated sediments which can contain free gas (e.g. Bangs *et al.* 1993; MacKay *et al.* 1994; Singh & Minshull 1994; Yuan *et al.* 1996). When the sediment permeability is too low to permit the transport of free gas away from the three-phase equilibrium boundary, a plane of weakness can develop at the BSR and lead to sediment failure.

While the three-phase equilibrium conditions determining the base of the hydrate stability field have received much attention and are fairly well known (e.g. Englezos & Bishnoi 1988; Sloan 1990; Dickens & Quinby-Hunt 1997), the two-phase equilibrium conditions which prevail above the base of the stability zone have been the subject of comparatively little research. If sufficient gas is available, we expect the pore space within the stability region to be occupied by a combination of solid hydrate and aqueous solution (Handa 1990). (In extremely gas-rich environments, it is possible to envision a two-phase equilibrium between hydrate and a mixture of free gas and water vapour, but this is not expected to be a common arrangement in the sea floor.) In a two-phase region with hydrate and aqueous solution, the equilibrium gas concentration dissolved in the water is a function of the *in situ* temperature and pressure. Experimental evidence and theoretical predictions show that the solubility of gas in water decreases when the system is brought further into the hydrate stability field by either increasing the pressure or reducing the temperature (Handa 1990; Yamane & Aya 1995; Zatsepina & Buffett 1997). (The presence of salts and multiple gas components will affect the value of the equilibrium gas concentration, but the basic form of the temperature and pressure dependence will be similar.) As shown in Fig. 2, this is the opposite of what occurs outside the hydrate stability field, where gas solubility is enhanced by increased pressure or decreased temperature (Fogg & Gerrard 1991). Because temperature largely controls the solubility behaviour under geological conditions, hydrate can crystallize from a gas-saturated aqueous solution without the presence of any free gas. Free gas would normally only be found beneath the gas hydrate zone. (The possibility that the effects of crystal growth kinetics may allow free gas to persist well into the hydrate stability zone will be discussed later.)

Governing equations

The interactions of several competing physical processes determine the rate of accumulation and spatial distribution of hydrate in a developing reservoir. As gas is removed from the interstitial fluid and incorporated into the hydrate structure, there is an associated latent heat release. The local hydrate accumulation rate is dependent on how quickly additional gas can be supplied and latent heat can be removed by advective and diffusive processes. By modelling the sediment–hydrate–liquid system as a continuum, we can use conservation principles to derive equations that describe these interactions. We restrict our attention to single-component hydrates in sediments which initially contain pure water, although our arguments may be easily modified to consider multi-component hydrates and the presence of salts. (The dynamics of heat and mass transport which control the hydrate accumulation process also govern the formation of mushy layers during the solidification of a binary melt. This related class of problems contains many fundamental similarities to the hydrate accumulation problem (e.g. Huppert & Worster 1985; Worster 1992).)

We consider the hydrate reservoir to consist of a uniform porous medium in which the void space ϕ is partitioned into hydrate, with pore-volume fraction h, and an aqueous solution with pore-volume fraction $1-h$ and gas concentration c (see Appendix 1 for a list of symbols and their definitions). The latent heat of formation L and the mass fraction of gas contained in the hydrate c_{h} are treated as constants; as are the densities ρ and specific heats C of each of the components (subscripts: s = sediment, h = hydrate, f = fluid). The bulk thermal conductivity $K(h)$, the bulk heat capacity $\bar{C}(h)$, and the combined diffusion–dispersion coefficient $D(h)$ all depend on the amount of hydrate present, h.

Heat and gas are transported both through the fluid by advection, and by diffusion and dispersion down the temperature and gas concentration gradients. Latent heat release provides a source of heat as hydrate is produced, and the gas which is incorporated into the hydrate structure reduces the aqueous solution's gas content. The conservation of energy and gas lead to the primary governing equations which describe how these processes change the temperature T and the mass fraction of gas in the fluid c (Rempel 1994; Rempel & Buffett 1997); (Appendix 2 contains a brief derivation)

$$\bar{C}(h)\frac{\partial T}{\partial t}+\mathbf{u}\cdot\nabla t=\nabla\cdot(\kappa(h)\nabla T)+\frac{\rho_{\mathrm{h}}\phi L}{\rho_{\mathrm{f}}C_{\mathrm{f}}}\frac{\partial h}{\partial t}$$

and

$$(1-h)\frac{\partial c}{\partial t}+\frac{1}{\phi}\mathbf{u}\cdot\nabla c=\nabla\cdot[(1-h)D(h)\nabla c] -\frac{\rho_{\mathrm{h}}}{\rho_{\mathrm{f}}}(c_{\mathrm{h}}-c)\frac{\partial h}{\partial t} \qquad (1)$$

where the effective thermal diffusivity

$$\kappa(h)\equiv\frac{K(h)}{\rho_{\mathrm{f}}C_{\mathrm{f}}}$$

is defined as the ratio of the bulk thermal conductivity to the heat capacity of the fluid, and the bulk heat capacity is

$$\bar{C}(h)\equiv\frac{\rho_{\mathrm{f}}C_{\mathrm{f}}\phi(1-h)+\rho_{\mathrm{h}}C_{\mathrm{h}}\phi h+\rho_{\mathrm{s}}C_{\mathrm{s}}(1-\phi)}{\rho_{\mathrm{f}}C_{\mathrm{f}}}.$$

The fluid velocity $\mathbf{u}$ is altered by the reduction in effective permeability $k(h)$ which must accompany the reduction in effective porosity when hydrate occupies a portion of the pore space. In addition, the density difference between hydrate and the pore fluid will cause a divergence in flow as hydrate forms. These effects are described by Darcy's law and the continuity condition

$$\mathbf{u}=\frac{k(h)}{\eta}\nabla P'$$

and

$$\nabla\cdot\mathbf{u}=\frac{\phi(\rho_{\mathrm{f}}-\rho_{\mathrm{h}})}{\rho_{\mathrm{f}}}\frac{\partial h}{\partial t} \qquad (2)$$

where $\nabla P'$ is the non-hydrostatic component of the pressure gradient, and η is the dynamic viscosity of the pore fluid.

Within the hydrate reservoir, the equilibrium condition gives a relationship between the fluid's gas concentration and the temperature. In principle the equilibrium concentration depends on both pressure and temperature, but in geological applications the temperature dependence is expected to be the controlling factor. The temperature dependence of c_{eq} may be expressed in the convenient form

$$c_{\mathrm{eq}}(T)=c_{\mathrm{eq}}(T_3)\exp[\alpha(T-T_3)] \qquad (3)$$

where T_3 is the temperature for three-phase equilibrium and $c_{\mathrm{eq}}(T_3)$ is the corresponding solubility. The two-phase equilibrium calculations of Zatsepina & Buffett (1997) show that $\alpha^{-1}\approx 10^{\circ}\mathrm{C}$ for the methane hydrate–water system in the range of temperature–pressure conditions that represent marine sediments.

Crystal growth kinetics can lead to departures from the equilibrium state. Kinetic effects have been the subject of numerous laboratory studies (e.g. Englezos *et al.* 1987; Sloan 1990), but many details remain poorly constrained. Given the uncertainties involved, we choose to model non-equilibrium effects by adopting a simple linear equation in which the rate of gas consumption is proportional to the concentration in excess of the equilibrium value

$$\frac{\mathrm{d}c}{\mathrm{d}t} = -\Re(c - c_{\mathrm{eq}}) \quad (4)$$

where $\Re$ is the reaction-rate constant.

Scaling arguments

To assess the relative importance of the different terms in equation (1) we compare coefficients in the corresponding dimensionless equations. We introduce the dimensionless temperature $\tilde{T} = (T - T_0)/\Delta T$ and gas concentration $\tilde{c} = (c - c_0)/\Delta c$, where ΔT and Δc are typical variations over the region of interest, and T_0 and c_0 are convenient reference values. Hydrate deposits are normally located in regions where the temperature and the equilibrium dissolved gas concentration change most rapidly in the vertical dimension. The fluid flow is also often oriented in a predominantly vertical direction. For simplicity then, it is reasonable to restrict our attention to one-dimensional problems and write the dimensionless form of equation (1) as

$$\bar{C}(h)\frac{\partial \tilde{T}}{\partial \tilde{t}} + Pe\frac{\partial \tilde{T}}{\partial \tilde{z}} = \frac{\partial}{\partial \tilde{z}}\left(\frac{\kappa(h)}{\kappa(0)}\frac{\partial \tilde{T}}{\partial \tilde{z}}\right) + S\frac{\partial h}{\partial \tilde{t}}$$

and

$$(1-h)\frac{\partial \tilde{c}}{\partial \tilde{t}} + \frac{1}{\phi}Pe\frac{\partial \tilde{c}}{\partial \tilde{z}}\varepsilon\frac{\partial}{\partial \tilde{z}}\left((1-h)\frac{D(h)}{D(0)}\frac{\partial \tilde{c}}{\partial \tilde{z}}\right) - \frac{\rho_{\mathrm{h}}}{\rho_{\mathrm{f}}}(\tilde{c}_{\mathrm{h}} - \tilde{c})\frac{\partial h}{\partial \tilde{t}} \quad (5)$$

where the dimensionless position is $\tilde{z} = z/l$, with l chosen as a typical length scale, and time is made dimensionless using the thermal diffusion time scale $\tilde{t} = t\kappa(0)/l^2$. Provided we choose the length scale l and the temperature and gas concentration scales, ΔT and Δc, appropriately, the derivatives in equation (5) will be roughly the same order of magnitude and the size of their dimensionless coefficients will indicate the relative importance of the various terms.

The new parameters introduced in equation (5) are the Peclet number, Pe the Stefan number, S, the Lewis number, ε, and the dimensionless hydrate gas concentration, $\tilde{c}_{\mathrm{h}}$. The Peclet number,

$$Pe \equiv \frac{ul}{\kappa(0)}$$

measures the speed with which heat can be transported through the fluid by advection, at fluid velocity u, relative to the speed with which it is conducted through the sediment–fluid mixture ($\kappa(0)$ is the thermal diffusivity in the absence of hydrate, i.e. $h = 0$). The Stefan number,

$$S \equiv \frac{\rho_{\mathrm{h}}\phi L}{\rho_{\mathrm{f}}C_{\mathrm{f}}\Delta T}$$

indicates the relative importance of the latent heat which is liberated as hydrate is produced compared to the heat required to change the temperature. The Lewis number,

$$\varepsilon \equiv \frac{D(0)}{\kappa(0)}$$

is defined as the ratio of the combined diffusion and dispersion coefficient for dissolved gas transport (with $h = 0$) to the thermal diffusion coefficient $\kappa(0)$. The dimensionless hydrate gas concentration,

$$\tilde{c}_{\mathrm{h}} \equiv \frac{c_h - c_0}{\Delta c}$$

is the ratio of the difference between the mass fraction of gas in the hydrate c_{h} and the reference gas concentration in the fluid c_0, to typical variations in fluid gas content Δc. The high gas storage capacity of hydrates ensures that c_{h} is always much larger than the amount of gas which can be dissolved in the fluid, so $c_{\mathrm{h}} \gg c_0$ and $\tilde{c}_{\mathrm{h}} \approx c_{\mathrm{h}}/\Delta c$.

The thermal conductivities of hydrate and water differ by about 10–20% depending on the hydrate former (Sloan 1990). As this difference is relatively small, and heat is also transported through the sediment and the pore fluid, we can expect the ratio of the bulk thermal diffusivity of hydrate-bearing sediments to the bulk thermal diffusivity of hydrate-free sediments $\kappa(h)/\kappa(0)$ to be of the order of 1. Thus, by comparing the coefficients in equation (5) we see that thermal diffusion dominates the heat transport if the Peclet number is much less than 1, and advective transport dominates if Pe is much greater than 1. On the Cascadia margin, where extensive hydrate deposits are found, the background fluid velocity is of the order of several millimetres per year (Hyndman *et al.* 1993). A typical vertical dimension for a hydrate deposit is roughly 10^2 m, and typical thermal diffusivities are on the order of $10^{-7}\,\mathrm{m^2\,s^{-1}}$ (Davis *et al.* 1990). In this case the Peclet number is much less than 1 ($Pe \approx (10^{-11}\,\mathrm{m\,s^{-1}})(10^{2}\,\mathrm{m})/(10^{-7}\,\mathrm{m^{2}\,s^{-1}}) \approx 10^{-2}$) and therefore advection of heat can be neglected when modelling the large-scale hydrate accumulation. In the same reservoir, however, fluid velocities in localized cracks and faults might be several orders of magnitude higher than the background field. In these local environments, where the Peclet number is large, it is essential to retain the advective term in models of hydrate accumulation.

In the equation for $\tilde{c}$ we compare the Peclet number and the Lewis number to determine the principal mechanism for transporting gas. The compositional diffusivities of dissolved hydrocarbons in sedimentary rocks vary over several orders of magnitude (Kroos & Leythaeuser 1988). In general we expect thermal diffusion to be more efficient than chemical diffusion so the Lewis number is usually very much less than 1. With the large uncertainty involved in estimating $D(0)$, however, it is difficult to ascertain whether advection or diffusion dominates gas transport. Flow in cracks and faults can enhance advective transport, but it also seems reasonable to expect diffusive transport to be important in reservoirs where the large scale fluid velocity is only a few millimetres or less per year.

In laboratory hydrate formation experiments, non-equilibrium effects have been shown to control the rate of gas consumption when vigorous stirring is used to enhance the transport of gas and heat. The methane hydrate formation experiments of Uchida & Narita (1996) give reasonable fits to the empirical first-order equation (4) when $\Re$ is of the order of $10^{-3}\,\mathrm{s^{-1}}$. When the time scale $\Re^{-1}$ for non-equilibrium effects is short compared to the shortest time scale for the transport of heat and gas, we can assume that the gas concentration in the two-phase region is close to the equilibrium value given by equation (3). In many situations the shortest time scale for transport is the thermal diffusion time $l^2/\kappa(0)$ which can be several orders of magnitude larger than $\Re^{-1}$. In these cases we infer that the equilibrium assumption is valid. In cracks and fractures, however, the advective time scale l/u might be comparable to $\Re^{-1}$, and the non-equilibrium effects described by equation (4) become more important. Non-equilibrium effects could also play an important role in laboratory simulations, where the fluid velocity is negligible but the length scale is short.

Hydrate accumulation models

Idealized physical models are useful for illustrating some of the important features of solutions to the governing equations. For example, a simple model can be obtained for the case of a uniform porous half-space cooled on its boundary to a temperature T_0 less than the equilibrium temperature T_{eq} (see Fig. 3). In this case, the fluid velocity, and hence the Peclet number, is initially zero. A divergence of flow is produced by the volume change due to hydrate formation, according to equation (2), but the associated velocities are small so the Peclet number remains close to zero and advective transport of heat and gas may be neglected. When the Lewis number is also small the diffusive gas transport is much slower than the diffusive heat transport. The temperature drops quickly relative to the speed of gas transport, so the gas cannot move appreciably before the temperature drops below the two-phase equilibrium temperature given by equation (3) and hydrate begins to form. Ther-

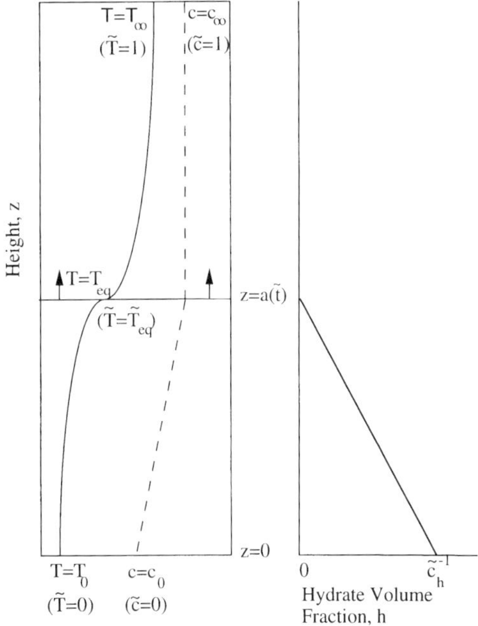

Fig. 3. A schematic representation of the temperature, fluid gas concentration and hydrate volume fraction profiles for a uniform porous half-space cooled on its boundary. Initially, the temperature is constant at T_∞ and the gas concentration is constant at c_∞. The base at $z = 0$ is then cooled to a temperature T_0 (where the corresponding equilibrium gas concentration is c_0). A hydrate layer develops with an advancing interface at $z = a(\tilde{t})$ along which the temperature is equal to the equilibrium value T_{eq}. The hydrate volume fraction in the layer is a function of both position and time, and decreases from a high of $\tilde{c}_h^{-1}$ at the base, to zero at the moving interface.

mal diffusion determines the position of the moving phase change interface separating the growing two-phase hydrate stability region from the warmer gas-rich aqueous solution. Given a constant initial temperature T_∞ and gas concentration c_∞, the problem reduces to become a member of the well-known class of solidification problems known as Stefan problems (e.g. Turcotte & Schubert 1982). The equations can be solved analytically to give expressions for the hydrate volume fraction and the growth rate of the layer (Rempel & Buffett 1997).

This model for hydrate formation in a porous half-space gives the hydrate volume fraction as approximately

$$h \approx \frac{1 - \tilde{c}}{\tilde{c}_h} \tag{6}$$

where $\tilde{c}$ is the dimensionless form of the two-phase equilibrium concentration and the values of c_0 and Δc used to define $\tilde{c}_h$ are the equilibrium gas concentration at the boundary and the difference between c_0 and the initial gas concentration c_∞. In practice, the large gas storage capacity of hydrates ensures that the dimensionless hydrate gas concentration $\tilde{c}_h$ is always much larger than the fluid gas concentration $\tilde{c}$, so the maximum value of h tends to be less than 1%. The hydrate distribution described by equation (6) decreases from a high of $\tilde{c}_h^{-1}$ at the chilled boundary to zero at the moving interface $a(\tilde{t})$ (see Fig. 3).

The position of the phase change interface is of the form $a(\tilde{t}) = 2\lambda\sqrt{\tilde{t}}$, where the constant λ is found by imposing energy conservation on the interface. An interesting feature of this problem is that the hydrate volume fraction at the interface goes to zero so there is no discontinuity in h at $a(\tilde{t})$; energy conservation implies that the temperature gradient must therefore be continuous across the interface. The most important parameters determining the value of λ, and hence the speed of the moving interface, are the Stefan number S, defined with $\Delta T = T_\infty - T_0$, and the dimensionless equilibrium temperature $\tilde{T}_{eq} \equiv (T_{eq} - T_0)/(T_\infty - T_0)$.The Stefan number measures the relative importance of the latent heat and the heat required to change the temperature. When S is high, the growth of the interface is slowed by the large heat release associated with the phase transition (see Fig. 4a). As S tends to zero, the latent heat release becomes negligible, and the only constraint on the growth rate of the layer is the time required to cool the sediments below the equilibrium temperature T_{eq}. When the initial temperature T_∞ is much warmer than T_{eq}, the dimensionless equilibrium temperature $\tilde{T}_{eq}$ is small; more heat must be extracted to cool the system, and the layer growth is slow (see Fig. 4b). When $\tilde{T}_{eq}$ is close to 1, the initial temperature is near the equilibrium temperature and the phase change interface moves rapidly.

In naturally occurring hydrate deposits, the manner in which the gas gets incorporated into the hydrate structure is a matter of some dispute. One theory holds that *in situ* biogenic gas production from buried organic material is the main gas source (Kvenvolden & Barnard 1983; Brooks *et al.* 1985). Others argue that there is insufficient organic material available to account for the large volumes of hydrate present, so there must be significant gas transport into the stability region from below (Hyndman & Davis 1992). Numerical solutions to the governing equations

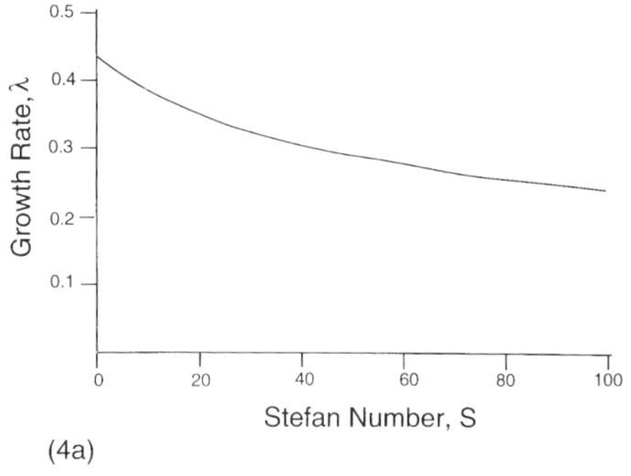

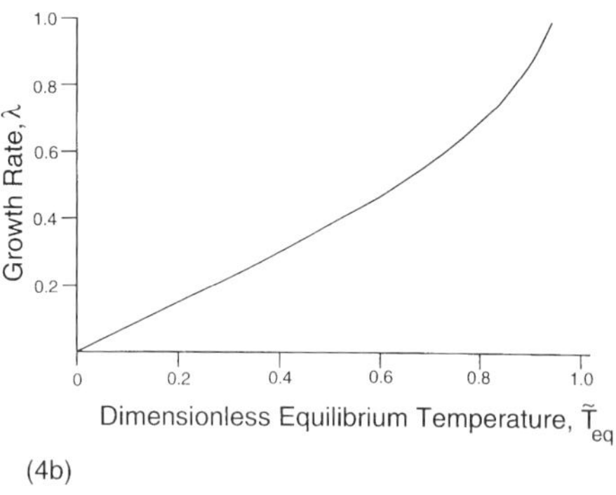

Fig. 4 (**a**) A plot of the growth-rate parameter λ as a function of the Stefan number $S \equiv \rho_h \phi L/(\rho_f C_f(T_\infty - T_0))$. At higher values of S, the growth rate is reduced by the need to remove more latent heat. (**b**) The dependence of λ on the dimensionless equilibrium temperature $\tilde{T}_{eq} \equiv (T_{eq} - T_0)/(T_\infty - T_0)$. When the equilibrium temperature T_{eq} equals the initial temperature T_∞, $\tilde{T}_{eq} = 1$ and layer growth is limited only by the need to remove latent heat.

can be used to explore the implications of these two models (Rempel & Buffett 1997).

For the gas flux model, we assume that the temperature profile is initially conductive and there is no gas in the hydrate zone. Beneath the level of the hydrate stability zone, the gas concentration is fixed at a constant value, and this gas is transported upwards by the combined effects of advection and diffusion. For typical parameter values (see Table 1) the temperature perturbations caused by hydrate production are insignificant, and the Peclet number is sufficiently small so that the temperature profile remains linear. After several thermal diffusion time cycles the gas concentration reaches a steady, equilibrium profile. When the Lewis number is small compared to the Peclet number the gas transport is dominated by advection and the hydrate accumulation rate is given by

$$\frac{\partial h}{\partial t} = -\frac{\rho_f \mathbf{u} \cdot \nabla c_{eq}}{\rho_h \phi (c_h - c_{eq})}. \quad (7)$$

Using the relationship from equation (3) between c_{eq} and temperature, we infer from equation (7) that the rate of hydrate accumulation decreases exponentially upwards into the stability zone. This prediction yields a hydrate saturation profile that is highest near the BSR, consistent with some estimates based on field observations (see Fig. 5) (e.g. Brown *et al.* 1996; Yuan *et al.* 1996). Using parameter values representative of conditions in marine sediments (see Table 1), we get an additive increase in hydrate volume of order 1% of the pore space in 10^5 years near the base of the hydrate stability field.

For the *in situ* gas production model we initially assume that the gas production rate is uniform throughout the entire sediment column. When the temperature profile is conductive, as in marine sediments, the gas solubility behaviour suggests that hydrate should form most readily where the system is furthest into the hydrate stability field at the top of the sediment column. This would lead to a hydrate saturation profile with a hydrate volume fraction that decreases with depth. If we allow the temperature conditions in the sediments to evolve due to the effects of continuing sedimentation, however, we get a much different result. The temperature at the base of the stability field increases and the position of the BSR migrates upwards as hydrate dissociates to liberate free gas (see Fig. 6). This gas rises back into the stability region by advection and compositional diffusion, and the new hydrate it forms is concentrated near the BSR. The overall hydrate saturation profile that

Table 1. *Parameter values*

Property	Nominal value	Units
ρ_f	1000	kg m^{-3}
ρ_h	930	kg m^{-3}
ρ_s	2650	kg m^{-3}
C_f	4200	J kg^{-1} K^{-1}
C_h	2080	J kg^{-1} K^{-1}
C_s	2200	J kg^{-1} K^{-1}
ϕ	0.5	–
L	430	kJ kg^{-1}
$\kappa(0)$	10^{-7}	m s^{-2}
$D(0)$	10^{-9}	m s^{-2}
$\mathbf{u}$	10^{-11}	m s^{-1}
c_h	0.13*	–
c_0	10^{-3}†	–
α	0.1	K^{-1}
$\Re$	10^{-3}	s^{-1}
G	0.04	K m^{-1}

* Assumes $CH_4 \cdot 6H_2O$; †solubility of methane at 6.06 MPa and 298 K (Fogg & Gerrard 1991). Other sources are Lide (1990), Sloan (1990), Hyndman & Davis (1992), and Uchida and Narita (1996).

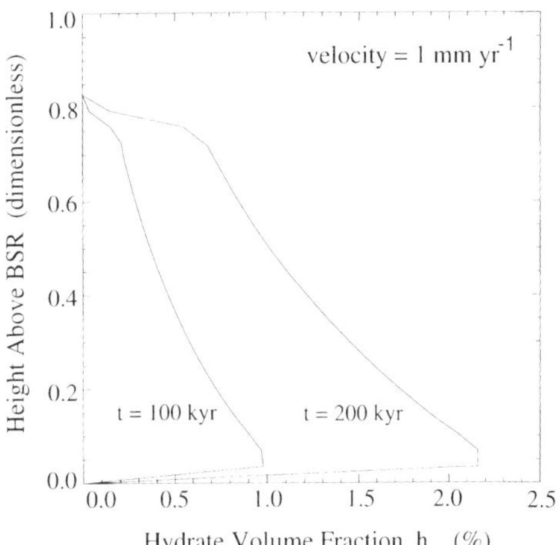

Fig. 5. The hydrate saturation profile for the gas flux model with a fluid velocity of 1 mm year^{-1}. The hydrate volume fraction decreases exponentially with height above the BSR. The ocean maintains the gas concentration at the sea water value at the sea floor (dimensionless height 1.0). This results in large gas concentration gradients near the sea floor, and compositional diffusion causes a sudden drop in hydrate volume fraction near dimensionless height 0.8. If the fluid velocity were higher, this drop in hydrate volume fraction would occur closer to dimensionless height 1.0.

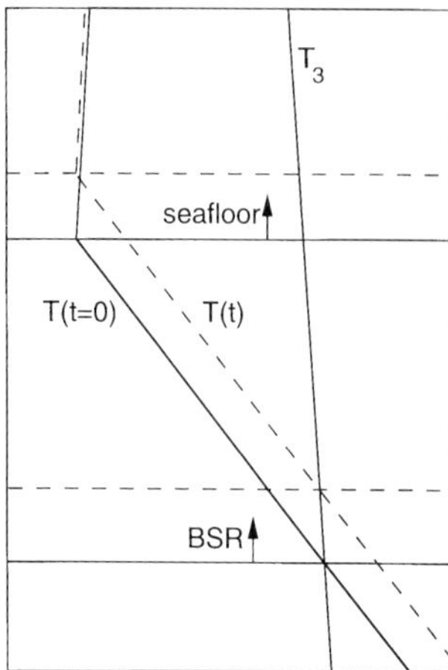

Fig. 6. A schematic diagram showing the effects of ongoing sedimentation on the temperature profile near the sea floor and on the position of the BSR. As time progresses and the sea-floor position migrates upwards, the temperature in the sediments changes from the solid line marked $T(t = 0)$ to the dashed line marked $T(t)$. The intersection of the geotherm and the three-phase equilibrium temperature T_3 migrates upwards, and hence the base of the stability region moves upwards as well.

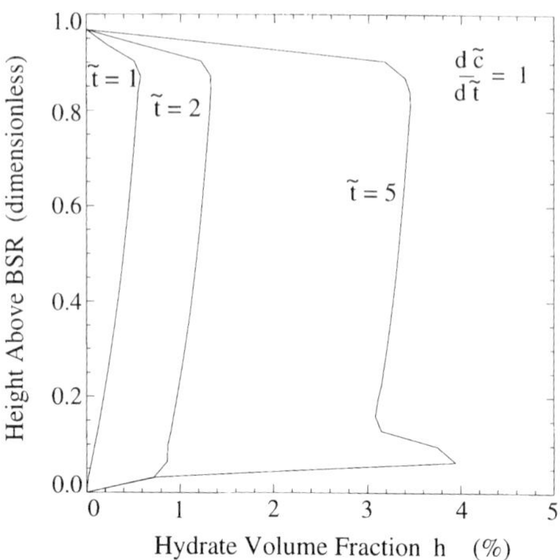

Fig. 7. The hydrate saturation level h as a function of dimensionless height above the initial BSR position for the *in situ* biogenic production model with the additional effects of sediment accumulation. The uniform dimensionless rate of biogenic gas production $d\tilde{c}/d\tilde{t}$ is 1. The sedimentation velocity is 0.01 so the new position of the BSR when $\tilde{t} = 5$ is 0.05. In this example the fluid velocity is zero so gas liberated from dissociated hydrate is transported back into the stability zone by compositional diffusion alone.

emerges from this scenario could contain two regions with an elevated hydrate volume fraction, similar to recent observations on the Blake ridge (Dickens *et al.* 1997). In Fig. 7, the upper region near the sea floor at dimensionless height 1.0 has an elevated hydrate volume fraction because of the decreased equilibrium gas concentration at lower temperatures. The peak in hydrate volume fraction near the BSR is caused by gas transported back into the stability region from hydrate that has dissociated because of the evolving thermal conditions. The time scale for this recycling of gas at the base of the stability zone depends on the rate of sedimentation and the rate of gas transport back into the stability zone, while the time scale for the total accumulation depends on the rate of biogenic gas production.

Discussion of model results

The observed vertical profiles of gas hydrate saturation in submarine deposits vary considerably from site to site. In addition, different methods of estimating hydrate saturation at the same location have yielded quite different results. For example, on the Cascadia margin, ODP Site 889, Yuan *et al.* (1996) found that velocity data from vertical seismic profiles and multi-channel seismic profiles suggest hydrate saturation levels of about 20% of the pore space immediately above the BSR, whereas chlorinity data indicate a pore saturation approaching 40% near the BSR. However, both data sets display the same general trend of increasing hydrate saturation levels with depth below the sea floor until a maximum is reached in the vicinity of the BSR. This is consistent with the predictions of the fluid-flux model of hydrate accumulation presented here (see Fig. 5). The fluid flux model suggests that hydrate saturation levels of order 10% of the pore space would require on the order of 10^6 years to accumulate.

Dickens *et al.* (1997) made direct measurements of methane content in recovered pressure core barrels at sites 995 and 997 of ODP Leg 164 on the Blake Outer ridge. They determined that the *in situ* methane concentration exceeds the two-phase hydrate–liquid equilibrium concentration in a region from 190 to 450 m below the sea floor (mbsf). The data indicate two peaks in hydrate saturation; the first at approximately 200 mbsf, and the second, more prominent peak immediately above the BSR at 450 mbsf. These data are qualitatively similar to the predictions of the *in situ* gas production model with the effects of sedimentation (see

Fig. 7). Dickens *et al.* (1997) report that the sediments from 0 to 190 mbsf and from 240 to 380 mbsf appear not to contain enough methane to support significant amounts of hydrate. These observations could also be explained by the *in situ* production model if we allow the rate of biogenic gas production to vary with depth. For example, near the sea floor bacterial gas production is not expected to occur above the base of the sulphate reduction zone (Claypool & Kaplan 1974). In addition, variations in the amount of organic material available at different depths could result in different rates of biogenic gas production. Over certain depth ranges this could lead to insufficient biogenic production of methane to enable the local gas concentration to exceed the two-phase equilibrium value.

Caution must be exercised when comparing the model results to field observations. For example, while the upper peak in hydrate saturation at ODP Leg 164 can be explained by the *in situ* gas production model, it could also have resulted from enhanced gas migration accompanying hydrate dissociation during past climate change. Alternatively, if sediment failure were to occur immediately above a BSR it would change the position of the sea floor and the thermal conditions in the sediment. The free gas previously beneath the initial location of the BSR could be converted to hydrate and lead to a region of enhanced hydrate saturation well above the subsequent depth of the base of the hydrate stability zone. The models presented here do not account for sea-level changes or sediment failure, which could both strongly influence the hydrate saturation profile. In principle, these effects would not alter the form of the governing equations (1), and the current models could be modified to consider such events. With the limited amount of data currently available, however, it could be difficult to determine which of the various model scenarios best explains the observed hydrate distribution.

Hydrate saturation profiles determined using bore hole techniques can be heavily influenced by anomalous zones containing segregated hydrate. Segregated hydrate can result from enhanced fluid transport in confined regions or from the influence of the sediment properties on hydrate formation (Clennell *et al.* 1995). Fractures can act as preferred fluid migration pathways where the rapid flow increases the gas flux. The hydrate accumulation models presented in this study have all been applied to uniform porous media. The actual geological settings where hydrate deposits are found can be much more complex. However, the general form of the equations in equation (1) should not be altered by features such as fractures. Hydrate accumulation is still controlled by balances between advection and diffusion of gas and heat, and the latent heat release and gas consumption associated with the phase change. When modelling hydrate accumulation in non-isotropic and inhomogeneous media, solutions to the governing equations, must be fully three dimensional. These complications are beyond the scope of the current paper, but the basic physical processes are described by equation (1).

Conclusions

We have presented a set of governing equations which describes the accumulation of hydrate in porous media. We use scaling arguments to show that simplifications to the governing equations are justified in a number of physical settings. This allows us to produce simple models that describe how hydrate deposits form.

We first examined the case of an isothermal porous half-space cooled on its boundary. We found that the growth rate of the stability zone is controlled by thermal diffusion, while the hydrate saturation profile is determined by the gas available in excess of the equilibrium concentration. Next we looked at the results of two numerical models of hydrate accumulation in a region with a conductive thermal profile similar to that found in marine sediments. In the fluid flux model, we considered the development of a hydrate layer as gas is advected and diffused into the layer from below. For the case where advection dominates the gas transport, we obtained a simple expression for the change in hydrate volume fraction with time after a steady-state equilibrium gas concentration profile has developed. The predicted rate of increase in hydrate saturation level is greatest at the base of the stability field, which is consistent with some indirect observations of hydrate deposits. For the *in situ* biogenic production model coupled with ongoing sedimentation, we obtained a similar saturation profile to that for the fluid flux model. One of the principal differences between the two models is the suggestion of two possible regions with elevated hydrate saturation in the biogenic model. The upper maximum is caused by the reduced gas solubility at lower temperatures in the hydrate zone, while the lower maximum is caused by the redistribution of gas from dissociated hydrate as the geotherm migrates due to ongoing sedimentation.

Hydrate reservoirs are actively maintained and the hydrate saturation profile is continuously evolving as additional gas is transported by diffusion and dispersion down the equilibrium concentration gradient, and by advection when there is a non-zero fluid velocity. Outgassing events such as those caused by sediment failure or sea-level changes are probably responsible for moderating the hydrate content of submarine reservoirs. Continued development and refinement of hydrate accumulation models using the results of laboratory experiments and field observations will improve our understanding of this important resource.

The authors would like to thank G. R. Dickens for thoughtful comments and suggestions that have undoubtedly improved the manuscript. A. W. Rempel thanks the organizers of the First Master Conference for their excellent work and hospitality.

Appendix 1. Nomenclature

c	mass fraction of gas in fluid
$\tilde{c}$	dimensionless gas concentration $\tilde{c} \equiv (c - c_0)/\Delta c$
c_h	mass fraction of gas in hydrate
$\tilde{c}_h$	dimensionless hydrate gas concentration $\tilde{c}_h \equiv (c_h - c_0)/\Delta c$
C	specific heat capacity at constant pressure
$\bar{C}(h)$	bulk heat capacity
$D(h)$	chemical dispersion–diffusion coefficient at hydrate pore-volume fraction h
$D(0)$	chemical dispersion–diffusion coefficient with no hydrate (i.e. $h=0$)
G	geothermal gradient
h	hydrate pore-volume fraction
H	specific enthalpy
$\bar{H}$	total enthalpy
$K(h)$	bulk thermal conductivity at hydrate pore-volume fraction h
$k(h)$	effective permeability to fluid flow at hydrate pore-volume fraction h.
l	length scale
L	latent heat of formation for hydrate
Pe	Peclet number $Pe \equiv ul/\kappa(0)$
Q	heat flux
$\Re$	reaction-rate constant for kinetic law, equation (4)
S	Stefan number $S \equiv \rho_h \phi L/(\rho_f C_f \Delta T)$
T	temperature
$\tilde{T}$	dimensionless temperature $\tilde{T} \equiv (T - T_0)/\Delta T$
t	time
$\tilde{t}$	dimensionless time $\tilde{t} \equiv t\kappa(0)/l^2$
$\mathbf{u}$	fluid velocity
z	vertical coordinate
$\tilde{z}$	dimensionless vertical coordinate $\tilde{z} \equiv z/l$
α	constant in empirical two-phase equilibrium equation (3)
ε	Lewis number $\varepsilon \equiv D(0)/\kappa(0)$
ϕ	porosity
η	dynamic viscosity of pore fluid
$\kappa(h)$	bulk thermal diffusivity $\kappa(h) \equiv K(h)/(\rho_f C_f)$
$\kappa(0)$	bulk thermal diffusivity with $h = 0$
λ	similarity coordinate for interface position
ρ	density

Subscripts

s	sediment
f	fluid
h	hydrate
0	reference value
eq	equilibrium

Appendix 2. Governing equations

The conservation equations for energy and gas are applied to a fixed volume in a continuum composed of sediment, pore fluid and hydrate. The heat equation relates the change in enthalpy H to the heat flux Q into the volume by

$$\frac{\partial \bar{H}}{\partial t} = Q. \tag{B1}$$

The total enthalpy $\bar{H}$ is written as

$$\bar{H} = \rho_f H_f \phi (1 - h) + \rho_h H_h \phi h + \rho_s H_s (1 - \phi), \tag{B2}$$

where H_f, H_h and H_s are the specific enthalpies of the fluid, hydrate and sediment components. The heat flux Q into the volume may be carried by advection and conduction

$$Q = -\nabla \cdot (\rho_f H_f \mathbf{u}) + \nabla \cdot (K(h) \nabla T) \tag{B3}$$

where the advective transport is through the fluid at velocity $\mathbf{u}$, and the bulk thermal conductivity is

$K(h)$. The specific heat at constant pressure is defined as

$$C = \left.\frac{\partial H}{\partial T}\right|_P$$

and the latent heat per unit mass of hydrate is

$$L = H_{\mathrm{f}} - H_{\mathrm{h}}$$

We treat each of the components as incompressible and substitute (B2) and (B3) into equation (B1) using the definitions for C and L and rearranging terms to get

$$\begin{aligned}[\rho_{\mathrm{f}} C_{\mathrm{f}} \phi(1-h) + \rho_{\mathrm{h}} C_{\mathrm{h}} \phi h + \rho_{\mathrm{s}} C_{\mathrm{s}}(1-\phi)] \frac{\partial T}{\partial t} \\ + \rho_{\mathrm{f}} C_{\mathrm{f}} \mathbf{u} \cdot \nabla T \\ + H_{\mathrm{f}} \left\{ \phi(\rho_{\mathrm{h}} - \rho_{\mathrm{f}}) \frac{\partial h}{\partial t} + \rho_{\mathrm{f}} \nabla \cdot \mathbf{u} \right\} \\ = \nabla \cdot (K(h) \nabla T) + \rho_{\mathrm{h}} \phi L \frac{\partial h}{\partial t}. \end{aligned} \qquad \text{(B4)}$$

The term in braces {} is equal to zero by conservation of mass. Dividing through by $\rho_{\mathrm{f}} C_{\mathrm{f}}$ we have

$$\begin{aligned}\bar{C}(h) \frac{\partial T}{\partial t} + \mathbf{u} \cdot \nabla T = \nabla \cdot (\kappa(h) \nabla T) \\ + \frac{\rho_{\mathrm{h}} \phi L}{\rho_{\mathrm{f}} C_{\mathrm{f}}} \frac{\partial h}{\partial t} \end{aligned} \qquad \text{(B5)}$$

where the bulk thermal diffusivity and the bulk heat capacity are

$$\kappa(h) \equiv \frac{K(h)}{\rho_{\mathrm{f}} C_{\mathrm{f}}} \quad \text{and}$$

$$\bar{C}(h) \equiv \frac{\rho_{\mathrm{f}} C_{\mathrm{f}} \phi(1-h) + \rho_{\mathrm{h}} C_{\mathrm{h}} \phi h + \rho_{\mathrm{s}} C_{\mathrm{s}}(1-\phi)}{\rho_{\mathrm{f}} C_{\mathrm{f}}}.$$

Changes in the total mass of gas in a fixed volume are due to the combined effects of advection, diffusion and dispersion. The mass of gas per unit volume is

$$\rho_{\mathrm{f}} \phi c(1-h) + \rho_{\mathrm{h}} \phi c_{\mathrm{h}} h \qquad \text{(B6)}$$

where the mass fraction of gas in the hydrate c_{h} may be treated as a constant. The advective gas flux into the volume through the fluid is

$$-\nabla \cdot (\rho_{\mathrm{f}} c \mathbf{u}). \qquad \text{(B7)}$$

We neglect the effects of the temperature gradient and the pressure gradient on the diffusive flux, and write the combined diffusive and dispersive flux as

$$\rho_{\mathrm{f}} \phi \nabla \cdot ((1-h) D(h) \nabla c). \qquad \text{(B8)}$$

The change in the mass of gas (B6) is equal to the sum of the transport terms, equations (B7) and (B8), so that

$$\begin{aligned}\rho_{\mathrm{f}} \phi(1-h) \frac{\partial c}{\partial t} + \rho_{\mathrm{h}} \phi(c_{\mathrm{h}} - c) \frac{\partial h}{\partial t} + \rho_{\mathrm{f}} \mathbf{u} \cdot \nabla c \\ + c \left\{ \rho_{\mathrm{f}} \nabla \cdot \mathbf{u} + \phi(\rho_{\mathrm{h}} - \rho_{\mathrm{f}}) \frac{\partial h}{\partial t} \right\} \\ = \rho_{\mathrm{f}} \phi \nabla \cdot ((1-h) D(h) \nabla c). \end{aligned} \qquad \text{(B9)}$$

Dividing through by $\rho_{\mathrm{f}} \phi$ and using conservation of mass to eliminate the term in braces {} we arrive at

$$\begin{aligned}(1-h) \frac{\partial c}{\partial t} + \frac{1}{\phi} \mathbf{u} \cdot \nabla c = \nabla \cdot [(1-h) D(h) \nabla c] \\ - \frac{\rho_{\mathrm{h}}}{\rho_{\mathrm{f}}} (c_{\mathrm{h}} - c) \frac{\partial h}{\partial t}. \end{aligned} \qquad \text{(B10)}$$

References

BANGS, N., SAWYER, D. & GOLOVCHENKO, X. 1993. Free gas at the base of the gas hydrate zone in the vicinity of the Chile triple junction. *Geology*, **21**, 905–908.

BROOKS, J., JEFFREY, A., MCDONALD, T., PFLAUM R. & KVENVOLDEN, K. 1985. Geochemistry of hydrate gas from Site 570, Deep Sea Drilling Project Leg 84. *In*: VON HUENE, R., AUBOUIN, J. *et al.* (eds) *Initial Reports of the Deep Sea Drilling Project*, Volume 84. US Government Printing Office, Washington, DC, 699–703.

BROWN, K., BANGS, N., FROELICH, P. & KVENVOLDEN, K. 1996. The nature, distribution, and origin of gas hydrate in the Chile Triple Junction region. *Earth and Planetary Science Letters*, **139**, 471–483.

CLAYPOOL, G. & KAPLAN, I. 1974. The origin and distribution of methane in marine sediments. *In:* KAPLAN, I.R. (ed.) *Natural Gases in Marine Sediments.* Plenum, New York, 99–139.

CLENNELL, M. B., HOVLAND, M., LYSNE, D. & BOOTH, J. S. 1995. Role of capillary forces, coupled flows and sediment-water depletion in the habitat of gas hydrate. *EOS Transactions of the American Geophysical Union*, **76**, S164–165.

DAVIS, E., HYNDMAN, R. & VILLINGER, H. 1990. Rates of fluid expulsion across the Northern Cascadia accretionary prism: Constraints from new heat flow and multichannel seismic reflection data. *Journal of Geophysical Research*, **95**, 8869–8889.

DICKENS, G. & QUINBY-HUNT, M. 1997. Methane hydrate stability in pore water: A simple theoretical approach for geophysical applications. *Journal of Geophysical Research*, **102**, 773–783.

——, PAULL, C., WALLACE, P. & THE ODP LEG 164 SCIENTIFIC PARTY. 1997. Direct measurement of in situ methane quantities in a large gas-hydrate reservoir. *Nature*, **385**, 426–428.

ENGLEZOS, P. & BISHNOI, P. 1988. Prediction of gas hydrate formation conditions in aqueous electrolyte solutions. *American Institute of Chemical Engineering Journal*, **34**, 1718–1721.

——, KALOGERAKIS, N., DHOLABHAI, P. & BISHNOI, P. 1987. Kinetics of formation of methane and ethane gas hydrates. *Chemical Engineering Science*, **42**, 2647–2658.

FOGG, P. & GERRARD, W. 1991. *Solubility of Gases in Liquids*. Wiley, Chichester, 113–159.

HANDA, Y. 1990 Effect of hydrostatic pressure and salinity on the stability of gas hydrates. *Journal of Physical Chemistry*, **94**, 2652–2657.

HOLBROOK, W., HOSKINS, H., WOOD, W., STEPHEN, R., LIZARRALDE, D. & THE ODP LEG 164 SCIENCE PARTY. 1996. Methane hydrate and free gas on the Blake Ridge from vertical seismic profiling. *Science*, **273**, 1840–1843.

HUPPERT, H. & WORSTER, M. 1985. Dynamic solidification of a binary melt. *Nature*, **314**, 703–707.

HYNDMAN, R. & DAVIS, E. 1992. A mechanism for the formation of methane hydrate and seafloor bottom-simulating reflectors by vertical fluid expulsion. *Journal of Geophysical Research*, **97**, 7025–7041.

—— & SPENCE, G. 1992. A seismic study of methane hydrate marine bottom simulating reflectors. *Journal of Geophysical Research*, **97**, 6683–6698.

——, WANG, K., YUAN, T. & SPENCE, G. 1993. Tectonic sediment thickening, fluid expulsion, and the thermal regime of subduction zone accretionary prisms: The Cascadia margin off Vancouver Island. *Journal of Geophysical Research*, **98**, 21, 865–21, 876.

KASTNER, M., KVENVOLDEN, K., WHITICAR, M., CAMERLENGHI, A. & LORENSON, T. 1995. Relation between pore fluid chemistry and gas hydrates associated with bottom-simulating reflectors at the Cascadia Margin, sites 889 and 992. *Proceedings of the Ocean Drilling Program, Scientific Results*, College Station, TX. Ocean Drilling Program, **146**, 175–187.

KROOS, B. & LEYTHAEUSER, D. 1988. Experimental measurements of the diffusion parameters of light hydrocarbons in water-saturated sedimentary rocks – II. Results and geochemical significance. *Organic Geochemistry*, **12**, 91–108.

KVENVOLDEN, K. 1993. Gas hydrates – Geological perspective and global change. *Reviews of Geophysics*, **31**, 173–187.

—— & BARNARD, L. 1983. Gas hydrates of the Blake Outer Ridge, Deep Sea Drilling Project Site 533, Leg 76. *In*: SHERIDAN, R., GRADSTEIN, F. *et al.* (eds) *Initial Reports of the Deep Sea Drilling Project*, Volume 75. US Government Printing Office, Washington, DC, 353–366.

LIDE, D. 1990. *CRC Handbook of Chemistry and Physics*, 71st edition. CRC, Boca Raton, FL 6–8.

MACKAY, M., JARRARD, R., WESTBROOK, G., HYNDMAN, R. & THE ODP LEG 146 SCIENCE PARTY. 1994. Origin of bottom-simulating reflectors: Geophysical evidence for the Cascadia accretionary prism. *Geology*, **22**, 459–462.

MINSHULL, T., SINGH, S. & WESTBROOK, G. 1994. Seismic velocity structure of a gas hydrate reflector, offshore western Columbia, from full waveform inversion. *Journal of Geophysical Research*, **99**, 4715–4734.

REMPEL, A. 1994. *Theoretical and Experimental Investigations into the Formation and Accumulation of Gas Hydrates*. M.Sc. thesis, University of British Columbia, Vancouver, Canada.

—— & BUFFETT, B. 1997. Formation and accumulation of gas hydrate in porous media. *Journal of Geophysical Research*, **102**, 10, 151–10, 164.

SELIM, M. & SLOAN, E. 1989. Heat and mass transfer during the dissociation of hydrates in porous media. *American Institute of Chemical Engineering Journal*, **35**, 1049–1052.

SINGH, S. & MINSHULL, T. 1994. Velocity structure of a gas hydrate reflector at ocean drilling program site 889 from a global seismic waveform inversion. *Journal of Geophysical Research*, **99**, 24, 221–24, 233.

SLOAN, E.D. 1990. *Clathrate Hydrates of Natural Gases*. Marcel Dekker, New York.

TURCOTTE, D. & SCHUBERT, G. 1982. *Geodynamics – Applications of Continuum Physics to Geological Problems*. Wiley, New York, 168–170.

TSYPKIN, G. 1991. Dissociation of gaseous hydrates in beds. *Journal of Engineering Physics*, **60**, 556–561.

—— 1992*a*. Appearance of two moving phase transition boundaries in the dissociation of gaseous hydrates in strata. *Soviet Physics Doklady*, **37**, 126–128.

—— 1992*b*. Effect of liquid phase mobility on gas hydrate dissociation in reservoirs. *Fluid Dynamics*, **26**, 564–572.

UCHIDA, T. & NARITA, H. 1996. Studies of formation and dissociation rates of methane hydrates in pure water – pure gas system. *In: Canada–Japan Joint Science and Technology Workshop on Gas Hydrates*, Victoria, Canada, 109–110.

WORSTER, M. 1992. The dynamics of mushy layers. *In*: DAVIS, S., HUPPERT, H. & MULLER, U. (eds) *Interactive Dynamics of Convection and Solidification*. Kluwer, Dordrecht, 113–138.

YAMANE, K. & AYA, I. 1995. Solubility of carbon dioxide in hydrate region at 30 MPa. *In: Proceedings of the MARIENV'95 Conference*, 911–917.

YUAN, T., HYNDMAN, R., SPENCE, G. & DESMONS, B. 1996. Seismic velocity increase and deep-sea gas hydrate concentration above a bottom-simulating reflector on the northern Cascadia continental slope. *Journal of Geophysical Research*, **101**, 13, 655–13, 671.

ZATSEPINA, O. & BUFFETT, B. 1997. Phase equilibrium of gas hydrate: Implications for the formation of hydrate in the deep seafloor. *Geophysical Research Letters*, **24**, 1567–1570.

Improvements in clathrate modelling II: the H_2O–CO_2–CH_4–N_2–C_2H_6 fluid system

R. J. BAKKER

Geologisch-Paläontologisches Institut, Universität Heidelberg, Im Neuenheimer Feld 234, D-69120 Heidelberg, Germany

Abstract: Clathrate stability conditions have been modelled for the H_2O–CO_2–CH_4–N_2–C_2H_6 fluid system based on all available experimental data. Optimum Kihara parameters are estimated for pure CO_2, CH_4, N_2 and C_2H_6 gas hydrates from using the most accurate calculation of other parameters involved in clathrate modelling, like fugacities, gas solubilities in H_2O and thermodynamic constants. For mixed gas hydrates of CO_2–CH_4, CH_4–N_2, and CH_4–C_2H_6 excess Gibbs free energy functions are introduced to obtain a good fit to experimental data. The excess Gibbs free energy is described according a modified Margules equation, which depends on mole fraction and temperature.

Gas hydrates are ice-like solids that form part of the clathrate family. They occur regularly in natural rock (e.g. Sloan 1990) and in fluid inclusions in single crystals (e.g. Roedder 1963). Gas hydrates are solid-solutions of H_2O and gases such as CO_2, CH_4, N_2, and C_2H_6, and its stoichiometry depends on temperature and pressure. Knowledge of the stability conditions of this clathrate phase as a function of salinity, temperature and pressure forms the basis of any conclusive predictions in clathrate-related oceanographic and atmospheric studies. Purely empirical best-fits to experimental data are the easiest way for modelling clathrate stability conditions, and this method has been frequently favoured throughout decades of clathrate research (e.g. Deaton & Frost 1946; Bozzo *et al.* 1975; Dholabhai *et al.* 1993). The non-stoichiometry and the infinite compositional combinations of clathrates, however, require a more thorough approach for interpolations and extrapolations of clathrate stability conditions in natural systems. Equilibrium thermodynamics is an important tool for a systematic approach of modelling clathrate melting conditions as a function of composition, temperature, pressure and salinity. Platteeuw & van der Waals (1958) and van der Waals & Platteeuw (1959) were the first to use this approach, which was only possible after thorough studies of the crystallographic structure of the solid clathrate phase by, for example, von Stackelberg & Müller (1954). Platteeuw & van der Waals (1958) and van der Waals & Platteeuw (1959) modified the absorption theory according to Langmuir (1918) to a three-dimensional generalization describing the physical interaction of a captured gas molecule in a cage of H_2O molecules. This model was the first to describe systematically clathrate stability conditions of pure and mixed gas clathrates.

The basic concept of the model from van der Waals & Platteeuw (1959) is the equality of chemical potential of one component in each phase in which it is present, which is the most important prerequisite for any equilibrium thermodynamics:

$$\mu^{\text{vap}}_{H_2O} = \mu^{\text{liq}}_{H_2O} = \mu^{\text{clath}}_{H_2O} \tag{1}$$

where the superscripts indicate the phases (vap is vapour, liq is liquid and clath is clathrate). The chemical potential of H_2O in the clathrate phase is obtained from statistical thermodynamics using the partition functions (van der Waals & Platteeuw 1959):

$$\mu^{\text{clath}}_{H_2O} = \mu^{\text{empty}}_{H_2O} + \mathrm{R}T\sum_i \nu_i \ln\left(1 - \sum_M y_{Mi}\right) \tag{2}$$

where $\mu^{\text{empty}}_{H_2O}$ is the chemical potential of a hypothetical gas-free (empty) clathrate, R is the gas constant ($=8.31439\,\mathrm{J\,mol^{-1}\,K^{-1}}$), T is temperature in Kelvin, ν_i is the number of cavities of type i per cage forming H_2O molecule and y_{Mi} is the probability of finding a gas molecule M in cavity of type i. van der Waals & Platteeuw (1959) related the probability y_{Mi} to Langmuir constants (equation (3)), which in turn are related to molecular cell potentials (equation (4))

$$y_{Mi} = \frac{C_{Mi} f_M}{1 + \sum_K C_{Ki} f_K} \tag{3}$$

$$C_{Mi} = \frac{4\pi}{kT}\int_0^\infty \exp\left(\frac{-w(r)}{kT}\right) r^2\,\mathrm{d}r \tag{4}$$

BAKKER, R. J. 1998. Improvements in clathrate modelling II: the H_2O-CO_2-CH_4-N_2-C_2H_6 fluid system. *In:* HENRIET, J.-P. & MIENERT, J. (eds) *Gas Hydrates: Relevance to World Margin Stability and Climate Change.* Geological Society, London, Special Publications, **137**, 75–105.

where C_{Mi} is the Langmuir constant for gas M in a cavity of type i, f_M is the fugacity of gas M, k is the Boltzman constant, r is the distance to the centre of the cavity and $w(r)$ is the spherically symmetrical potential function describing the intermolecular potential between a gas molecule at the centre of the cage and an H_2O molecule of the cavity wall. Equations (2)–(4) are a combination of Langmuir's isotherm, which is characteristic for localized adsorption without interaction of the adsorbed molecules, and a generalization of Raoult's law for the properties of the solvent in a solution where solute–solute interaction can be neglected.

In the scope of this study it is important to focus on the major postulates originally proposed by van der Waals & Platteeuw (1959), apart from their axiom of classical statistics:

(a) the mode of occupation of the cavities does not affect the contribution of the H_2O molecules to the total chemical potential of the host lattice;
(b) a cavity can never hold more than one encaged molecule, which is always localized in the cavities;
(c) the mutual interaction of the solute molecules is neglected;
(d) the solute molecules can rotate freely in their cavities, i.e. the rotational partition function for the motion in the cavity is the same as that in the perfect gas;
(e) the potential energy of a solute molecule when at a distance r from the centre of its cage is given by the spherically symmetrical potential $w(r)$ proposed by Lennard-Jones & Devonshire (1937, 1938) for dense gases. This method is equally valid for the cavities in a clathrate, as long as one restricts oneself to first neighbour interactions.

By that time, van der Waals & Platteeuw (1959) already indicated the limitations of these postulates. Encaged molecules with larger dimensions, like CO_2, may distort host lattice and damage assumption (a). Postulate (c) can never be strictly true because numerical calculations and experimental evidence show that the contribution of solute–solute interaction to the configurational energy is at most a few per cent of the energy of binding of the solute molecules in their cages. Non-spherical molecules, like O_2 and N_2, will not be free to rotate in the entire cavity. When such a molecule comes close to the wall of its cage it will have to orientate itself parallel to this wall. The relative contribution of second and third neighbour solvent molecules to $w(r)$ in hydrate structures is only of the order of a quarter of that in the much denser face-centred cubic lattice. Despite the indicated limitations of the postulates (a)–(e), van der Waals & Platteeuw (1959) were able to give a clear relation between the equilibrium vapour pressure, composition, and chemical potential of the solvent in a clathrate. The individual postulates (a)–(e) have frequently been modified (e.g. McKoy & Sinanoglu 1963; Saito *et al.* 1964; Parrish & Prausnitz 1972; Ng & Robinson 1976; Holder *et al.* 1980; Dharmawardhama *et al.* 1980; John *et al.* 1985; Munck *et al.* 1988; Dubessy *et al.* 1992; Bakker *et al.* 1996) to increase the physical integrity at molecular scale and to obtain a better fit to increasingly available experimental data.

Modifications proposed by Bakker *et al.* (1996) for pure CO_2 clathrate with various salts are further developed in this study for a complex fluid system including H_2O–CO_2–CH_4–N_2–C_2H_6. The aim of this study is to present a model which is based on all available experimental data within this fluid system, and which is able to reproduce these data accurately over a wide range of TP conditions. In addition, we have tried to avoid increasing the complexity of the model, which has no experimental justification, and to show that some original proposed principles are eminently useful in describing this fluid system accurately.

Topology of H_2O–CO_2–CH_4–N_2–C_2H_6 fluids near clathrate melting conditions

Qualitative thermodynamic equilibria analysis, based on Gibbs phase rule, constrains possible configurations for this multi-component fluid system (Fig. 1). Five phases, i.e. clathrate, water, gas mixtures (liquid- and vapour-like) and ice form the basic ingredients for this analysis. Equilibrium between three phases is illustrated by univariant lines, which intersect at invariant quadruple points Q_1 and Q_2 (e.g. Fig. 1d), where four phases form a stable configuration. Quadruple point Q_2 can only be present if the immiscibility field of gas mixtures and pure gases, i.e. the dew-point curve and bubble-point curve, intersects the limits of the clathrate stability field (Fig. 1c and d). Quadruple point Q_1 remains a point for this multi-component system because the coexisting gas mixture is always vapour-like. Q_2 is transformed in a line segment for ternary and higher-order fluid systems. Clathrate melting conditions for binary gas–H_2O systems are represented by Fig. 1a

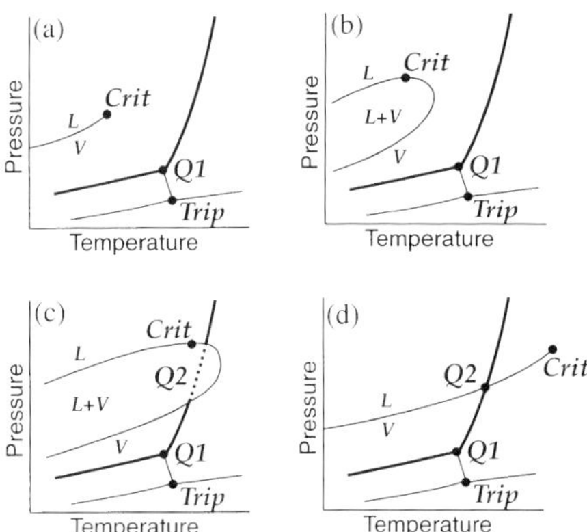

Fig. 1. Schematic P–T diagrams illustrating the position of the clathrate stability field (shaded areas) compared to the position of liquid–vapour equilibria of pure gases, (**a**) and (**d**), and gas mixtures, (**b**) and (**c**). L, liquid; V, vapour; Crit, critical point; Q_1, quadruple point 1; Q_2, quadruple point 2; Trip, triple point of pure H_2O.

and d. These topologies form the basis of the systematic clathrate modelling in this study.

Model modifications

Improvements and simplifications of the clathrate model as proposed by Bakker *et al.* (1996) are further developed for the multi-component H_2O–CO_2–CH_4–N_2–C_2H_6 fluid system.

Thermodynamic properties of H_2O in the liquid phase

The thermodynamic properties of pure H_2O in liquid and solid phase are intensively studied fluid parameters in the literature (e.g. Franks 1972). Several recently published equations of state for pure H_2O (e.g. Haar *et al.* 1984; Hill 1990) are able to describe accurately all measurable properties of this fluid from a unified Helmholtz function, i.e. saturation properties, densities, pressures, speed of sound, specific heats and virial coefficients. In general, these studies do not include metastable regions for the H_2O phase. Although many parameters for H_2O are accurately known, most clathrate stability models include oversimplified values for density and specific heat. For example, constant values have been used for these parameters by, for example, Munck *et al.* (1988) and Dubessy *et al.* (1992). Bakker *et al.* (1996) introduced a more accurate calculation of these parameters, using published equations of state that reproduce accurately experimental data, even extended into metastable regions. The molar volume of liquid H_2O is calculated from the equation of state as proposed by Kell & Whalley (1965) and Kell (1967) up to 100 MPa. At higher pressures, the equation from Haar *et al.* (1984) is used. The heat capacity of liquid H_2O is obtained from Osborne *et al.* (1939). At temperatures below 273.15 K, a polynomal best-fit (Bakker *et al.* 1996, p. 1662, equation 6) was obtained through data on metastable liquid H_2O from Angell *et al.* (1973).

van der Waals & Platteeuw (1959) concentrated on clathrate equilibrium in the presence of pure ice and a gas phase. They did not consider explicitly an equilibrium involving a liquid solution, because solubilities of the solute in the liquid phase should be taken into account. The chemical potential of H_2O in the liquid phase was, in general, not known at that time. The effect of gas solubilities in an ideal liquid solution was introduced by Saito *et al.* (1964) and Parrish & Prausnitz (1972). According to relatively simple thermodynamics, the chemical potential of H_2O in the liquid solution ($\mu^{\text{liq}}_{\text{H}_2\text{O}}$) is defined by:

$$\mu^{\text{liq}}_{\text{H}_2\text{O}} = \mu^{\text{pure}}_{\text{H}_2\text{O}} + \text{R}T\ln(a_{\text{H}_2\text{O}}) \qquad (5)$$

where $\mu^{\text{pure}}_{\text{H}_2\text{O}}$ is the chemical potential of pure H_2O at a selected temperature and pressure, and $a_{\text{H}_2\text{O}}$ is the activity of H_2O. The activity coefficient for H_2O is assumed to be equal to 1 due to low gas solubilities, and, therefore, activities can be replaced by mole fraction in real solutions. However, the activity coefficient is noticeably affected in salt-bearing solutions and it should, therefore, be obtained from theoretical consideration by Debye & Hückel (1923) and Pitzer (1992) on dissolved electrolytes, as illustrated for some selected liquid solutions by, for example, Englezos & Bishnoi (1988), Dubessy *et al.* (1992) and Bakker *et al.* (1996).

Thermodynamic properties of gas hydrates

To illustrate directly equilibrium conditions for clathrates as a function of Langmuir constants van der Waals & Platteeuw (1959) defined the difference between the chemical potentials of H_2O in the two modifications, i.e. empty clathrate and ice. The quotient of this difference ($\Delta\mu_{\text{H}_2\text{O}}$) and R$T$ was directly related to measurable parameters such as enthalpy differences and volume changes:

$$\frac{\Delta\mu_{H_2O}(T,P)}{RT} = \frac{\Delta\mu^{ref}_{H_2O}}{RT_0} - \int_{T_0}^{T}\left[\frac{\Delta h_{H_2O}}{RT^2}\right]dT + \int_{P_0}^{P}\left[\frac{\Delta v_{H_2O}}{RT}\right]dP \quad (6)$$

where $\Delta\mu^{ref}_{H_2O}$ is the chemical potential difference at standard conditions T_0 and P_0, Δh_{H_2O} and Δv_{H_2O} are the difference in enthalpy and molar volume, respectively, which may be obtained from the Clapeyron equation and structural data on crystal lattices. This difference was considered a standard value for a given clathrate structure, because it is not related to the type of encaged gas molecule. The standard value for $\Delta\mu^{ref}_{H_2O}$ could have been obtained from direct analysis of the composition of clathrates at 0°C. However, determination of the composition of clathrate compounds with direct chemical analysis seems to be problematic, and many significantly different values have been published (Sloan 1990, p. 231, table 5-5). Bakker *et al.* (1996) give thorough arguments for the adoption of the values given by Dharmawardhama *et al.* (1980) as standard values at 273.15 K and 0.1 MPa. The standard chemical potentials $\mu^{empty}_{H_2O}(ref)$ and enthalpy $h^{empty}_{H_2O}(ref)$ of the hypothetical empty clathrate Structure I and II (Table 1) are obtained after addition of the standard values for the well-defined thermodynamic constants for the liquid solution or ice to $\Delta\mu^{ref}_{H_2O}$ from equation (6). The standard value for entropy $s^{empty}_{H_2O}(ref)$ is deduced from these values according to classical thermodynamics:

$$\mu^{empty}_{H_2O}(ref) = h^{empty}_{H_2O}(ref) - T_0 \times s^{empty}_{H_2O}(ref) \quad (7)$$

where T_0 is 273.15 K. Equation (6) has been extensively used in literature and it has been built up to extreme proportions. For example, the use of classical thermodynamics for this arbitrarily defined chemical potential difference lead to very complex notations (e.g. Parrish & Prausnitz 1972). A better organization of these formulae improves the survey of the model. In fact, there is no need to use this difference because the values for one component, i.e. ice and water, are well defined in the literature (see previous section). Simplifying thermodynamic notations, Bakker *et al.* (1996) used straightforward thermodynamics to calculated the chemical potential of empty clathrates:

$$\mu^{empty}_{H_2O}(T,P) = \mu^{empty}_{H_2O}(ref) - \int_{T_0}^{T} s^{empty}_{H_2O}\,dT + \int_{P_0}^{P} v^{empty}_{H_2O}\,dP \quad (8)$$

where $s^{empty}_{H_2O}$ and $v^{empty}_{H_2O}$ are entropy and molar volume of the empty clathrate, respectively. In most studies, the temperature and pressure dependency of thermodynamical properties of hypothetical empty clathrates were set equal to some properties of pure ice. In this study, the temperature dependency of entropy is obtained from the heat capacity equation for pure ice (Equation 9), which was fitted to experimental data from Giauque & Stout (1936) between 240 and 270 K (Bakker *et al.* 1996, p. 1661, equation 5). This linear function has to be extrapolated to temperatures well above pure ice melting conditions (up to 320 K) for clathrate stability conditions at higher pressures. Equation (9a) is assumed to be appropriate for both clathrate structures and ice

$$c^{empty}_{P_{H_2O}} = -0.3840 + 0.1406 \times T \quad (9a)$$

$$s^{empty}_{H_2O}(T) = s^{empty}_{H_2O}(ref) + \int_{T_0}^{T}\left(\frac{c_P}{T}\right)dT. \quad (9b)$$

The volumetric properties of the hypothetical empty clathrate are obtained from the so-called equation of state for pure ice. This equation combines the isothermal compressibility (equation (10a)) according to Bakker *et al.* (1996),

Table 1. *Thermodynamic properties of water in four phases at standard conditions of 273.15 K and 0.1 MPa*

Phase	μ_0 (kJ mol^{-1})	h_0 (kJ mol^{-1})	s_0 (J mol^{-1})	v_0 (kJ mol^{-1})
Liquid	−303.935	−286.634	63.34	18.01
Ice	−303.916	−292.645	41.26	19.65
Clathrate Structure I (empty)	−302.638	−291.256	41.67	22.35†
Clathrate Structure II (empty)	−302.961	−291.620	41.52	22.57†

† Avlonitis (1994).

and thermal expansion of clathrate structures I and II (equations (10b) and (10c)) according to Avlonitis (1994). Equation (10) gives, therefore, a complete description of volumetric properties of the empty clathrate structure at selected temperatures and pressures

$$v_P = v_0 \times \exp[-K\Delta P] \tag{10a}$$

$$v_{H_2O}^{\text{empty I}} = v_P(1.0 + 3.1075 \times 10^{-4}\Delta T + 5.9537 \times 10^{-7}\Delta T^2 + 1.3707 \times 10^{-10}\Delta T^3) \tag{10b}$$

$$v_{H_2O}^{\text{empty II}} = v_P(1.0 + 1.9335 \times 10^{-4}\Delta T + 21768 \times 10^{-7}\Delta T^2 - 1.4786 \times 10^{-10}\Delta T^3) \tag{10c}$$

where Roman numerals I and II refer to clathrate Structure I and II, v_0 is the molar volume at reference conditions (Table 1), K is the isothermal compressibility ($=10^{-4}$ MPa), ΔP is the difference between the pressure (in MPa) of interest and the standard pressure (=0.1 MPa), and ΔT is the difference between the temperature (in K) of interest and the standard temperature (=273.15 K).

Clathrate stability predictions appear to be highly sensitive to the size of cavities (e.g. Lundgaard & Mollerup 1992). Most studies have adopted a constant value for the cell radius (R) of spherical symmetrical cavities, and volumetric properties were only used for the chemical potential difference calculations in equation (6). Bakker *et al.* (1996) introduced a variable cell radius which is in a straightforward relation to volumetric properties from equation (10). Changes in molar volume resulting from variable temperature and pressure can be characterized by the total differential of equation (10) (equation (11a)). Isotropic behaviour of the clathrate structure relates the alterations in cavity radius to the cubic root of this changes in molar volume of the clathrate phase (equation (11b))

$$\mathrm{d}V = \left(\frac{\partial V}{\partial T}\right)_P \mathrm{d}T + \left(\frac{\partial V}{\partial P}\right)_T \mathrm{d}P \tag{11a}$$

$$\mathrm{d}R \equiv \sqrt[3]{\mathrm{d}V}. \tag{11b}$$

Fugacity coefficients of gases

Calculation of gas fugacities is applied to all phases which appear in equation (1). Fugacity coefficients are calculated from equations of state, which describe accurately the relation between P–T–V–X properties of fluids, according to the theoretical considerations from Prausnitz *et al.* (1986). In addition, the accuracy of fugacity calculations is dependent on the ability of equations of state to represent volumetric data for the entire pressure range from 0 MPa to the pressure of interest.

There is a rich supply of equations of state from the literature, and the selection is based, first, on the temperature and pressure of interest to clathrate stability conditions. Secondly, the equation of state should be able to describe CO_2–CH_4–N_2–C_2H_6 mixtures, which also include very small amounts of H_2O. Bakker *et al.* (1996) have demonstrated this selection procedure for pure CO_2 gas hydrates. This selection procedure has been omitted in most other studies.

Recently, unified Helmholtz energy functions have been developed for many pure gases, such as H_2O (Haar *et al.* 1984; Hill 1990), CO_2 (Span & Wagner 1996), CH_4 (Setzmann & Wagner 1991), N_2 (Angus *et al.* 1979) and C_2H_6 (Friend *et al.* 1991). These equations of state accurately describe P–T–V properties of pure gases over a wide range of temperatures and pressures. Unfortunately, these function are only defined for pure components and they are not available for gas mixtures. Therefore, another family of equations of state, such as the modifications to the equation of state according to van der Waals (1873), Benedict *et al.* (1940) or Carnahan & Starling (1969) have to be used in this study. These equations have a limited range of application and should be carefully selected. First, the clathrate melting pressure is described as a polynome in temperature (equation (12) and Table 2), which is fitted to experimental data for pure gas hydrates

$$\log_{10}(P) = A + BT_C + CT_C^2 + DT_C^3 \tag{12}$$

where T_C is temperature in °C, and P is pressure in MPa. Equation (12) is of direct use to pure gas hydrates, and can be applied for the temperature range indicated in Table 2. Along this polynome several equations of state are compared to fugacity calculations for the pure gases (Fig. 2), which are obtained from the previously mentioned unified Helmholtz energy function. Comparison to pure gases is justified because the gas mixture contains in general less than 0.1 mol% H_2O. The equation of states according to Redlich & Kwong (1949), Chueh & Prausnitz (1967), Soave (1972), Lee & Kesler (1975), Holloway (1977, 1981), Flowers (1979) and Duan *et al.* (1992*a*, *b*, 1996) are used for comparison (Fig. 2).

Table 2. *Constants in equation (12) to calculate clathrate stability for pure gas hydrates*

Gas hydrate	A	B	C	D	Temperature interval (°C)
CO_2	0.087925	0.049309	0.00068487	–	−1.5 to 9.9
	−8.3417	1.6697	−0.093782	0.0018506	9.9–20
CH_4	0.43115	0.034	0.0010333	-1.7994×10^{-5}	−0.3 to 40
N_2	1.208	0.046288	−0.00017311	–	−1.3 to 27
C_2H_6	−0.30931	0.046365	0.00072746	–	−0.1 to 14.6
	−8.1526	0.87596	−0.019077	–	14.6–18

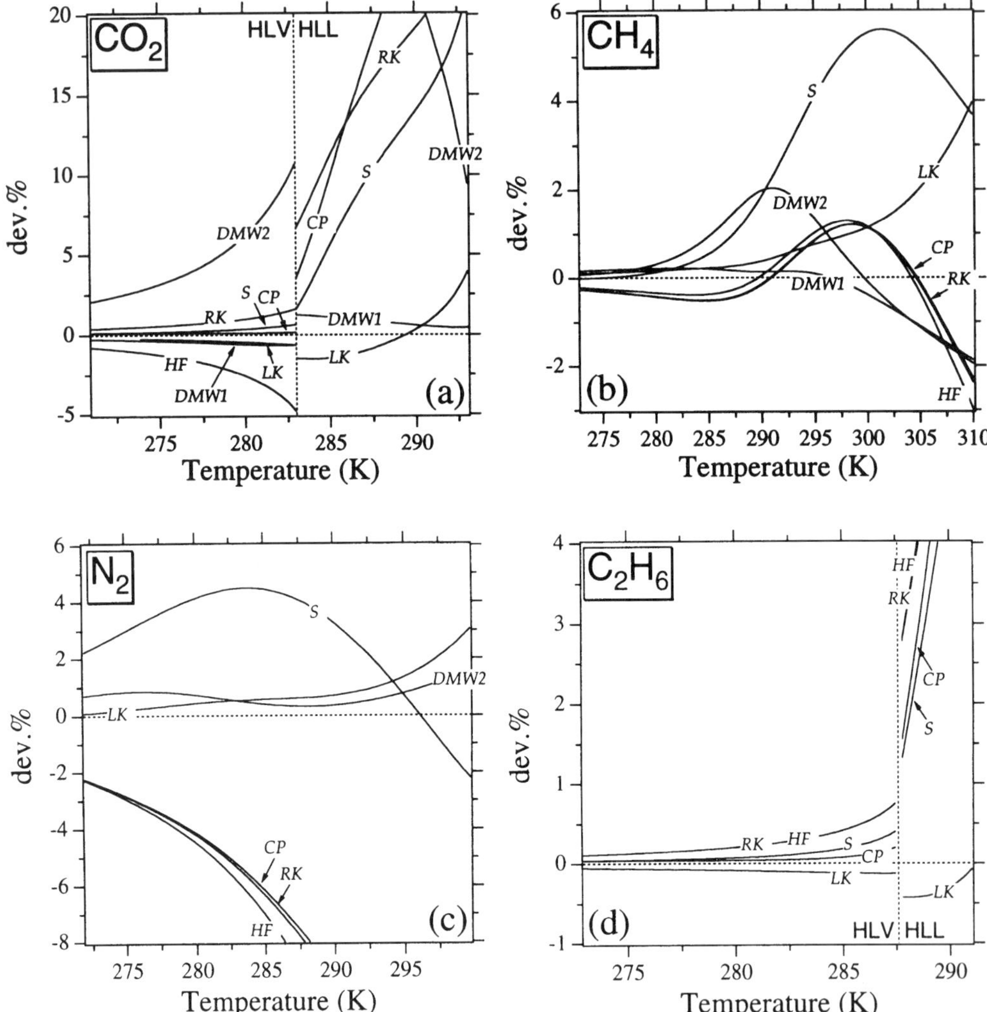

Fig. 2. Comparison of fugacity calculation for pure gases (**a**) CO_2; (**b**) CH_4; (**c**) N_2; (**d**) C_2H_6, obtained from several equations of state (RK, Redlich & Kwong 1949; CP, Chueh & Prausnitz 1967; S, Soave 1972; LK, Lee & Kesler 1975; HF, Holloway 1977, 1981; and Flowers 1979; DMW1, Duan *et al.* 1992*a*, *b*; DMW2, Duan *et al.* 1996), which can be used for gas mixtures including small amounts of H_2O. Deviation (dev.%) is defined as a percentage from fugacities obtained from unified Helmholtz energy functions. The vertical dashed line for CO_2 (**a**) and C_2H_6 (**d**) marks the boundary between clathrate melting in equilibrium with a liquid-like gas-rich phase (HLL) and a vapour-like gas-rich phase (HLV).

Table 3. *Selected equations of state for fugacity calculations at clathrate melting conditions*

Gas	EOS
CO_2	Duan *et al.* (1992*a*, *b*)
CH_4	Duan *et al.* (1992*a*, *b*)
N_2	Duan *et al.* (1996)
C_2H_6	Lee & Kesler (1974)

Table 3 gives selected equations of state which reproduce most accurately the $P-T-V-X$ properties, in particular fugacity, of the fluid at clathrate melting conditions for pure gas hydrates. The equation of state according to Duan *et al.* (1992*a*, *b*) has been chosen for fugacity calculations of pure CO_2 gas hydrate and pure CH_4 gas hydrate. Duan *et al.* (1996) provides the equation of state for pure N_2 gas hydrates, and Lee & Kesler (1975) for C_2H_6. The accuracy in fugacity calculations of these equations remains within 2%.

Solubility of gases

Although the solubility of gases like CH_4 and N_2 in H_2O is very low (e.g. Culberson & McKetta 1951), a small difference in H_2O activity may seriously affect modelled clathrate stability conditions. The solubility of gases can be theoretically predicted with equations of state that accurately describe both coexisting liquid and vapour phases at relative low-temperature conditions (e.g. Lundgaard & Mollerup 1991). This model is based on the equality of chemical potential of all components in each coexistent phase (equation (13)), which are obtained from one equation of state

$$\begin{cases} \mu_{H_2O}^{liq} = \mu_{H_2O}^{vap} \\ \mu_M^{liq} = \mu_M^{vap} \end{cases} \quad (13)$$

where subscript M refers to any type of dissolved gas. Some equations of state accurately describe liquid–vapour equilibria for H_2O–free gas mixtures (e.g. Thiéry *et al.* 1994), however, the liquid-like coexisted phase in clathrate stability calculations is H_2O-rich. The previously mentioned equations of state for the fugacity calculation accurately describe only a part of binary H_2O-gas mixtures. For example, the experimental data on CH_4 solubility in H_2O from Culberson & McKetta (1961) are accurately reproduced by Duan *et al.* (1992*a*, *b*). However, the coexisting composition of the vapour phase is increasingly in error at higher temperatures (Fig. 3). Furthermore, the method of liquid–vapour equilibria calculation from a single equation of state is a lengthy operation due to the numerical–iterative approach. Therefore, Henry's law has been used to replace the liquid phase part of equation (13). Henry's law appears to represent more accurately the low solubility data of gases in H_2O at temperature-pressure conditions relevant to clathrate stability con-

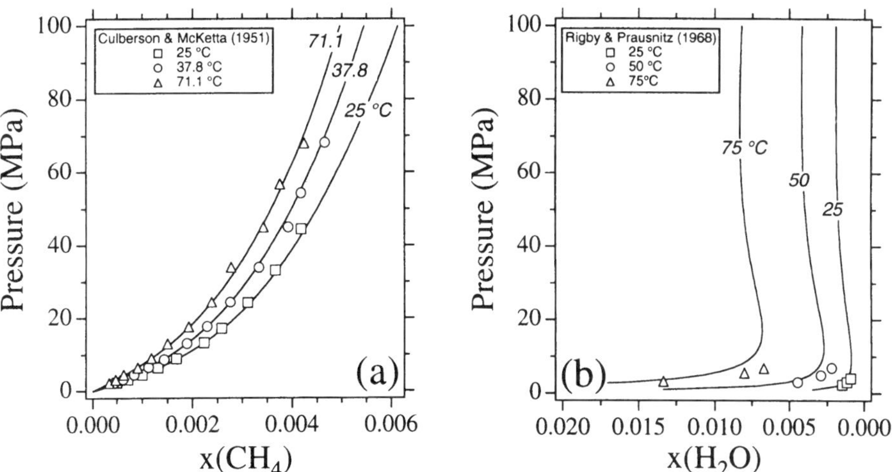

Fig. 3. P–X diagrams for comparison of solubility data of (**a**) CH_4 in H_2O liquid from Culberson & McKetta (1951) and (**b**) H_2O in CH_4 vapour from Rigby & Prausnitz (1968) to calculated coexisting liquid and vapour compositions from the equation of state according to Duan *et al.* (1992*a*, *b*) at selected temperatures.

ditions. Bakker *et al.* (1996) illustrated the careful choice of the equation according to Carroll *et al.* (1991) and Carroll & Mather (1992) for the CO_2 solubility in H_2O. Using a similar approach, Henry's constants of CH_4 and C_2H_6 from Rettich *et al.* (1981), and of N_2 from Benson & Kraus (1976), are chosen to calculate their solubility in H_2O. A polynome in reciprocal temperature is the general form to describe the temperature dependence of these Henry's constants:

$$\ln(H) = a_0 + \frac{a_1}{T} + \frac{a_2}{T^2} + \frac{a_3}{T^3} + \frac{a_4}{T^4} \qquad (14)$$

where H is Henry's constant in MPa, and a_i are the coefficients for individual gases (Table 4). It is important to realize that Henry's constants are directly related to gas fugacities according to equation (13), and the originally measured mole fraction of gas in solutions can only be reproduced exactly if a similar equation of state is used to calculated fugacities of components in the vapour phase. Fortunately, fugacity calculations from different equations of state do not differ significantly at these relatively low-temperature conditions. At higher pressures, adjustments of Henry's constant are described according to the correction from Krichevsky & Kasarnowsky (1935):

$$\ln(H)_P = \ln(H)_{P_{\mathrm{ref}}} + \int_{P_{\mathrm{ref}}}^{P} \left(\frac{v_{\mathrm{gas}}^{\infty}}{\mathrm{R}T}\right) \mathrm{d}P \qquad (15)$$

where $v_{\mathrm{gas}}^{\infty}$ is the partial molar volume of a gas at infinite dilution (Table 4), and equation (14) is used to calculate Henry's constants at the reference pressure of 1 atmosphere.

Intermolecular potential

Platteeuw & van der Waals (1958) and van der Waals & Platteeuw (1959) used the cell theory according to Lennard-Jones & Devonshire (1937, 1938) to describe the molecular interaction between an encaged gas molecule and an H_2O molecule from the cavity wall, which is directly related to lowering the chemical potential of H_2O in the clathrate phase according to equations (2)–(4). In the context of determination of interatomic forces of gases, Lennard-Jones & Devonshire (1937) were astonished by the power of the equation of state from van der Waals (1873) but noticed its failure for dense gases. They introduced a cell theory by means of statistical mechanics for gas molecules whose average position is something like an atom in a liquid or a crystal. Platteeuw & van der Waals (1958) applied this theory to describe molecular

Table 4. *Henry's constants for equation (14) and partial molar volume of gases at infinite dilution in H_2O used in equation (15)*

Gas	a_0	a_1	a_2	a_3	a_4	$v_{\mathrm{gas}}^{\infty}$ ($\mathrm{cm^3\,mol^{-1}}$)
CO_2 (Ref. 1)	−6.8346	1.2817×10^4	-3.7668×10^6	2.997×10^8	–	$58.9 - 0.08 \times T$*
CH_4 (Ref. 2)	9.881539	9864.392	-2.150256×10^6	8.810241×10^7	–	$\exp(3.205 + 0.00123 \times T)$
N_2 (Ref. 3)	−4.857	1.87007×10^4	9.674615×10^5	–	–	35.7†
C_2H_6 (Ref. 2)	−205.6101	2.624901×10^5	-1.128303×10^8	2.159875×10^{10}	-1.569604×10^{12}	$\exp(3.5808 + 0.00122 \times T)$

References: 1. Carroll *et al.* (1991); 2. Rettich *et al.* (1981); 3. Benson & Kraus (1976).
* From Bakker *et al.* (1996).
† From Moore *et al.* (1982).

forces in clathrate cavities, using the Lennard-Jones 12–6 potential function (Jones 1924). McKoy & Sinanoglu (1963) compared several molecular potential function and they concluded that the Kihara potential (Kihara 1953) predicts better dissociation pressures for gas hydrates of rod-like molecules, such as CO_2, N_2, and C_2H_6. Unlike the Lennard-Jones 12–6 potential, the Kihara potential takes into account the shape and the size of encaged molecules. Unfortunately, the Kihara potential is not the best potential function to describe the properties of H_2O molecules, the indispensable counterpart of molecular interaction in clathrates. However, this potential is eminately useful for gases like CO_2, CH_4, N_2 and C_2H_6. The spherical form of the Kihara potential (Tee *et al.* 1966) was used solely in later studies of clathrate modelling:

$$\Gamma(r) = 4\varepsilon_{ij}\left\{\left(\frac{\sigma_{ij}-2a_{ij}}{r-2a_{ij}}\right)^{12} - \left(\frac{\sigma_{ij}-2a_{ij}}{r-2a_{ij}}\right)^{6}\right\} \quad (16)$$

where Γ is intermolecular potential as a function of the distance (r) between molecules i and j, ε is the minimum potential energy, σ is the distance between two molecules for zero potential energy and a is the radius of impenetrable molecule cores. Bakker *et al.* (1996) noted the necessity for clarification of the application of the Kihara potential functions in clathrate modelling because many different notation methods for equal formulae have been used in the literature (e.g. Kihara 1953; McKoy & Sinanoglu 1963; Tee *et al.* 1966; Parrish & Prausnitz 1972; Avlonitis 1994). The transparency of clathrate stability modelling was seriously mystified by this variation. The parameters in equation (16) are all defined from the centre of molecules, which is a direct and clear method to describe potentials and which avoids any double standards.

The interaction parameters between unalike molecules, i.e. gas and H_2O, were originally obtained from classical mixing rules:

$$\varepsilon_{ij} = \sqrt{\varepsilon_i\varepsilon_j}$$
$$\sigma_{ij} = \frac{\sigma_i+\sigma_j}{2} \qquad (i \neq j). \qquad (17)$$
$$a_{ij} = \frac{a_i+a_j}{2}$$

The Kihara parameter values for H_2O (Table 5) were originally based on theoretical considerations and indirect obtained values from argon gas hydrate (van der Waals & Platteeuw 1959). The core radius for a H_2O molecule is assumed to be 0 (McKoy & Sinanoglu 1963), which reduces the potential function to a Lennard-Jones 12–6 function. Holder *et al.* (1980) and John *et al.* (1985) obtained empirical values for ε and σ for H_2O molecules from experimental data on CH_4 gas hydrates (Table 5) using the values for pure CH_4 and the mixing rules from equation (17). These values are markedly different to the theoretical values from van der Waals & Platteeuw (1959), which illustrates the diversity of describing the molecular potential of H_2O molecules. The Kihara parameters for CO_2, CH_4, N_2, and C_2H_6 molecules (Table 5) are obtained directly both from second virial coefficient and from viscosity data from Tee *et al.* (1966). They give several sets of values for these parameters based on different independent data sets, and they give smoothed values from unified correlations, which are favoured in most clathrate modelling literature (e.g. Avlonitis 1994). However, unsmoothed parameters which are based on a larger data set, and which are also obtained from independent sources, are preferred in this study.

Table 5. *Kihara parameters for similar molecule interactions*

Gas	ε/k (K)	σ (pm)	a (pm)
H_2O (Ref. 1)	167.0	249.45	0
(Ref. 2)	115.5	375.1	0
(Ref. 3)	102.134	356.438	0
CO_2 (Ref. 4)	469.73	350.1	68.05
CH_4 (Ref. 4)	232.20	350.5	38.34
N_2 (Ref. 4)	97.31	366.1	0
C_2H_6 (Ref. 4)	425.32	397.7	56.51

References: 1. van der Waals & Platteeuw (1959); 2. Holder *et al.* (1980); 3. John *et al.* (1985); 4. Tee *et al.* (1966).

The molecular potential $w(r)$ in spherically symmetrical cells in clathrate structures, which McKoy & Sinanoglu (1963) designed for the Kihara potential was modified by Bakker *et al.* (1996) to do justice to equation (16):

$$w(r) = 2Z\varepsilon_{ij}\left\{\frac{(\sigma_{ij}-2a_{ij})^{12}}{rR^{11}}\left(\delta^{10}+\frac{2a_{ij}}{R}\delta^{11}\right) - \frac{(\sigma_{ij}-2a_{ij})^{6}}{rR^{5}}\left(\delta^{4}+\frac{2a_{ij}}{R}\delta^{5}\right)\right\} \quad (18a)$$

$$\delta^n = \frac{1}{n}\left\{\left(1-\frac{r}{R}-\frac{2a_{ij}}{R}\right)^{-n} - \left(1+\frac{r}{R}-\frac{2a_{ij}}{R}\right)^{-n}\right\} \quad (18b)$$

Table 6. *Coordination number* Z *and approximate radius* R *of small and large spherical cavities in clathrate Structure I and II*

	Structure I		Structure II	
	Small cavity (Pentagonal dodecahedron)	Large cavity (Tetrakaidecahedron)	Small cavity (Pentagonal dodecahedron)	Large cavity (Hexakaidecahedron)
Z	20	24	20	28
R (pm)*	387.5	430.0	387.0	470.3

* From John & Holder (1981).

where r is the distance from the centre of the cavity, Z and R are the coordination number and the cell radius of a type of cavity (Table 6), respectively, n is either 4, 5, 10 or 11. The cell radii are obtained from John & Holder (1981) who extensively studied deviations from the cavity symmetry according to van der Waals & Platteeuw (1959). The coordination number Z is set equal to the vertices of the cavities, unlike the estimation of effective coordination numbers by John & Holder (1981). John & Holder (1981) and John *et al.* (1985) noted that the binary interaction parameters in equations (16) and (18) calculated with the arbitrary mixing rules from equation (17) do not result in an accurate reproduction of clathrate stability conditions. They incorporate second neighbour molecular interactions and the asymmetry of such interactions, modifying postulate (e) from van der Waals & Platteeuw (1959).

A different approach to clathrate modelling is chosen in this study. The Kihara parameters of binary gas–H_2O interactions in clathrate cavities can be directly obtained from minimizing differences between model predictions and experimental data on clathrate melting conditions. This method was introduced by Saito *et al.* (1964) for CH_4, N_2 and argon gas hydrates. Subsequently, they used the previously mentioned mixing rules (equation (17)) and the theoretical values for a H_2O molecule to estimate the molecular potential parameters for pure gases according to the Lennard-Jones 12–6 potential. This method has been adopted by, for example, Parrish & Prausnitz (1972), Anderson & Prausnitz (1986), Dubessy *et al.* (1992) and Bakker *et al.* (1996) for the Kihara potential function. A very important implication of this method is that all irregularities and deviations in principle assumptions for other parameters are projected onto these Kihara parameters. The first step of this method avoids uncertainties in the arbitrarily defined mixing rules and in the molecular potential function for pure H_2O in the clathrate phase. Prausnitz *et al.* (1986) give a general overview of many variations on arbitrarily mixing rules like equation (17), which reinforces its preferred omission in this study. Consequently, this study has two significant uncertain steps less than models which include equation (17) to calculate gas–H_2O interactions in clathrate phases (e.g. John *et al.* 1985; Avlonitis 1994). In this study, only the core radius for unalike molecule interaction is obtained from the geometrical mean (equation (17)) of values given in Table 5 for pure components.

Optimum Kihara parameters in the clathrate phase

Unsmoothed experimental data on clathrate phase equilibria for pure CO_2, CH_4, N_2, and C_2H_6 gas hydrates (Table 7) are used to determine optimum Kihara parameters ε and σ between an encaged gas molecule and a cage-forming water molecule. The large amount of available data allows a careful discrimination of those data which appear to be inconsistent with the system. For example, data on C_2H_6 gas hydrate from Roberts *et al.* (1940) are not used in this study because they are not consistent with other data sets in the same range of clathrate melting temperatures. Data on clathrate melting above 210 MPa are also excluded in this study, because many other parameters, like gas solubility and fugacity, that play an important role in clathrate modelling cannot be accurately calculated at higher pressures. For example, experimental data on CH_4 and N_2 gas hydrates above 210 MPa from Marshall *et al.* (1964) are excluded for this reason.

Optimum Kihara parameters are found after a lengthy search within ε/k–σ diagrams (Fig. 4 and Table 8). These parameters are systematically varied, and for each set of ε/k–σ values the average deviation between the model and all experimental data has been calculated. The

Table 7. *Available experimental data*

Gas hydrate	Literature	Amount of data
CO_2	Deaton & Frost (1946)	19
	Unruh & Katz (1949)	5
	Larson (1955)	38
	Takenouchi & Kennedy (1965)	15
	Robinson & Metha (1971)	7
	Ng & Robinson (1985)	9
	Adisasmito *et al.* (1991)	9
	Dholabhai *et al.* (1993)	4
	Englezos & Hall (1994)	6
CH_4	Deaton & Frost (1946)	13
	Kobayashi & Katz (1949)	4
	McLeod & Campbell (1961)	10
	Marshall *et al.* (1964)	20
	Jhaveri & Robinson (1965)	8
	Galloway *et al.* (1970)	4
	Verma *et al.* (1975)	7
	De Roo *et al.* (1983)	9
	Adisasmito *et al.* (1991)	11
	Dickens & Quinby–Hunt (1994)	7
N_2	van Cleeff & Diepen (1960)	39
	Marshall *et al.* (1964)	14
	Jhaveri & Robinson (1965)	8
C_2H_6	Roberts *et al.* (1940)	20
	Deaton & Frost (1946)	20
	Reamer *et al.* (1952)	4
	Galloway *et al.* (1970)	3
	Holder & Grigoriou (1980)	7
	Holder & Hand (1982)	7
	Ng & Robinson (1985)	8
	Avlonitis (1988)	10
	Englezos & Bishnoi (1991)	6

numerical approach of this method takes into account the chemical potential of both clathrate structures, I and II. Obviously, the most stable configuration has the lowest chemical potential. The boundary between clathrate Structure I and II is defined by certain σ value and it is independent upon ε/k values (Fig. 4). Clathrate Structure II appears to be stable at lower σ values than Structure I for CO_2, CH_4 and N_2 gas hydrates. This phase change is also temperature dependent, therefore only the boundary near Q_1 conditions is illustrated in Fig. 4. The boundary moves only several pm to lower σ values at higher temperatures. These estimations imply that a clathrate structure with larger cavities is more stable if encaged gas molecules can get closer to the wall of the cavities. C_2H_6 gas hydrate has a clathrate Structure I over the entire range of σ and ε/k values illustrated in Fig. 4d. Contours of equal deviation have been draw in these diagrams (Fig. 4). The very irregular shape of the 5% deviation contour indicates the complexity of a systematic approach. The pattern in Fig. 4 is an interference between best-fits in both clathrate structures, which is clearly demonstrated by the discontinuity at the phase boundary. Minima in average deviation are present in both parts of the diagrams. The intention of the method is to choose the lowest minimum as a solution for optimum Kihara parameters. The position of minima for CO_2, N_2 and C_2H_6 (Table 8) are consistent with their observed clathrate structures, however, that for CH_4 appears to be in the Structure II field, close to the phase boundary. These thermodynamic considerations imply that CH_4 has a clathrate Structure II at Q_1 conditions, and changes to clathrate Structure I above 292 K. However, it is well known that relatively small CH_4 molecules form a clathrate Structure I, which has been measured from X-ray diffraction data by von Stackelberg & Müller (1954). Consequently, the minimum in the clathrate Structure I field has to be chosen for CH_4 gas hydrate (Table 8), which does not differ very much in magnitude from the lowest minimum. Small differences in chemical potential between clathrate Structure I and Structure II may allow the formation of the metastable phase due to small growth irregularities, or the presence of impurities. The diagram for N_2 gas hydrates has a similar pattern with three minima (Fig. 4c), but the middle minimum in the clathrate Structure II field has a significant lower average deviation than the other minima. The interpretation of these thermodynamic calculations are consistent with the findings of Davidson *et al.* (1987) who were the first to identify a clathrate Structure II for N_2 gas hydrate from powder diffraction patterns. C_2H_6 gas hydrate has the least pronounced minimum, which is directly related to the relatively small amount of experimental data used for the optimization.

The low average deviation for all types of pure gas hydrates (Fig. 4 and Table 8) illustrates the excellent fit between this clathrate model and the experimental data. The error values illustrated in Fig. 4 are average values, which may differ from error values for individual data points. Therefore, the deviation of each individual experimental data point is given in Fig. 5. Most individual data are located within 2% accuracy for CO_2 (Fig. 5a), 4% for CH_4 (Fig. 5b), 2% for N_2 (Fig. 5c) and 4% for C_2H_6 (Fig. 5d) from model predictions. This accuracy is improved even further if the error in measurements of individual datum is taken into account.

The results of these optimalization are strongly dependent on other parameters used in this clathrate stability model. A thorough error

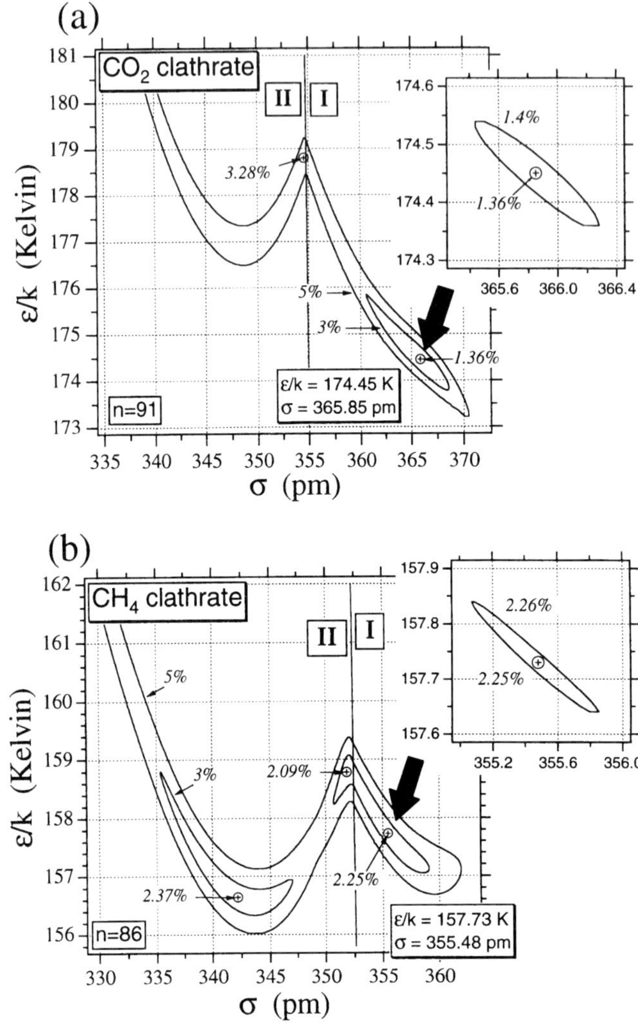

Fig. 4. $\sigma-\varepsilon/k$ diagrams for (**a**) CO_2, (**b**) CH_4, (**c**) N_2, and (**d**) C_2H_6 with contours of equal average deviation between the clathrate stability model and experimental data. I is clathrate Structure I; II is clathrate Structure II; n is the amount of used experimental data.

analysis by Bakker *et al.* (1996) illustrates that mainly uncertainties in the Kihara core parameter a and chemical potential of the empty clathrate phase at standard conditions $\mu_{H_2O}^{empty}(\mathrm{ref})$ influence the position of the best solution for the two other Kihara parameters ε and σ. The main variation is obtained in ε/k values (Bakker *et al.* 1996, p. 1667, Fig. 12). An error indication of approximately 2 pm is obtained for σ values, and approximately 4 K for ε/k values. This error indications are only slightly improved if more data are available.

Best-fit estimations of clathrate melting conditions in the presence of a liquid-like gas-rich phase for clathrates with a Q_2 point (see Fig. 1), like CO_2 and C_2H_6, are separated from those in the presence of a vapour-like gas-rich phase, below Q_2 conditions. Deviations between calculated clathrate melting pressures and observed pressures at selected temperatures are, in general, very large above Q_2 conditions due to the relatively steep slope of the melting curve. This apparently bad fit is an artefact, and the inverse method, i.e. comparison of clathrate melting temperatures at selected pressures, gives much better results at these conditions.

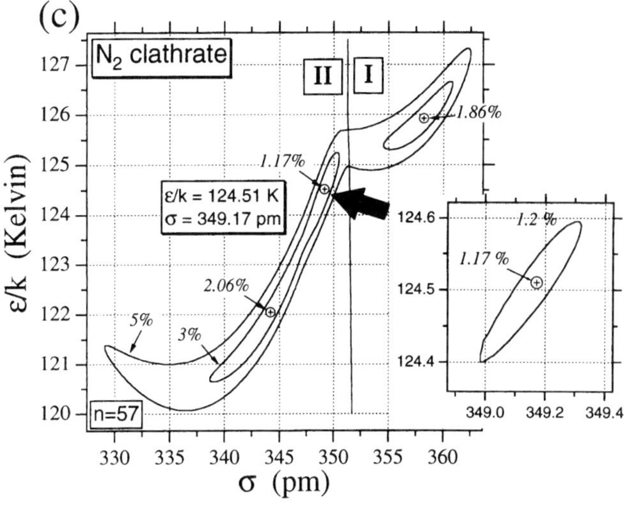

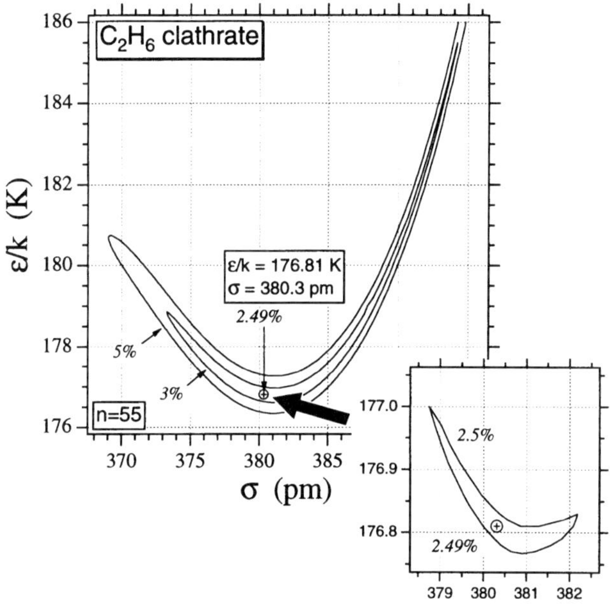

Fig. 4. (*cont.*)

Unfortunately, experimental data above Q_2 conditions are highly inconsistent for CO_2 gas hydrates (Takenouchi & Kennedy 1965; Ng & Robinson 1985) and C_2H_6 gas hydrates (Roberts *et al.* 1940; Ng & Robinson 1985). Using the optimum Kihara parameters obtained from below Q_2 conditions for CO_2 gas hydrate results in overestimated pressures at selected temperatures above Q_2 conditions. For CO_2, the best solution is obtained from $\sigma = 364.74$ pm (Table 8), while ε/k remains unchanged. Differences between modelled and measured clathrate melting pressures at selected temperatures above Q_2 conditions may exceed 30% (Fig. 6a). The fit of

Table 8. *Best-fit solutions for Kihara parameters ε/k and σ for gas–H_2O molecular interactions in clathrate structures,* n *is the amount of experimental data used in the fitting procedure*

Gas hydrate	ε/k (K)	σ (pm)	n	Average deviation
CO_2	174.45	365.85	91	1.36%
		364.74*	19	–
CH_4	157.73	355.48	86	2.25%
N_2	124.51	349.17	57	1.17%
C_2H_6	176.81	380.3	55	2.49%

* Above Q_2 conditions.

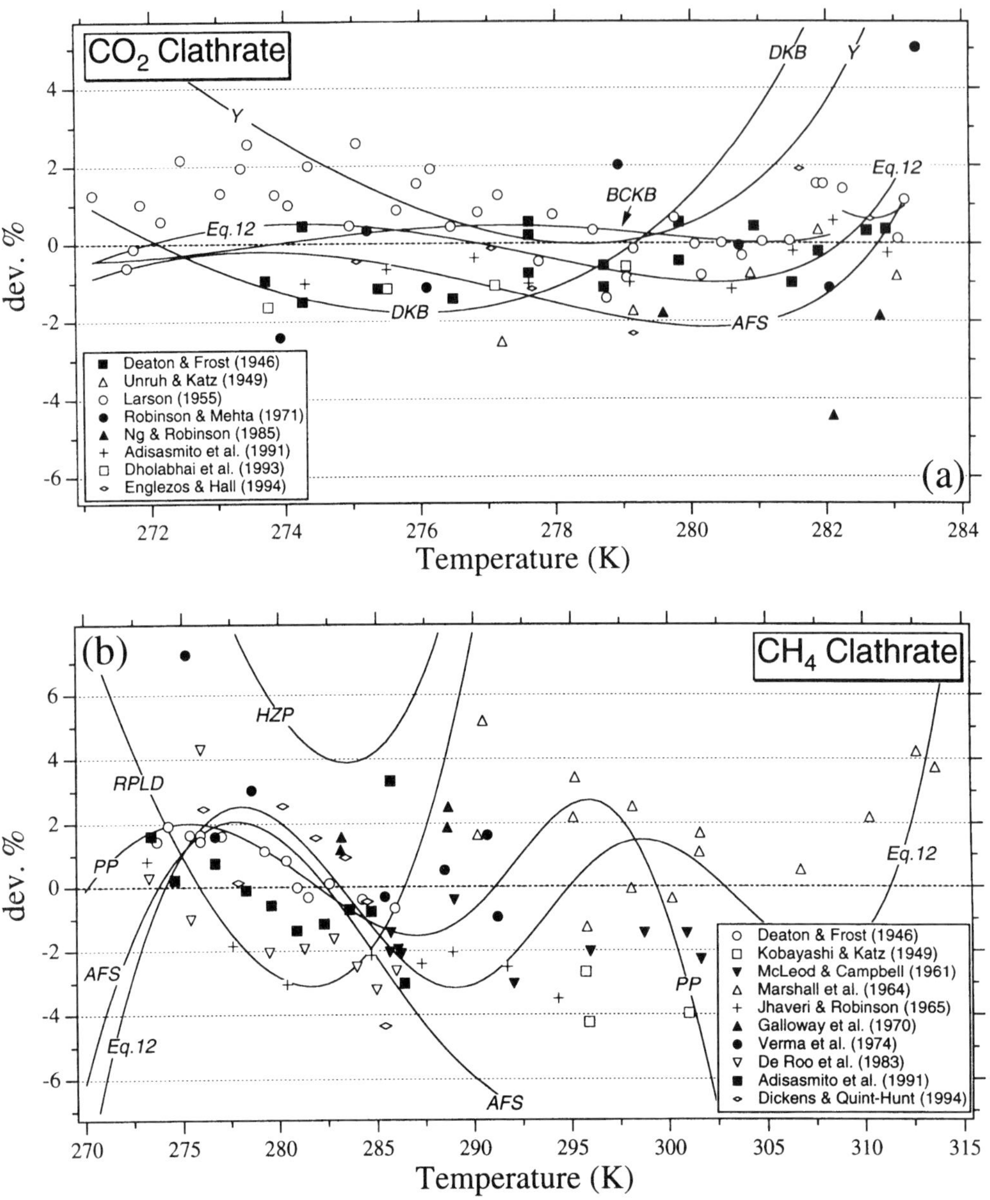

Fig. 5. Deviation of individual experimental data on clathrate stability conditions from (**a**) CO_2, (**b**) CH_4, (**c**) N_2, and (**d**) C_2H_6 gas hydrates to model predictions according to the estimated optimum Kihara parameters σ and ε/k in Fig. 4. Solid lines are comparisons for available purely empirical equations for clathrate stability conditions: BCKB, Bozzo *et al.* (1975); HZP, Holder *et al.* (1988); AFS, Adisasmito *et al.* (1991); Y, Yerokin (1993); PP, Parrish & Prausnitz (1975); RPLD, De Roo *et al.* (1983); Eq. 12, equation (12) from this study.

the model to experimental data is improved at higher temperatures. The model produces the midpoint between experimental data sets, which makes it an acceptable mediator. The experimental data for C_2H_6 gas hydrate above Q_2 conditions are adequately reproduced by the σ value obtained at lower pressures (Fig. 6b), although the scatter in data from Roberts *et al.* (1940) do not allow any conclusive interpretation.

The quadruple points for each pure gas

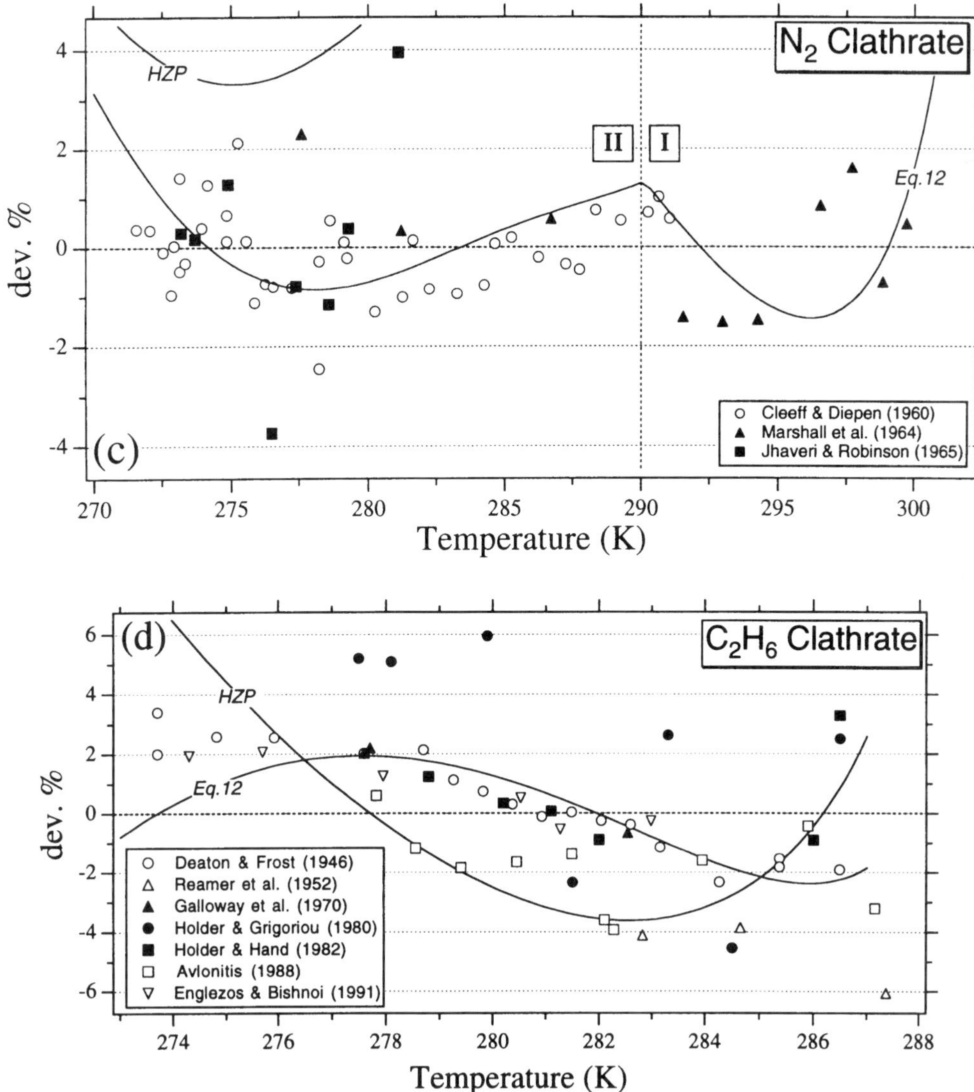

Fig. 5. (*cont.*)

hydrate have been calculated from these optimum Kihara parameters (Table 9). Q_2 points are obtained from the intersection of the clathrate melting curve and saturation curve for pure gases. Q_1 points are calculated from the intersection of clathrate melting curves in equilibrium with ice and liquid.

Comparison to literature

Bakker *et al.* (1996) already illustrated large differences for Kihara parameters for CO_2–H_2O molecular interactions in clathrate structures. At first sight, reported Kihara parameters for gas–water interactions are astonishingly different in the literature (Table 10). To eliminate the effect of the core radius a, molecule interactions are compared from the edge of the impenetrable core as a reference distance ($\sigma - 2a$). The pioneering study from McKoy & Sinanoglu (1963) show large deviations from this work, in particular ε/k values are extremely overestimated in their study. The second large anomaly is the study from John *et al.* (1985), whose ε/k value for CO_2 gas hydrate differs

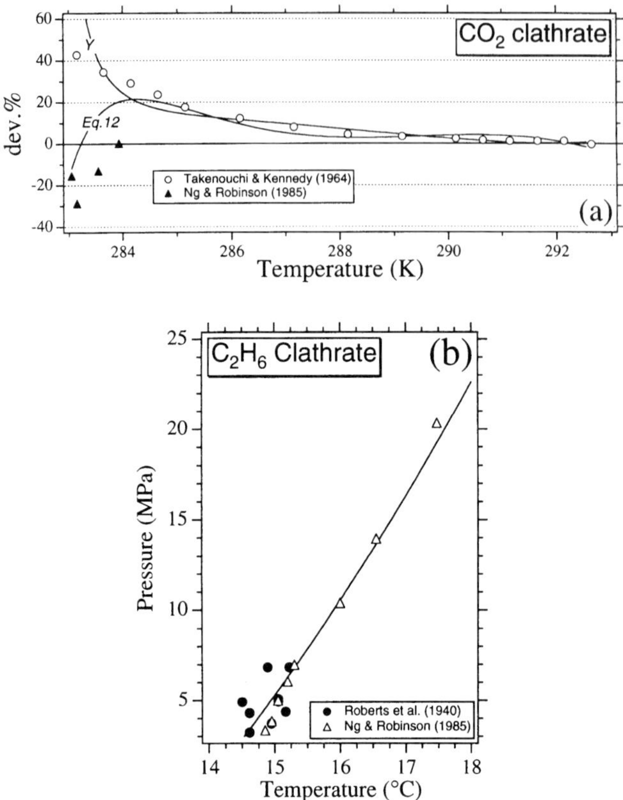

Fig. 6. (**a**) Deviation of individual experimental data from CO_2 gas hydrate at high pressures, above Q_2 conditions. Solid lines are comparisons with purely empirical equations from Yerokin (Y) and this study (Eq. 12). (**b**) *T–P* diagram with a modelled clathrate stability curve for C_2H_6 gas hydrate (solid line) and experimental data from Roberts *et al.* (1940) and Ng & Robinson (1985), above Q_2 conditions.

by up to 30% from this study. This variation is expected because John *et al.* (1985) used a different approach to clathrate modelling and included second neighbour interactions, as previously described. For the other studies, the relative deviation in $\sigma - 2a$ and ε/k values are, in general, less 10%, which still greatly exceeds the absolute error range estimated in this study. This variation corresponds perfectly to the amount of experimental data used to fit these parameters. The scatter of experimental data in Fig. 5 indicates a maximum difference of *c.* 12%, which is similar to the variation found for Kihara parameters in several studies (Table 10). Thus, the ambiguous disagreement, which has been recognized in the literature (e.g. Sloan 1990; Avlonitis 1994), is merely a reflection of the use of an insufficient amount of data. This interpretation justifies the necessity to use the most accurate values for gas solubilities, fugacities and thermodynamic constants, in addition to a large amount of independently obtained experimental data, as demonstrated in this study.

Differences in optimum Kihara parameters for N_2 gas hydrates may result from its only recently discovered clathrate Structure II (Davidson *et al.* 1987), which is in contradiction to the classical size–structure rule of von Stackelberg & Müller (1954). Only Lundgraad & Mollerup (1991) and Avlonitis (1994) modelled the N_2 according to Structure II clathrates. For example, Parrish & Prausnitz (1972), Anderson & Prausnitz (1986) and Dubessy *et al.* (1992) modelled N_2 according to Structure I clathrates, which obviously must result in different values for the optimized Kihara parameters.

Several purely empirical equations are compared to model predictions and experimental data in Figs 5 and 6. Most empirical equations for pure CO_2 gas hydrate are accurately repro-

Table 9. *Quadruple points Q_1 and Q_2 for pure CO_2, CH_4, N_2, and C_2H_6 gas hydrates*

Gas hydrate	Q_1		Q_2	
	T (K)	P (MPa)	T (K)	P (MPa)
CO_2	271.66	1.041	282.961	4.481
CH_4	272.85	2.583	–	–
N_2	271.85	14.253	–	–
C_2H_6	273.13	0.483	287.798	3.348

ducing experimental data. The equation from Bozzo *et al.* (1975) is very similar to our model predictions, and, therefore, is most accurately producing clathrate stability conditions between Q_1 and Q_2. The equation from Dholabhai *et al.* (1993) and Yerokhin (1993) increasingly deviate from both this model and the experimental data at temperatures above 280 K and near Q_1 conditions. Above Q_2 conditions (Fig. 6a) the equation from Yerokhin (1993) accurately reproduces the data from Takenouchi & Kennedy (1965). The empirical equations for pure CH_4 gas hydrates are, in general, less accurate. The equation according to Parrish & Prausnitz

Table 10. *Comparison of Kihara parameters $\sigma - 2a$ and ε/k for gas–H_2O interactions*

Gas hydrate	Literature	ε/k (K)	Deviation (%)	$\sigma - 2a$ (pm)	Deviation (%)
CO_2	MS	309	77	299.3	0.5
	PP	169.09	−3.1	296.81	−0.3
	JPH	227.39	30	280.869	−5.7
	AP	169.44	−2.9	295.61	−0.7
	LM	162.37	−6.9	335.43	12.6
	DTC	170.0	−2.6	295.8	−0.7
	A	172.0	−1.4	290.4	−2.5
	This study	174.45		297.8	
CH_4	MS	205	30	298.5	−5.9
	PP	153.17	−2.9	323.98	2.2
	HCP	148.88	−5.6	343.45	8.3
	JPH	141.987	−10	327.269	3.2
	AP	153.39	−2.8	325.78	2.7
	LM	151.41	−4.0	326.6	3.0
	DTC	152.0	−3.6	327.0	3.1
	A	153.8	−2.5	325	2.5
	This study	157.73		317.16	
N_2	MS	124	−0.4	309.0	−11.5
	PP	127.95	2.8	361.42	3.5
	JPH	127.422	2.3	316.319	−9.4
	AP	128.57	3.3	326.77	−6.4
	LM	125.94	1.1	331.16	−5.2
	DTC	126.25	1.4	328.0	−6.1
	A	127.0	2.0	318.3	−8.8
	This study	124.51		349.17	
C_2H_6	MS	609	244	230.7	−29
	PP	174.97	−1.0	331.80	2.4
	JPH	200.397	13.4	322.619	−0.4
	AP	175.94	−0.5	343.81	6.1
	LM	176.89	0.1	347.32	7.2
	DTC	175.0	−1.0	344.0	6.2
	A	184.1	4.1	343.0	5.9
	This study	176.79		323.99	

MS, McKoy & Sinanoglu (1963); PP, Parrish & Prausnitz (1972); HCP, Holder *et al.* (1980); JPH, John *et al.* (1985); AP, Anderson & Prausnitz (1986); LM, Lundgaard & Mollerup (1991); DTC, Dubessy *et al.* (1992); A, Avlonitis (1994).

(1972) is in good agreement with our model predictions and remains within 3% accuracy from Q_1 up to 300 K. The equations from De Roo *et al.* (1983) and Adisasmito *et al.* (1991) are only applicable up to 288 K. Near Q_1 conditions none of these equations are able to reproduce the experimental data. The equation according to Holder *et al.* (1988) is unable to reproduce any experimental data at all, and extensively overestimates the model predictions. Equation (12) from this study is the sole empirical equation which can be used up to 313 K for CH_4 gas hydrate (Fig. 5b). For N_2 gas hydrates (Fig. 5c) the equation from Holder *et al.* (1988) is, again, not related to any experimental data and unable to reproduce model predictions. Equation (12) from this study accurately reproduces experimental data between Q_1 and 300 K, and is in close agreement with the model. Both equation (12) and the equation from Holder *et al.* (1988) accurately reproduce experimental data of C_2H_6 gas hydrate (Fig. 5d). Although, purely empirical equations can be easily applied by solving a simple mathematical equation, these comparisons illustrate its limited use. In fact, most of them can only be used for pure gas hydrates and in limited ranges of clathrate stability conditions.

Clathrate equilibria of binary guest mixtures

It has been demonstrated that the previously described model very accurately reproduces clathrate stability conditions for pure gas hydrates over a wide range of temperatures, including all available experimental data. The definitions for this clathrate stability model (equations (2)–(4)) allow application to calculate any type of gas mixture in equilibrium with clathrate. Therefore, experimental data of binary guest mixtures are compared to the model prediction. An unexpected large difference between the calculated and modelled clathrate melting pressure at selected temperatures is observed, which greatly exceeds those predictions for pure gas hydrates. This result is similar to prediction from other classical clathrate stability models, which are unable to accurately reproduce mixed gas hydrate stability conditions.

A method has been developed in this study to describe this discrepancy systematically. Available data on gas mixture composition of the vapour phase and melting temperature and pressure of the clathrate phase (Table 11) allow calculation of the amount of energy that is missing in the original equilibrium formula (equation (1)). This missing chemical potential

Table 11. *Experimental data for binary gas hydrates*

Binary gas hydrate	Literature	Amount of data
CO_2–CH_4	Unruh & Katz (1949)	17
	Adisasmito *et al.* (1991)	42
CH_4–N_2	Jhaveri & Robinson (1965)	63
CH_4–C_2H_6	Deaton & Frost (1946)	24
	McLeod & Campbell (1961)	16
	Holder & Grigoriou (1980)	15

is attributed to the clathrate and is suggested to be an excess Gibbs free energy function (equation (19)) which depends on the gas mixture composition and temperature

$$\mu_{H_2O}^{clath} = \mu_{H_2O}^{empty} + RT\sum_i v_i \ln\left(1 - \sum_M y_{Mi}\right) + G^{excess}(x, T). \qquad (19)$$

Physically, the G^{excess} function is suggested to result from non-ideal interactions between cavities which are occupied by different types of guest molecules. Thompson (1967) introduced the Margules equations (Margules 1895) to characterize non-ideality for binary crystalline solution in petrological and geochemical research. The Margules model describes an excess thermodynamic function as a power series expressed in mole fractions of the components involved. Thompson (1967) used symmetrical (equation (20a)) and asymmetrical (equation (20b)) excess functions to obtain the position of solvi for immiscible mineral solutions. He noted that most real solutions are asymmetric to some degree:

$$G^{excess} = X_1 X_2 W_G \qquad (20a)$$

$$G^{excess} = X_1 X_2 (W_{G1} X_2 + W_{G2} X_1) \qquad (20b)$$

where W_G are interchange energies which depend only on temperature, not on composition. In this study the strong asymmetry for clathrate phases needs a modification of equation (20) to be able to characterize clathrate excess functions:

$$G^{excess} = X_1 X_2 (W_{G1} X_2^2 + W_{G2} X_1^2). \qquad (21)$$

The temperature dependence of W_G is described according to a second-order polynome:

$$W_G = w_0 + w_1 T_C + w_2 T_C^2 \qquad (22)$$

where T_C is temperature in °C, and w_0, w_1 and w_2 are constant values for selected binary gas hydrates (Table 12). The derivation of these functions from experimental data is illustrated in detail in the next paragraph using CO_2–CH_4

Table 12. *Constants for the modified Margules formulae in equation (22)*

Mixed gas hydrate		w_0	w_1	w_2
CO_2–CH_4	$W_G^{CO_2}$	119.43	16.792	–
	$W_G^{CH_4}$	33.699	7.5437	–
CH_4–N_2	$W_G^{CH_4}$	140.45	−128.58	5.9032
	$W_G^{N_2}$	286.32	−34.053	0.42591
CH_4–C_2H_6	$W_G^{CH_4}$	2057.7	−43.155	−1.7486
	$W_G^{C_2H_6}$	−1756.6	86.956	−0.70654

mixed gas hydrate as an example. The summation over all binary excess functions can be used in ternary and higher-order guest mixture gas hydrates.

Several complications are linked to the application of the excess functions (equations (21) and (22)). First, the composition of gas hydrates is difficult to obtain by direct chemical analysis (e.g. Dharmawardhama *et al.* 1980). Hydrate composition is subject to large experimental uncertainties. Parrish & Prausnitz (1972) mentioned that only the vapour phase composition, which is in equilibrium with the clathrate phase, is important in process design application. In this study, the filling of the clathrate structures is theoretically obtained with the sophisticated treatment from van der Waals & Platteeuw (1959). To avoid any systematic errors in modelling and large experimental uncertainties, it is preferable to compare the calculated excess function for the clathrate phase to the mole fraction in the vapour phase. Distribution coefficients between both phases must be used in order to obtain true excess functions for the clathrate phase. Second, in the previous section two values are given for the optimized Kihara parameter σ for pure CO_2 gas hydrate (Table 8), which were obtained from equilibria with a CO_2-rich gas phase at high densities (above Q_2 conditions) and low densities (below Q_2 conditions). Clathrate melting in equilibrium with a CO_2-rich fluid phase with intermediate densities, between 51 and 332 cm^3/mol^{-1}, do not occur, which is a direct consequence of the intersection with the saturation curve for pure CO_2. Addition of small amounts of CH_4 or N_2 reduces the liquid–vapour equilibria of the fluid phase to lower temperatures (see Fig. 1), and subsequently, these intermediate densities do occur in mixed gas hydrates stability calculations. Therefore, a transition zone has to be defined for the σ parameter from CO_2–H_2O interactions to describe its alteration as a function of molar volume of the gas-rich phase:

$$\sigma_{CO_2} = X_V \times \sigma_{liq} + (1 - X_V) \times \sigma_{vap} \quad (23a)$$

$$X_V = \frac{(332 - V_M)}{(332 - 51)} \quad (23b)$$

where V_M is the molar volume of the gas-rich phase, X_V varies between 0 and 1. X_V is 1 for molar volumes lower than 51 $cm^3\,mol^{-1}$ (high densities), and 0 for molar volumes higher than 332 $cm^3\,mol^{-1}$ (low densities). An important implication of the use of excess functions is the possibility of the demixing of mixed gas hydrates into pure gas hydrates at lower temperatures. Although this has not yet been described in the literature, it is proposed as a possible process in natural gas hydrates.

CO_2–CH_4 gas hydrate

CO_2–CH_4 gas hydrate forms a Structure I clathrate over the entire compositional range, which is obviously inherited from the pure end-member CO_2 and CH_4 gas hydrates. The equation of state according to Duan *et al.* (1992*a*, *b*) is selected to calculate the fugacities of the components in CO_2–CH_4 mixtures with small amounts of H_2O. This equation of state was previously selected for both pure CO_2 gas hydrate and pure CH_4 gas hydrate.

Experimental data from Unruh & Katz (1949) and Adisasmito *et al.* (1991) are compared to model predictions as presented in the previous section, without an excess energy function. The difference between calculated and measured clathrate melting pressure at selected temperatures approaches 20% for mixtures containing 40 mol% CO_2, which greatly exceeds the accuracy of the pure end-members. The amount of missing energy is calculated for each individual data point (Table 13 and Fig. 7) in order to be able to perform a systematic analysis of these differences. Most data from Unruh & Katz (1949) are compatible with those from Adisasmito *et*

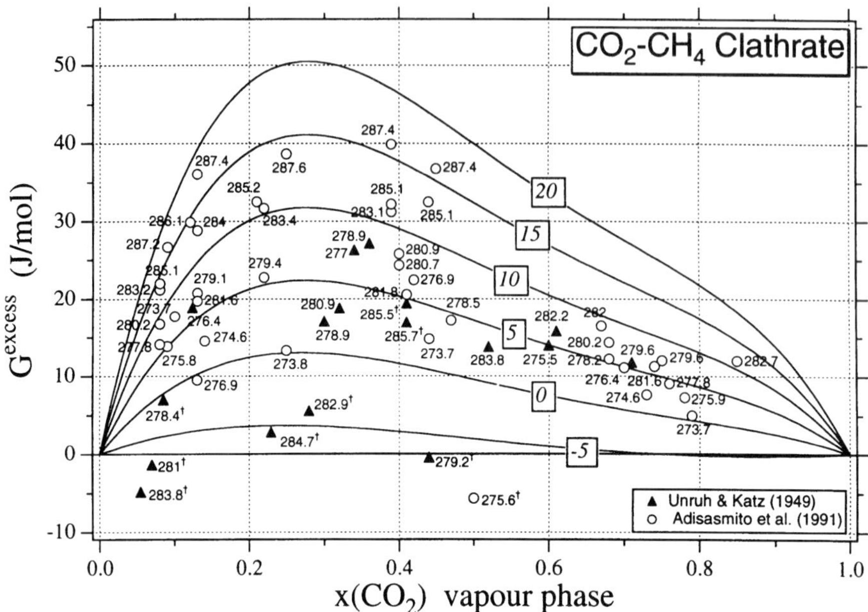

Fig. 7. $G^{\text{excess}}-X$ diagram for CO_2–CH_4 mixed gas hydrate. Numbered temperatures for individual data points are in Kelvin. Curved solid lines are modelled G^{excess} values according to equations (21) and (22) at selected temperatures (squared numbers in °C). Experimental data points marked with an † are excluded from the estimation of interchange energies W_G.

al. (1991), although their scatter is much larger. Figure 7 demonstrates an increase in excess energy at higher temperatures and intermediate compositions. First, a $G^{\text{excess}}-T$ diagram is constructed (Fig. 8a) with isopleths from the composition of the vapour phase. Data close to a selected isopleth are projected. This diagram clearly illustrates individual data points that are inconsistent with the general trend outlined by most data, and it allows a careful selection procedure. For example, calculated G^{excess} values of data from Unruh & Katz (1949) for the 0.42 isopleth (Fig. 8a) are too low, and, therefore, they are not included in a best-fit line estimation to describe the temperature dependency of G^{excess}. For the 0.73 isopleth (Fig. 8a), all data are consistent. The best-fit method smooths calculated data and allows interpolation at selected temperatures. Second, a $G^{\text{excess}}-X$ diagram is constructed with these smoothed values for several vapour compositions at selected temperatures (Fig. 8b). An equation in the form of equation (21) is fitted to this data set by constantly arbitrarily varying the values for interchange energies W_G, until an acceptable fit is obtained. Third, these interchange energies for selected temperatures are plotted in a W_G-T diagram (Fig. 8c) to estimate a polynomal fit for the temperature dependency of W_G in the form of equation 22. The interchange energies for both CO_2 and CH_4 appear to increase linearly with temperature (Table 12). The asymmetry of the calculated isotherms in Fig. 7 is strongly dragged towards CH_4-rich compositions. Analysis based solely on one data set may result in very dissimilar estimated excess functions.

The CO_2 gas hydrate has two quadruple points at two defined temperatures and pressures, where four different phases are in stable configuration (Q_1 and Q_2), whereas CH_4 gas hydrate has only a Q_1 point (Table 9). Addition of small amounts of CH_4 to a CO_2-rich vapour phase in equilibrium with the clathrate phase transforms its Q_2 point into a limited line in a *TP*-diagram (Fig. 9), according to the classical phase rule. This line is defined by the intersection of the immiscibility field of a CO_2-rich liquid phase and a CH_4-rich vapour phase (e.g. Thiéry *et al.* 1994; Bakker 1997) and the clathrate melting curve. This immiscibility will retreat to lower temperatures as the amount of CH_4 in the gas mixture increases, whereas the clathrate melting curve moves to a much lesser extent. For a gas mixture of about 79 mol% CO_2 and 21 mol%

Table 13. *Measured clathrate stability conditions and calculated* G^{excess} *and mole fractions in the clathrate phase for binary* CO_2*–*CH_4 *gas hydrate*

Literature	xCO_2 (vap)	Measured values T (K)	P (MPa)	G^{excess} (J mol^{-1})	Calculated values xCO_2(cla)	xCH_4(cla)
Adisasmito *et al.* (1991)	0.10	273.7	2.52	17.73	0.032	0.108
	0.09	275.8	3.10	13.86	0.029	0.111
	0.08	277.8	3.83	14.09	0.026	0.115
	0.08	280.2	4.91	16.73	0.026	0.116
	0.08	283.2	6.80	21.16	0.025	0.117
	0.08	285.1	8.40	22.00	0.025	0.119
	0.09	287.2	10.76	26.67	0.027	0.117
	0.14	274.6	2.59	14.52	0.041	0.099
	0.13	276.9	3.24	9.53	0.038	0.102
	0.13	279.1	4.18	20.77	0.038	0.103
	0.13	281.6	5.38	19.73	0.038	0.105
	0.13	284.0	7.17	28.80	0.037	0.106
	0.12	286.1	9.24	29.89	0.034	0.110
	0.13	287.4	10.95	36.06	0.035	0.109
	0.25	273.8	2.12	13.31	0.061	0.077
	0.22	279.4	3.96	22.74	0.055	0.086
	0.22	283.4	6.23	31.65	0.053	0.089
	0.21	285.2	7.75	32.51	0.051	0.093
	0.25	287.6	10.44	38.66	0.054	0.090
	0.44	273.7	1.81	14.78	0.085	0.051
	0.42	276.9	2.63	22.41	0.083	0.056
	0.40	280.7	4.03	24.33	0.079	0.092
	0.39	283.1	5.43	31.21	0.076	0.065
	0.39	285.1	6.94	32.15	0.074	0.068
	0.39	287.4	9.78	39.87	0.070	0.074
	0.50	275.6	1.99	−5.67	0.091	0.045
	0.47	278.5	2.98	17.17	0.087	0.051
	0.40	280.9	4.14	25.77	0.079	0.062
	0.41	281.8	4.47	20.55	0.079	0.061
	0.44	285.1	9.84	32.49	0.079	0.063
	0.45	287.4	9.59	36.77	0.076	0.068
	0.73	274.6	1.66	7.64	0.111	0.023
	0.70	276.4	2.08	11.07	0.108	0.027
	0.68	278.2	2.58	12.20	0.107	0.030
	0.68	280.2	3.28	14.33	0.106	0.032
	0.67	282.0	4.12	16.47	0.104	0.035
	0.79	273.7	1.45	5.00	0.115	0.018
	0.78	275.9	1.88	7.27	0.115	0.020
	0.76	277.8	2.37	9.10	0.113	0.023
	0.75	279.6	2.97	12.01	0.112	0.025
	0.74	281.6	3.79	11.24	0.111	0.028
	0.85	282.7	4.37	11.96	0.120	0.018
Unruh & Katz (1949)	0.34	277.0	2.84	26.29	0.073	0.066
	0.30	278.9	3.46	17.05	0.067	0.073
	0.36	278.9	3.43	27.13	0.075	0.065
	0.32	280.9	4.24	18.81	0.069	0.072
	0.28	282.9	5.17	5.55	0.063	0.079
	0.23	284.7	6.47	2.81	0.055	0.088
	0.60	275.5	1.99	13.94	0.100	0.036
	0.44	279.2	3.08	−0.45	0.084	0.055
	0.123	276.4	3.20	18.88	0.037	0.104
	0.085	278.4	3.95	6.98	0.027	0.114
	0.07	281.0	5.10	−1.41	0.023	0.119
	0.055	283.8	6.89	−4.86	0.019	0.124
	0.71	279.6	3.00	11.75	0.109	0.029
	0.61	282.2	4.27	15.78	0.099	0.041
	0.52	283.8	5.27	13.77	0.089	0.052
	0.41	285.5	6.89	19.37	0.076	0.066
	0.41	285.7	7.00	16.93	0.076	0.066

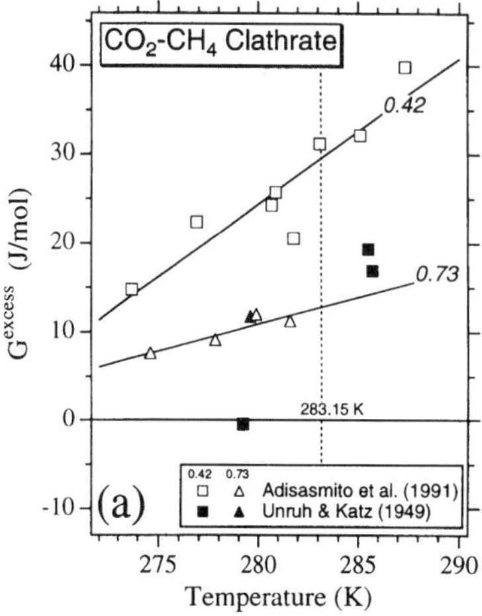

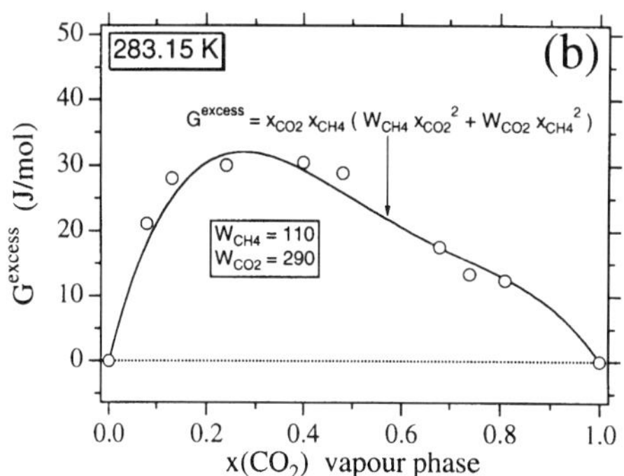

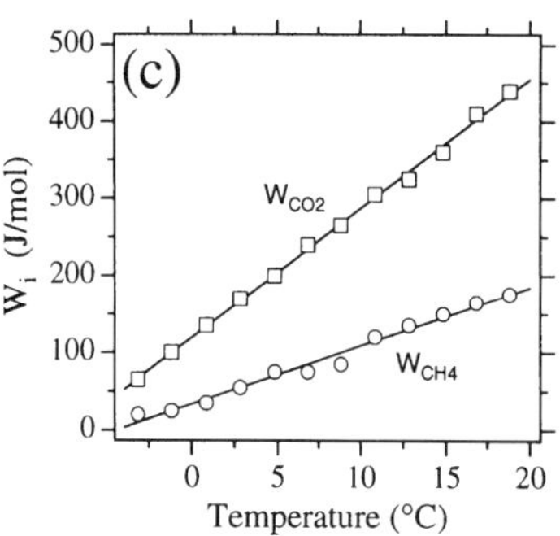

Fig. 8. (**a**) $T-G^{\mathrm{excess}}$ diagram for CO_2–CH_4 gas hydrate with isopleths for 0.42 and 0.73 mole fraction CO_2; (**b**) $X-G^{\mathrm{excess}}$ diagram for CO_2–CH_4 gas hydrate at 283.15 K; and (**c**) $T-W_G$ diagram with the interchange energies for CO_2 and CH_4 in CO_2–CH_4 gas hydrate.

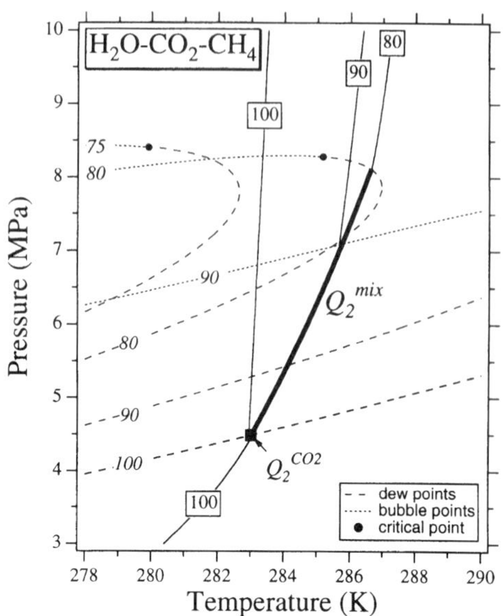

Fig. 9. T–P diagram for CO_2–CH_4 mixed gas hydrate near Q_2 melting conditions. Curved dashed lines illustrate the position of immiscibility for CO_2–CH_4 fluids, numbers are mol% CO_2. Thin solid lines are clathrate stability curves for several selected vapour compositions (squared numbers are mol% CO_2). The thick solid line is the line segment which contains all possible clathrate melting at Q_2 conditions.

CH_4 the clathrate melting curve no longer intersects the immiscibility curve, and Q_2 disappears. For higher mole fractions of CH_4 the clathrate phase melts in the presence of a supercritical fluid. For 90 mol% CO_2 mixture, the gas-rich phase at the lower Q_2 point is vapour-like, while the higher Q_2 point has a liquid-like gas-rich phase. The 80 mol% CO_2 mixture has a vapour-like gas-rich phase at both the lower and upper Q_2 point (Fig. 9). The position of the critical point of this mixture is located within the clathrate stability field, while the dew-point line intersects this boundary twice.

The distribution of CO_2 and CH_4 between the clathrate phase and the fluid is illustrated in Fig. 10. Within a small temperature range, between 283 and 293 K, this fluid system has an azeotrope. The clathrate phase in equilibrium with a CH_4-rich fluid is relatively enriched in CO_2, and a clathrate phase in equilibrium with a CO_2-rich fluid is strongly enriched in CH_4. The azeotropic point moves from CO_2-rich compositions near 283 K to CH_4-rich compositions at higher temperatures. The 284 K isotherm in Fig. 10 intersects the immiscibility field for CO_2–CH_4 fluids, which defines Q_2. The relative

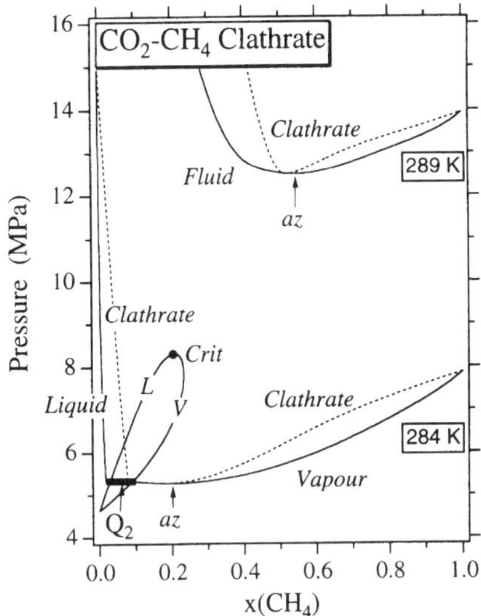

Fig. 10. *P–X* diagram for CO_2–CH_4 mixed gas hydrate with two isotherms (284 and 289 K). Mole fraction CH_4 is calculated on a water-free basis. L, liquid phase; V, vapour phase; Crit, critical point at 284 K, az, azeotropic point.

amount of CH_4 in the clathrate compared to CO_2 is similar to the composition of the vapour phase, while the coexisting liquid phase is slightly enriched in CO_2.

CH_4–N_2 gas hydrate

The hydrate formers CH_4 and N_2 have different clathrate structures, I and II, respectively. Therefore, mixed CH_4–N_2 gas hydrates can form both type of structures, which is mainly defined by their composition. Neither pure CH_4 nor pure N_2 gas hydrates have a Q_2 point, and clathrate melting always occurs in the presence of a supercritical fluid. The equation of state according to Duan *et al.* (1996) is selected to calculate fugacities in CH_4–N_2 mixtures. Although this equation of state was originally not used for pure CH_4 gas hydrate, it has a similar accuracy to the equation of state from Duan *et al.* (1992*a*, *b*) (see Fig. 2b).

Jhaveri & Robinson (1965) provide the sole experimental data set in this fluid system. Therefore, the estimation of excess functions must be regarded as a first approach. The difference between measured and modelled clathrate melting pressures without excess energy may exceed 30% for gas mixtures of *c.* 30 mol% CH_4, which, again, greatly exceeds the accuracy of the pure end-members. The missing energy G^{excess} in equation (19) has been calculated for each individual experimental data point (Fig. 11 and Table 14). A simple dependence on mole fraction and temperature, like that for CO_2–CH_4 gas hydrates, is not easily observed

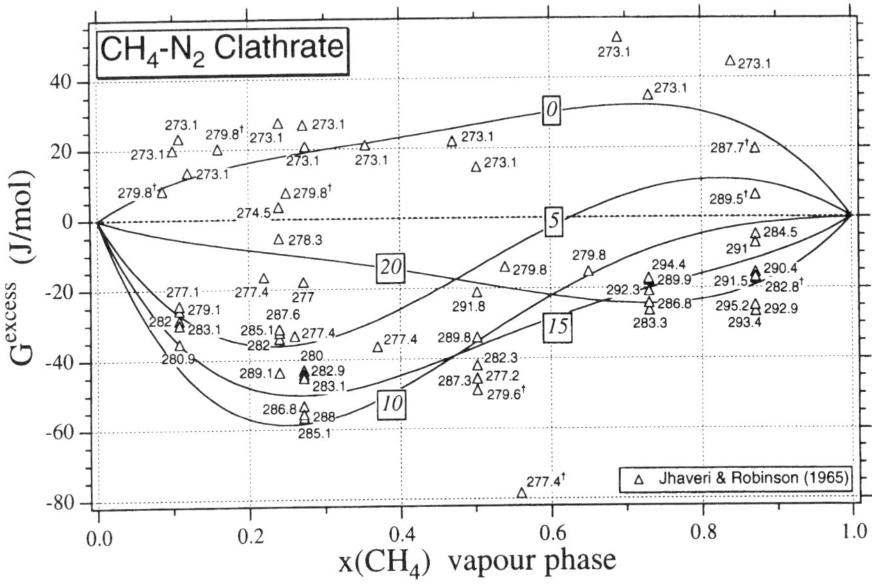

Fig. 11. G^{excess}–X diagram for CH_4–N_2 mixed gas hydrate. Numbered temperatures for individual data points are in Kelvin. Curved solid lines are modelled G^{excess} values according to equations (21) and (22) at selected temperatures (squared numbers in °C). Experimental data points marked with an † are excluded from the estimation of interchange energies W_G.

Table 14. *Measured clathrate stability conditions and calculated* G^{excess} *and mole fraction in the clathrate phase for the binary* CH_4–N_2 *gas hydrate*

Literature	Measured values				Calculated values	
	xCH_4 (vap)	T (K)	P (MPa)	G^{excess} ($J\,mol^{-1}$)	xCH_4 (cla)	xN_2 (cla)
Jhaveri & Robinson (1965)	0.873	282.76	7.40	−17.46	0.138	0.005
		284.54	9.31	−4.94	0.139	0.005
		287.71	14.52	19.61	0.139	0.006
		289.48	17.11	6.43	0.139	0.006
		290.37	17.49	−15.67	0.139	0.006
		290.98	19.53	−7.27	0.139	0.006
		291.48	19.99	−16.99	0.139	0.006
		292.87	22.94	−26.83	0.139	0.007
		293.43	24.66	−26.73	0.138	0.007
		295.21	31.31	−25.02	0.138	0.007
	0.731	273.15	3.90	34.86	0.132	0.009
		283.32	8.95	−26.49	0.132	0.011
		286.82	13.22	−24.37	0.132	0.012
		289.93	19.55	−17.91	0.131	0.014
		292.32	25.99	−20.91	0.130	0.015
		294.43	34.34	−17.31	0.129	0.017
	0.5025	273.15	4.96	14.57	0.119	0.022
		277.21	6.13	−45.65	0.118	0.023
		279.65	7.77	−48.93	0.118	0.024
		282.32	10.49	−41.90	0.117	0.026
		287.32	17.90	−45.70	0.115	0.029
		289.82	24.99	−33.93	0.114	0.031
		291.82	33.19	−21.18	0.112	0.033
	0.272	273.15	7.96	26.93	0.094	0.048
		277.04	10.16	−17.92	0.092	0.050
		279.98	12.64	−43.06	0.091	0.051
		282.87	17.04	−44.34	0.089	0.054
		283.15	17.50	−45.28	0.089	0.054
		285.09	20.72	−56.69	0.088	0.056
		286.76	25.14	−53.21	0.087	0.057
		287.98	28..49	−55.58	0.086	0.058
	0.24	273.15	8.62	27.60	0.088	0.053
		274.54	9.15	3.51	0.088	0.054
		278.26	12.96	−5.42	0.086	0.056
		282.04	17.44	−33.93	0.084	0.059
		285.09	24.34	−32.27	0.082	0.062
		287.59	31.99	−31.31	0.080	0.065
		289.09	35.96	−43.60	0.079	0.066
	0.108	273.15	12.55	23.22	0.055	0.087
		277.15	15.86	−25.78	0.053	0.089
		279.09	19.39	−24.44	0.052	0.091
		280.93	22.52	−35.28	0.051	0.092
		282.04	25.82	−28.60	0.050	0.093
		283.15	28.79	−30.10	0.050	0.094
Jhaveri & Robinson (1965)	0.840	273.15	3.56	44.47	0.136	0.005
	0.690		4.31	51.36	0.130	0.011
	0.470		5.35	22.00	0.116	0.025
	0.355		6.55	21.00	0.105	0.036
	0.275		7.75	20.759	0.094	0.047
	0.185		10.64	46.98	0.077	0.065
	0.120		11.65	13.48	0.059	0.083
	0.100		12.77	19.81	0.052	0.090
	0.075		13.32	9.56	0.031	0.104
	0.060		14.59	22.92	0.026	0.111
	0.560	277.43	5.20	−78.49	0.122	0.019
	0.370		8.11	−36.43	0.106	0.036
	0.260		10.34	−33.23	0.090	0.052
	0.220		12.06	−16.52	0.083	0.059
	0.650	279.82	7.14	−15.39	0.128	0.014
	0.540		8.37	−13.91	0.121	0.022
	0.250		15.55	7.57	0.087	0.056
	0.160		20.67	20.14	0.067	0.077
	0.086		25.23	8.32	0.043	0.101

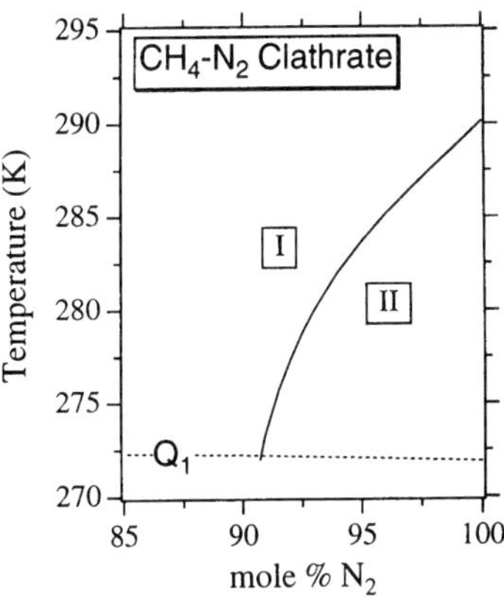

Fig. 12. *T–X* diagram for CH_4–N_2 mixed gas hydrate with the boundary between clathrate Structure I and II. Mole fraction N_2 is calculated on a water-free basis.

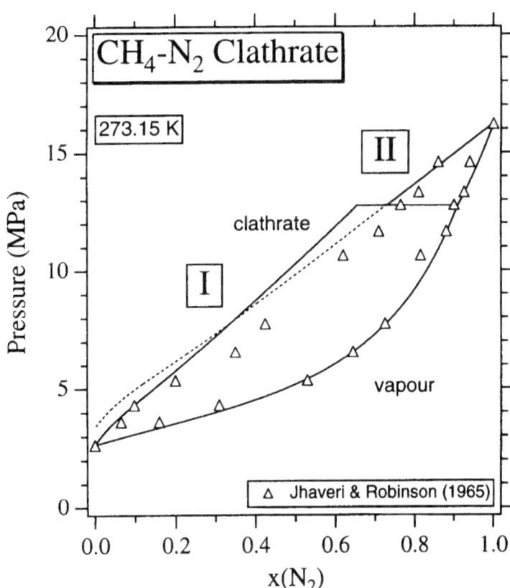

Fig. 13. *P–X* diagram for CH_4–N_2 mixed gas hydrate at 273.15 K. Mole fraction N_2 is calculated on a water-free basis. Iand II are the stability fields for clathrate Structure I and Structure II, respectively.

from Fig. 11. Some experimental data do not follow the general trend in temperature dependence at selected compositions. However, relatively simple equations for temperature and mole fraction dependency of G^{excess} and interchange energies (Table 12) are estimated after careful analysis of Fig. 10. The non-ideality switches from positive excess energies at 273 K to negative values at higher temperatures (>278 K). At temperatures higher than 283 K, G^{excess} switches back in the direction of positive values only for CH_4-rich compositions, while N_2-rich compositions continue to decrease in value. The asymmetry changes from most intensively non-ideal for CH_4-rich compositions, similar to CO_2–CH_4 gas hydrates, to N_2-rich compositions at higher temperatures. In general, the calculated chemical potential for clathrate Structure II exceeds that for Structure I clathrates for all experimental data in Fig. 11 and Table 14. Consequently, clathrate Structure I is the most stable configuration, even for N_2-rich compositions. The stability of clathrate Structure II is limited to a small range at high N_2 mole fractions (>0.91) and relatively low temperature (Fig. 12).

The distribution of N_2 and CH_4 between the clathrate phase and the vapour phase at 273.15 K is illustrated in Fig. 13. The clathrate phase is systematically enriched in CH_4. Theoretical considerations already suggest that denser phases are enriched in less volatile components which have higher critical points. Modelled compositions of the vapour phase at selected pressures are consistent with the experimental data from Jhaveri & Robinson (1965). However, the coexisting composition of the clathrate phase is very dissimilar, which may result from large uncertainties in measured values. Experimental data which are located in the modelled stability field of clathrate Structure I are more likely to correspond to a modelled clathrate Structure II (Fig. 13). Small difference in chemical potential between Structure I and Structure II may ease the presence of a metastable phase during experiments. This example illustrates the advantage of presenting the clathrate phase G^{excess} as a function of the vapour phase composition, which is experimentally verifiable.

CH_4–C_2H_6 gas hydrate

The CH_4–C_2H_6 gas mixture forms clathrate Structure I over the total compositional range. Fugacities in gas mixtures are calculated using the equation of state according to Lee & Kesler (1975). This equation accurately reproduces fugacity coefficients for pure CH_4, C_2H_6 and mixtures up to 300 K (Fig. 2b and d).

Experimental data from Deaton & Frost (1946), McLeod & Campbell (1961) and Holder & Grigoriou (1980) are compared to model predictions. Again, the model without a G^{excess} function is unable to reproduce stability conditions of the mixed gas hydrate. Table 15 and Fig. 14 illustrate the amount of missing energy in the model to obtain a perfect fit to experimental data. Although three independent data

Table 15. *Measured clathrate stability conditions and calculated* G^{excess} *and mole fractions in the clathrate phase for binary* CH_4–C_2H_6 *gas hydrate*

Literature	Measured values			Calculated values		
	xCH$_4$ (vap)	T (K)	P (MPa)	G^{excess} (J mol^{-1})	xCH$_4$ (cla)	xC$_2$H$_6$ (cla)
Deaton & Frost (1946)	0.988	274.8	2.86	−7.45	0.131	0.009
	0.988	277.6	3.81	−3.72	0.132	0.009
	0.988	280.4	5.09	−0.47	0.134	0.008
	0.978	274.8	2.36	−53.46	0.122	0.017
	0.978	277.6	3.23	−41.57	0.125	0.016
	0.978	280.4	4.41	−30.92	0.127	0.015
	0.978	282.6	5.67	−24.78	0.129	0.014
	0.978	283.1	6.09	−20.94	0.129	0.013
	0.971	274.8	2.16	−72.33	0.117	0.021
	0.971	277.6	2.96	−59.24	0.120	0.020
	0.971	280.4	4.03	−49.57	0.122	0.019
	0.950	274.8	1.84	−92.89	0.104	0.0329
	0.950	277.6	2.53	−79.97	0.107	0.032
	0.950	280.4	3.45	−71.38	0.110	0.030
	0.950	283.1	4.77	−60.53	0.113	0.028
	0.904	274.8	1.52	−95.61	0.084	0.051
	0.904	277.6	2.10	−84.38	0.088	0.050
	0.904	280.4	2.89	−73.77	0.091	0.048
	0.904	283.1	3.96	−67.13	0.094	0.046
	0.534	274.8	0.94	−14.54	0.032	0.097
	0.564	277.6	1.29	−11.88	0.035	0.096
	0.564	280.4	1.76	−10.47	0.038	0.095
	0.564	283.1	2.43	−6.71	0.041	0.094
McLeod & Campbell (1961)	0.944	302.0	68.43	24.61	0.134	0.013
	0.944	301.1	62.23	22.11	0.134	0.013
	0.944	299.0	48.23	16.52	0.133	0.013
	0.944	296.5	34.44	9.55	0.132	0.014
	0.945	293.6	24.24	20.58	0.130	0.016
	0.945	289.6	13.89	21.22	0.124	0.020
	0.945	287.9	10.45	8.29	0.121	0.023
	0.946	284.9	6.93	−3.83	0.116	0.027
	0.807	304.0	68.57	−0.76	0.108	0.038
	0.807	303.1	61.95	−2.53	0.109	0.038
	0.807	301.3	48.64	−15.28	0.108	0.038
	0.807	299.0	35.61	−19.68	0.107	0.039
	0.808	296.4	23.48	−26.19	0.103	0.042
	0.808	293.3	13.89	−27.17	0.094	0.050
	0.808	291.7	10.45	−29.44	0.088	0.055
	0.808	288.8	7.00	−23.43	0.081	0.061
Holder & Grigoriou (1980)	0.983	283.9	1.81	−453.53	0.122	0.012
	0.983	285.7	2.31	−431.05	0.124	0.011
	0.983	286.6	2.71	−406.97	0.125	0.011
	0.983	287.8	3.08	−403.81	0.126	0.011
	0.952	279.4	0.99	−455.32	0.098	0.031
	0.952	281.5	1.34	−425.97	0.101	0.030
	0.952	283.3	1.71	−405.25	0.104	0.029
	0.952	285.3	2.17	−392.90	0.106	0.029
	0.952	286.4	2.51	−381.28	0.108	0.028
	0.952	287.6	2.99	−363.40	0.110	0.028
	0.822	281.6	1.42	−269.34	0.064	0.069
	0.822	283.3	1.77	−255.824	0.066	0.098
	0.822	284.8	2.14	−245.67	0.068	0.067
	0.822	286.2	2.66	−233.81	0.071	0.066
	0.822	287.0	2.96	−217.32	0.072	0.065

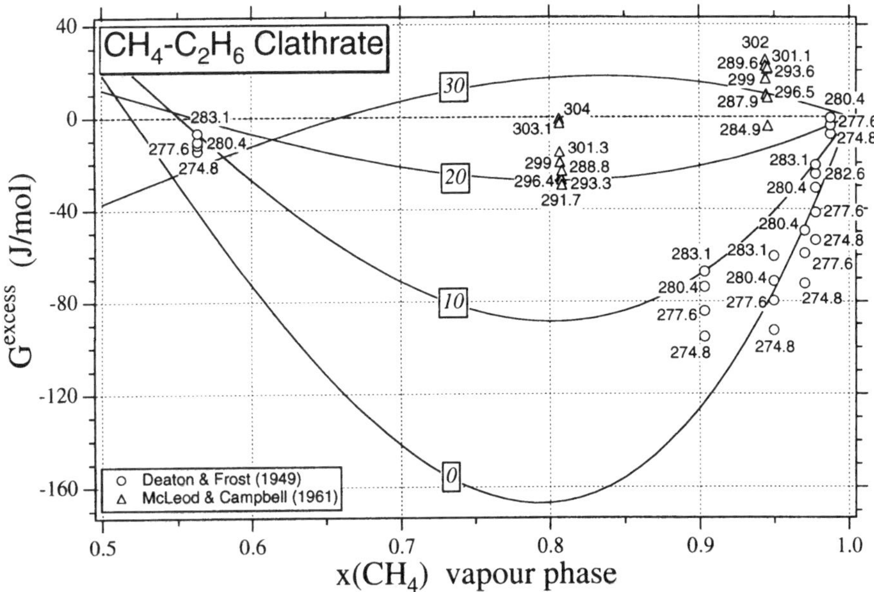

Fig. 14. $G^{excess}-X$diagram for CH_4–C_2H_6 mixed gas hydrate. Numbered temperatures for individual data points are in Kelvin. Curved solid lines are modelled G^{excess} values according to equations (21) and (22) at selected temperatures (squared numbers in °C).

sets are available, Holder & Grigoriou (1980) present data which are completely inconsistent with the other data sets. For example, clathrate melting pressure for a 95 mol% CH_4 mixture at 283.2 K is 4.77 MPa according to Deaton & Frost (1946), while Holder & Grigoriou (1980) have measured it at 1.71 MPa. In addition, the extremely high calculated G^{excess} values from their data are regarded as unrealistic. Therefore, experimental data from Holder & Grigoriou (1980) are excluded from the estimation of G^{excess} and interchange energies W_G. Figure 14 illustrates a strong asymmetry with high negative G^{excess} values for about 80 mol% CH_4 and relatively low temperatures. All data are consistent with a general trend towards positive values for increasing temperatures. Unfortunately, experimental data are not available for C_2H_6-rich fluids (>50 mol%), but the continuation of the asymmetry for CH_4-rich compositions suggests a knot at about 50 mol% and strong positive G^{excess} values at relatively low temperatures.

CO_2–N_2 gas hydrate

Experimental data are not available for this fluid system. Therefore, the excess Gibbs free energy function for CO_2–N_2 gas hydrates cannot be calculated. In a first approximation it is assumed that both interchange energies for CO_2 and N_2 are zero. However, thermodynamic considerations on the position of Q_2 for CO_2-rich gas mixtures (>75 mol%) suggest the existence of excess energy in this fluid system. Similar to the CO_2–CH_4 system (Fig. 9), a quadruple point Q_2 for pure CO_2 gas hydrate is transformed into a limited line if small amounts of N_2 are added to the fluid system (Fig. 15). The lower Q_2 point for a mixture of 90 mol% CO_2 in Fig. 15 is defined by the intersection of the dew-point line and the clathrate melting curve in equilibrium with a vapour-like CO_2-rich phase. Likewise, the upper Q_2 point is defined by the intersection of the bubble-point line and the clathrate melting curve in equilibrium with a liquid-like CO_2-rich phase. However, the coexisting vapour-like CO_2-rich phase at this point does not give an equal intersection point (Fig. 15). This gap can be closed if an G^{excess} value of 60 J mol^{-1} is used.

Computer programmes

Calculation of clathrate stability conditions in this study have been performed with the computer code CLATHRATE in C++ developed by Bakker (1997). The package of programs was developed for fluid inclusion research,

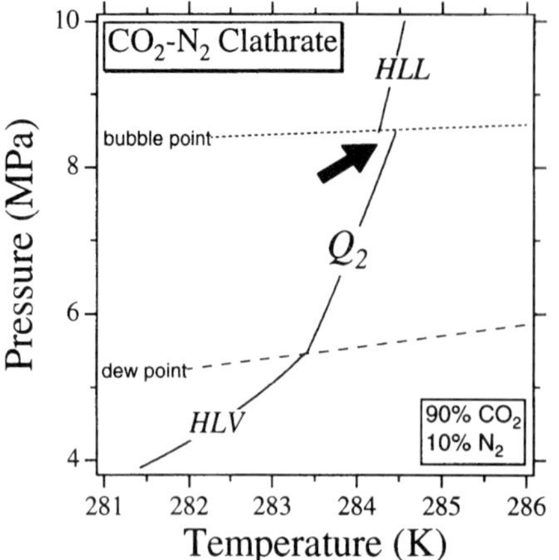

Fig. 15. *T–P* diagram for CO_2–N_2 gas hydrate near Q_2 conditions. The fluid has 90 mol% CO_2 and 10 mol% N_2. HLV and HLL are the clathrate melting curves in equilibrium with a vapour-like gas mixture and a liquid-like gas mixture, respectively. The arrow indicates the discontinuity in the clathrate melting curve at the bubble point.

where clathrate melting temperatures are frequently measured. The package is extended with the program CURVES, which is able to calculate clathrate melting curves for a chosen fluid composition, including salinities, and can which calculate clathrate melting pressure at a selected temperature or the inverse. Furthermore, the programs calculate the composition of the clathrate phase at a selected temperature and pressure, and the composition of each phase that is present during clathrate melting. The program can run on an IBM (compatible) computer and is readily available on request.

Conclusions

Using the most accurate estimations for fugacities, gas solubilities in H_2O and thermodynamic constants for the clathrate, liquid and vapour phase, optimum Kihara parameters σ and ε for gas–H_2O interactions are obtained for pure CO_2, CH_4, N_2, and C_2H_6 gas hydrates. A large amount of experimental data from independent sources for the fitting procedure are necessary to obtain a true best-fit. The postulates from the original clathrate stability model from van der Waals & Platteeuw (1959) are not violated in this optimization procedure. These Kihara parameters differ from previous estimations found in the literature, which can be directly related to the use of an insufficient amount of data. Experimental data for mixed gas hydrates can only be reproduced accurately after the introduction of an excess Gibbs free energy function, which must have resulted from an interaction between cavities with different filling molecules. A modified Margules equation describes the composition and temperature dependence of the excess energy for CO_2–CH_4, CH_4–N_2, and CH_4–C_2H_6 gas hydrates.

This work is financially supported by the Deutsche Forschungsgemeinschaft: Graduiertenkollegs 'Wirkung fluider Phasen auf Locker- und Festgesteine'.

References

ADISASMITO, S., FRANK, R. J. & SLOAN, E. D. Jr. 1991. Hydrates of carbon dioxide and methane mixtures. *Journal of Chemical Engineering Data*, **36**, 68–71.

ANDERSON, F. E. & PRAUSNITZ, J. M. 1986. Inhibition of gas hydrates by methanol. *American Institute of Chemical Engineering Journal*, **32**, 1321–1333.

ANGELL, C. A., SHUPPERT, J. & TUCKER, J.C. 1973. Anomalous properties of supercooled water. Heat capacity, expansivity, and proton magnetic resonance chemical shift from 0 to −38°. *Journal of Physical Chemistry*, **77**, 3092–3099.

ANGUS, S., DE REUCK, K. M. & ARMSTRONG, B. 1979. *International Thermodynamic Tables of the Fluid State: 6. Nitrogen*. Pergamon, Oxford.

AVLONITIS, D. 1988. *Multiple Equilibria in Oil-water Hydrate Forming Systems*. MSc thesis, Heriot-Watt University, UK.

—— 1994. The determination of Kihara potential parameters from gas hydrate data. *Chemical Engineering Science*, **49**, 1161–1173.

BAKKER, R. J. 1997. Clathrates: computer programmes to calculate fluid inclusion V–X properties using clathrate melting temperatures. *Computers & Geosciences*, **23**, 1–18.

——, DUBESSY, J. & CATHELINEAU, M. 1996. Improvements in clathrate modeling. I: The H_2O–CO_2 system with various salts. *Geochimica et Cosmochimica Acta*, **60**, 1657–1681.

BENEDICT, M., WEBB, G. B. & RUBIN, L. C. 1949. An empirical equation for thermodynamic properties of light hydrocarbons and their mixtures I. Methane, ethane, propane and n-butane. *Journal of Chemical Physics*, **8**, 334–345.

BENSON, B. B. & KRAUS D., JR. 1976. Empirical laws for dilute aqueous solutions of nonpolar gases. *Journal of Chemical Physics*, **64**, 689–709.

BOZZO, A. T., CHEN, H.-S., KASS, J. R. & BARDUHN, A. J. 1975. The properties of the hydrates of chlorine and carbon dioxide. *Desalination*, **16**, 303–320.

CARNAHAN, N. F. & STARLING K. E. 1969. Equation of state for non-attracting rigid spheres. *Journal of Chemical Physics*, **51**, 635–636.

CARROLL, J. J. & MATHER, A. E. 1992. The system

carbon dioxide–water and the Krichevsky–Kasarnovsky equation. *Journal of Solution Chemistry*, **21**, 607–621.

——, SLUPSKY, J. D. & MATHER, A. E. 1991. The solubility of carbon dioxide in water at low pressure. *Journal of Physical Chemistry Reference Data*, **20**, 1201–1209.

CHUEH, P. L. & PRAUSNITZ, J. M. 1967. Vapor–liquid equilibria at high pressures. Vapor-phase fugacity coefficients in nonpolar and quantum–gas mixtures. *Industrial and Engineering Chemistry Fundamentals*, **6**, 492–498.

CULBERSON, O. L. & MCKETTA, J. J., JR. 1951. Phase equilibria in hydrocarbon–water systems III: The solubility of methane in water at pressures to 10,000 psia. *Petroleum Transactions, American Institute for Mechanical Engineers*, **192**, 223–226.

DAVIDSON D. W., GOUGH, S. R., HANDA, Y. P., RATCLIFFE, C. I., RIPMEESTER, J. A. & TSE, J. S. 1987. Some structural studies of clathrate hydrates. *Journal de Physique C1*, **3**, 537–542.

DEATON, W. M. & FROST, E. M., JR. 1946. *Gas Hydrates and their Relation to Operation of Natural-gas Pipe Lines*. United States Bureau of Minerals Monograph **8**, 1–108.

DEBYE, P. & HÜCKEL, E. 1923. Zur Theorie der Electrolyte I: Gefrierpunktserniederung und verwandte Erscheinungen. *Physikalisch Zeitschrift*, **24**, 185–206.

DE ROO, J. L., PETERS, C. J., LICHTENTHALER, R. N. & DIEPEN, G. A. M. 1983. Occurence of methane hydrate in saturated and unsaturated solutions of sodium chloride and water in dependence of temperature and pressure. *American Institute of Chemical Engineering Journal*, **29**, 651–657.

DHARMAWARDHAMA, P. B., PARRISH, W. R. & SLOAN, E. D. JR. 1980. Experimental thermodynamic parameters for the prediction of natural gas hydrate dissociation conditions. *Industrial Engineering and Chemical Fundamentals*, **19**, 410–414.

DHOLABHAI, P. D., KALOGERAKIS, N. & BISHNOI, P. R. 1993. Equilibrium conditions for carbon dioxide hydrate formation in aqueous electrolyte solutions. *Journal of Chemical Engineering Data*, **38**, 650–654.

DICKENS, G. R. & QUINBY-HUNT, M. S. 1994. Methane hydrate stability in seawater. *Geophysical Research Letters*, **21**, 2115–2118.

DUAN, Z., MØLLER, N. & WEARE, J. H. 1992*a*. An equation of state for the CH_4–CO_2–H_2O system: I. Pure systems from 0 to 1000°C and 0 to 8000 bar. *Geochimica et Cosmochimica Acta*, **56**, 2605–2617.

——, —— & —— 1992*b*. An equation of state for the CH_4–CO_2–H_2O system: II. Mixtures from 50 to 1000°C and 0 to 1000 bar. *Geochimica et Cosmochimica Acta*, **56**, 2619–2631.

——, —— & —— 1996. A general equation of state for supercritical fluid mixtures and molecular dynamics simulation of mixture PVTX properties. *Geochimica et Cosmochimica Acta*, **60**, 1209–1216.

DUBESSY, J., THIÉRY, R. & CANALS, M. 1992. Modeling of phase equilibria involving mixed gas clathrates: application to the determination of molar volume of the vapour phase and salinity of aqueous solution in fluid inclusions. *European Journal of Mineralogy*, **4**, 873–884.

ENGLEZOS, P. & BISHNOI, P. R. 1988. Prediction of gas hydrate formation conditions in aqueous electrolyte solutions. *American Institute for Chemical Engineering Journal*, **34**, 1718–1721.

—— & —— 1991. Experimental study on the equilibrium ethane hydrate formation conditions in aqueous electrolyte solutions. *Industrial and Engineering Chemistry Research*, **30**, 1655–1659.

—— & HALL, S. 1994. Phase equilibrium data on carbon dioxide hydrate in the presence of electrolytes, water soluble polymers and montmorillonite. *Canadian Journal of Chemical Engineering*, **72**, 887–893.

FLOWERS, G. C. 1979. Correction of Holloway's (1977) adaption of the modified Redlich-Kwong equation of state for calculation of the fugacities of molecular species in supercritical fluids of geologic interest. *Contributions to Mineralogy and Petrology*, **69**, 315–319.

FRANKS, F. 1972. *Water, A Comprehensive Treatise*. Plenum, New York.

FRIEND, D. G., INGHAM, H. & ELY, J. F. 1991. Thermophysical properties of ethane. *Journal of Physical Chemistry Reference Data*, **20**, 275–347.

GALLOWAY, T. J., RUSKA, W., CHAPPELEAR, P. S. & KOBAYASHI, R. 1970. Experimental measurements of hydrate numbers for methane and ethane and comparison with theoretical values. *Industrial Engineering and Chemistry Fundamentals*, **9**, 237–243.

GIAUQUE, W. F. & STOUT, J. W. 1936. The entropy of water and the third law of thermodynamics. Heat capacity of ice from 15 to 273 K. *Journal of the American Chemical Society*, **58**, 1144–1150.

HAAR, L., GALLAGHER, J. S. & KELL, G. S. 1984. *NBS/NRC Steam Tables*. Hemisphere, Washington, DC.

HILL, P. G. 1990. A unified fundamental equation for the termodynamic properties of H_2O. *Journal of Physics and Chemistry Reference Data*, **19**, 1233–1274.

HOLDER, G. D. & GRIGORIOU, G. C. 1980. Hydrate dissociation pressures of (methane + ethane + water). Existence of a locus minimum pressure. *Journal of Chemical Thermodynamics*, **A150**, 1093–1104.

—— & HAND, J. H. 1982. Multiple-phase equilibria in hydrates from methane, ethane, propane, and water mixtures. *American Institute for Chemical Engineering Journal*, **28**, 440–447.

——, CORBIN, G. & PAPADOPOULOS, K. D. 1980. Thermodynamic and molecular properties of gas hydrates from mixtures containing methane, argon, and krypton. *Industrial Engineering and Chemistry Fundamentals*, **19**, 282–286.

——, ZETTS, S. P. & PRADHAN, N. 1988. Phase behaviour in systems containing clathrate hydrates, a review. *Reviews in Chemical Engineering*, **5**, 1–70.

HOLLOWAY, J. R. 1977. Fugacity and activity of molecular species in supercritical fluids. *In*:

FRASER, D. G. (ed.) *Thermodynamics in Geology*, 161–182.

—— 1981. Compositions and volumes of supercritical fluids in the earth's crust. *In*: HOLLISTER, L. S. & CRAWFORD, M. L. (eds) *Short Course in Fluid Inclusions: Application to Petrology*, 13–38.

JHAVERI, J. & ROBINSON, D. B. 1965. Hydrates in the methane-nitrogen system. *Canadian Journal of Chemical Engineering*, **43**, 75–78.

JOHN, V. T. & HOLDER, G. D. 1981. Langmuir constants for spherical and linear molecules in clathrate hydrates. Validity of the cell theory. *Journal of Physical Chemistry*, **89**, 3279–3285.

——, PAPADOPOULOS, K. D. & HOLDER, G. D. 1985. A generalized model for predicting equilibrium conditions for gas hydrates. *American Institute of Chemical Engineering Journal*, **31**, 252–259.

JONES, J. E. 1924. On the determination of molecular fields II: From the equation of state of a gas. *Proceedings of the Royal Society of London*, **A106**, 463–477.

KELL, G. S. 1967. Precise representation of volumetric properties of water at one atmosphere. *Journal of Chemical Engineering Data*, **12**, 66–69.

—— & WHALLEY, E. 1965. The PVT properties of water. I. Liquid water in the temperature range 0 to 150°C and at pressures up to 1 kb. *Philosophical Transactions of the Royal Society of London*, **258A**, 565–614.

KIHARA, T. 1953. Virial coefficients and models of molecules in gases. *Reviews in Modern Physics*, **25**, 831–843.

KOBAYASHI, R. & KATZ, D. L. 1949. Methane hydrate at high pressures. *Petroleum Transactions, American Institute for Mechanical Engineers*, **1**, 66–70.

KRICHEVSKY, I. R. & KASARNOWSKY, J. S. 1935. Thermodynamical calculations of solubilities of nitrogen and hydrogen in water at high pressures. *Journal of the American Chemical Society*, **57**, 2168–2172.

LANGMUIR, I. 1918. The adsorption of gases on plane surfaces of glass, mica and platinum. *Journal of the American Chemical Society*, **40**, 1361–1403.

LARSON, S. D. 1955. *Phase Studies of Two-component Carbon Dioxide–Water System Involving the Carbon Dioxide Hydrate*. PhD thesis, University of Illinois.

LEE, B. I. & KESLER, M. G. 1975. A generalized thermodynamic correlation based on three-parameter corresponding states. *American Institute of Chemical Engineering Journal*, **21**, 510–527.

LENNARD-JONES, J. E. & DEVONSHIRE, A. F. 1937. Critical phenomena in gases I. *Proceedings of the Royal Society of London*, **163**, 53–70.

—— & —— 1938. Critical phenomena in gases. II: Vapour pressures and boiling points. *Proceedings of the Royal Society of London*, **165**, 1–11.

LUNDGAARD, L. & MOLLERUP, J. 1991. The influence of gas phase fugacity and solubility on correlations of gas-hydrate formation pressure. *Fluid Phase Equilibria*, **70**, 199–213.

—— & —— 1992. Calculation of phase diagrams of gas-hydrates. *Fluid Phase Equilibria*, **76**, 141–149.

MARGULES, M. 1895. Über die Zusammensetzung der gesättigten Dämpfe von Mischungen. *Sitzungberichte der Wiener Akademie*, **104**, 1243–1278.

MARSHALL, D. R., SAITO, S. & KOBAYASHI, R. (1964) Hydrates at high pressure, part I: methane–water, argon–water, and nitrogen–water system. *American Institute of Chemical Engineering Journal*, **10**, 202–205.

MCKOY, V. & SINANOGLU, O. 1963. Theory of dissociation pressures of some gas hydrates. *Journal of Chemical Physics*, **38**, 2946–2956.

MCLEOD, H. O. & CAMPBELL, J. M. 1961. Natural gas hydrates at pressures to 10,000 psia. *Journal of Petroleum Technology*, **222**, 590–594.

MOORE, J. C., BATTINO, R., RETTICH, T. R., HANDA, Y. P. & WILHELM, E. 1982. Partial molar volumes of 'gases' at infinite dolution in water at 298.15 K. *Journal of Chemical Engineering Science*, **27**, 22–24.

MUNCK, J., SKJOLD-JORGENSEN, S. & RASMUSSEN, P. 1988. Computations of the formation of gas hydrates. *Chemical Engineering Data*, **27**, 22–24.

NG, H. J. & ROBINSON, D. B. 1976. The measurement and prediction of hydrate formation in liquid hydrocarbon–water systems. *Industrial and Engineering Chemistry Fundamentals*, **15**, 293–298.

—— & —— 1985. Hydrate formation in systems containing methane, ethane, propane, carbon dioxide or hydrogen sulphide in the presence of methanol. *Fluid Phase Equilibria*, **21**, 145–155.

OSBORNE, N. S., STIMSOM, H. F. & GINNINGS, D. C. 1939. Measurements of heat capacity and heat of vaporization of water in the range of 0° to 100°C. *Journal of Research National Bureau of Standards*, **23**, 145–155.

PARRISH, W. R. & PRAUSNITZ, J. M. 1972. Dissociation pressure of gas hydrates formed by gas mixtures. *Industrial and Engineering Chemistry, Processes, Design, and Development*, **11**, 26–35.

PITZER, K. S. 1992. Ion interaction approach: theory and data correlation. *In*: PITZER, K. S. (ed.) *Activity Coefficient in Electrolyte Solutions*. CRC, Boca Raton, FL, 76–153.

PLATTEEUW, J. C. & VAN DER WAALS, J. H. 1958. Thermodynamic properties of gas hydrates. *Molecular Physics*, **1**, 91–96.

PRAUSNITZ, J. M., LICHTENTHALER, R. N. & DE AZEVEDO, E.G. 1986. *Molecular Thermodynamics of Fluid-phase Equilibria*. Prentice-Hall, Englewood Cliffs, NJ.

REAMER, H. H., SELLECK, F. T. & SAGE, B. H. 1952. Some properties of mixed paraffinic and olefinic hydrates. *Petroleum Transactions, American Institute for Mechanical Engineers*, **195**, 197–205.

REDLICH, O. & KWONG, J. N. S. 1949. On the thermodynamics of solutions V: an equation of state. Fugacities of gaseous solutions. *Chemical Reviews*, **44**, 233–244.

RETTICH, T. R., HANDA, Y. P, BATTINO, R. & WILHELM, E. 1981. Solubility of gases in liquids 13. High-precision determination of Henry's constants for methane and ethane in liquid water at 275 to 328 K. *Journal of Physical Chemistry*, **85**, 3230–3237.

RIGBY M. & PRAUSNITZ, J. M. 1968. Solubility of water

in compressed nitrogen, argon, and methane. *Journal of Physical Chemistry*, **72**, 330–334.

ROBERTS, O. L., BROWNSCOMBE, E. R. & HOWE, L. S. 1940. Methane and ethane hydrates. *Oil & Gas Journal*, **39**, 37–42.

ROBINSON, D. B. & METHA, B. R. 1971. Hydrates in the propane–carbon dioxide–water system. *Journal of Canadian Petroleum Technology*, **10**, 33–35.

ROEDDER, E. 1963. Studies of fluid inclusions. II: Freezing data and their interpretation. *Economical Geology*, **58**, 167–211.

SAITO, S., MARSHALL, D. R. & KOBAYASHI, R. 1964. Hydrates at high pressures, part II: Application of statistical mechanics to the study of the hydrates of methane, argon, and nitrogen. *American Institute of Chemical Engineering Journal*, **10**, 734–740.

SETZMANN, U. & WAGNER, W. 1991. A new equation of state and tables of thermodynamic properties for methane covering the range from the melting line to 625 K at pressure up to 1000 MPa. *Journal of Physical Chemistry Reference Data*, **20**, 1061–115.

SLOAN, E. D., JR. 1990. *Clathrate Hydrates of Natural Gases*. Chemical Industries, Volume 39., Marcel Dekker, New York.

SOAVE, G. 1972. Equilibrium constants from a modified Redlich–Kwong equation of state. *Chemical Engineering Science*, **27**, 1197–1203.

SPAN, R. & WAGNER, W. 1996. A new equation of state for carbon dioxide covering the fluid region from the triple-point temperature to 1100 K at pressures up to 800 MPa. *Journal of Physical Chemistry Reference Data*, **25**, 1509–1596.

TAKENOUCHI, S. & KENNEDY, G. C. 1965. Dissociation of the phase CO_2–5.75H_2O. *Journal of Geology*, **73**, 383–390.

TEE, L. S., GOTOH, S. & STEWARD, W. E. 1966. Molecular parameters for normal fluids, Kihara potential with spherical core. *Industrial and Engineering Chemistry Fundamentals*, **5**, 363–367.

THIÉRY, R., VAN DEN KERKHOF, A. M. & DUBESSY, J. 1994. VX properties of CH_4–CO_2 and CO_2–N_2 fluid inclusions: modeling for $T < 31°C$ and $P < 400$ bars. *European Journal of Mineralogy*, **6**, 753–771.

THOMPSON, J. B. Jr. 1967. Thermodynamic properties of simple solutions. *In:* ABELSON, P. H. (ed.) *Researches in Geochemistry*, Volume 2. Wiley, New York, 340–361.

UNRUH, C. H. & KATZ, D. L. 1949. Gas hydrates of carbon dioxide–methane mixtures. *Journal of Petroleum Technology*, **1**, 83–86.

VAN CLEEFF, A. & DIEPEN, G. A. M. 1960. Gas hydrates of nitrogen and oxygen. *Recueil des Travaux Chiniques des Pays-Bas*, **79**, 582–586.

VAN DER WAALS, J. D. 1873. *De continuiteit van den gas- en vloeistoftoestand*. Academisch Proefschrift Leiden, Sijthof.

—— & PLATTEEUW, J. C. 1959. Clathrate solutions. *Advances in Chemical Physics*, **2**, 1–57.

VERMA, V. K., HAND, J. H., KATZ, D. L. & HOLDER, G. D. 1975. Denuding hydrocarbon liquids of natural gas constituents by hydrate formation. *Journal of Petroleum Technology*, **27**, 223–226.

VON STACKELBERG, M. & MÜLLER, H. R. 1954. Feste Gashydrate II: Structur und Raumchemie. *Zeitschrift fr Electrochemie*, **58**, 25–39.

YEROKHIN, A.M. 1993. A new method of determining CO_2 density and solution concentration in H_2O–CO_2–NaCl inclusions from the gas-hydrate melting point. *Geochemistry International*, **30**, 107–129.

Synthesis of CO_2 hydrate in various CH_3CO_2Na/CH_3CO_2H pH buffer solutions

H. LU & R. MATSUMOTO

Geological Institute, University of Tokyo, Hongo 7-3-1, Tokyo 113, Japan

Abstract: CO_2 hydrates were synthesized in CH_3CO_2Na/CH_3CO_2H buffer solutions with various pH values and different concentrations to investigate whether the change of pH value can influence CO_2 hydrate formation. The results imply that the change in pH value has a minimum influence on the equilibrium condition of CO_2 hydrate. The equilibrium condition of CO_2 hydrate in the various pH buffer solutions is seemingly controlled by the total concentration of the acetic acid and acetate anion, while Na^+ cation plays a minimum role.

It has been suggested for years to store CO_2 in the form of hydrate on and under the deep sea floor to release its green house effect (Marchetti 1977; Ohsumi 1995), but this may cause new environmental problems without thoroughly knowing the fate of CO_2 after being buried. Detailed information about the stable condition of CO_2 hydrates in the deep sea environment is crucial to this problem. It has been suggested that CO_2 hydrates have possibly played a significant role in determining the atmospheric composition in geological history (Matsumoto 1987). Natural CO_2 hydrate has been recognized in the backarc basin of the Mid-Okinawa Trough (Sakai *et al.* 1990). Detailed information on the stable condition of CO_2 hydrates is also important in understanding the carbon cycle on Earth. The phase diagrams of CO_2 hydrate have been well established in pure water and sodium chloride solutions (Larson 1955). Below the sea floor, however, CO_2 hydrate should form in the pores of sediments with interstitial water, the chemistry of which is different to pure water, or even sea water. Moreover, the interstitial water chemistry changes with depth in the course of burial diagenesis. As reported by Matsumoto (1992), the pH of the interstitial water of Japan Sea sediments can vary from 5 to 8. However, the effect of these changes on the formation of CO_2 hydrate has as yet not been well documented. The present research is designed to investigate whether a change of pH value can influence the CO_2 hydrate formation by synthesizing CO_2 hydrate in various pH buffer solutions.

Experimental methods

A stainless steel autoclave with a volume of 1 litre was employed for the described experiments (Fig. 1). It is designed to bear a pressure of up to $90\,kgf\,cm^{-2}$. The temperature in the autoclave can be adjusted between $-20°C$ and $80°C$ by a liquid coolant circulating in a jacket which is connected with a programmable chiller unit. Two thermometers are installed to monitor the temperatures of gas and solution; the pressure is detected by a pressure transducer. The precision of temperature and pressure readings are $\pm0.1°C$ and $\pm0.1\,kgf\,cm^{-2}$, respectively. A stirrer is used to agitate the solution when necessary. The pressure and temperature data were collected using a PC.

The pH buffer solutions used for this research were prepared using the method of Perrin & Dempsey (1974) with $CH_3CO_2Na \cdot 3H_2O$ (superfine, Wako Pure Chemical Industries, Ltd) and CH_3CO_2H acid (EL grade, Junsei Chemical Co. Ltd) as described in Table 1.

The autoclave was filled with about 430 ml pH buffer solution for each run of the experiment. After evacuating the whole system, CO_2 gas was injected. To avoid liquifying of the CO_2 gas, the maximum pressure was kept lower than $40\,kgf\,cm^{-2}$. The experiment started by lowering the temperature from 10°C to less than 4°C at a rate of $1.5°C\,h^{-1}$; then the temperature was kept at a constant value. CO_2 hydrate formed after an induction period of 8.5 hours–17 days for different solutions. When the pressure and temperature reached a stable condition, the system was supposed to be in a state of equilibrium. Then the temperature of the autoclave was increased to 10°C at a rate of $1.5°C\,h^{-1}$. The already formed CO_2 hydrate started to decompose consequently until its complete disappearance. The data were recorded during the dissociation of CO_2 hydrate.

Results and discussion

As the solubility of CO_2 gas is about 0.8–$1.0\,mol\,kg^{-1}$ at a pressure of $20–40\,kgf\,cm^{-2}$

LU, H. & MATSUMOTO, R. 1998. Synthesis of CO_2 hydrate in various CH_3CO_2Na/CH_3CO_2H pH buffer solutions. *In:* HENRIET, J.-P. & MIENERT, J. (eds) *Gas Hydrates: Relevance to World Margin Stability and Climate Change.* Geological Society, London, Special Publications, **137**, 107–111.

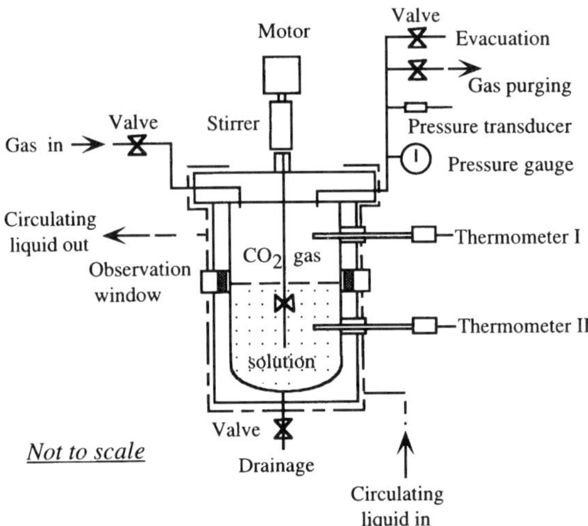

Fig. 1. The autoclave used for gas hydrate synthesizing.

Table 1. *Preparation (mixing) method for pH buffer solutions*

pH value	0.2 M CH_3CO_2Na (ml)	0.2 M CH_3CO_2H (ml)	1 M CH_3CO_2Na (ml)	1 M CH_3CO_2H (ml)
3.1	37.5	462.5	37.5	462.5
5.2	452.5	47.5	–	–

and a temperature of 0–5°C (Ebinuma unpublished data), and the K_1 of the CO_2-water system is only $10^{-6.58}$–$10^{-6.52}$ at a temperature of 0–5°C (Plummer & Bussenberg 1982), the amount of H^+ produced by the dissolution of CO_2 gas should be much less than the buffer capacity of the solutions (Table 2), and therefore the pH values of the buffer solutions should show a minimum change.

Because the solutions were mixed using specific ratios of CH_3CO_2Na to CH_3CO_2H and hydrate formed by the reaction of CO_2 gas and water, there were only two independent components in the studied systems: CO_2 gas and the buffer solution. Of the three phases of gas, solution and hydrate present, only one variable could be changed freely during hydrate dissociation according to the Gibbs Phase Rule. This means that each temperature will correspond to a specific pressure upon hydrate decomposition. Thus, the data recorded during hydrate dissociation represent the stable condition of three phases of CO_2 hydrate–gas–solution.

However, as gas hydrates are constructed only of water and suitably sized gas molecules, no other ions can be accommodated, and therefore the concentration of the buffer solution becomes concentrated in the course of CO_2 hydrate formation. The increased concentration was calculated roughly referring to the pressure change during hydrate formation. The concentrations of the applied buffer solutions were about 5–6% higher than the original ones upon the maximum hydrate formation. The data points on each line of Fig. 2 thus correspond to the equilibrium condition for the solution with the composition at that time. Owing to the special property of the buffer solution, the concentration change during hydrate formation did not cause a variation in the pH value.

As shown in Fig. 2, the data for the equilibrium boundary of CO_2 hydrate–gas–water are consistent with the data of Larson (1955). Compared with pure water, the $m_{Ac} = 0.2$ M, pH = 3.1 and 5.2 buffer solutions cause a shift of the equilibrium line by about −0.5°C. In the interval between 20.5 and 25.5 kgf cm^{-2} the data points for the two solutions overlap, although the pH values and the ratios of CH_3CO_2Na/CH_3CO_2H differ from each other, and their concentrations change somewhat during hydrate formation, as mentioned above.

Table 2. *Specifications of the solutions and their effects on the equilibrium condition of CO_2 hydrate*

	pH = 3.2–3.3* Pure water	pH = 3.1 Mixed using 0.2 M CH_3CO_2Na and CH_3CO_2H	pH = 5.2 Mixed using 1 M CH_3CO_2Na & CH_3CO_2H	pH = 3.1 Mixed using 1 M CH_3CO_2Na and CH_3CO_2H
Σm_{Ac} (mol l^{-1})†‡	0	0.2	0.2	1
m_{Na^+} (mol l^{-1})‡§	0	0.015	0.181	0.075
Buffer capacity (mol)	–	0.115	0.115	0.576
ΔT (°C)‖	–	−0.5	−0.5	−3.5 to −4.5
Concentration change (%)¶	–	5	6	6

*Calculated value at P_{CO_2} = 20–32 kgf cm^{-2}.
†Total concentration of $CH_3CO_2^-$ anion and CH_3CO_2H.
‡Concentrations refer to the original values before hydrate formation.
§Concentration of Na^+ cation.
‖Shift of equilibrium boundary of CO_2 hydrate–gas–water.
¶The concentration change at most caused by CO_2 hydrate formation.

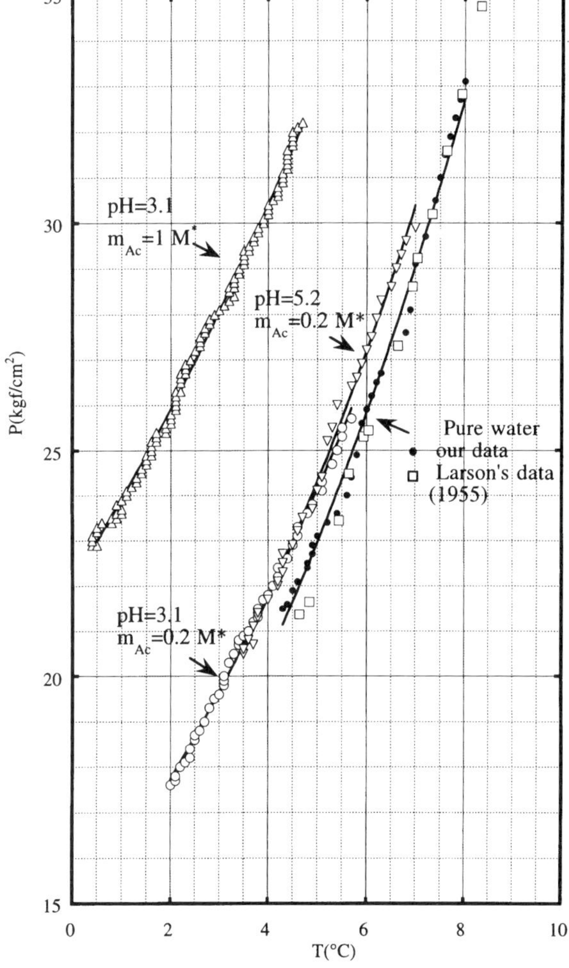

Fig. 2. Dissociation lines of CO_2 hydrate in CH_3CO_2Na/CH_3CO_2H pH buffer solutions (* concentration refers to the original solutions – it increased 5–6% at most with hydrate formation).

The $m_{Ac} = 1$ M, pH = 3.1 solution causes the equilibrium line to shift by −3.5 to −4.5°C. Because solutions with the same pH but with different concentrations cause a different shift in the equilibrium conditions of CO_2 hydrate, while those with different pH values cause a similar shift, it is reasonable to conclude that the change of pH value has an insignificant influence on the equilibrium condition of CO_2 hydrate.

CO_2 hydrate is formed by CO_2 gas filling the Structure I cage, made of water molecules, by forming hydrogen bonds, and therefore the factors which affect the clustering of water molecules to the cage will result in a change of the equilibrium condition (Christiansen & Sloan 1994; Lederhos *et al.* 1996). There are two such factors in a solution: one is the dissociation of water molecules, which can be described as pH value; another is the interaction between water molecule and the solute. Because pH is defined as $-\log(a_{H^+})$, the concentrations of H^+ in pH = 3.1 and 5.2 solutions are only $10^{-3.1}$ and $10^{-5.2}$ M, respectively, so the water dissociation in those solutions should be less than 0.1% and 0.001%. This minimum dissociation should have little influence on the equilibrium condition of CO_2 hydrate.

As discussed by Sloan (1990), the inhibition mechanism stems from the increased competition for the water molecules by the dissolved inhibitor molecule or ion. In the employed pH buffer solutions, both the CH_3CO_2H molecule and the $CH_3CO_2^-$ anion can form a hydrogen bond with water molecules. They will compete for water molecules with hydrate clusters and consequently cause the CO_2 hydrate to form under more subcooling conditions. The Na^+ cation in solution interacts with water molecules using a much stronger bond than the van der Waals' forces which cause clustering around the dissolved CO_2 molecules (Sloan 1990), and this effect will also make the CO_2 hydrate formation difficult. However, Table 2 shows that the pH = 3.1 and 5.2 solutions, which were prepared using 0.2 M CH_3CO_2H and 0.2 M CH_3CO_2Na solutions, have the same total concentration of CH_3CO_2H molecules and $CH_3CO_2^-$ anions, whereas the latter is 11 times more concentrated than the former in the Na^+ cation solution (even when the solution became more concentrated due to hydrate). They, however, cause almost the same shift in the equilibrium condition. This may imply that CH_3CO_2H molecules and $CH_3CO_2^-$ anions have a much stronger ability to influence CO_2 hydrate formation than the Na^+ cation. This is clearly demonstrated by the result of the pH = 3.1 buffer solution which is made up of 1 M CH_3CO_2H and 1 M CH_3CO_2Na solutions. Even if its Na^+ cation concentration was increased by about 6%, it was still less than half of the above pH = 5.2 solution, so the only factor resulting in a much greater shift is its high total concentration of CH_3CO_2H molecules and $CH_3CO_2^-$ anions, which is 4 times higher than the latter.

Conclusions

- The change in pH value of the solutions has a minimum influence on the equilibrium condition of CO_2 hydrate.
- The concentration of acetic acid and acetate anions in the studied pH buffer solutions is the main factor influencing the equilibrium condition of CO_2 hydrate due to the hydrogen bond formation between CH_3CO_2H molecules, $CH_3CO_2^-$ anions and water molecules. The Na^+ cation can also be an inhibitor but, compared with the above factor, its ability is small.

We would like to express our sincere thanks to J.-P. Henriet for his great help in completing this manuscript. We are also grateful to the reviewers for helpful comments on the initial manuscript. This research was partly supported by Tsuji Asia Scholarship and the Sasakawa Scientific Research Grant from the Japan Science Society to H. Lu, as well as Grand-in-Aid from the Ministry of Education and Culture (No. 07559004) to R. Matsumoto. The technical advice of T. Ebinuma of Hokkaido National Industrial Research Institute is greatly appreciated by the authors.

References

CHRISTIANSEN, R. & SLOAN, E. 1994. Mechanisms and kinetics of hydrate formation. *Annals of the New York Academy of Sciences*, **715**, 283–305.

LARSON, S. 1955. *Phase Studies of the Two-component Carbon Dioxide–Water System, Involving the Carbon Dioxide Hydrate*. PhD thesis, University of Illinois.

LEDERHOS, J., LONG, J., SUM, A. *et al.* 1996. Effective kinetic inhibitors for natural gas hydrates. *Chemical Engineering Science*, **51**(8), 1221–1229.

MARCHETTI, C. 1977. On engineering the CO_2 problem. *Climate Change*, **1**, 59–68.

MATSUMOTO, R. 1987. Nature and occurrence of gas hydrates and their implications to geological phenomena. *Journal of Geological Society of Japan*, **93**(8), 597–615 (in Japanese with an English abstract).

——1992. Causes of the oxygen isotopic depletion of interstitial waters from Sites 798 and 799, Japan Sea, Leg 128. *Proceedings of the Ocean Drilling Program*, **127/128**, 697–703.

OHSUMI, T. 1995. Feasibility study on CO_2 storage in the deep sea. *Proceedings of International Conference on Technologies for Marine Environment Preservation*, **2**, 867–874

PERRIN, D. & DEMPSEY, B. 1974. *Buffers for pH and Metal Ion Control*. Chapman & Hall, London.

PLUMMER, L. & BUSSENBERG, E. 1982. The solubilities of calcite, aragonite and vaterite in CO_2–H_2O solutions between 0 and 90°C, and an evaluation of the aqueous model for the system $CaCO_3$–CO_2–H_2O. *Geochimica et Cosmochimica Acta*, **46**, 1011–1040.

SAKAI, H., GAMO, T., KIM, E.-S. *et al.* 1990. Venting of carbon dioxide-rich fluid and hydrate formation in Mid-Okinawa Trough backarc basin. *Science*, **248**, 1093–1096.

SLOAN, E.D. JR 1990. *Clathrate Hydrates of Natural Gases*. Marcel Dekker, New York.

Major occurrences and reservoir concepts of marine clathrate hydrates: implications of field evidence

J. S. BOOTH[1], W. J. WINTERS[1], W. P. DILLON[1], M. B. CLENNELL[2] & M. M. ROWE[3]

[1] *U.S. Geological Survey, 384 Woods Hole Road, Woods Hole, MA 02543, USA*
[2] *Department of Earth Sciences, University of Leeds, Leeds LS2 9JT, UK*
[3] *High Tech, Inc., 1390 29th Avenue, Gulfport, MS 39501, USA*

Abstract: Questions concerning clathrate hydrate as an energy resource, as a factor in modifying global climate and as a triggering mechanism for mass movements invite consideration of what factors promote hydrate concentration, and what the quintessential hydrate-rich sediment may be. Gas hydrate field data, although limited, provide a starting point for identifying the environments and processes that lead to more massive concentrations. Gas hydrate zones are up to 30 m thick and the vertical range of occurrence at a site may exceed 200 m. Zones typically occur more than 100 m above the phase boundary. Thicker zones are overwhelmingly associated with structural features and tectonism, and often contain sand. It is unclear whether an apparent association between zone thickness and porosity represents a cause-and-effect relationship. The primary control on the thickness of a potential gas hydrate reservoir is the geological setting. Deep water and low geothermal gradients foster thick gas hydrate stability zones (GHSZs). The presence of faults, fractures, etc., can favour migration of gas-rich fluids. Geological processes, such as eustacy or subsidence, may alter the thickness of the GHSZ or affect hydrate concentration. Tectonic forces may promote injection of gas into the GHSZ. More porous and permeable sediment, as host sediment properties, increase storage capacity and fluid conductivity, and thus also enhance reservoir potential.

Evaluating the energy resource potential of gas hydrate in local areas, such as in the Blake Ridge area off the south-eastern United States (Dillon *et al.* 1995) or in the Japan offshore (Okuda 1996), is the first step towards developing an overall picture of clathrates as a viable energy source. How closely the assessments of these areas truly represent the population of hydrate-bearing strata is unknown. The framework of possible effects that methane hydrate dissociation might have on global climate has been constructed by Nisbet (1990), Englezos & Hatzikiriakos (1994) and Paull *et al.* (1991). As a potent 'greenhouse' gas, being 10 times more effective than CO_2 on a molar basis (Lashof & Ahuja 1990) but having a short residence time in the atmosphere (about 9 years), it is the rapid release of massive quantities of CH_4 that may well govern its relative importance on global climate. Without knowledge of what may constitute a major gas hydrate deposit, however, the possible impact of these effects cannot be prudently evaluated. The potential of gas hydrates and attendant processes to act as an important geological agent in transporting sediment and controlling continental slope and rise morphology depends on estimates of local hydrate concentration. Popenoe *et al.* (1993) and Booth *et al.* (1994) presented evidence of possible associations between gas hydrates and marine landslides. Gas hydrates were a major part of a sea-floor collapse event off the south-eastern United States (Dillon *et al.*, 1998). In addition, Kayen & Lee (1991) examined possible consequences of sea-level fall on slope stability if excess pore pressures accompany hydrate dissociation. Further effects may become apparent in conjunction with general geotechnical engineering and, if the case, with extraction of gas hydrate itself. The considerations of offshore engineering, and the possible magnitude of slope failures and their significance in the geological record, cannot yet be determined.

Accordingly, there are three first-order questions regarding marine gas hydrate that constitute the essence of our research:

(1) Can gas hydrate be an energy resource?
(2) Can there be places where the quantity of gas hydrate is such that its rapid dissociation would have a consequential effect on climate?
(3) Can gas hydrate deposits be large enough to serve as triggering mechanisms for major marine landslides and sea-floor collapses?

The answers to these questions necessarily involve a fundamental characterization of the

BOOTH J. S., WINTERS W. J., DILLON W. P., CLENNELL M. B. & ROWE M. M. 1998. Major occurrences and reservoir concepts of marine clathrate hydrates: implications of field evidence. *In:* HENRIET, J.-P. & MIENERT, J. (eds) *Gas Hydrates: Relevance to World Margin Stability and Climate Change*. Geological Society, London, Special Publications, **137**, 113–127.

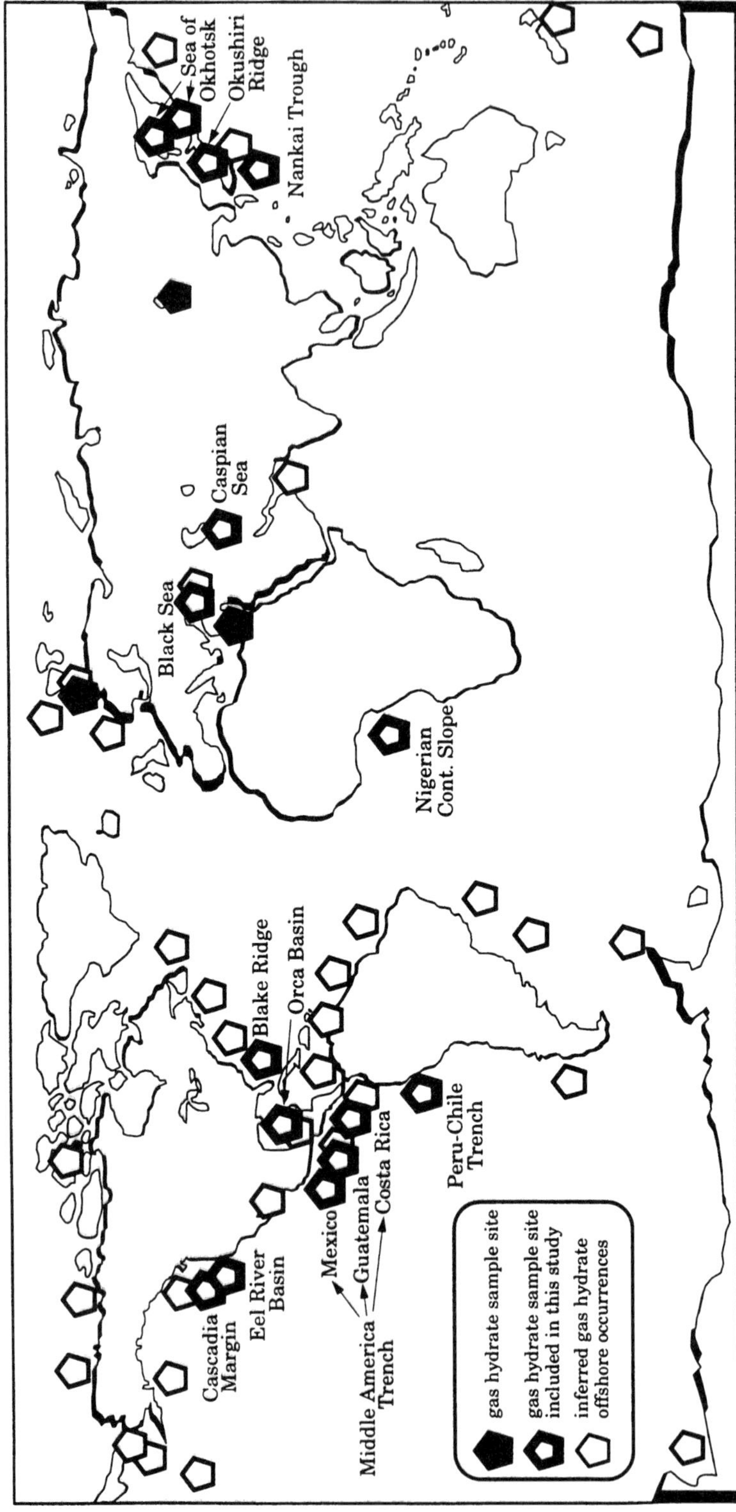

Fig. 1. World-wide distribution of confirmed or inferred offshore areas of gas hydrate-bearing sediments. Most areas contain more than one sample site. References to sites not specifically mentioned in the text may be found in Booth *et al.* (1996).

quintessential hydrate-rich sediment. Attempts have not yet been made to search for the most impressive hydrate 'reservoirs'; i.e. the thickest, most concentrated zones of clathrate hydrate that may have resource potential, may be vulnerable to sudden breakdown and release of massive quantities of CH_4 to the atmosphere, or may harbour enough gas to cause large-scale slope failures and mass movements.

A database of marine gas hydrates samples compiled by Booth *et al.* (1996) permits analysis of the gamut of natural occurrences and the characteristics associated with each. The data represent Deep Sea Drilling Program (DSDP) and Ocean Drilling Project (ODP) downhole samples, and samples from sea-floor gas hydrates (see Booth *et al.* (1996) for a complete list of references). The database comprises 15 areas (Fig. 1) and represents approximately 100 samples of gas hydrate. Because it is based solely on field samples and thus does not include inferred occurrences, it is conservative. It is also biased because of the nature of site selection, its data are often clustered (e.g. 20 of 28 downhole samples are from the Middle America Trench region) and somewhat qualitative, and the few data incorporated within it may not constitute a valid sampling of the marine gas hydrate population. Despite these limitations, the database serves as a starting point toward understanding the environments and processes that lead to the more concentrated and/or conspicuous deposits of marine gas hydrates. It reveals some plausible tendencies and associations, and permits preliminary characterization. In turn, inferences about geological settings, geological processes and gas hydrate host sediment properties that favour concentration of gas hydrates and development of thick zones of hydrate-bearing sediment may be deduced.

Summary of pertinent marine gas hydrate characteristics

Prominent samples, zones and ranges

Although the size of most pure gas hydrate samples is expressed in terms of millimetres or centimetres, a sample collected in the vicinity of the Middle America Trench off Guatemala (DSDP Site 570) was 1.05 m thick and may have come from a section of pure gas hydrate that is as much as 3–4 m thick (Shipboard Scientific Party Leg 84 1985). Also, a 14 cm thick sample of pure gas hydrate was taken from ODP Site 997 in the Blake Ridge area (off the south-eastern United States) (Shipboard Scientific Party Leg 164 1996*b*). A 9.5 m thick bed of coarse sand completely cemented by gas hydrate (i.e. 100% occupancy of pores) was also discovered off Guatemala (Shipboard Scientific Party Leg 66 1982*b*) and two other, similar, sub-seabed gas hydrate-cemented sands were cored that were about 0.5 m thick.

Individual samples often constitute a part of thicker zones of gas hydrate-bearing strata, where a zone is a unit in which there is a conspicuous and essentially uninterrupted presence of gas hydrate. Seven zones identified from DSDP and ODP data are greater than 1 m thick and four zones are greater than 10 m thick. Several sea-floor sites are more than 1 m thick as well. Gas hydrate zones vary in thickness up to as much as 30 m (zone in Okushiri Ridge, Sea of Japan (Shipboard Scientific Party Leg 127 1990). Other zones more than 10 m thick are located in Orca Basin (Gulf of Mexico) – 20 m, the Cascadia margin (Pacific–US) – 17 m and the Middle America Trench region (Pacific–Guatemala) – 15 m. A zone about 33 m thick may exist at the last site listed (Shipboard Scientific Party Leg 84 1985).

'Range' is the sub-seabed depth over which clathrate occurs. Often, this is the vertical distance spanned by multiple zones at one site. Near the Middle America Trench off Guatemala (DSDP Site 570) six zones of hydrate exist from 192 to 338 mbsf (146 m range) (Shipboard Scientific Party Leg 84 1985). Off Mexico in the same region (DSDP Site 490) four zones are present from 140 to 364 mbsf (224 m range) (Shipboard Scientific Party Leg 66 1982*a*).

The relative frequencies of occurrence of these three levels of hierarchy for different size/thickness categories are shown in Fig. 2. The inference is that grains, or thin veins or laminae, of pure gas hydrate may be common in many zones but that typically they may be only a minor constituent (small percentage of total sediment volume) of the thicker zones in which they occur. Some zones may show substantial concentrations, however. Concentrations of clathrate tend to be in zones less than 10 m thick, although the inset (Fig. 2) shows that more than a quarter of the zones and most of the ranges have greater thicknesses.

Tendencies in spatial distribution

The primary controls of hydrate occurrence at a given site in the marine environment are: (1) an ample source of methane (references to a gas herein mean methane gas; marine hydrates are typically ≈99% methane) which can exist

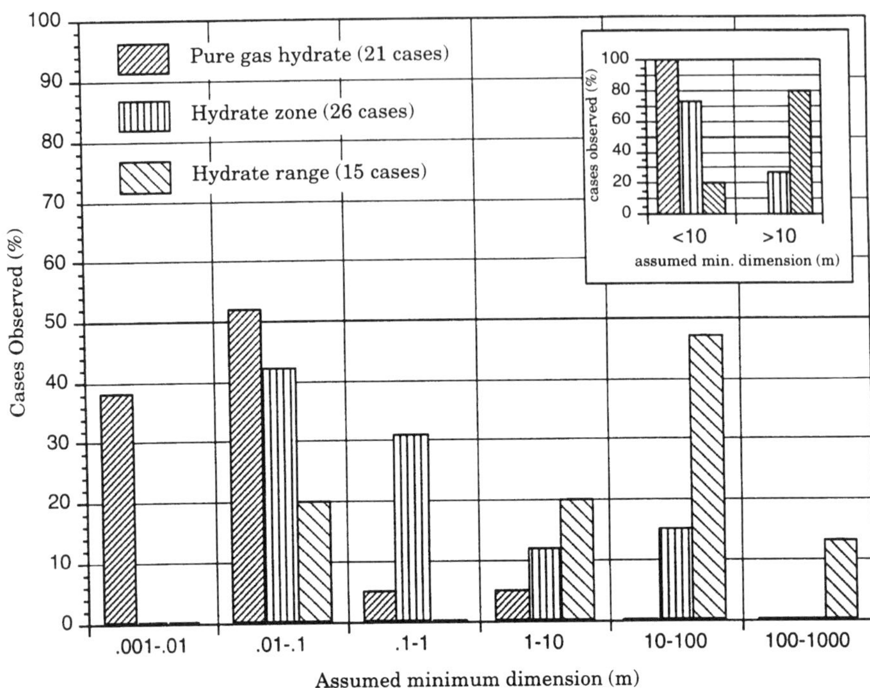

Fig. 2. Frequency distribution of presumed minimum dimension of pure gas hydrate samples, thickness of gas hydrate-bearing zones and vertical range at a sample site over which gas hydrate occurs (may include multiple zones). Inset shows the frequency of occurrence of each category using 10 m as the sorting criterion.

almost anywhere where organic matter and suitable reducing conditions or sufficient heat are juxtaposed (Demaison & Moore 1980; Hunt 1995), along with appropriate pathways for fluid migration; and (2) a specific range of pressure and temperature conditions (generally >5 MPa (≈500 m water depth) and <25°C) within which gas hydrates are stable; i.e. a gas hydrate stability zone (GHSZ). Such combinations of pressure and temperature are pervasive in world ocean sediments to sub-seabed depths of 100 to perhaps 1000 m or more, depending upon the geothermal gradient. Accordingly, gas hydrate sites are geographically widespread but not ubiquitous. Continental margins or areas near land masses, which tend to develop thick sedimentary sections that are relatively rich in organic matter, favour hydrate accumulation (Fig. 1).

In addition to the bias in global surface distribution, there is a vertical bias imposed by the phase boundary: that is, with reference to Fig. 3a, even at extremely low geothermal gradients the rise in temperature with increasing sub-bottom depth will ultimately yield a P–T combination that precludes gas hydrate formation. Because this temperature is associated with a sub-bottom depth, this depth is the base of the GHSZ. The position of this 'floor' establishes a site's absolute potential to develop a 'rich' gas hydrate deposit because it defines the vertical range over which a hydrate may exist in the sediment column. A general model asserts that clathrates form when an ample supply of gas has migrated upward from some source, encountered the phase boundary (base of the GHSZ), and been encaged there. This implies that gas hydrates occur in a somewhat narrow band at or proximal to the base of the GHSZ.

Most gas hydrate samples taken from DSDP and ODP drill holes show that *in situ* they were situated well above the base of the GHSZ (Fig. 3a). No samples were found below the calculated position of this base at any site. The average position of a sample was approximately 300 m above the base, and about three-quarters of the samples came from more than 100 m above the base's assumed position (Fig. 3b). However, drilling to the bottom simulating reflector (BSR), which may indicate the presence of a gas hydrate layer at the base of the stability zone, was prohibited in DSDP operations as a safety precaution.

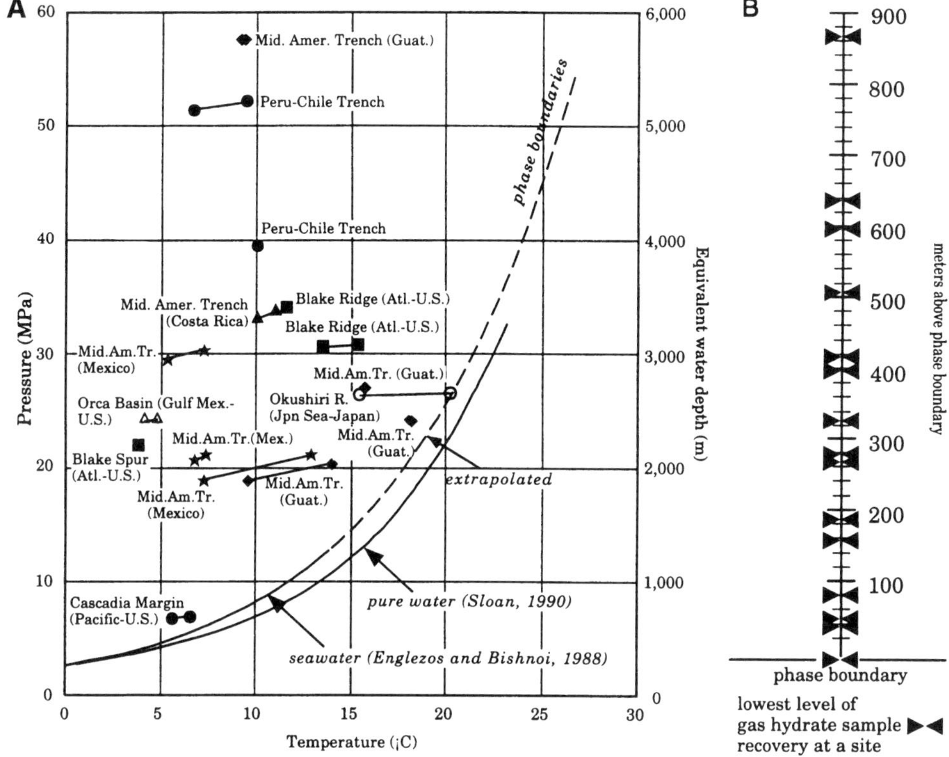

Fig. 3. (**A**) Approximate positions of gas hydrate samples with respect to the methane clathrate hydrate phase boundary. Two symbols connected by a line indicate the *in situ P–T* range from which gas hydrate samples were recovered from one site. Only borehole data were plotted. (**B**) Difference between borehole gas hydrate sample recovery depth and calculated depth of the regional phase boundary.

Borehole sampling was thus skewed away from the boundary, and this alone would tend to mask the frequency of hydrate occurrence at that boundary if there were such an association. This not withstanding, at five of six gas hydrate-bearing sites, in which corers penetrated the assumed depth of the base of the GHSZ, there was no evidence of gas hydrates. In addition, an analysis of sites based on standardizing geothermal gradients (i.e. comparing distances of gas hydrate occurrence above the base of the GHSZ if all sites had the same geothermal gradient) showed that there was no obvious relationship between the position of hydrate zones and the base of the GHSZ. Not only does the base of the GHSZ make an unreliable 'containment' zone, but it does not necessarily make a good criterion for locating hydrate-bearing sediments either.

Analysis of the gas hydrate zone and the range of vertical distribution shows no conspicuous tendencies. Of the hydrate-bearing zones thicker than 1 m and the ranges of occurrence greater than 10 m, there may be a slight tendency for them to exist relatively high (toward the sea floor) in the sedimentary column.

Associations with structural features

There is a clear association between gas hydrate occurrence and fault zones, as well as other tectonically related features, as has been observed by Hyndman & Davis (1992), Soloviev & Ginsburg (1994), and others. Approximately three-quarters of the gas hydrate-bearing sites in the database (including both seabed and sub-seabed sites) are within or proximal to features that may promote upward migration of gas or gas-rich fluids. All but one of the 15 drill hole sites in which hydrates were present lie within tectonically active continental margin environments. The Blake Ridge off the south-eastern United States is the only site in a passive

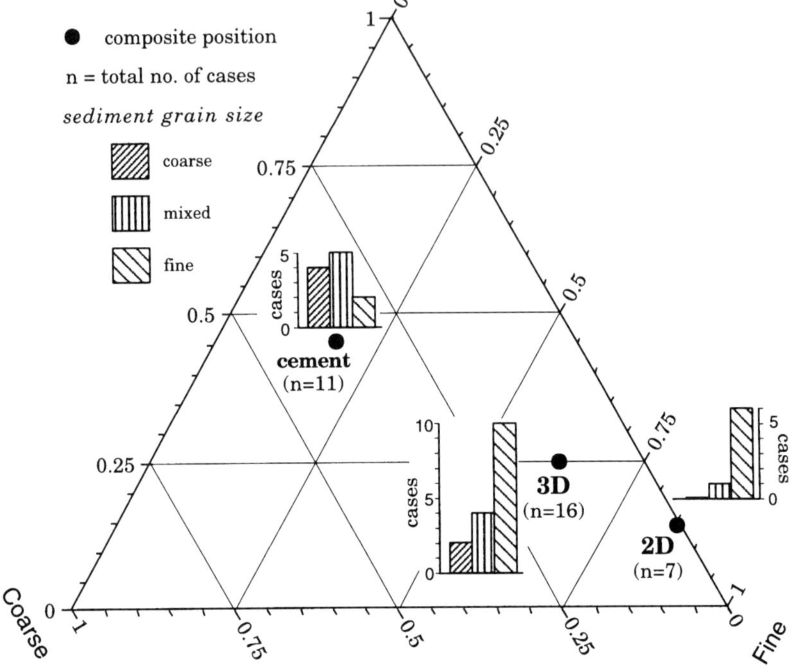

Fig. 4. Composite relative frequency of occurrence of different gas hydrate forms (habit) with respect to sediment type (borehole samples only). Insert shows the specific grain size associations for each gas hydrate habit. '2D' refers to planar form hydrates (e.g. laminae, lenses, layers); '3D' refers to hydrates described as grains, particles, blebs, nodules, massive, etc.; 'cement' are hydrates that act as a matrix for included sediment particles. Sediment types are: coarse – sand or larger grain sizes (includes volcanic ash); fine – silt and clay grain sizes (includes fine ash); mixed – both coarse and fine sediment mixed or juxtaposed where the hydrate sample was found.

margin. It is nonetheless characterized by the presence of extensive fault zones (Dillon *et al.* 1994; Rowe & Gettrust 1994). The presence of vents, seeps and mud volcanoes near many sea-floor gas hydrate sites (e.g. Brooks *et al.* 1986; Soloviev & Ginsburg 1994; Shipboard Scientific Party Leg 164 1996*b*; Basov *et al.* 1996) verifies that gas(es) have moved up conduits and locally breached the sea floor at these locales.

Associations with sediment properties

Natural gas hydrates typically exist as individual grains or particles disseminated throughout the sediment, but also commonly exist as cements, nodules and as laminae or layers. Figure 4 shows the relationships between habit and grain size for sub-seabed samples. The drill hole data show that no particular category (habit) was dominant. As indicated by Fig. 4, almost all of the two-dimensional samples (layers, laminae, etc.) were associated with fine sediment, either intrinsically or in fractures. Three-dimensional samples (granules, nodules, etc.) were numerous in fine sediments as well, but were also identified in sediment of more than one basic grain size. Clathrate cements clearly are more associated with coarser material. The two largest pure samples (1.05 m apparent layer from DSDP Site 570 and 14 cm thick borehole sample from ODP Site 997) formed at the contact between an indurated dolomite and a mudstone, and in a silty clay, respectively. As stated, a sand layer 9.5 m thick was cemented with clathrate hydrate in DSDP Site 498.

Porosities determined proximal to the hydrate samples in drill holes ranged from just above 40% at 364 mbsf (DSDP Site 491 in the Middle America Trench (Shipboard Scientific Party Leg 66 1982*b*)) to nearly 75% at 161 mbsf (ODP Site 688 in the Peru–Chile Trench (Shipboard Scientific Party Leg 112 1988)). An average porosity for sediments above and below gas hydrate zones is about 55%. The sediments from which the two largest samples of pure gas hydrate were recovered had porosities slightly higher than the average. Because the data are too few and there are other shortcomings in the database, a rigorous statistical analysis of porosity values above and below gas hydrate zones was not attempted. Nonetheless, DSDP/ODP data show that thicker gas hydrate zones are associated with more porous sediment. Five of the seven thickest zones (1.5–29.3 m thick) have the five highest porosities among the borehole gas hydrate zones for which there are data. Values are close to or in excess of 60%. The porosity of the sixth zone (20–40 mbsf in Orca Basin, Gulf of Mexico) is estimated to be similar. Porosities were not determined by the site investigators for the remaining (seventh) zone, which is the 9.5 m thick zone discovered at DSDP Site 498 (Middle America Trench). Figure 5 is a plot of zone thickness vs porosity for zones >0.1 m thick.

Given that in two of the three outliers the host sediment are sands, which characteristically have much lower porosities than unconsolidated cohesive sediment, the apparent correlation between porosity/permeability and hydrate concentration is noteworthy. However, because this apparent relationship may represent an effect (of gas expansion in the sediment once on deck) as well as a cause, and because of other factors, it is considered more of a justification for research than it is a prospecting criterion.

Field data and gas hydrate reservoir concepts

Potential gas hydrate 'reservoirs' are defined by their storage capacity and capability to concentrate clathrates, as well as being within the GHSZ. Whereas petroleum traps have to leak to function as a trap – a petroleum trap must allow water to be expelled (Roberts 1980) – water must be present for a gas hydrate to form. Also, gas hydrate reservoirs can be self-sealing. The filling of sediment interstices with clathrate can prevent further migration of the source fluid and thus set up further production of clathrate hydrates. We focus on three main factors which bear on reservoir potential: geological setting, geological processes and host sediment properties. Discussion is limited to

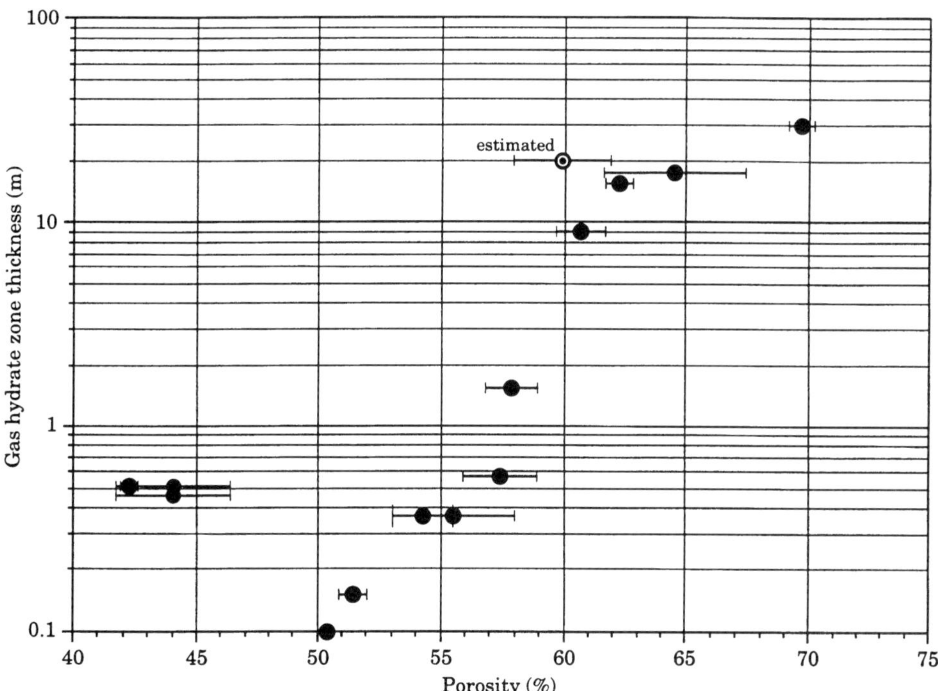

Fig. 5. Plot of porosity vs zone thickness for zones ≥0.1 m thick. The range shown is based on porosities measured above and below, but not within, the gas hydrate-bearing zone. Solid circles are the range midpoints. The estimated porosity range is for the Orca Basin site (Gulf of Mexico). The gas hydrate zone at this site is 20–40 mbsf. Not included: porosities associated with 9.5 m thick zone of gas hydrate-cemented ash in the Middle America Trench off Guatemala. Porosities were not determined near this zone.

concepts for which data can be presented in the context of existing gas hydrate sites.

Geological setting

This factor includes water depth and geothermal gradient, which together are the primary controls of the absolute thickness of the GHSZ, and hence, without known variability in lateral dimensions, the absolute storage capacity of the site. The thickness of the GHSZ is consequential because it sets the overall site potential for hydrate concentration and may affect the probability of including more favourable host sediment within the GHSZ. Figure 6 is a plot of the relationship between water depth and thickness of the GHSZ for specific geothermal gradients. As shown, the GHSZ thickens with increasing water depth and decreasing geothermal gradient. The DSDP/ODP sites are also plotted in Fig. 6 to show the thicknesses of their GHSZ and so that site comparison is possible. The sites with the two largest ranges and the most gas hydrate zones (DSDP sites 490 and 570) have GHSZ thicknesses of more than 500 m. Both sites are also relatively shallow (<2000 m), but, given the non-linear (approximately exponential) relationship between *P* and *T* that sets the phase boundary, they are deep enough to avoid significant constraints on the thickness of their GHSZs. Figure 6 also shows that although a greater GHSZ thickness may be inherently more desirable, it may not be required for a substantial gas hydrate deposit to accumulate. The Okushiri Ridge and Cascadia margin sites have two of the thicker gas hydrate zones described, but have the two thinnest GHSZs.

Fault zones, joint and fracture systems, and other features that are potential pathways for upward migration of methane or other gases can be an important facet of a possible gas hydrate 'reservoir', as has been previously recognized in gas hydrate field studies (e.g. Brooks *et al.* 1986, 1994; Soloviev & Ginsburg 1994). Such conduits additionally provide a means for transporting methane considerable distances into the GHSZ. In some of these systems, fluids can be injected into more horizontal pathways as well, but such flow must be hydraulically driven to counter the strong vertical component of movement associated with free methane due to its buoyancy. Data from field studies can support these possibilities, as is the case historically for the numerous studies on secondary oil and gas migration (e.g. see Roberts & Cordell 1980; England & Fleet 1991; Hunt 1995). Rowe & Gettrust (1994) and Dillon *et al.* (1994) present high-resolution seismic profiles from the Blake Ridge gas hydrate area that show indications of gas hydrate penetration of fault zones. In the Cascadia margin gas hydrate area, Zwart *et al.* (1996) have thermal evidence that fault zones are acting as conduits for warm fluids. Seabed sites are direct evidence of the migration of gas and may well reflect processes of clathrate formation in zones penetrated by these conduits before they reach the sea floor (Soloviev & Ginsburg 1994; Ginsburg & Soloviev 1997). Veins, ash layers (sands) and silty sands are prevalent in the Middle America Trench area, and their connection to the widespread occurrence of gas hydrate there has been examined (Taylor & Bryant 1985).

There are several other geological features that affect hydrate accumulation within a GHSZ, but they are not inherently a factor in hydrate occurrence. Among these are diapirs. During piercement, diapirs often generate fault networks and cause radial fracturing. Sea-floor gas hydrates in the Gulf of Mexico (Brooks *et al.* 1986), the Blake Ridge area (Paull *et al.* 1995), and the Black, Caspian and Okhotsk seas (Soloviev & Ginsberg 1994) are linked to diapirs.

Geological processes

That geological processes can shape a potential gas hydrate reservoir is evident by the explanations required for why gas hydrates occupy positions well above their present regional phase boundary. If phase boundary gas hydrates (gas hydrates at or proximal to the present base of the GHSZ) are truly subordinate in frequency to internal gas hydrates (gas hydrates that are well above the present base of the stability zone), why is this latter category dominant? Four possible types of explanations for an internal gas hydrate are: (1) it formed as a consequence of localized pressure, temperature or pore water chemistry effects, rather than as a consequence of regional geothermal and hydrostatic conditions; (2) it formed by site-specific gas enrichment, either in the absolute or relative sense; (3) it formed at a time when the base of the GHSZ was shallower; it is relict; and/or (4) it formed elsewhere and was transported to its present location by mass movement processes; it is allochthonous.

A comprehensive discussion of the possibilities associated with each of these types is beyond the scope of this paper. Rather, two separate scenarios based on these possibilities

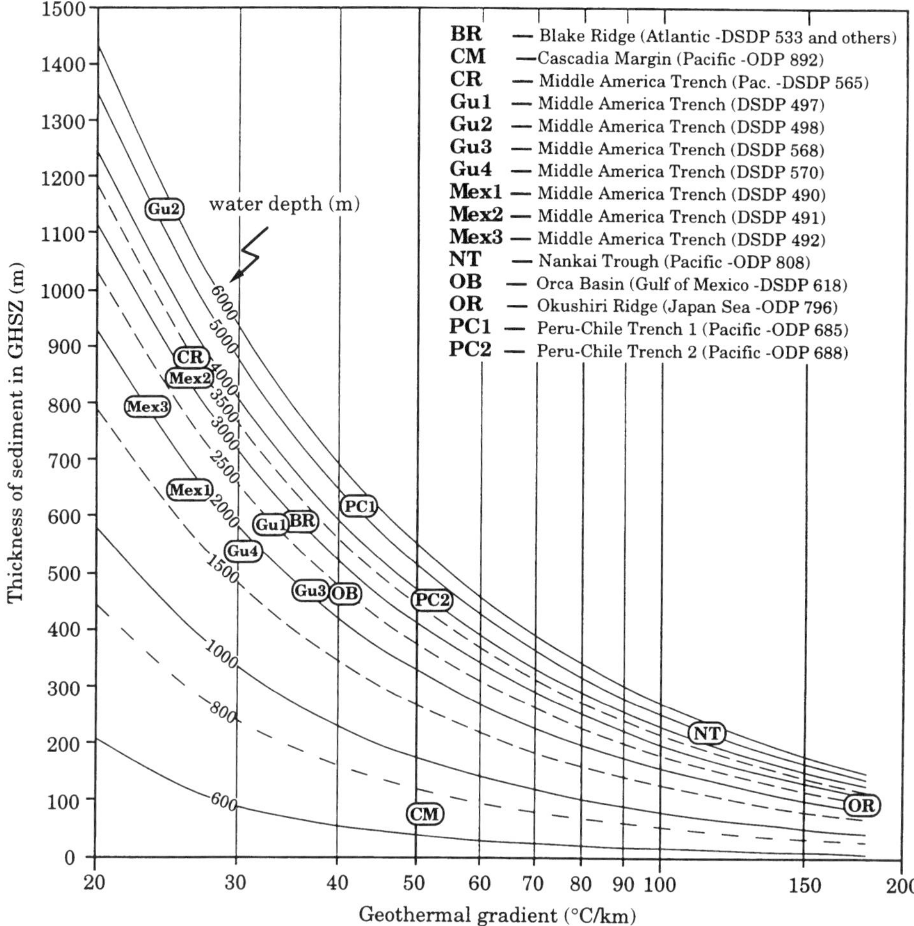

Fig. 6. Thickness of sediment in the gas hydrate stability zone (sediment height above the phase boundary) as a function of pressure (water depth + sub-bottom depth, assuming normal consolidation) and temperature (geothermal gradient). Thicknesses of GHSZs and DSDP/ODP sites are plotted for comparison.

will be examined to show their relevance in a potential hydrate enrichment process. The first one involves Type 1 ('localized phase boundary') internal hydrates. Field evidence indicates that anomalously warm, presumably methane-rich fluids can rise in conduits in gas hydrate areas (Zwart *et al.* 1996). These fluids ultimately produce clathrates when they achieve thermal equilibrium with respect to regional isotherms; that is, they create their own localized, moving phase boundary front until they dissipate their excess heat. If this fails to happen, as near Paramushir Island in the Sea of Okhotsk, the gas escapes into the overlying water column (Basov *et al.* 1996). The Type 3 'relict base of the GHSZ' scenario applies to, for example, the last major eustatic event, subsidence or a decrease in the geothermal gradient (e.g. due to moving away from a hot spot or spreading centre). The effect of such processes could make a pre-existing hydrate the top of an ever-thickening zone as the base of the GHSZ moved lower in the section, or it could 'strand' a pre-existing hydrate zone with respect to the new regional position of the GHSZ base. For predicting sites that may have more potential to bear hydrate-rich sediment, it is again noteworthy that the P–T relationship approximates an exponential curve. A consequence of this is that a eustatic rise or seabed subsidence, which would both cause an increase in pressure at a given sub-surface depth, could result in a significant thickening of a shallow-water GHSZ, but only a slight thickening in deep-water GHSZ

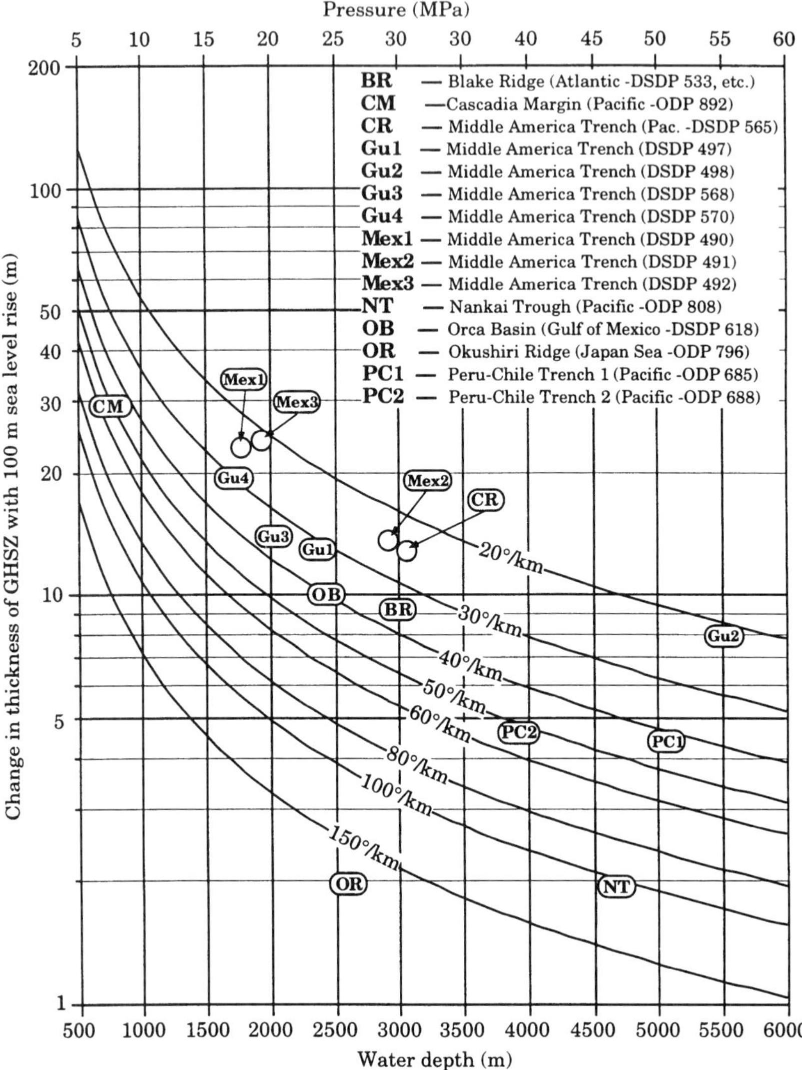

Fig. 7. Increase in GHSZ thickness with increase in pressure equivalent to a 100 m sea-level rise. Deeper gas hydrate zones or those associated with high geothermal gradients are marginally affected; shallow sites can show a significant increase.

(Fig. 7). Conversely, a decrease in the geothermal gradient could substantially increase the thickness of a hydrate in deep water, but have much less an effect on a shallow-water hydrate (Fig. 7). A plot of the ODP data in Fig. 6 shows which sites have the greater potential to change their GHSZ thicknesses based on Type 3 possibilities. The GHSZ on the Cascadia Margin, for example, would have almost doubled its thickness in response to last major eustatic event; conversely, the Japanese sites (Okushiri Ridge and Nankai Trough) and the Peru sites would have been virtually unaffected.

The potential reservoir thickness of a site can be analysed in a similar way for depositional and erosional processes. Paull *et al.* (1994) show how the phase boundary alone could be a zone of increasing concentration of gas hydrate through depositional processes. Similarly, slope failures and mass movements may also induce changes in reservoir thickness, even if only temporarily.

Geological processes not only promote access of gas to the entire GHSZ, but can alter the thickness of the potential reservoir. Tectonism, besides its possible role in establishing favourable attributes in the geological environment and in aiding gases to bypass the base of the GHSZ or in enlarging the GHSZ, also can improve the reservoir potential of a site by creating forces that actively inject gases into it. Hyndman & Davis (1992) propose that large-scale compression associated with tectonically induced thickening of sediment wedges or subduction can expel fluids into the upper parts of the sediment section. They infer that convergent margins are not only typified by the presence of pathways for fluid movement, but provide mechanisms to move the fluids as well.

Host sediment

If we assume a basic proportionality between porosity and permeability (the Kozeny–Carman equation), then higher porosity would also suggest a greater capability to transport fluids. Simplistically, high porosity and permeability favour the accumulation of clathrates. Because permeability tends to be anisotropic (more permeable normal to the direction of compaction) in consolidating sediments (e.g. Vasseur *et al.* 1995), more porous beds may also favour clathrate accumulation through fluid pathways that may trend toward horizontal. Extrapolation of the findings of Clennell *et al.* (1995) infers that when a pore throat (capillary) diameter fails to exceed a threshold

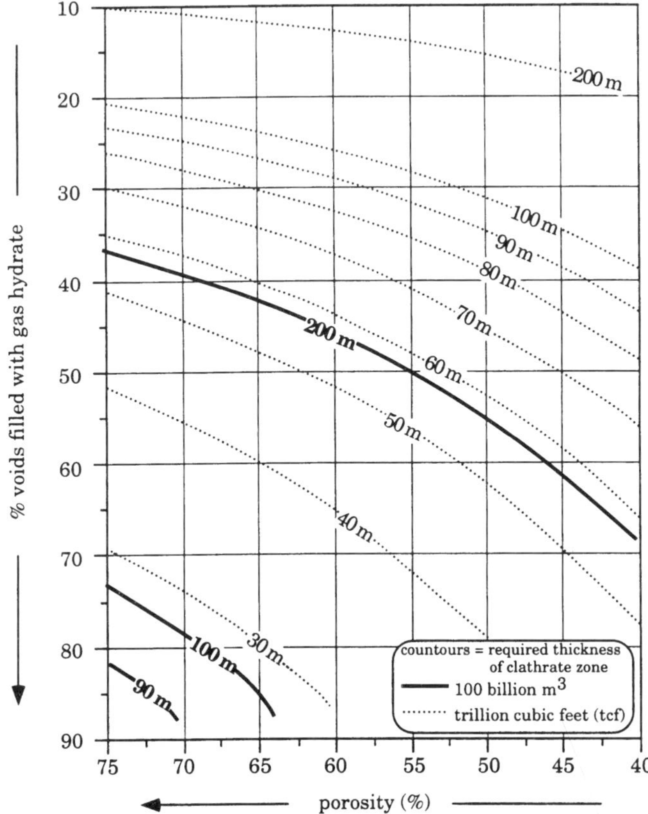

Fig. 8. Thickness of the zone required to yield 100 billion cubic metres or a trillion cubic feet (tcf) of methane for different combinations of porosity and percentage of voids filled with gas hydrate. All values of porosity, percentage of voids filled and possible thicknesses are within boundaries established by field data. Calculations are based on a cylindrical gas hydrate zone with a radius of 2000 m. For a typical porosity associated with gas hydrates (55%), if 50% of the voids are filled, the gas hydrate would have to be present over a vertical range of 200 m to yield 100 bcm, or about 57 m thick to yield a tcf of natural gas.

Table 1. *Gas hydrate zone facts*

Area	Site identification	Zone thickness (m)	Site near land mass?	Porosity $\gtrsim 60\%$?*	Proximal to phase boundary?	Tectonic features present?	Sand present?	Habit
Okushiri Ridge	Sea of Japan ODP Site 796	29.3	yes	yes	yes	yes	yes	mixed
Orca Basin	Gulf of Mexico DSDP Site 618	≈20	yes	yes†	no	yes	yes	grains
Cascadia Margin	Pacific–US ODP Site 892	≈17	yes	yes	no	yes	no‡	mixed
Middle American Trench	Pacific–Guatemala DSDP Site 570 (zone b)	≈15	yes	yes	no	yes	no	discrete
								10 m
Middle American Trench	Pacific–Guatemala DSDP Site 489	9.5	yes	–	no	yes	yes	cement
Middle American Trench	Pacific–Guatemala DSDP Site 570 (zone c)	≈9	yes	yes	no	yes	no	discrete
Middle American Trench	Pacific–Guatemala DSDP Site 568	1.5	yes	yes	no	no	yes	discrete
								1 m
Blake Ridge	Atlantic–US ODP Site 997	0.55	yes	yes	no	yes	no	discrete
Middle American Trench	Pacific–Mexico DSDP Site 490 (zone d)	{ 0.5	yes	no	no	no	yes	cement
Middle American Trench	Pacific–Mexico DSDP Site 491 (zone c)	0.5 }	yes	no	no	yes	no	mixed
Middle American Trench	Pacific–Mexico DSDP Site 491 (zone b)	0.45	yes	no	no	yes	yes	cement
Middle American Trench	Pacific–Mexico DSDP Site 490 (zone c)	{ 0.35	yes	no	no	no	yes	cement
Middle American Trench	Pacific–Mexico DSDP Site 490 (zone b)	0.35 }	yes	no	no	no	yes	cement
Middle American Trench	Pacific–Mexico DSDP Site 492 (zone a)	0.15	yes	no	no	no	yes	cement
Middle American Trench	Pacific–Mexico DSDP Site 491 (zone a)	0.1	yes	no	no	yes	no	discrete
								0.1 m

* Based on values above and below zone; † estimated; ‡ sand rare.
Habits: mixed – two or more of grains, 2D, 3D, cement; grains – ≈equant forms ≤1 cm; discrete – 2D or 3D forms with mic. $D > 1$ cm; cement – matrix for sediment particles.

size (perhaps less than 1 μm), intergranular hydrate growth can be greatly inhibited. Clennell *et al.* (1995) argue that under these conditions growth is a function of the ability of the clathrate to displace the surrounding sediment grains that form the pore space. Constraints on growth are also discussed by Harrison & Curiale (1982), who suggest that a pore diameter of 100 Å may be a limiting size to hydrate formation. We speculate that if there is a limiting capillary diameter with respect to hydrate growth, and if there is a limit to the capability of a nucleated hydrate to displace sediment, then gas hydrate growth would be precluded in some sediments, regardless of gas supply, in all but the most severe hydraulic gradients.

Pure sand layers facilitate fluid flow but generally have lower porosities than fine-grained sediments in the upper few hundred metres of a section; i.e. above the position of most regional phase boundaries. These sand layers also tend to be thin, reducing their overall potential to hold large volumes of gas hydrate compared to the much more prevalent cohesive sediment. The 9.5 m thick cemented sand (DSDP 498) was the thickest zone of sand discovered and was unique among the zones thicker than 1 m.

Work by Clennell *et al.* (1995) also lays a thermodynamic foundation for pore-geometry control of gas hydrate growth behaviour and infers that nodules or interstratal occurrences of gas hydrate would tend to form in finer sediments (typified by smaller pores), whereas intergranular growth (cements) may be more likely in sands or sediments with larger pore sizes. We also speculate, therefore, that more massive, pure gas hydrates would tend to be located either relatively deep in a sediment column or in overconsolidated sediment. Grains of gas hydrates or gas hydrate cement could occur at any level within the same sediment column.

Implications with respect to massive gas hydrate occurrence

Based on the field evidence and reservoir concepts, what could be expected in terms of gas hydrate concentration and what would the attributes of the site be? Estimates of the potential of a gas hydrate to concentrate to the degree that such a deposit would achieve the label 'significant' would be both tenuous and variable, depending upon whether in the context of energy resource, global climate or slope stability. Despite this, a hypothetical scenario provides a starting point for discussion. Figure 8 is an attempt to construct a framework for scenarios that may be possible. It shows, for different combinations of porosity and gas hydrate infilling of the pores (% voids filled), the thickness of the hydrate zone that would yield 100 billion cubic metres (100 bcm) or trillion cubic feet (tcf) of CH_4. The boundary conditions were set using borehole data from gas hydrate sample sites. Porosities range from 40 to 75%. Zones ranged to 30 m or more in thickness and ranges exceeded 200 m. Infilling ranged to apparently 100% in some zones and some samples were pure gas hydrate. The model used for Fig. 8 assumes a borehole site is the axis of a cylinder of gas hydrate-bearing sediment that is 2000 m in radius. This radius is well within the lateral extent of gas hydrate occurrence implied by site clusters in several regions (Blake Ridge, Middle America Trench regions off Guatemala and off Mexico).

Regardless of whether a gas hydrate deposit could attain 'significant' volumes, a profile of the more impressive deposits yet to be discovered may be foreshadowed in Table 1. It is a summary of common attributes among the more prominent gas hydrate zones. Most, particularly those greater than 10 m thick, possess many of the attributes of the conceptual 'gas hydrate reservoir'. All of the zones greater than 10 m thick are located within or proximal to features that may promote fluid migration and are in tectonically active areas. Interestingly, sites not associated with tectonic features have some percentage of sand in the clathrate-bearing zone. There are no gas hydrates among the top four that are characterized as cements. All but one zone are relatively high in the sedimentary column with respect to their GHSZ base.

In considering the types and amount of data available, Table 1, rather than taken as research findings, is meant to be viewed as a source of research directions: it shows common characteristics, not innate criteria.

Many discussions at the workshop helped shape this report; it evolved from them. We therefore gratefully acknowledge the contributions of the participants. I. A. Pecher and E. D. Sloan provided specific helpful comments and discussion. We especially thank J.-P. Henriet for his support and patience. This work was a part of the US Geological Survey marine gas hydrates research programme.

References

Basov, E., van Weering, T., Gaedike, C. *et al.* 1996. Seismic facies and specific character of the bottom simulating reflector on the western margin of Paramushir Island, Sea of Okhotsk. *Geo-Marine Letters*, **16**, 297–304.

BOOTH, J., ROWE, M. & FISCHER, K. 1996. *Offshore Gas Hydrate Sample Database*. US Geological Survey Open File Report, **96**, 272.

——, WINTERS, W. & DILLON W. 1994. Circumstantial evidence of gas hydrate and slope failure associations on the U.S. Atlantic continental margin. *In*: SLOAN, E., JR., HAPPEL, J. & HNATOW, M. (eds) *International Conference on Natural Gas Hydrates*. Annals of the New York Academy of Sciences, **715**, 487–489.

BROOKS, J. M., COX, H., BRYANT, W. *et al.* 1986. Association of gas hydrates and oil seepage in the Gulf of Mexico. *Organic Geochemistry*, **10**, 221–234.

BROOKS, J., ANDERSON, A., SASSEN, R. *et al.* 1994. Hydrate occurrence in shallow subsurface cores from continental slope sediments. *In*: SLOAN, E., JR., HAPPEL, J. & HNATOW, M. (eds) *International Conference on Natural Gas Hydrates*. Annals of the New York Academy of Sciences, **715**, 381–391.

CLENNELL, M., HOVLAND, M., LYSNE, D. & BOOTH, J. 1995. Role of capillary forces, coupled flows and sediment–water depletion in the habitat of gas hydrate. (Abstract.). *EOS, Transactions of the American Geophysical Union*, **76** (17 Supplement), 164–165.

DEMAISON, G. J. & MOORE, G. T. 1980. Anoxic environments and oil source bed genesis. *Organic Geochemistry*, **2**, 9–31.

DILLON, W. P., DANFORTH, W. W., HUTCHINSON, D. R., DRURY, R. M., TAYLOR, M. H. & BOOTH, J. S. 1998. Evidence for faulting related to dissociation of gas hydrate and release of methane off the south-eastern United States. *This volume*.

——, FEHLHABER, K., COLEMAN, D. *et al.* 1995. *Maps Showing Gas Hydrate Distribution Off the East Coast of the United States*. US Geological Survey Miscellaneous Field Investigations Map, **MF2268**.

——, LEE, M. & COLEMAN, D. 1994. Identification of marine hydrates *in situ* and their distribution off the Atlantic coast of the United States. *In*: SLOAN, E., JR., HAPPEL, J. & ANATOW, M. (eds) *International Conference on Natural Gas Hydrates*. Annals of the New York Academy of Sciences, **715**, 364–380.

ENGLAND, W. A. & FLEET, A. J. (eds) 1991. *Petroleum Migration*. Geological Society, London, Special Publications, **59**.

ENGLEZOS, P. & BISHNOI, P. R. 1988. Prediction of gas hydrate formation conditions in aqueous electrolyte solutions. *American Institute of Chemical Engineers Journal*, **34**, 1718–1721.

—— & HATZIKIRIAKOS, S. G. 1994. Environmental aspects of clathrate hydrates. *In*: SLOAN, E., JR., HAPPEL, J. & HNATOW, M. (eds) *International Conference on Natural Gas Hydrates*. Annals of the New York Academy of Sciences, **715**, 270–282.

GINSBURG, G. & SOLOVIEV, V. 1997. Methane migration within the submarine gas-hydrate stability zone under deep-water conditions. *Marine Geology*, **137**, 49–57.

——, SOLOVIEV, V., CRANSTON, R., LORENSON, T. & KVENVOLDEN, K. 1993. Gas hydrates from the continental slope, offshore Sakhalin Island, Okhotsk Sea. *Geo-Marine Letters*, **13**, 41–48.

HARRISON, W. & CURIALE, J. 1982. Gas hydrates in sediments of holes 497 and 498A. *In*: AUBOUIN, J., VON HUENE, R. *et al.* (eds) *Initial Reports of the Deep Sea Drilling Project*, Volume 67. US Government Printing Office, Washington, DC, 591–594.

HUNT, J. M. 1995. *Petroleum Geochemistry and Geology*. Freeman, New York.

HYNDMAN, R. & DAVIS, E. 1992. A mechanism for the formation of methane hydrate and seafloor bottom simulating reflectors by vertical fluid expulsion. *Journal of Geophysical Research*, **97**(B5), 7025–7041.

LASHOF, D. A. & AHUJA, D. R. 1990. Relative contributions of greenhouse gas emissions to global warming. *Nature*, **344**, 529–531.

KAYEN, R. E. & LEE, H. J. 1991. Pleistocene slope instability of gas hydrate-laden sediment on the Beaufort sea margin. *Marine Geotechnology*, **10**, 125–141.

NISBET, E. G. 1990. The end of the ice age. *Canadian Journal of Earth Science*, **27**, 148–157.

OKUDA, Y. 1996. Research on gas hydrates for resource assessment in relation to national drilling programme in Japan. *In: Proceedings of the 2nd International Conference on Natural Gas Hydrates*, Toulouse, France, 633–639.

PAULL, C. K., USSLER, W., III & BAROWSKI, W. S. 1994. Sources of biogenic methane to form marine gas hydrate. *In:* SLOAN, E., JR., HAPPEL, J. & HNATOW, M. (eds) *International Conference on Natural Gas Hydrates*. Annals of the New York Academy of Sciences, **715**, 392–409.

——, USSLER, W., III & DILLON, W. P. 1991. Is the extent of glaciation limited by marine gas hydrates? *Geophysical Research Letters*, **18**, 432–434.

——, SPIESS, F. N., USSLER, W., III & BOROWSKI, W. S. 1995. Methane-rich plumes on the Carolina continental rise: associations with gas hydrates. *Geology*, **23**, 89–92.

POPENOE, P., SCHMUCK, E. A. & DILLON, W. 1993. The Cape Fear Landslide: Slope failure associated with salt diapirism and gas hydrate decomposition. *In*: SCHWAB, W., LEE, H. & TWICHELL, D. (eds) *Submarine Landslides: Selected Studies in the U.S. Exclusive Economic Zone*. US Geological Survey Bulletin, **2002**, 40–53.

ROBERTS, W. H., III. 1980. Design and function of oil and gas traps. *In:* ROBERTS, W., III & CORDELL, R. (eds) *Problems of Petroleum Migration*. American Association of Petroleum Geologists, Studies in Geology, **10**, 217–240.

—— & CORDELL, R. J. (eds). 1980. *Problems of Petroleum Migration*. American Association of Petroleum Geologists, Studies in Geology, **10.**

ROWE, M. M. & GETTRUST, J. F. 1994. Methane hydrate content of Blake Outer Ridge sediments. *In*: SLOAN, E., JR., HAPPEL, J. & HNATOW, M. (eds) *International Conference on Natural Gas Hydrates*. Annals of the New York Academy of Sciences, **715**, 492–494.

SHIPBOARD SCIENTIFIC PARTY LEG 66. 1982*a*. Site 490. *In*: WATKINS, J., MOORE, J. *et al.* (eds) *Initial Reports of the Deep Sea Drilling Project*, Volume 66. US Government Printing Office, Washington, DC, 151–217.

—— 1982*b*. Site 491. *In*: WATKINS, J., MOORE, J. *et al.* (eds) *Initial Reports of the Deep Sea Drilling Project*, Volume 66. US Government Printing Office, Washington, DC, 219–287.

—— LEG 84. 1985. Site 570. *In*: VON HUENE, R., AUBOUIN, J. *et al.* (eds.) *Initial Reports of the Deep Sea Drilling Project*, Volume 67. US Government Printing Office, Washington, DC, 283–336.

—— LEG 112. 1988. Site 688. *In*: SUESS, E., VON HUENE, R. *et al.* (eds) *Proceedings of the Ocean Drilling Program, Initial Reports*, College Station, TX. Ocean Drilling Program, **112,** 873–1004.

—— LEG 127. 1990. Site 796. *In*: TAMAKE, K., PISCIOTTO, K., ALLEN, J. *et al.* (eds) *Proceedings of the Ocean Drilling Program, Initial Reports*, College Station, TX. Ocean Drilling Program, **127,** 247–322.

—— LEG 164. 1996*a*. Site 996. *In*: PAULL, C., MATSUMOTO, R., WALLACE, P. *et al.* (eds) *Proceedings of the Ocean Drilling Program, Initial Reports*. College Station, TX. Ocean Drilling Program, **164,** 241–276.

—— 1996*b*. Site 997. *In*: PAULL, C., MATSUMOTO, R.,WALLACE, P. *et al.* (eds.) *Proceedings of the Ocean Drilling Program, Initial Reports*, College Station, TX. Ocean Drilling Program, **164**, 277–336.

SLOAN, E. D., JR. 1990. *Clathrate Hydrates of Natural Gases*. Marcel Dekker, New York.

SOLOVIEV, V. & GINSBURG, G. 1994. Formation of submarine gas hydrates. *Bulletin of the Geological Society of Denmark*, **41**, 86–94.

TAYLOR, E. & BRYANT, W. 1985. Geotechnical properties of sediments from the Middle America Trench and Slope. *In*: VON HUENE, R., AUBOUIN, J. *et al.* (eds) *Initial Reports of the Deep Sea Drilling Project*, Volume 67. US Government Printing Office, Washington, DC, 745–766.

VASSEUR, G., DJERAN-MAIGRE, I., GRUNBERGER, D. *et al.* 1995. Evolution of structural and physical parameters of clays during experimental compaction. *Marine and Petroleum Geology*, **12**, 941–954.

ZWART, G., MOORE, J. & COCHRANE, G. 1996. Variations in temperature gradients identify active faults in the Oegon accretionary prism. *Earth and Planetary Science Letters*, **139**, 485–495.

Detection of gas hydrates using downhole logs

D. GOLDBERG & S. SAITO

Borehole Research Group, Lamont-Doherty Earth Observatory, RT 9W, Palasides, NY 10964, USA

Abstract: Downhole logs have proven to be critical in quantifying natural gas hydrates found in marine sediments and the seismic signature associated with free gas below. In recent drilling on the Blake Ridge, the Ocean Drilling Program recorded *in situ* velocity and resistivity logs that reveal an increasing amount of hydrate with depth in the pressure-temperature stability window. The associated increase in shear rigidity (decrease in V_p/V_s ratio) and decrease in acoustic amplitude over this interval is attributed to the cementation of sediment grains by hydrate, which may also explain the prevalent seismic blanking across the Blake Ridge. In future drilling campaigns, the use of 'logging-while-drilling' (LWD) sensors placed just above the drill bit will improve the *in situ* estimation of porosity, lateral variability and hydrocarbon reservoir potential of natural gas hydrates.

Downhole logs in sedimentary basins can reveal *in situ* properties that are difficult to measure in laboratory samples, such as those in gas hydrates and gas-rich zones which dissociate or severely alter during the process of coring. Downhole logs have been recorded previously in several gas hydrate environments (e.g. Mathews 1986; Collett 1993; Prensky 1995; Goldberg 1997). Velocity logs in particular have proven to be critical in quantifying the nature of the seismic signature associated with hydrates; velocity decreases due to free gas below the hydrate layer and the generation of a large seismic reflection (e.g. Mackay *et al.* 1994; Hyndman *et al.* 1996; ODP Leg 164 Shipboard Scientific Party 1996). Other logs, such as porosity, can be used by proxy to quantify the geophysical properties that are required for detection of natural gas hydrates using surface seismic methods. If the extent of hydrates and free gas can then be predicted from seismics, more reliable and robust estimates of the volume of gas and methane hydrate could be made with limited drilling.

Porosity is a fundamental measurement for the *in situ* investigation of gas hydrates. A major limitation in determining porosity from velocity is our poor understanding of their relationship in hydrates as well as in high porosity sediments. Because of differences in lithology and styles of hydrate formation, our knowledge of this relationship would improve significantly if field data were acquired at several, varied, study sites. The ocean drilling program has recorded downhole logs in gas hydrates at a few such locations, most recently on the Blake Ridge off the US eastern coast (ODP Leg 164 Shipboard Scientific Party 1996). Because the physical properties of the sediment play a critical role in gas migration and the stability of hydrate formation, accurate *in situ* measurements are essential. On the Blake Ridge, the logs reveal that compressional velocity and resistivity increase through the depth interval of gas hydrate stability, then decrease below due to the increased presence of free gas (ODP Leg 164 Shipboard Scientific Party 1996). The V_p/V_s ratio decreases in the hydrates due to increasing shear rigidity (Goldberg *et al.* 1997), and the log amplitudes are generally low compared to the sedimentary interval below. Such changes in the elastic properties by hydrate cementation may explain the seismic blanking observed across the Blake Ridge, a common feature which has been previously attributed to the lithological homogeneity of these sediments (Holbrook *et al.* 1996).

Drilling and logging results on the Blake Ridge and at other locations, such as the Cascadia accretionary prism, provide excellent opportunities to compare continuous records of porosity and velocity as a function of depth. These data significantly improve our understanding of the *in situ* physical properties of gas hydrates and may help to determine their hydrocarbon reservoir potential. Because marine drilling conditions are difficult, however, the use of conventional logging technology may fail to deliver complete and continuous profiles from the sea floor through to the base of a gas hydrate layer. In particular, the interval immediately below the sea floor, where porosity reduction is the greatest, has never been logged at a location where the base of gas hydrate stability is shallow.

Logging-while-drilling

Over the last 5–10 years, new technology has been developed to measure *in situ* properties in

GOLDBERG, D. & SAITO, S. 1998. Detection of gas hydrates using downhole logs. *In:* HENRIET, J.-P. & MIENERT, J. (eds) *Gas Hydrates: Relevance to World Margin Stability and Climate Change*. Geological Society, London, Special Publications, **137**, 129–132.

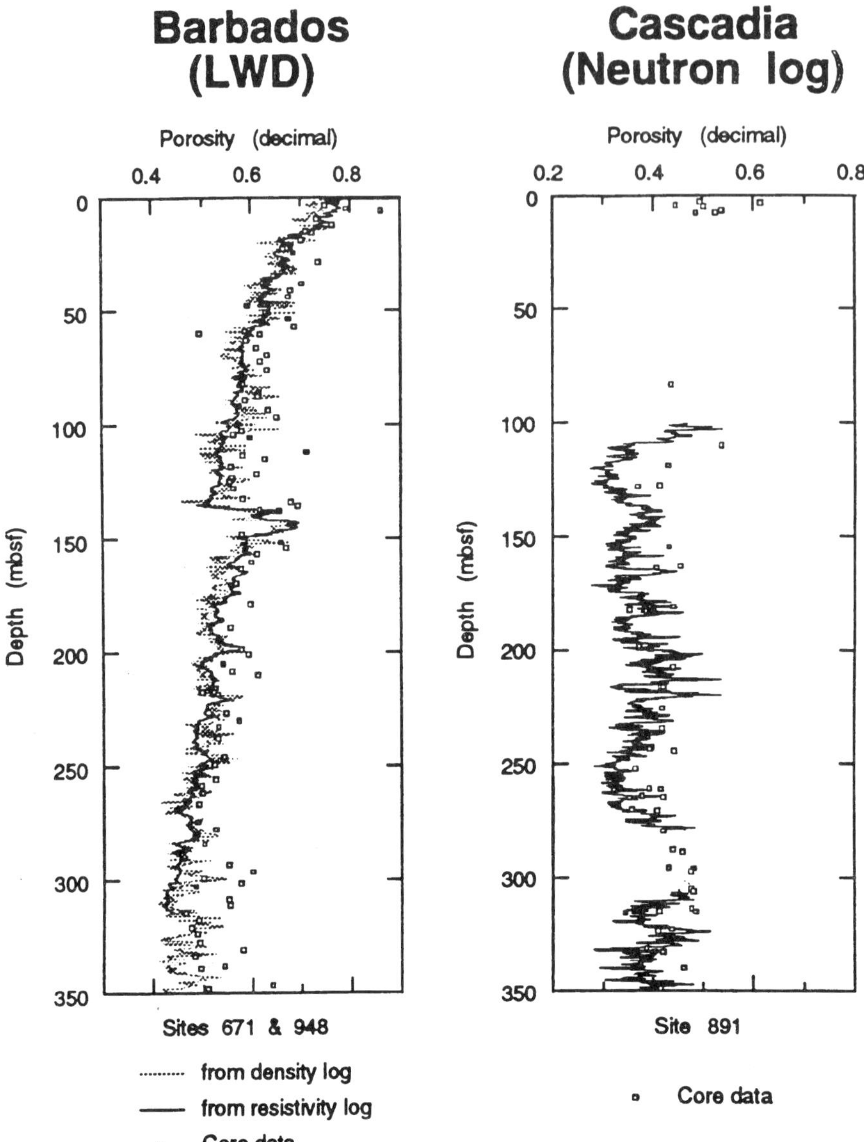

Fig. 1. Porosity logs through accretionary prisms in ODP Hole 948A (Barbados) and 889C (Cascadia) recorded using LWD and wireline logging tools, respectively. The LWD logs indicate continuous porosity profiles derived from density and resistivity measurements from the sea floor to 350 m depth. The wireline logs are limited to depths greater than 80 m below the sea floor and are of lesser quality due to poor borehole conditions after drilling. Measurements of porosity from laboratory core tests are shown from both holes for comparison.

oil industry holes that are drilled horizontally. In this environment, conventional logging with a flexible wireline is impossible. This innovative technology is called 'logging-while-drilling' (LWD) and uses sensors placed just above the drill bit that allow measurements of porosity, resistivity, density and natural gamma radiation, among others, to be made minutes after the drill bit cuts through the formation (e.g. Allen *et al.* 1989; Bonner *et al.* 1992; Murphy 1993). Transient physical properties can then be measured before they change significantly. Sonic velocity measurements, however, can only be acquired while drilling when formation velocities exceed approximately 2.0 km s^{-1} (Aron *et al.* 1994), precluding many marine environments in which natural gas hydrates occur. In drilling hydrates, existing LWD technology would provide several key advantages: (1) high-quality data acquired in typically unstable marine holes; (2) data acquired over the entire drilled interval, particularly in the critical shallow section immediately below the sea floor; (3) data acquired immediately after the borehole is opened, not allowing time for porosity and permeability to change significantly; and (4) data acquired continuously with depth to correct sampling bias that results from less than 100% core recovery.

LWD has been used once previously in a deep marine environment by the Ocean Drilling Program, where data were recorded in the Barbados accretionary prism, but no gas hydrates were observed (Shipley *et al.* 1995). A comparison of two LWD porosity logs from Barbados with a conventional wireline porosity log from the Cascadia accretionary prism (where hydrates were recovered) is shown in Fig. 1. Measurements of porosity from laboratory core tests and the wireline log are more sparse and less representative of the complete porosity structure. The porosity profiles from the Barbados prism cover the entire depth range, including the critical top 100 m, and correspond well with the laboratory data. When calibrated, LWD and core data have been used jointly to compute continuous pore pressure and effective stress profiles vs depth (Moore *et al.* 1995). Determining variations in effective stress, both downhole and laterally, is also critical to understanding the spatial homogeneity and failure criteria of sediments containing hydrates on continental slopes.

The differences in measurement technologies, however, must be taken into consideration whether computing effective stress profiles, seismic properties or simply comparing log profiles from wireline and LWD tools (e.g. Evans 1991). The resolution of LWD sensors is similar to that of wireline logging tools; porosity measurements have a vertical resolution of about 30 cm, while density and gamma-ray measurements have a vertical resolution of 15 cm, depending in part on consistent drilling rates. The position of these sensors above the drill bit allows data to be recorded from 4–30 min after the hole is cut by the bit for a penetration rate of 0.5 m min^{-1} (the maximum recommended for high-quality data). New instruments have recently been developed to record oriented electrical images with 5-cm resolution using sensors that scan a full 360° around the drill bit (Lovell *et al.* 1995). These images resemble wireline data from the Formation Microscanner (FMS) tool with somewhat poorer resolution, but, unlike the FMS, borehole wall coverage is complete and the data are recorded before borehole conditions deteriorate. The structural fabric and formation anisotropy of gas hydrates could be observed using image logs. We recommend use of LWD tools to detect the occurrence and accurately measure the physical properties of gas hydrates *in situ* at sites on the Blake Ridge, Cascadia Margin and in other locations.

References

ARON, J., CHANG, S. K., DWORAK, R. *et al.* 1994. Sonic compressional measurements while drilling. *In: Transactions of the SPWLA 35th Annual Logging Symposium*, paper SS.

ALLEN, D., BERGT, D., BEST, D. *et al.* 1989. Logging while drilling. *Oilfield Review*, **1**, 4–17.

BONNER, S., CLARK, B., HOLENKA, J. *et al.* 1992. Logging while drilling – a three-year perspective. *Oilfield Review*, **4**, 4–21.

COLLETT, T. S. 1993. Natural gas hydrates of the Prudhoe Bay and Kuparuk River area, North Slope, Alaska. *AAPG Bulletin*, **77**, 793–812.

EVANS, H. B. 1991. Evaluating differences between wireline and MWD systems. *World Oil*, **212**, 51–61.

GOLDBERG, D. 1997. The role of downhole measurements in marine geology and geophysics. *Review of Geophysics*, **35**(3), 315–342.

——, GUERIN, G., MELTSER, A. & THE ODP LEG 164 SHIPBOARD SCIENTIFIC PARTY. 1997. Dipole sonic logs and velocity–porosity relationships in gas hydrates. *Transactions of the AAPG Bulletin*, **6**, A41.

HOLBROOK, W., HOSKINS, H., WOOD, W., STEPHEN, R., LIZARRALDE, D. & THE ODP LEG 164 SHIPBOARD SCIENTIFIC PARTY. 1996. Methane hydrate and free gas on the Blake Ridge from vertical seismic profiling. *Science*, **273**, 1840–1842.

HYNDMAN, R., SPENCE, G. YUAN, T. & DESMONS, B. 1996. Gas hydrates on the continental slope off Vancouver Island. *In: Proceedings of the 2nd International Conference on Natural Gas Hydrates*, Toulouse, France, 485–490.

LOVELL, J., YOUNG, R., ROSTHAL, R., BUFFINGTON L. & ARCENEAUX, C., JR. 1995. Structural interpretation of resistivity-at-the-bit images. In: *Transactions of the SPWLA 36th Annual Logging Symposium*, Paris, France, paper TT.

MACKAY, M., JARRARD, R., WESTBROOK, G. & HYNDMAN, R. 1994. Origin of bottom-simulating reflectors – geophysical evidence from the Cascadia accretionary prism. *Geology*, **22**, 459–462.

MATHEWS, M. A. 1986. Logging characteristics of methane hydrate. *Log Analyst*, **27**, 26–63.

MOORE, J. C. *et al.* 1995. Abnormal fluid pressures and fault-zone dilation in the Barbados accretionary prism: Evidence from logging while drilling. *Geology*, **23**, 605–608.

MURPHY, D. P. 1993. What's new in MWD and formation evaluation. *World Oil*, **214**, 47–52.

OCEAN DRILLING PROGRAM LEG 164 SHIPBOARD SCIENTIFIC PARTY. 1996. Methane gas hydrates drilled at Blake Ridge. *EOS, Transactions of the American Geophysics Union*, **77(23)**, 219.

PRENSKY, S. 1995. A review of gas hydrates and formation evaluation of hydrate-bearing reservoirs. *In*: *Transactions of the SPWLA 36th Annual Logging Symposium*, Paris, France, paper GGG.

SHIPLEY, T., OGAWA, Y., BLUM, P. *et al.* 1995. *Proceedings of the Ocean Drilling Program, Initial Reports*, College Station, TX. Ocean Drilling Program, **156**.

Tomographic seismic studies of the methane hydrate stability zone in the Cascadia Margin

J. W. HOBRO[1], T. A. MINSHULL[1] & S. C. SINGH[2]

[1] *Bullard Laboratories, Department of Earth Sciences, University of Cambridge, Madingley Road, Cambridge CB3 0EZ, UK*

[2] *BIRPS, Bullard Laboratories, Madingley Road, Cambridge CB3 0EZ, UK*

Abstract: A seismic study in the Cascadia Margin in June 1993 focused upon the bottom simulating reflector (BSR) around Hole 889B of the Ocean Drilling Program. Extensive wide-angle and normal-incidence data were collected during two deployments of five ocean bottom hydrophones (OBHs). We have applied a new two-dimensional travel-time inversion method to data from one line within the survey. Data from four OBHs and normal-incidence arrivals are inverted simultaneously, and a distribution of P-wave velocities above the BSR is obtained. These velocities are found to be slightly higher than those given by vertical seismic profile (VSP) and sonic log data from Hole 889B, with velocities of 1.83–1.95 $\mathrm{km\,s^{-1}}$ occurring immediately above the BSR. Estimates of hydrate concentration derived using two different methods range from 2 to 24% of the pore space. The velocity model provides some support for the existence of a correlation between BSR strength and hydrate concentration above the BSR in this region.

The detection of methane hydrates beneath the deep seabed has been primarily through observation of a strong bottom simulating reflector (BSR) in seismic reflection profiles (Shipley *et al.* 1979), which parallels the seabed and commonly cuts across the stratigraphy. The BSR is thought to mark the base of the hydrate stability zone, because this phase boundary locally follows sub-seabed isotherms that approximately parallel the seabed. Recent waveform inversion studies and Ocean Drilling Program (ODP) drilling have shown that high-amplitude BSRs are normally underlain by free methane gas (e.g. Singh *et al.* 1993; MacKay *et al.* 1994; Holbrook *et al.* 1996), but the distribution of hydrate above the BSR is not well known. Anomalously high velocities immediately above the BSR have been inferred on several margins, and these are interpreted to represent zones of increased hydrate concentration up to 50 m thick (e.g. Wood *et al.* 1994). An important unresolved question is how, if at all, the strength and continuity of the BSR may be linked to the continuity and concentration of the hydrate deposit. Inversion of wide-angle and normal-incidence data allows the P-wave velocity distribution of the sediment above the BSR to be inferred, thus enabling estimation of vertical and lateral variations in the hydrate-induced velocity anomaly, and hence in the hydrate content of the sediments.

In June 1993 a collaborative study between the University of Victoria (British Columbia, Canada) and the University of Cambridge was undertaken in which extensive single-channel normal-incidence and wide-angle seismic data were collected. The survey was centred upon the site of Hole 889B, drilled in ODP Leg 146, and the survey geometry was constructed to give good three-dimensional ray coverage of the sediments surrounding the drill site, in addition to providing a number of lines of data suitable for two-dimensional analysis (Fig. 1). Spence *et al.* (1995) and Whitsett (1996) have performed analyses of four such lines (7A/6B, 8A, 2B and 3B in Fig. 1) from wide-angle data using the travel-time inversion method of Zelt and Smith (1992), interpreting P-wave velocities between 1.75 and 1.93 $\mathrm{km\,s^{-1}}$ immediately above the BSR and estimating hydrate saturations along these lines of between 10 and 28% of the pore space. In a study of BSR and seabed reflection coefficients, Fink (1995) correlated strong BSR reflections with topographic highs in the seabed, and used the BSR reflection coefficients to interpret above-BSR velocities which ranged from 1.6 to 2.2 $\mathrm{km\,s^{-1}}$ (suggesting hydrate concentrations varying between 8 and 45% of the pore space).

This paper presents results from a combined two-dimensional analysis of wide-angle and normal-incidence data from line 2A/1B (Fig. 1), using the new simultaneous travel-time inversion method of McCaughey & Singh (1997). In this method, a velocity distribution is obtained that contains the minimum structure required to model the seismic data satisfactorily. The inversion algorithm is currently being extended to

HOBRO, J. W., MINSHULL, T. A. & SINGH, S. C. 1998. Tomographic seismic studies of the methane hydrate stability zone in the Cascadia Margin. *In:* HENRIET, J.-P. & MIENERT, J. (eds) *Gas Hydrates: Relevance to World Margin Stability and Climate Change*. Geological Society, London, Special Publications, **137**, 133–140.

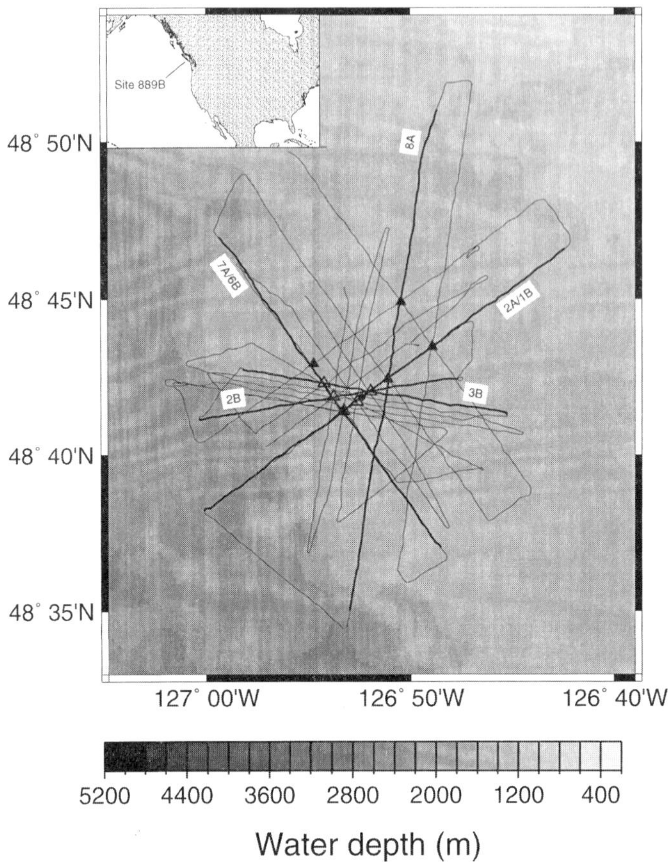

Fig. 1. Acquisition lines followed during OBH deployments A and B (light solid lines) and lines referred to in the text (heavy solid lines). Solid triangles mark OBH locations in deployment A, and outline triangles those in deployment B. The circle indicates the position of ODP Site 889B.

three dimensions. We model the vertical P-wave velocity distribution and determine the extent of lateral variations in the velocity anomaly above the BSR that are required to fit the normal-incidence and wide-angle travel-time data from this line.

Data acquisition and processing

The seismic data were collected in June 1993 on the *John P. Tully*. A total of ~1160 km of single-channel seismic (SCS) data were recorded digitally, most of which were acquired using a 120 in.^{3} airgun source firing at an interval of 14 s (corresponding to ~35 m). Dominant frequencies were 60–90 Hz for this source, which was used to obtain all the data used in this analysis. The SCS data were collected in two grids at different resolutions, with nominal line separations of 200 and 100 m. A more detailed description of the survey is given by Spence *et al.* (1995). Wide-angle data from the 120 in.^{3} airgun were collected on a set of five digital ocean bottom hydrophones (OBHs) from the University of Cambridge in two consecutive deployments. In deployment A, the instruments were placed in a grid at a spacing of 3.3 km, and in deployment B at a spacing of 1 km. The location of all lines in the survey, which were recorded out to shot-receiver offsets of up to 13 km, are shown in Fig. 1.

Processing applied to the data before travel times were picked focused upon the removal of the strong bubble pulses produced by the single gun source. Data from the 120 in.^{3} airgun exhibited bubble pulses at a period of 100 ms which, particularly in the OBH data, made the isolation of the various ray phases difficult. Signature deconvolution was used successfully

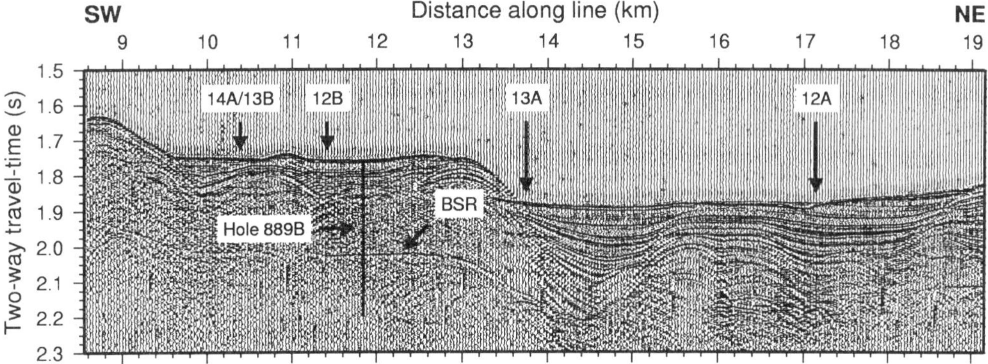

Fig. 2. Single-channel seismic section along line 2A showing the BSR and the locations of OBHs supplying data used in this inversion (processing includes a bandpass filter and signature deconvolution).

on the normal incidence section (Fig. 2), and with varying degrees of success on different sections of the OBH data. For this preliminary study, travel times were picked for seabed and BSR reflections from the normal-incidence data, and for BSR reflections and turning rays above the BSR from the OBH data (Fig. 3).

Tomographic inversion method

The travel-time inversion was performed using the new method described in McCaughey & Singh (1997). This method achieves joint inversion of wide-angle and normal-incidence travel times to obtain a detailed velocity and interface model of the subsurface. The model is described as a series of layers each containing a mesh of velocity nodes which describe a two-dimensional velocity distribution. The layers are separated by interfaces, whose depths are sampled at regular intervals. The interface sample spacing and velocity mesh density are chosen so that the inversion problem is over-parameterized. Smooth and continuous interfaces and velocity fields are constructed from the model parameters by two- and three-dimensional interpolation using B-splines.

The forward problem is solved using the ray-tracing method of Farra (1990) and a scheme of interpolation to obtain source–receiver travel times for a particular model. An iterative linearized inversion approach is used, which performs successive small modifications

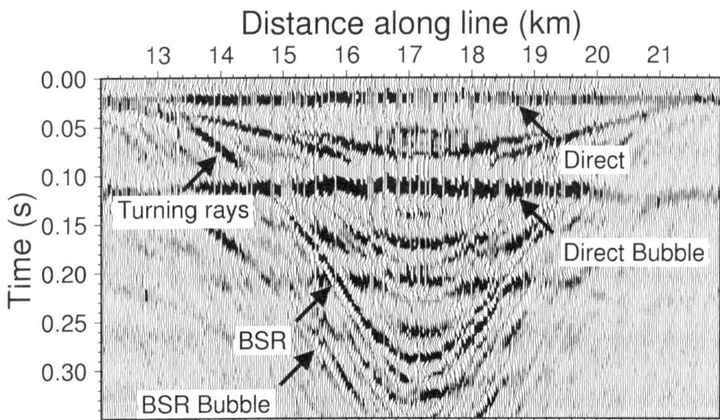

Fig. 3. Bandpass filtered data from OBH 12A, illustrating the strength of the direct arrival bubble pulse and a strong BSR. This section could not be deconvolved due to clipping during acquisition, but the main phases can still be accurately picked. Normal moveout correction has been applied to condense the section, flattening the direct arrival through the water column.

to model parameters in order to minimize the misfit between a set of synthetic ray-traced travel times and the set of travel times picked from normal-incidence and wide-angle data. Regularization of the evolving model is achieved by minimizing the two-term misfit function, Ψ,

$$\Psi(\delta\mathbf{m}) = \delta\mathbf{t}^{\mathrm{T}}\mathbf{C}_{\mathrm{D}}^{-1}\delta\mathbf{t} + \lambda\big((\mathbf{m}_0 + \delta\mathbf{m})^{\mathrm{T}} \cdot \mathbf{C}_{\mathrm{M}}^{-1}(\mathbf{m}_0 + \delta\mathbf{m})\big) \quad (1)$$

where $\mathbf{m}_0$ is a reference model, $\delta\mathbf{m}$ represents a change in model parameters and is the parameter over which the minimization is performed, $\delta\mathbf{t}$ is the set of travel-time residuals associated with model $\mathbf{m}_0$, $\mathbf{C}_{\mathrm{D}}^{-1}$ is the data covariance matrix which, in this case, weights each travel-time residual by the error in the corresponding picked travel-time $\mathbf{C}_{\mathrm{M}}^{-1}$ is the model covariance matrix that describes the regularization, and λ is the regularization strength. The model covariance matrix is constructed in order to minimize the second spatial derivatives of interface and velocity parameters, thus constraining the roughness of the model. Partial minimization is achieved at each iteration using the conjugate gradient method (Scales 1987). The regularization strength is gradually decreased throughout the course of the inversion, enabling the best fit with a smooth model to be achieved before greater levels of detail are allowed to emerge. The dimensionless parameter χ^2 is defined,

$$\chi^2 = \frac{1}{n}(\delta\mathbf{t}^{\mathrm{T}}\mathbf{C}_{\mathrm{D}}^{-1}\delta\mathbf{t}) \quad (2)$$

where n is the number of travel-time residuals in the set $\delta\mathbf{t}$. The value of χ^2 is monitored as the inversion progresses; a value of unity signifies that the data have been matched to within the estimated level of noise, and at this point the inversion process is halted.

The objective of this inversion method is to produce the simplest and smoothest possible interface and velocity model that is able to explain the travel-time data. It is expected that the final model will be unique within the realm of these criteria.

Inversion procedure

A total of 978 reflection picks were made from the SCS section along line 2A, 739 from the seabed reflection and 239 from the BSR. These picks were assigned errors of ± 2 ms – an estimate of the travel-time error due to the bandwidth of the recorded signal – as both interfaces were easily identified in the seismic section. In total, six OBHs were deployed along line 2A/1B. One of these yielded no data, and one location was duplicated. Thus, there were four independent OBH sites from which useful data could be obtained, and a total of 675 travel-time picks were obtained (365 reflection picks and 310 turning ray picks). These picks were initially assigned errors of ± 10 ms, reflecting the uncertainties in isolating the BSR reflections and turning rays from other arrivals in the OBH data. Normal-incidence reflection picks were obtained for the BSR in the vicinity of these sites and their two-way travel times were used as guides in interpreting the OBH data.

The seismic velocity of the water column was modelled from temperature and salinity data (Wilson 1960), and the seabed reflection data were inverted to produce a detailed model of the seabed using 800 depth parameters (a slightly over-parameterized model) spread evenly over the model span of 26 km. This process is equivalent to a depth migration of the seabed reflection. The water column was then substituted by an optimized constant velocity layer at P-wave velocity 1.4825 km s^{-1} which introduced a negligible systematic error of ± 0.5 ms into synthetic wide-angle travel times, but significantly decreased the time required to run each inversion step. A starting velocity model of the layer between the seabed and the BSR was created, with a constant vertical velocity gradient and horizontal velocity contours. The gradient in this model was chosen to give velocities of ~ 1.5 km s^{-1} at the seabed, and ~ 1.8 km s^{-1} at the BSR at Site 889B.

Two inversions were performed. In both, the model was parameterized slightly more densely than would be required to map the anticipated level of model complexity. The level of parameterization would have been increased had more detail been required to explain the data. In inversion 1, a fixed BSR was created with identical topography to the seabed, at a depth of 225 m, and the data were inverted for the velocities above the BSR. The BSR depth was derived from a VSP taken at Site 889B, and is the depth at which the VSP travel time matched that of the BSR on seismic sections (Westbrook *et al.* 1994, pp. 208–212); thus, it is effectively a physical measurement of the BSR depth, and is not dependent upon any velocity estimates. The velocity layer was modelled as a grid of 48×40 velocity parameters, covering a line of 26 km in length with a thickness of 500 m. During the inversion, turning-ray arrivals quickly established a velocity profile whose contours closely followed the seabed topography, increasing our confidence in the uniqueness of the final model. Before the inversion was complete, however, a tension developed between the wide-angle and

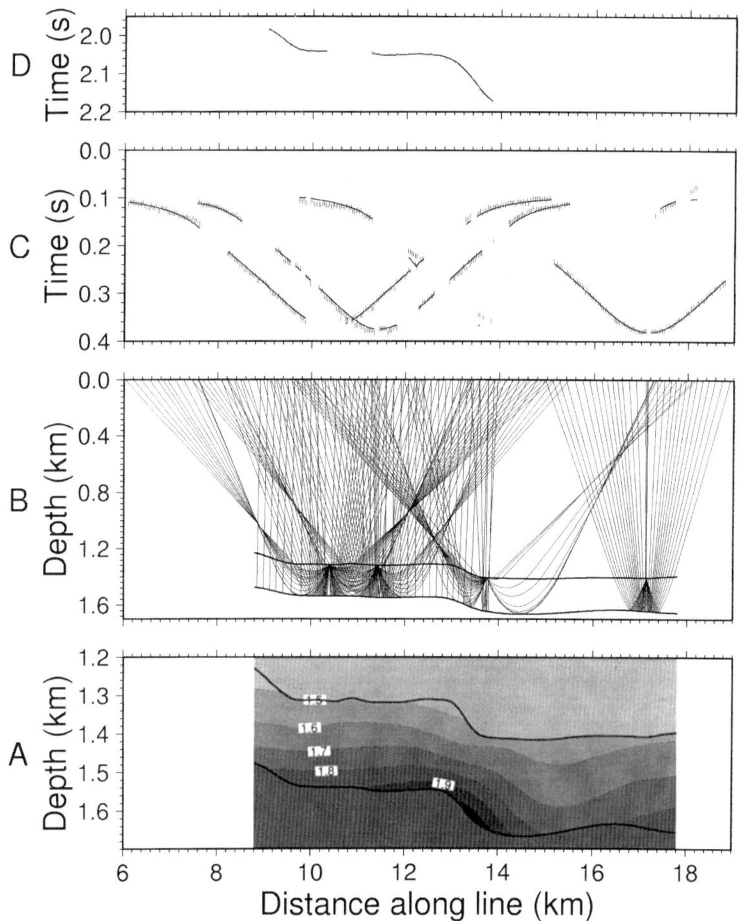

Fig. 4. (**A**) Final model from inversion 2, with a χ^2 of 1.36. Velocity contours are in km s^{-1}. (**B**) Ray coverage for final model, showing all rays used for interpolation to produce synthetic travel-time data. (**C**) Comparison of real and synthetic OBH travel times, with data from all four instruments superimposed. Travel times are reduced using a normal-moveout correction. (**D**) Comparison of real and synthetic normal-incidence travel times.

normal-incidence data in the regions of the model with uneven seabed topography. This could only be resolved by allowing the depth of the BSR to change, an unsurprising result as isotherms in the sediment would be expected to smooth out with depth and so would follow an uneven seabed profile only approximately. A partially inverted velocity model from inversion 1 was smoothed and used as a starting model for inversion 2, a joint BSR depth and velocity inversion in which the BSR was described by 136 depth parameters. When it became evident that no incompatibility existed between travel-time data for different ray phases, the errors in the OBH travel-time picks were reduced to 5 ms to reflect confidence in the correct identification of these phases. A horizontal–vertical velocity smoothing ratio of 10 was used throughout. This value was picked to allow the horizontal velocity field to follow the seabed topography before a significant vertical structure emerged.

Results and discussion

The final model from inversion 2 is shown in Fig. 4, along with the final ray coverage and travel-time residuals. Ray coverage in the region 8.8–13.8 km is good, but it is poor from 13.8 to 17.8 km (Fig. 4B). This, coupled with the uneven topography of the seabed and BSR between 13 and 14 km, caused problems for the

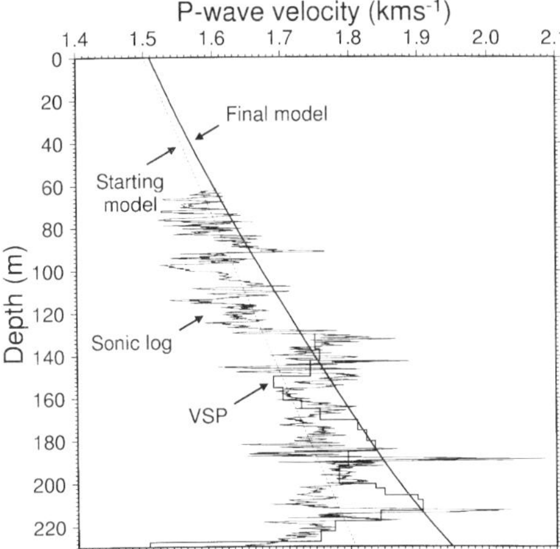

Fig. 5. Velocity–depth profiles at Site 889B for the starting model and final model, and comparison with profiles from downhole measurements.

inversion algorithm. The regularization strength could not be reduced sufficiently to explain the data fully without causing instabilities in the region of poor ray coverage. A modification in the algorithm to allow variations in regularization strength across the model span should enable this, or other similar inversions to proceed to completion. The model shown was inverted to a χ^2 of 1.36, the point at which all travel-time picks for rays passing through the region from 8.8 to 13.0 km were matched satisfactorily (Fig. 4C and D). The velocity field is over-smoothed at around 13 km where the seabed slopes sharply, and the OBH reflection and turning-ray data are not well matched at this point, where the surface velocity in the model is too low and the velocity gradient too high. Velocities at the BSR in the well-constrained region range from 1.87 km s^{-1} at 9.5 km along the line, to 1.96 km s^{-1} at 12.4 km. Velocity gradients at the BSR range from 2.2 to 2.8 s^{-1}, and at the seabed from 1.0 to 1.4 s^{-1}, measured at the same two locations. The velocity structure in the upper 100 m of sediment appears to some extent to follow stratigraphic features visible in the SCS data, with a syncline in the 1.6 km s^{-1} contour at 9.8 km and an anticline at 10.8 km both matching similar structural features in Fig. 2. In the 100 m above the BSR these features are replaced by a gradual velocity increase across the section, as indicated in the measurements above.

There is a correlation throughout the model between high velocities at the BSR and prominent BSR reflections in the SCS data (Fig. 2). The degree of this correlation was assessed by sampling the model velocity directly above the BSR and plotting this alongside BSR amplitude derived from the SCS and MCS data within the well-constrained region of the velocity model. The BSR amplitude was evaluated by measuring the RMS amplitude in a time window constructed around the BSR arrival in each seismic trace. The windows used in the MCS data were of length 40 ms, and those in the SCS data of length 20 ms. A running-average filter was applied to remove high-frequency variations and noise, the spatial length of which was ~1850 m in both cases. The result (Fig. 6) reveals a general trend in BSR amplitude that follows the trend in velocity above the BSR, although analysis of a larger dataset would be necessary to demonstrate this correlation unequivocally, due to the current uncertainties in model velocities. These uncertainties were estimated from comparisons between models generated at different stages during the inversion. A conservative estimate of the uncertainty in velocities at the BSR is ±0.05 km s^{-1}.

A vertical velocity profile, extracted from this model at the location of Site 889B, follows the general velocity trend suggested by the sonic log and VSP at this site (Fig. 5), although the synthetic velocities near the BSR are up to 0.04 km s^{-1} higher than the trend in the measured values. The profiles are, however, consistent within the estimated margin of error. These discrepancies may be due in some degree to anisotropy in the sediment, with slightly higher horizontal velocities. In general, P-wave velocities at the BSR along this line correlate reasonably well with those obtained in the other analyses performed on this seismic dataset. In comparisons of BSR velocities at the points of intersection with the lines modelled using the method of Zelt & Smith (1992) – lines 2B, 3B, 8A and 7A/6B in Fig. 1 – the values were consistently slightly higher in this model with the difference ranging from 0.03 to 0.14 km s^{-1}. However, most of these discrepancies lie within the margins of uncertainty in the results.

From our P-wave velocity model we estimate hydrate saturation using the methods of Lee *et al.* (1996), who use a weighted equation based on the three-phase time-average and Wood equations, and that of Hyndman & Spence (1992), in which it is assumed that the effect on seismic velocity of porosity reduction due to hydrate saturation is the same as that due to compaction with depth. These estimates rely upon the assumption that the reference (unhydrated) velocity profile across the model is constant. Although the presence of lateral structural velocity variations is suggested by the

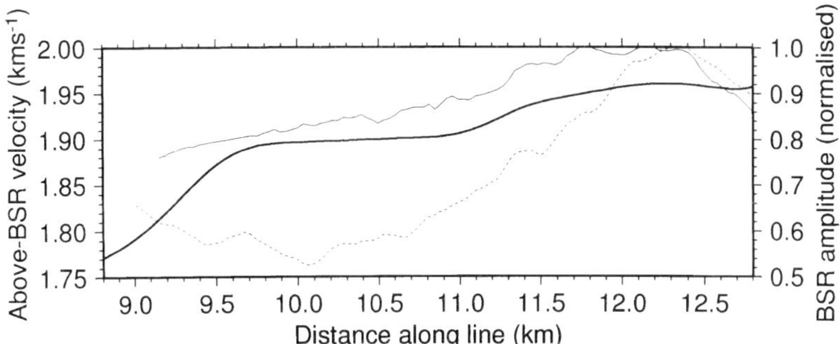

Fig. 6. A comparison between model velocity immediately above the BSR (heavy solid line) and BSR amplitudes derived from SCS (light solid line) and MCS (light dashed line) reflection profiles, showing a correlation between the general trends in velocity and BSR strength.

tendency of the shallow velocity contours to follow the local structure evident in Fig. 2, these variations in reference velocity are small enough not to introduce a significant error into the saturation estimates.

In the 50 m above the BSR, the P-wave velocity at Site 889B increases steadily from 1.83 to 1.95 km s^{-1}. With a matrix velocity of 4.37 km s^{-1} and a porosity of 50% (Westbrook *et al.* 1994, pp. 206–210), the method of Lee *et al.* (1996) yields a hydrate saturation in this region ranging from 2 to 12% of the pore space, or 1–6% of the sediment volume. Using a reference P-wave velocity of 1.65 km s^{-1} for unhydrated sediment (Yuan *et al.* 1996), we obtain a velocity anomaly of 11–18%. From the semi-quantitative porosity relation of Hyndman & Spence (1992), a hydrate saturation of 9–12% of the sediment volume, or 18–24% of the pore space, would be required to produce such an anomaly.

Conclusions

In summary, we conclude the following:

- The inversion method of McCaughey & Singh (1997) was applied successfully, yielding a satisfactory two-dimensional P-wave velocity model in an area of good ray coverage. The model contained some sedimentary structure and mapped lateral variations in the hydrate velocity anomaly in one well-constrained region.
- Velocities directly above the BSR ranged from 1.87 to 1.96 km s^{-1}, with a steady increase along the line towards the north-eastern end. The velocity gradient at the BSR was approximately 2.5 s^{-1}. Porosity reduction estimates suggest that 18–24% of the pore space is filled by hydrate. A weighted time-average and Wood equation estimate yields a concentration of 2–12% of the pore space.
- Velocities in the model were slightly higher than those obtained from other two-dimensional modelling in the region, and from VSP and sonic log data from ODP Hole 889B. However, our results were consistent with these within the estimated margin of error.
- A correlation was observed between the variations in P-wave model velocity above the BSR and BSR strength. This may provide evidence of a correlation between BSR strength and hydrate concentration above the BSR in this region.

Data acquisition was coordinated by G. D. Spence (University of Victoria, British Columbia), and was funded by NSERC in Canada and the Natural Environment Research Council (NERC) in the UK. J. Hobro was supported by a NERC research studentship and T. Minshull by a Royal Society University Research Fellowship. BIRPS is funded by NERC and BIRPS Industrial Associates (Amerada-Hess Ltd, ARCO British Ltd, BP Exploration Co. Ltd, Chevron UK Ltd, Conoco (UK) Ltd, Lasmo North Sea Plc, Mobil North Sea Ltd, Shell UK Exploration and Production, Statoil (UK) Ltd). We thank I. A. Pecher and an anonymous reviewer for their constructive comments. This paper is a University of Cambridge, Department of Earth Sciences contribution 4950.

References

FARRA, V. 1990. Amplitude computation in heterogeneous media by ray perturbation theory: a finite element approach. *Geophysical Journal International*, **103**, 341–354.

FINK, C. 1995. *Methane Hydrate Distribution Offshore Vancouver Island From Detailed Single Channel Seismic Studies*. MSc thesis, School of Earth and Ocean Sciences, University of Victoria, British Columbia.

HOLBROOK, W., HOSKINS, H., WOOD, W., STEPHEN, R., LIZARRALDE, D. & LEG 164 SCIENCE PARTY. 1996. Methane hydrate and free gas on the Blake Ridge from vertical seismic profiling. *Science*, **273**, 1840–1843.

HYNDMAN, R. & SPENCE, G. 1992. A seismic study of methane hydrate marine bottom simulating reflectors. *Journal of Geophysical Research*, **97**, 6683–6698.

LEE, M., HUTCHINSON, D., COLLETT, T. & DILLON, W. 1996. Seismic velocities for hydrate-bearing sediments using weighted equation. *Journal of Geophysical Research*, **101**, 20 347–20 358.

MACKAY, M., JARRARD, R., WESTBROOK, G. & HYNDMAN, R. 1994. Origin of bottom-simulating reflectors: geophysical evidence from the Cascadia accretionary prism. *Geology*, **22**, 459–462.

MCCAUGHEY, M. & SINGH, S. 1997. Simultaneous velocity and interface tomography of normal-incidence and wide-aperture seismic travel-time data. *Geophysical Journal International*, **131**, 87–99.

SCALES, J. A. 1987. Tomographic inversion via the conjugate gradient method. *Geophysics*, **52**, 179–185.

SHIPLEY, T., HOUSTON, M., BUFFLER, R. *et al.* 1979. Seismic reflection evidence for the widespread occurrence of possible gas-hydrate horizons on continental slopes and rises. *AAPG Bulletin*, **63**, 2204–2213.

SINGH, S., MINSHULL, T. & SPENCE, G. 1993. Velocity structure of a gas hydrate reflector. *Science*, **260**, 204–207.

SPENCE, G., MINSHULL, T. & FINK, C. 1995. Seismic studies of methane gas hydrate, offshore Vancouver Island. *In:* CARSON, B., WESTBROOK, G., MUSGRAVE, R. & SUESS, E. (eds) *Proceedings of the Ocean Drilling Program, Scientific Results*, College Station, TX. Ocean Drilling Program, 163–174.

WESTBROOK, G., CARSON, B., MUSGRAVE, R. *et al.* 1994. *Proceedings of the Ocean Drilling Program, Initial Reports*, College Station, TX. Ocean Drilling Program, **146**(1).

WHITSETT, J. 1996. *Seismic Modelling of Methane Hydrates Found in the Northern Cascadia Accretionary Prism off Vancouver Island*. BSc thesis, School of Earth and Ocean Sciences, University of Victoria, British Columbia.

WILSON, W. 1960. Equation for the speed of sound in sea water. *Journal of the Acoustical Society of America*, **32**, 13–57.

WOOD, W., STOFFA, P. & SHIPLEY, T. 1994. Quantitative detection of methane through high-resolution seismic velocity analysis. *Journal of Geophysical Research*, **99**, 9681–9695.

YUAN, T., HYNDMAN, R., SPENCE, G. & DESMONS, B. 1996. Seismic velocity increase and deep-sea gas hydrate concentration above a bottom-simulating reflector on the northern Cascadia continental slope. *Journal of Geophysical Research*, **101**, 13,655–13,671.

ZELT, C. & SMITH, R. 1992. Seismic traveltime inversion for 2-D crustal velocity structure. *Geophysical Journal International*, **108**, 16–34.

Seismic tomography study of a bottom simulating reflector off the South Shetland Islands (Antarctica)

U. TINIVELLA, E. LODOLO, A. CAMERLENGHI & G. BOEHM

Osservatorio Geofisico Sperimentale, Dipartimento di Geofisica della Litosfera, P.O. Box 2011, 34016 Trieste, Italy

Abstract: Reflection tomography techniques have been applied to two multi-channel seismic profiles, acquired across the accretionary prism of the South Shetland margin, in order to reconstruct the velocity field associated with gas hydrate and free gas layers in the sedimentary sequence. Data show the presence of a strong bottom simulating reflector (BSR), running along the slope in water depths ranging from 1000 to 4600 m, locally underlain by a weak normal polarity reflector about 80 ms deeper in the section. The analysis indicates a velocity trend from the sea floor to the BSR generally consistent with that of normally compacted marine sediments, with an abrupt decrement between the BSR and the underlying reflector, indicating the presence of free gas in the sediment pore spaces. The calculated thickness of this gas-bearing layer is approximately 50 m. Local increments of tomographic velocity above the BSR can be related either to gas hydrate abundances in normally compacted slope basin sediments or to overcompaction in accreted sediments, as imaged by the pre-stack depth migrated sections. We conclude that clathrates and free gas distribution on the South Shetland continental slope are strongly controlled by the structural setting of the accretionary prism, where faults act as conduits for migration of natural gas towards the surface. A brief description of the adopted tomography method is also presented.

Detailed knowledge of the vertical acoustic velocity distribution in marine environments is a key factor in revealing the presence of gas hydrate-bearing layers and/or free gas-bearing layers. Gas hydrate-bearing sediments have high compressional wave velocity compared to water-saturated sediments (Stoll *et al.* 1971), whereas free gas-bearing sediments present low compressional wave velocity (Domenico 1977). The effect of a strong acoustic impedance contrast between sediments containing gas hydrates and sediments with free gas in the pore space is evident in marine seismic profiles as bottom simulating reflectors (BSRs; e.g. Shipley *et al.* 1979).

In a first approximation, we may obtain the acoustic velocity field adopting standard processing techniques on seismic data: the root-mean-square velocities (RMS) can show anomalous velocity gradients vs depth for sedimentary layers containing gas hydrates and/or free gas, but the seismic acquisition geometry (large spatial average corresponding to the spread length) and the physical behaviour of the sonic waves travelling through the Earth (high-frequency absorption) intrinsically limit the resolution of this velocity analysis, both horizontally and vertically.

Miller *et al.* (1991) used synthetic seismograms to estimate the thickness of the free gas zone present below the hydrate-bearing sediments, by comparing the observed and modelled waveforms: they obtained values of 5–17 m for the free gas zone in the Peru accretionary prism. Vertical seismic profiling (VSP) through the BSR of the Cascadia Margin (Ocean Drilling Program, Leg 146) provided indications of the thickness of the free gas zone as at least 50 and 15 m in the Oregon and Vancouver margins, respectively (MacKay *et al.* 1994). Thicknesses of gas-bearing layers calculated from another VSP experiment performed in the three Blake Ridge drill holes (Ocean Drilling Program, Leg 164; Holbrook *et al.* 1996), give a value of at least 250 m. Singh & Minshull (1994) utilized full waveform analyses for calculating low-velocity layer thickness. The positive seismic velocity anomaly generated by the presence of clathrates in the sediment pore spaces is less evident than the negative anomaly produced by the presence of free gas. In addition, many authors (Scholl & Hart 1993; Lee *et al.* 1994; Wood *et al.* 1994) found that the hydrate zone is characterized by significant lateral velocity anomalies.

In this paper, acoustic reflection tomography is applied to multi-channel seismic data acquired on the South Shetland margin (Antarctic Peninsula), where a BSR has been found (Lodolo *et al.* 1993), in order to obtain a local velocity field associated with the presence of gas hydrates and/or free gas in the sedimentary layers. We will show that this technique increases the vertical and lateral resolution of the velocity

Tinivella, U., Lodolo, E., Camerlenghi, A. & Boehm, G. 1998. Seismic tomography study of a bottom simulating reflector off the South Shetland Islands (Antarctica). *In:* Henriet, J.-P. & Mienert, J. (eds) *Gas Hydrates: Relevance to World Margin Stability and Climate Change*. Geological Society, London, Special Publications, **137**, 141–151.

structure associated to the BSR, allowing the determination of the thickness of the gas-bearing layer below the BSR. Moreover, we will demonstrate that the geological background of the South Shetland accretionary prism and the gas hydrate and free gas distribution, inferred from the velocity field anomalies, are strictly related as evidenced by the pre-stack depth migrated sections obtained with the tomographic velocity field.

Regional geological setting

The South Shetland margin belongs to a wide and complex continental margin (Pacific Margin of the Antarctic Peninsula), which extends from approximately the Bellingshausen Sea to the northern tip of the Antarctic Peninsula in the Drake Passage (Fig. 1).

The complexity of the Pacific Margin of the Antarctic Peninsula is related to a long tectonic history of axial ridge crest–trench collision that progressed, from the southwest to the northeast, from the Eocene until the Pliocene (Barker 1982; Larter & Barker 1991). Recent multi-channel seismic surveys conducted in the area (Maldonado *et al.* 1994; Kim *et al.* 1995) have delineated the main structural features of the convergent margin, and described the depositional regime of the trench–slope system. The South Shetland trench represents the zone of convergence between the continental domain of the South Shetland microplate and the oceanic floor of the Phoenix microplate to its northwest. Although the Phoenix ridge is commonly believed to have ceased spreading about 4 Ma ago (it should thus be considered inactive), an extremely low spreading rate may have persisted until younger times.

The oceanic side of the trench is characterized by normal faulting in the basement, which produces a marked horst-and-graben tectonic style. The inner portion of the trench is characterized by folding and reverse faulting of the sediments, and a décollement surface can be identified along a strong high-amplitude–low-frequency reflector. A wedge of accreted sediments is identified on the continental slope by chaotic reflectors alternating landward-dipping,

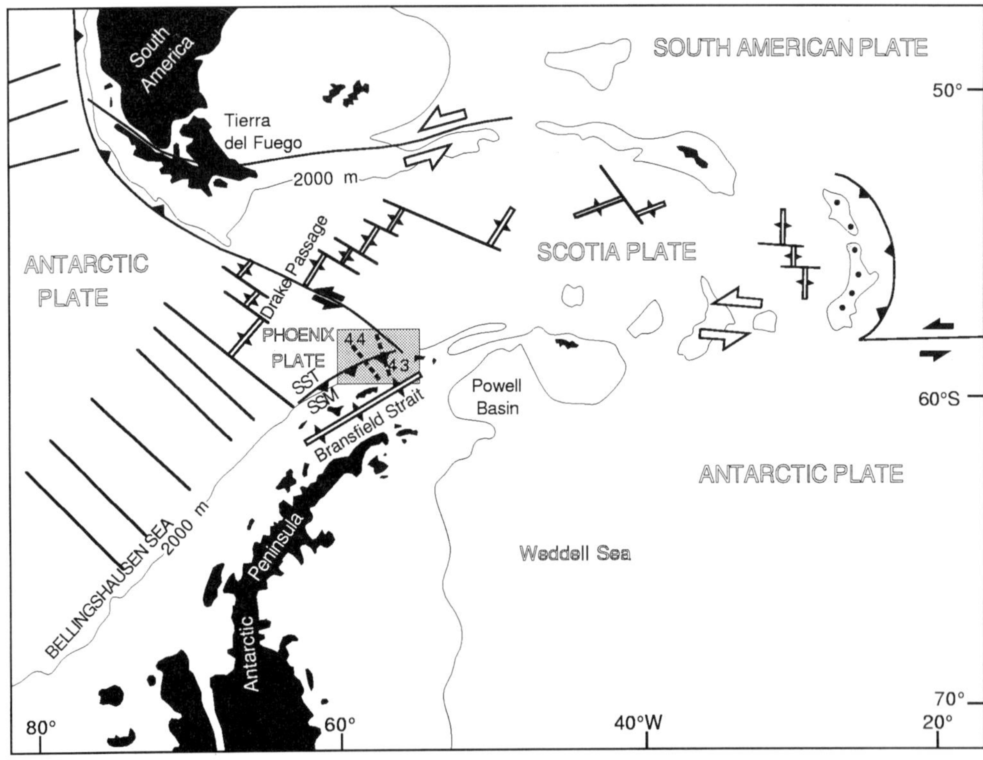

Fig. 1. Simplified map of the Scotia Sea region, with the main tectonic elements. Dashed box refers to the survey area. SSM, South Shetland margin; SST, South Shetland trench; 43, MCS line IT9043; 44, MCS line IT9044.

high-amplitude–low-frequency reflectors. In the upper slope a part of a forearc sedimentary basin has been affected by the deformation of the accretionary complex, with the occurrence of folding and faulting.

The tectonic activity presently observed in the trench on the accretionary prism and forearc basin, and the onset of the extensional regime of the Bransfield Strait, a backarc basin located behind the South Shetland continental platform, are believed to be produced by the gravitational sinking and roll-back of the oceanic lithosphere that followed the cessation (or marked reduction) of the Phoenix ridge push (4 Ma) (Larter & Barker 1991).

Seismic data

Strong, high-amplitude sub-bottom reflections have been identified on two multi-channel seismic reflection profiles (IT9043 and IT9044) acquired by the R/V *OGS-Explora* (Austral summer 1989–1990) on the continental slope of the South Shetland Islands (Lodolo *et al.* 1993) (Fig. 2). The energy source was two air-gun arrays of 15 guns each with a total volume of 45 l. Thirty-fold data were recorded using a 120-channel, 3000 m long analogue streamer, with a hydrophone group interval of 25 m, and shot spacing of 50 m. The sampling interval was 4 ms. The conventional seismic processing applied on the data included: (1) spherical divergence and absorption compensation; (2) trace editing and despiking; (3) shot gather and adjacent trace sum applying a differential normal move-out; (4) normal move-out and common depth-point stack using stacking velocities derived from semblance analyses and constant velocity-stack analyses; (5) deconvolution after stack; and (6) trace mixing and time-variant filtering. After reconstruction of the velocity field from reflection tomography (see next paragraphs), we applied the pre-stack depth migration on raw field data (unfiltered and undeconvolved) in order to enhance the seismic image of the structures associated to the presence of the BSR.

General character of the South Shetland BSR

The BSR is almost always a single symmetrical pulse, and it crosses horizons that reflect the position and orientations of sedimentary layers. It must therefore be younger and superimposed upon the acoustic structure of these sediments. Sub-bottom two-way travel time (TWTT) of the BSR on line IT9043 varies from 520 ms at the shallowest water depth surveyed (about 1000 m) to 900 ms at a depth of 4800 m in the vicinity of the South Shetland trench. On line IT9044, the BSR is less evident and continuous, and it lies in water depths ranging between 1350 and 4500 m, with sub-bottom TWTT of 500 and 800 ms, respectively. Local discontinuities of the BSR in both seismic profiles could be caused by structural discontinuities and/or disturbances at the base of gas hydrates stability field (sudden changes in pressure–temperature values).

True amplitude traces (Fig. 3) indicate a consistent negative polarity of the BSR with respect to the sea-floor reflection which testifies the presence of a significant acoustic impedance contrast. The amplitude of the BSR is larger than any other sub-bottom reflector, and has a laterally continuous character. The reflection coefficient of the sea floor, calculated by comparing the amplitude of the sea-floor reflection with the amplitude of the sea-floor multiple on near-normal incidence traces (to avoid complications due to variation in amplitude with offset), gives values ranging from 35 to 40% (Lodolo *et al.* 1993).

Acoustic velocity field calculations

Conventional stacking velocity analysis

Conventional stacking velocity analyses estimate vertical velocity changes, if the layers are nearly horizontal and the incidence angles are small. In this case, we can simply apply Dix's formula to obtain the interval velocities, which is the information we need to identify anomalous variations of the compressional waves velocity. This approach is certainly acceptable in the area comprised between CDP 250 and 400 of line IT9043 (Fig. 4), where the BSR is flat and nearly horizontal. The layer above the BSR is characterized by lateral variations of the average interval velocity between 1900 and 2200 m s^{-1}. Below the BSR, a continuous low-velocity zone of average interval velocity comprised in the range of 1300 m s^{-1} $\pm 15\%$ suggests the presence of gas-bearing sediments, between the BSR and the first detectable reflector about 80 ms below it, corresponding in depth to a thickness of 60–70 m.

In stacking velocity determinations the resolution in space and time of the velocity spectra is quite limited. In fact, a sort of spatial average

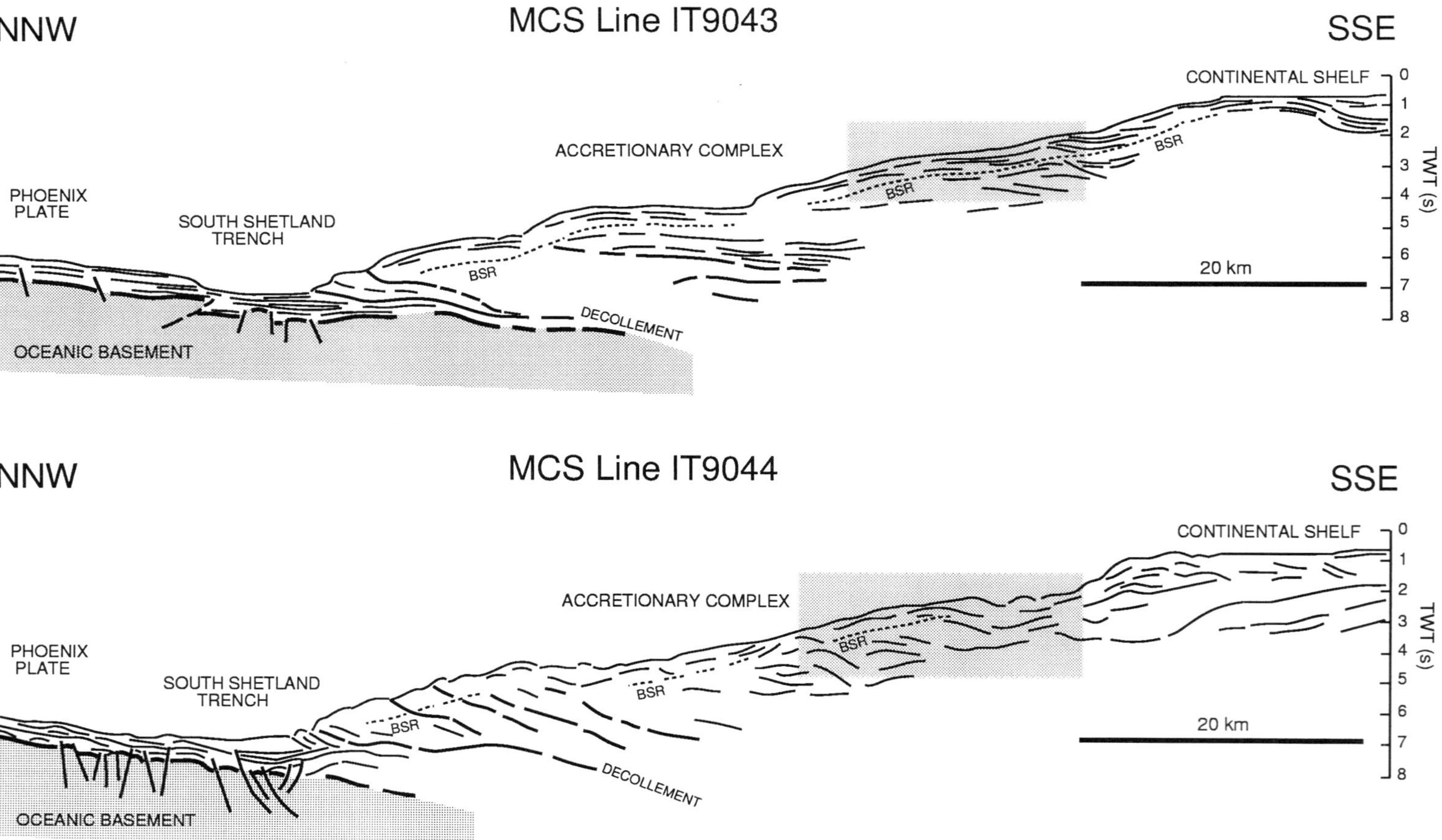

Fig. 2. Line drawings of the seismic profiles IT9043 (top) and IT9044 (bottom). Dashed boxes on the profiles refer to the parts of seismic data where the tomographic analysis has been applied.

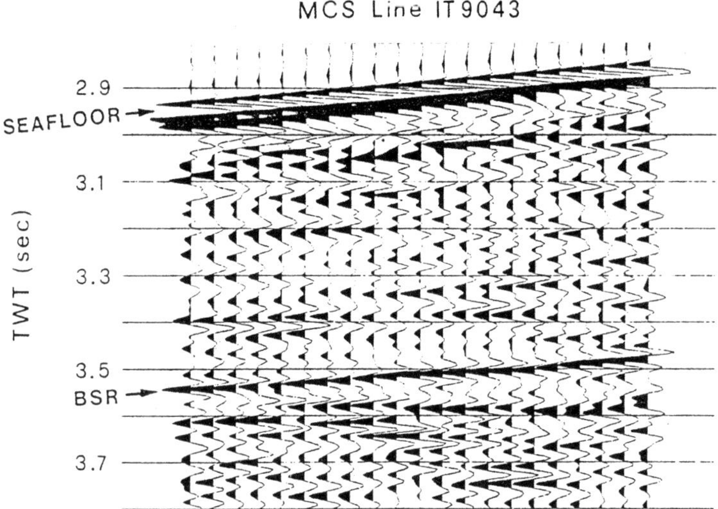

Fig. 3. Reflection amplitude display (corrected for geometric spreading) of part of the sea floor and BSR on line IT9043. Note the consistent negative polarity of the BSR with respect to the sea-floor reflection.

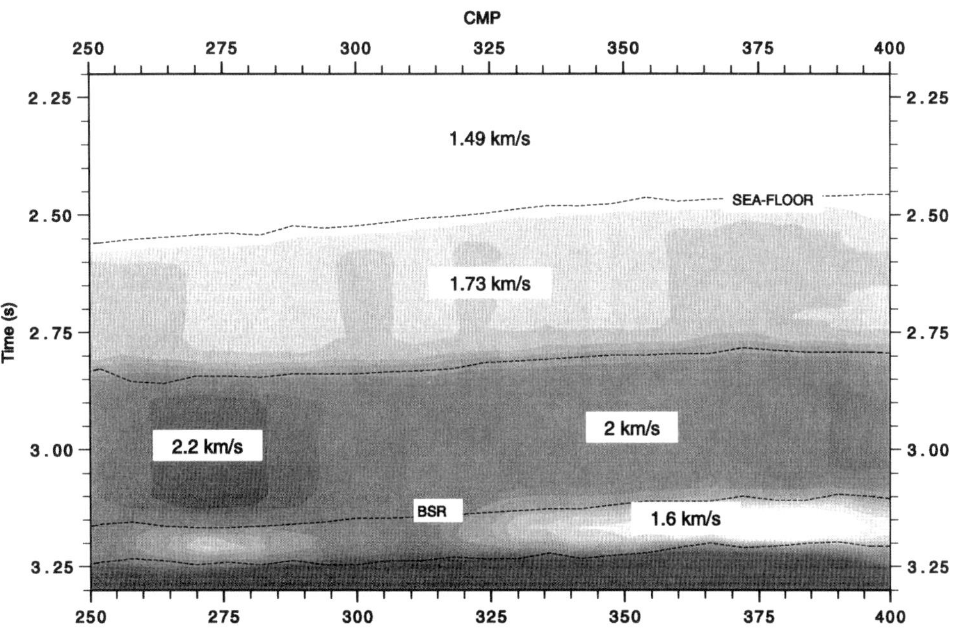

Fig. 4. Interval velocity field associated to the BSR on line IT9043, showing the low-velocity zone below the BSR. Lateral velocity variations have been revealed in the layer above the BSR (from 1.9 to 2.2 km s^{-1}).

is carried out in stacking velocity analysis over the acquisition spread, whose length, in our case, is 2670 m. As a consequence, we are not able to measure sharp lateral velocity variations, but only to observe (and in some case invert) the spatially organized oscillations induced in the continuous velocity spectra of the underlying horizons (Loinger 1983; Harlan 1989; Carlini *et al.* 1989). Along the time dimension we have to consider that the coherency values are averaged usually within time windows, whose length is comparable with that of the seismic wavelets.

In our case, at the target depth, this length is about 40 ms, which means, converted in depth, about 50 m. This figure is again larger than the sought anomaly, and therefore could be a serious drawback. For these reasons we also considered velocity spectra without time windows. Because the signal to noise ratio is good in the considered seismic data this test produced fair results, but unfortunately not significantly different from those displayed above.

Reflection tomography velocity analysis

Tomographic inversion of travel times is able to significantly increase the resolution, for two main reasons: (1) the picking of seismic reflections transforms them practically into punctual events, and (2) there is no spatial average, as each event preserves its individual contribution to the velocity estimation during the whole inversion procedure.

The method basically consists of two steps:

- identification of acoustic discontinuities (horizons) and determination of relative travel times for various source–receiver positions (picking of horizons);
- local velocity model estimations by iterative procedure (data inversion).

Picking of seismic horizons is a spiking process, which concentrates an event into a single point, the picking accuracy is limited by various causes for the noise presence, the interference with other events, phase rotations at supercritical incidence angles, etc. We have, therefore, to consider this point in a statistical sense and this means that some averaging process is still necessary for our velocity estimate. In addition, we can say that the expected spatial resolution is of the order of the trace spacing, which is 25 m in our case; this value is different by 2 orders of magnitude with respect to that provided by stacking velocity analysis, which is of the order of the spread length, as discussed.

The reflection tomography algorithm applied on the South Shetland seismic data is described in detail by Carrion (1991) and Carrion *et al.* (1993). Here we briefly recall its basic concepts, which can be useful in interpreting the velocity diagrams presented later. The space is discretized by pixel's, which are zones where the propagation velocity of seismic waves is assumed to be constant. The pixel's shape is dictated by the available resolution with a surface acquisition geometry, which allows a good estimation of the lateral gradients but is poor for the vertical variations comprised between two reflectors. In this way, a vertically averaged velocity is estimated between the upper and lower reflector. The iterative inversion procedure starts with any initial model, which can be very far from the true solution. Rays are traced to simulate the seismic wave's propagation in that velocity field and the reflector's position. At each step, the velocity distribution is updated first, obtaining an estimate for the depth location of the reflection point for each source–receiver position and for each considered reflector. Then, observing the pattern and the dispersion of these reflection points, we can formulate a new guess for the reflector's structure and change the local velocity; in fact, the possible presence of lateral velocity gradients is revealed by a spatially organized dip of reflection points as a function of the offset between source and receiver. The final solution is obtained by iterating alternatively these two steps, until a minimum dispersion of the reflection points bound to different shots and offsets is obtained for all horizons.

Velocity field results

Travel times, as described, are provided by the picking of seismic reflections in common-shot and common offset gathers. The analysis has been conducted along 16 km of the seismic profile IT9043 that crosses a slope sediment basin where the BRS is particularly well defined, in water depths ranging from 1400 to 2300 m (see Fig. 2 for location). On profile IT9044, the analysis has been performed along 20 km of data, in depths ranging from 1300 to 2500 m (see Fig. 2). For both profiles IT9043 and IT9044, we inverted the arrival times of 11 contiguous hydrophone groups with spacing of 25 m (group numbers 35–45 of a total 120) providing a minimum offset of 1000 m. The selected groups were found to provide the optimum reflectivity for picking horizons on common shot gathers. Six reflectors were picked on line IT9043: the sea floor, the BSR, three reflectors between the two, and one reflector consistently found about 80 ms TWTT below the BSR, called the BGR (base of the free gas layer). No additional continuous reflectors suitable for picking were found further below. On line IT9044, the picking of horizons was more difficult because of a lack of continuity of some reflectors, in particular for those horizons comprised between the sea floor and the BSR. The number of picked horizons varies from three to seven, depending on the seismic signature of the section.

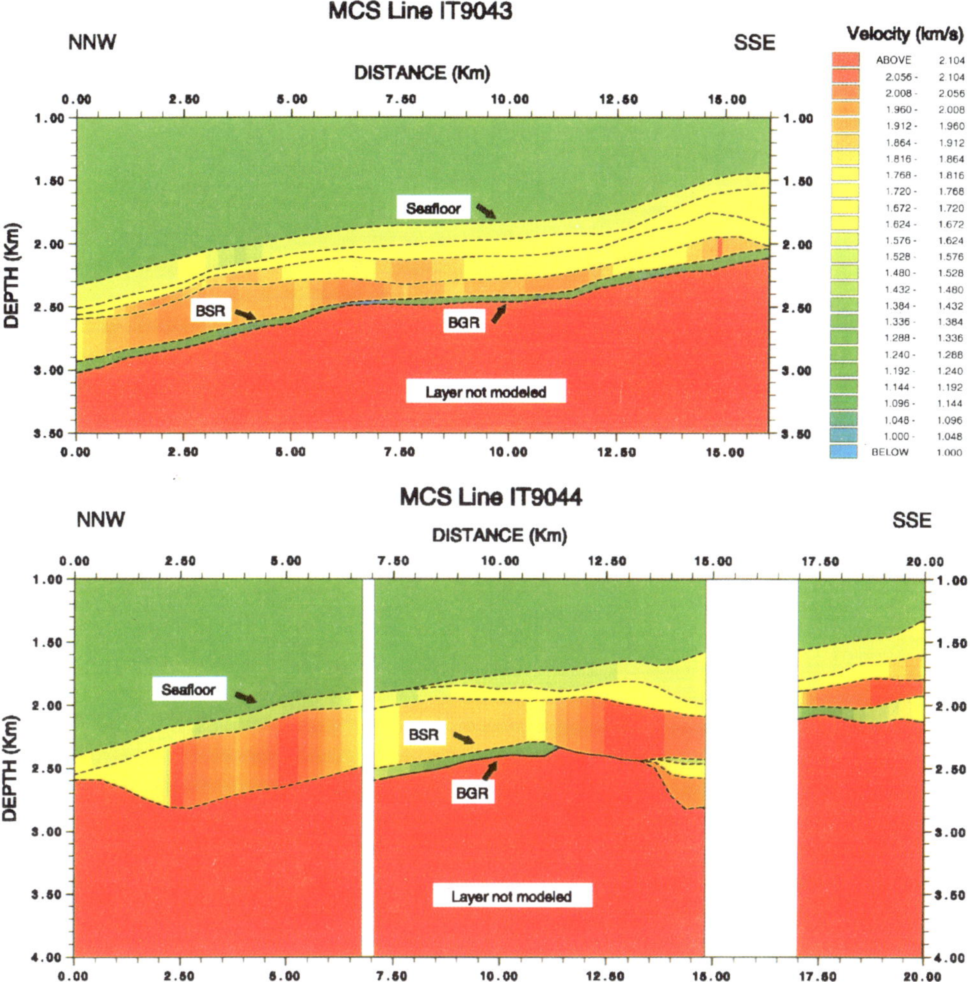

Fig. 5. Velocity fields obtained from the tomography inversion of travel times of part of the MCS lines IT9043 (top) and IT9044 (bottom) (see Fig. 2 for location).

The results of tomographic analysis (Fig. 5) indicate a velocity field from sea floor to the BSR generally consistent with that of a normally compacted terrigenous sequence (Hamilton 1978), increasing from 1750 to 2200 m s^{-1} on line IT9043, and from 1600 to 2350 m s^{-1} on line IT9044. Between the BSR of line IT9043 and the reflector 80 ms below, a drop of the interval velocity to 1200–1400 m s^{-1} is found (1200–1700 m s^{-1} for line IT9044), which indicates the presence of free gas in the sedimentary sequence. The picked horizon below the BSR represents the base of a gas-bearing layer as it displays a normal polarity and so do numerous ghost reflections produced by ringing between upper and lower boundaries of the gas layer. The tomographic inversion of these ghosts in fact produced velocities close to that of the gas layer. Conversely, the interval velocities found in the layers picked above the BSR cannot univocally support the presence of high-velocity hydrate-bearing sediment layers. However, lateral velocity gradients as high as +350 m s^{-1} with respect to the adjacent sediments, have been found.

Pre-stack depth migration

With the obtained tomography velocity field, pre-stack depth migration (Kirchhoff method) has been performed (Figs 6 and 7). The higher

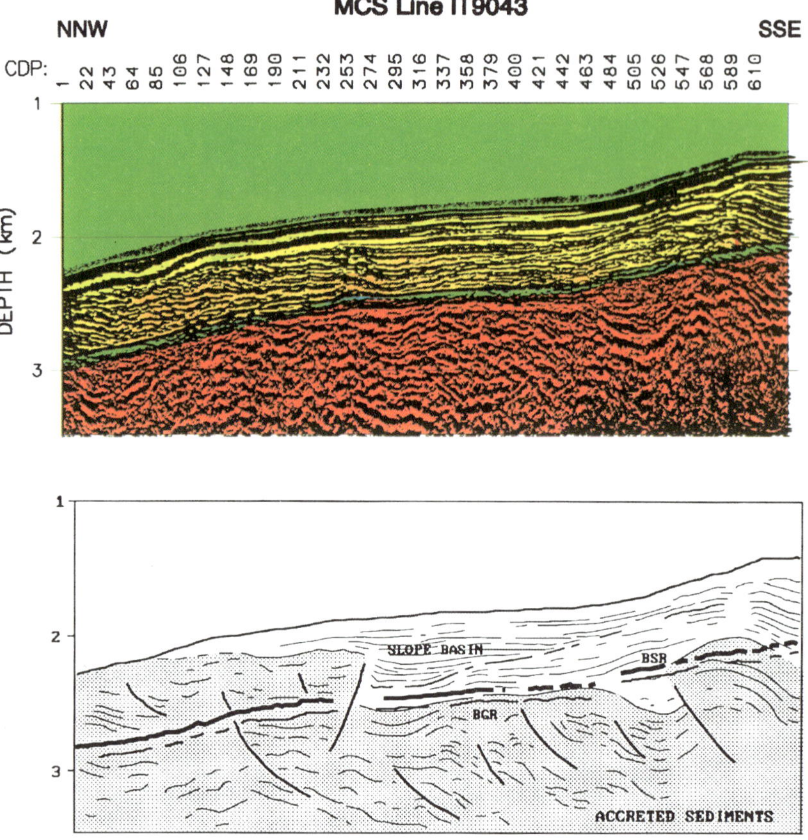

Fig. 6. Pre-stack depth migrated version (top) and interpreted section (bottom) of MCS line IT9043 (see Fig. 5 for colour scale).

resolution velocity structure enhanced the amplitudes of reflections from the base of the gas reflector and provided an estimate of the thickness of the gas layer, varying between 50 and 100 m. Furthermore, it provided a better constraint for the relationship between the deformed zone and the high-velocity anomalies. The good results of migration confirm that the velocity structure is reliable, and outline the presence of geological structures such as reverse faults and folds in the accreted sediments (Kim *et al.* 1996). In one case, the strongest BSR and the associated low-velocity zone below the BSR are coincident with the downslope side of a faulted anticline (Figs 6 and 7).

Discussion

In order to better constrain the velocity anomalies, we compare the velocity structure obtained with the tomographic analysis to a reference velocity profile in a terrigenous sedimentary sequence (Fig. 8a and b) according to Hamilton (1978). The tomographic velocity profile is consistent, within the error of the inversion technique, with the Hamilton's profile in the flat-lying undisturbed turbiditic sequence of the South Shetland trench (Fig. 8a), where there is no evidence of BSR. On the accretionary complex instead, the deviation of the tomographic velocity profile from the Hamilton's profile is remarkable (Fig. 8b). The pronounced negative velocity between the BSR and the BGR can be produced by a variable quantity of free gas in the sediment pore spaces, from a few per cent to several tens of per cent of the pore spaces, because of the very steep velocity–gas concentration curve (Domenico 1977). Above the BSR, the positive velocity anomaly may indicate that gas hydrate concentrations in the pore spaces significantly affect the velocity of the

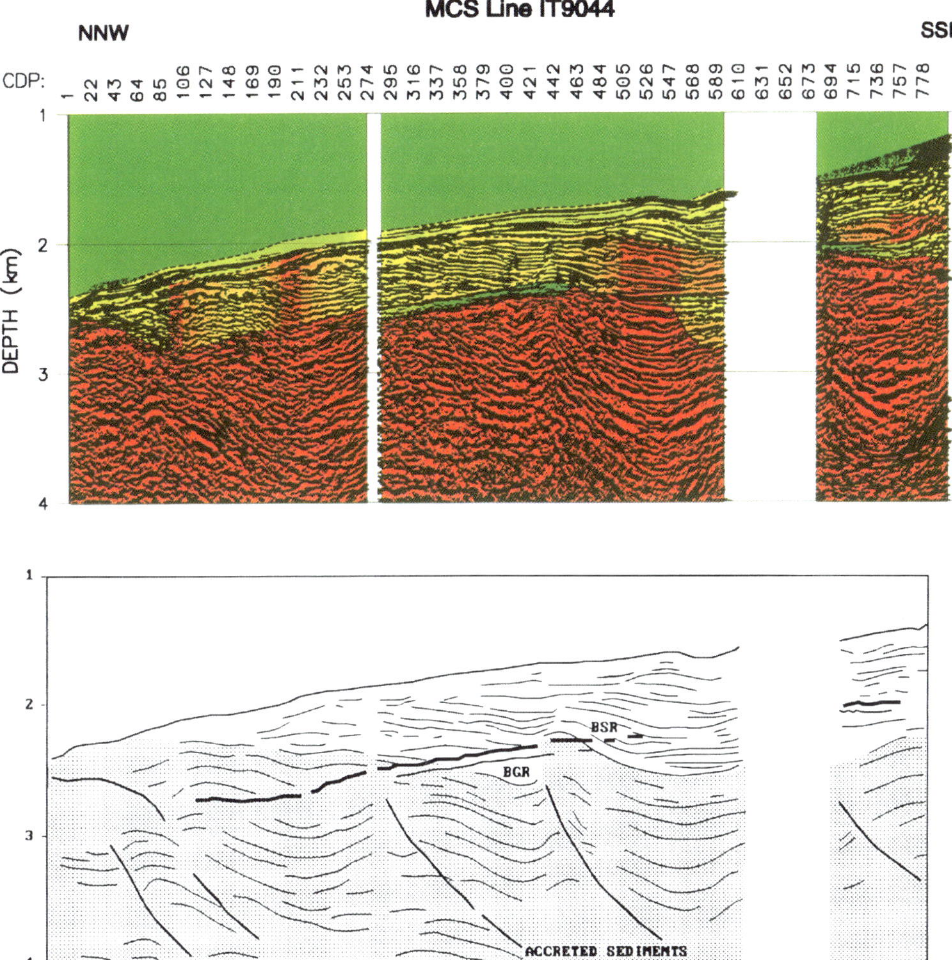

Fig. 7. Pre-stack depth migrated version (top) and interpreted section (bottom) of MCS line IT9044 (see Fig. 5 for colour scale).

bulk sediment. According to the velocity vs clathrate concentration relation obtained by Lee *et al.* (1996), the observed anomaly is potentially produced by approximately 5–20% of the pore spaces occupied by clathrates.

On line IT9043, the velocity field shows local increments above the BSR in different geological situations (see Fig. 6). In the NNW part of the section (CDP 1–250) accreted sediments are present at shallow depths below the sea floor and therefore are consistently present above the BSR. At the SSE end of the section (CDP 550–610) a faulted anticline brings accreted sediments of its crest above the BSR. In both cases the presence of deformed, probably low-porosity and overconsolidated sediments above the BSR is reflected in positive tomographic velocity anomalies with respect to the Hamilton's reference curve. Therefore, it is difficult to discriminate between contributions of gas hydrate crystals and tectonic history to the velocity of the bulk sediments. In the central part of the profile, the presence of a slope basin, characterized by flat lying and normally consolidated turbidites, unconformably resting on the accreted sediments above the BSR suggests that the local positive deviations of the tomographic velocity from the Hamilton's curve may be produced by gas hydrate abundance in the sediment pore spaces. Around CDP 450 the BSR is offset; this could be due to the proximity of a reverse fault that may be acting as a conduit for migration of

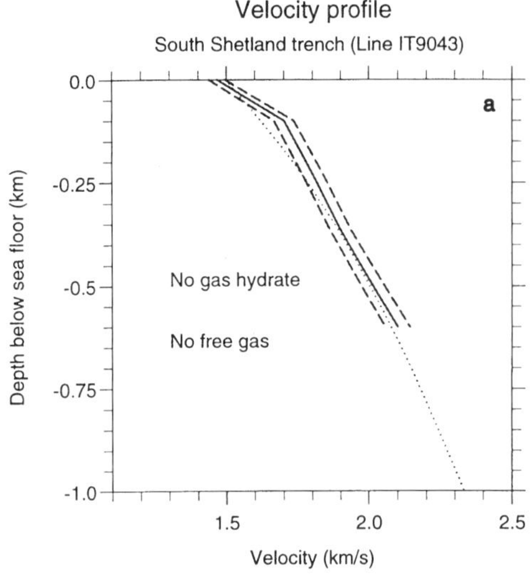

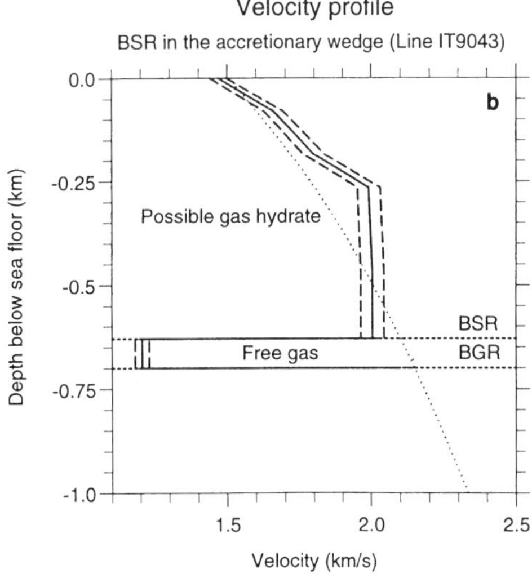

Fig. 8. Comparison between reference velocity profile (dotted line) in marine sequences (Hamilton 1978) and tomographic velocity profiles (solid line) obtained from MCS line IT9043. An error of ±2% (dashed line) has been estimated for the tomographic velocity. (**a**) Layered and undisturbed sediments of the South Shetland trench, where there is no evidence of BSR. (**b**) Slope basin sediments overlying accreted sediments on the South Shetland accretionary complex, where the BSR is stronger.

gas and other fluids towards the surface. The layer just below the BSR, whose averaged thickness is about 70 m and averaged velocity is 1300 m s^{-1}, is clearly associated with the presence of free gas. Because the free gas layer is not associated with any particular stratigraphic or structural element, we must conclude that the gas is trapped by a permeability reduction produced by the occurrence of clathrates above, the seismic evidence of which is not always and conclusively found.

On line IT9044 (see Fig. 7) the lateral continuity of seismic horizons is generally lower than that on line IT9043, and in some cases the relationship between tomographic velocity and the structural and geological setting is more difficult to assess. In the NNW part of the section (CDP 1–270) lateral velocity gradients are found throughout the section above the BSR in both accreted and tilted overlying apron sediments. Erosional truncations near the sea floor suggest that sediments are overconsolidated, therefore a higher velocity than in a normally consolidated terrigenous sequence can be expected. The high velocity found in a small mid-slope basin from CDP 210 to CDP 340 may be produced by local clathrate accumulations. The very sharp velocity change at CPD 100 is an artifact due to the noise in the picked reflectors. A very gradual and steady lateral velocity increase towards SSE from CDP 300 to CDP 600 in the gently dipping layers cropping out at the sea floor above the BSR may also reflect the distribution of gas hydrates, possibly accumulated in coarser, higher porosity sediments that produce higher amplitude reflectors towards the SSE. The low-velocity zone below the BSR is not as continuous as in line IT9043. From CDP 1 to CDP 270 the BGR is not seen, therefore no reflectors below the BSR have been picked. Free gas is therefore probably not present in this part of the margin. Evidence of a nearly 90 m thick free gas layer below the BSR is strong between CDP 270 and CDP 440, while towards the SSE (CDP 550–CDP 600) gas appears to be structurally trapped in larger thickness within a piggy-back basin.

Conclusions

Seismic tomographic inversion applied to marine multi-channel seismic reflection profiles provides a valuable tool in reconstructing acoustic velocity fields and revealing lateral velocity anomalies that can be linked to local abundances of gas hydrates and free gas across BSRs. Because of the sharp seismic velocity decrease produced by even small amounts of free gas in the sediments, we have shown that it is possible to identify the top and bottom of a low-velocity zone below the BSR that provides information on thickness and lateral extent of the free gas layer associated to the formation of gas hydrates. In geologically complex areas, such as the South Shetland margin, local increments of seismic velocity above the BSR can only tentatively be related to occurrence of gas hydrates. Local seismic velocity anomalies can also be produced by overcompacted sediments of the accretionary complex.

The velocity model obtained with the tomographic inversions appears to be effective in producing pre-stack depth migrated sections that help recognize the structural architecture of the sediments and therefore a discrimination between gas hydrates and structurally related velocity anomalies can be attempted. In our case, we relate the presence of gas hydrate-related velocity anomalies to the presence of faults that probably act as conduits for the upward migration of fluids.

We would like to thank the scientists and the technicians participating in the cruise during data acquisition. We are also grateful to M. Marchi and J. Centonze for preliminary seismic data processing. A. Vesnaver contributed with valuable discussions. Support for this work was provided by the Italian Programma Nazionale di Ricerche in Antartide (PNRA).

References

BARKER, P. F. 1982. The Cenozoic subduction history of the Pacific margin of the Antarctic Peninsula: ridge crest-trench interaction. *Journal of the Geological Society, London*, **139**, 787–801.

CARLINI, A., VESNAVER, A., BOEHM, G. & HARLAN, W. 1989. Inversione tomografica di anomalie nella velocità di Stack. *In: Atti del VIII Convegno GNGTS*, Roma, 695–704.

CARRION, P. 1991. Dual tomography for imaging complex structures. *Geophysics*, **56**, 1395–1404.

——, BOEHM, G., MARCHETTI, A., PETTENATI, F. & VESNAVER, A. 1993. Reconstruction of lateral gradients from reflection tomography. *Journal of Seismic Exploration*, **2**, 55–67.

DOMENICO, S. N. 1977. Elastic properties of unconsolidated porous sand reservoirs. *Geophysics*, **42**, 1339–1368.

HAMILTON, E. L. 1978. Sound velocity gradients in marine sediments. *Journal of the Acoustical Society of America*, **65**, 909–922.

HARLAN, W. 1989. Tomographic estimation of seismic velocities from reflected raypaths. *Expanded Abstracts of 59th SEG Meeting*, Dallas, 922–924.

HOLBROOK, W., HOSKINS, H., WOOD, W., STEPHEN, R. & THE LEG 164 SCIENCE PARTY 1996. Methane hydrate, bottom-simulating reflectors and gas bubbles: results of vertical seismic profiles on the Blake Ridge. *Science*, 273, 1840–1843.

KIM Y., SAMUELSEN, C. & HAUGE, T. 1996. Efficient velocity model building for prestack depth migration. *The Leading Edge*, **15**, 751–753.

——, KIM, H.-S., LARTER, R., CAMERLENGHI, A., GAMBÔA, L. & RUDOWSKI, S. 1995. Tectonic deformation in the upper crust and sediments at the South Shetland trench. *In*: COOPER, A., BARKER, P. & BRANCOLINI, G. (eds) *Geology and Seismic Stratigraphy of the Antarctic Margin*. Antarctic Research Series, American Geophysical Union, **68**, 157–166.

LARTER, R. & BARKER, P. 1991. Effects of ridge crest-trench interaction on Antarctic–Phoenix spreading: forces on a young subducting plate. *Journal of Geophysical Research*, **96**, 19 583–19 607.

LEE, M., HUTCHINSON, D., AGENA, W., DILLON, W., MILLER J. & SWIFT, B. 1994. Seismic character of gas hydrates on the southeastern U.S. Continental Margin. *Marine Geophysical Researches*, **16**, 163–184.

——, ——, COLLETT, T. & DILLON, W. 1996. Seismic velocities for hydrate-bearing sediments using weighted equation. *Journal of Geophysical Research*, 101, 20 347–20 358.

LODOLO, E., CAMERLENGHI, A. & BRANCOLINI, G. 1993. A bottom simulating reflector on the South Shetland Margin, Antarctic Peninsula. *Antarctic Science*, **5**, 207–210.

LOINGER, E. 1983. A linear model for velocity anomalies. *Geophysical Prospecting*, **31**, 98–118.

MACKAY, N., JARRARD, R., WESTBROOK, G. & HYNDMAN, R. 1994. Origin of bottom simulating reflectors: geophysical evidence from the Cascadia accretionary prism. *Geology*, **22**, 459–462.

MALDONADO, A., LARTER, R. D. & ALDAYA, F. 1994. Forearc tectonic evolution of the South Shetland Margin, Antarctic Peninsula. *Tectonics*, **13**, 1345–1370.

MILLER, J., LEE, M. & VON HUENE, R. 1991. An analysis of a seismic reflection from the base of a gas hydrate zone, offshore Peru. *AAPG Bulletin*, **75**, 910–924.

SCHOLL, D. & HART, P. 1993. Velocity and amplitude structures on seismic reflection profiles. Possible massive gas hydrate deposits and underlying gas accumulations in the Bering Sea Basin. *In*: HOWELL, D. (ed.) *The Future of Energy Gases*. US Geological Survey Professional Paper, **1570**, 331–351.

SHIPLEY, T., HOUSTON, M., BUFFLER, R., SHAUB, F. *et al.* 1979. Seismic evidence for widespread possible gas hydrate horizons on continental slopes and rises. *AAPG Bulletin*, **63**, 2204–2213.

SINGH, S. & MINSHULL, T. 1994. Velocity structure of a gas hydrate reflector at Ocean Drilling Program Site 889 from a global seismic waveform inversion. *Journal of Geophysical Research*, **99**, 24,221–24,233.

STOLL, R., EWING, J. & BRYAN, G. 1971. Anomalous wave velocities in sediments containing gas hydrates. *Journal of Geophysical Research*, **76**, 2090–2094.

WOOD, W., STOFFA, P. & SHIPLEY, T. 1994. Quantitative detection of methane hydrate through high-resolution seismic velocity analysis. *Journal of Geophysical Research*, **99**, 9681–9695.

Marine gas hydrate inventory: preliminary results of ODP Leg 164 and implications for gas venting and slumping associated with the Blake Ridge gas hydrate field

C. K. PAULL, W. S. BOROWSKI, N. M. RODRIGUEZ & THE ODP LEG 164 SHIPBOARD SCIENTIFIC PARTY

Geology Department, UNC-CH, Chapel Hill, NC 27599-3315, USA

Abstract: Deep ODP holes on the Blake Ridge (sites 994, 995 and 997; Leg 164) show that sediments within the interval ~200–450 metres below sea floor (mbsf) contain 1–4% gas hydrate based on interstitial geochemistry, pressure core samples, well logs and borehole seismic data. In addition, a considerable methane gas reservoir also exists below the gas hydrate stability zone in the form of free and dissolved methane. Addition of methane from these deep reservoirs to overlying ocean water (and ultimately to the atmosphere) could have profound geological, geochemical and climatic effects. Pervasive flux of methane to the sea floor via diffusion is unlikely because methane is almost completely consumed by reaction with sulphate at the sulphate-methane interface. Further, sulphate and $^{87}Sr/^{86}Sr$ profiles suggest that upward advection of potential methane-containing fluids is also unlikely. However, point sources for methane advection to the sea floor do occur. Active methane transport along faults occurs at Site 996, and the methane ^{13}C isotopic signature indicates derivation from the gas hydrate-bearing and free gas zones below. ^{14}C data from sediments indicate that the frequency of sediment slumping on the Blake Ridge is higher during the sea-level lowstand associated with the last ice age. Whereas there is no direct evidence that methane exchange is associated with slump events, nor that gas hydrates are necessarily related to the increased slumping frequency, the association between slumps and faults suggest that the amounts of gas escape should be greater during sea-level lowstands.

The sensitivity of the marine gas hydrate reservoir to pressure and temperature conditions has lead to speculation about potential connections between the decomposition of gas hydrates, venting of greenhouse gases to the atmosphere, global carbon isotope shifts and increased slumping frequency during sea-level lowstands. Essential steps toward evaluating these assertions include assessing the size of the gas hydrate reservoir, evaluating the potential mechanisms by which gases may escape from gas hydrate-related deposits and acquiring data on the timing of slumping in areas associated with gas hydrates.

Summary of ODP Leg 164 and assessing the size of the Blake Ridge gas hydrate deposits

ODP Leg 164 was devoted to investigating the amount and *in situ* characteristics of gas hydrates stored in marine sediments. Sites 994, 995 and 997 were drilled on the Blake Ridge (Fig. 1), a sediment drift that is composed of homogeneous and rapidly deposited (350 m Ma^{-1}) nannofossil-rich clays. The holes extend to 700–750 mbsf, penetrating through the depth (~450 mbsf) of the bottom simulating reflector (BSR) into sediments below (Fig. 2), which is taken to be equivalent to the base of gas hydrate stability. Minimal lithological variation occurs near the BSR.

Finely disseminated gas hydrates occupy more than 1% of the sedimentary section between 200 and 450 mbsf at all three sites, regardless of whether a BSR is present (sites 995 and 997) or not (Site 994). Some solid gas hydrate nodules also occur, the largest of which was a >30 cm thick horizon of massive gas hydrate. Pore water profiles from the three sites show progressive freshening to depths of ~200 mbsf. From 200–450 mbsf chloride concentrations are highly variable, and characterized by local, anomalous excursions toward fresher values (Fig. 3). These anomalies indicate variations of up to 14% in the amount of gas hydrate contained in adjacent samples throughout this zone (Fig. 4). Well logs show distinct zones of higher electrical resistivity and sonic velocity that are coincident with the chloride anomaly zones. Vertical seismic profiles indicate that the velocities of the sediments above the BSR are not significantly elevated above normal sediment velocities. However, velocities as low as 1400 m s^{-1} were measured immediately beneath the BSR

PAULL, C. K., BOROWSKI, W. S., RODRIGUEZ, N. M. & ODP LEG 164 SHIPBOARD SCIENTIFIC PARTY 1998. Marine gas hydrate inventory: preliminary results of ODP Leg 164 and implications for gas venting and slumping associated with the Blake Ridge gas hydrate field. *In:* HENRIET, J.-P. & MIENERT, J. (eds) *Gas Hydrates: Relevance to World Margin Stability and Climate Change*. Geological Society, London, Special Publications, **137**, 153–160.

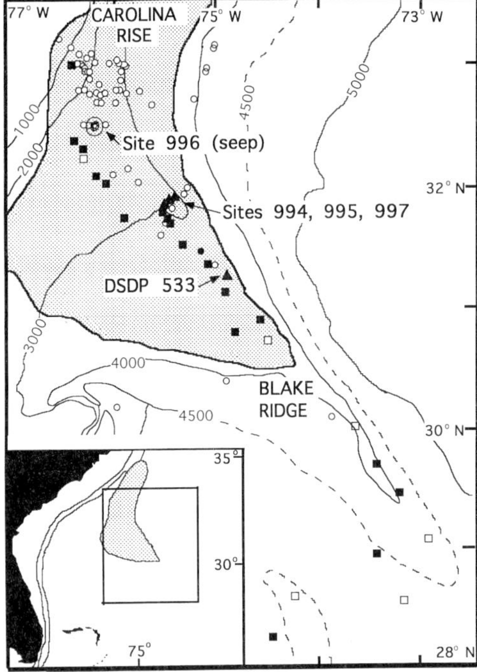

Fig. 1. Map showing the location of piston core sites and Leg 164 drill sites of the Carolina Rise and Blake Ridge. Traditional piston cores (<16 m long) from five different cruises (CH-15-91, CH-11-92, CH-31-92, Kr-140-1 and Kr-140-2) are shown with circles. Deeper-penetrating (>16-m long), jumbo piston cores (JPCs, cruise Kr-140-2) are shown with squares. Triangles indicate DSDP and ODP Sites. Filled symbols highlight cores with adequate amounts of methane for sampling and isotopic measurements (see Fig. 8). Bathymetric contours are in metres. The outline of gas hydrates (stippled) is based on the mapping of seismic reflectors by Dillon & Paull (1983). Map after Borowski *et al.* (1997).

at Site 997 (Holbrook *et al.* 1996). Methane gas volumes from a pressure core sampler are in excess of the *in situ* methane saturation, thus demonstrating that free gas exists intermittently throughout the sedimentary section below the base of gas hydrate stability. In fact, data from the PCS (Dickens *et al.* 1997) indicate that the amount of free gas in the sediments beneath the base of gas hydrate stability may exceed the volume stored in the hydrates above. The results of ODP Leg 164 confirm that sedimentary gas hydrates, plus the associated dissolved and gaseous phases, represent a major methane reservoir.

Assessing the potential for diffusive and/or advective venting from the gas hydrate reservoir

For the gas hydrate reservoir to have implications for either the atmospheric gas composition or for the isotopic composition of the global carbon pool, transport mechanisms need to be established between the methane that is stored within the sub-seafloor reservoirs and the overlying water column and atmosphere. Upward gas transport could either occur as a pervasive process throughout the sedimentary section or be concentrated along conduits which act as point sources for gas discharge.

Pervasive methane discharge

The importance of pervasive upward methane transport from the Blake Ridge gas hydrate deposits can be constrained in two ways:

(1) Detailed pore water profiles were measured in sediments retrieved from piston cores from sediments overlying the Carolina Rise and Blake Ridge gas hydrate field (Fig. 1). Sulphate gradients in this region are linear (Fig. 5), implying that sulphate depletion is dominated by anaerobic methane oxidation ($CH_4 + SO_4^{2-} \rightarrow HCO_3^- + HS^- + H_2O$) localized at the base of the sulphate-bearing zone. Apparently, downward diffusion of sulphate from the overlying sea water is required to balance the upward methane flux (Borowski *et al.* 1996) because the methane is consumed by reaction with sulphate. Thus, linear sulphate gradients can be used to quantify and assess the *in situ* methane flux (Fig. 6), which is a function of the methane inventory below. The methane flux, calculated from sulphate profiles from the Carolina Rise and Blake Ridge sediments, varies by at least a factor of 16, which suggests lateral variation in methane concentrations at depth. The occurrence of a downward flux of sea water sulphate limits the significance of upward advection within these sediments and prevents pervasive escape of methane through the sea floor.

(2) The concentration and isotopic composition of strontium also indicate that the chemical gradients in the pore waters within the upper ~350 m of the sediment column (Fig. 7A) are controlled by diffusion. Variations in strontium isotopes of sea water are well known for the Neogene, and provide a basis from which to assess whether the pore waters are 'younger' or 'older' than their host sediments. Pore waters down to >200 mbsf contain strontium with

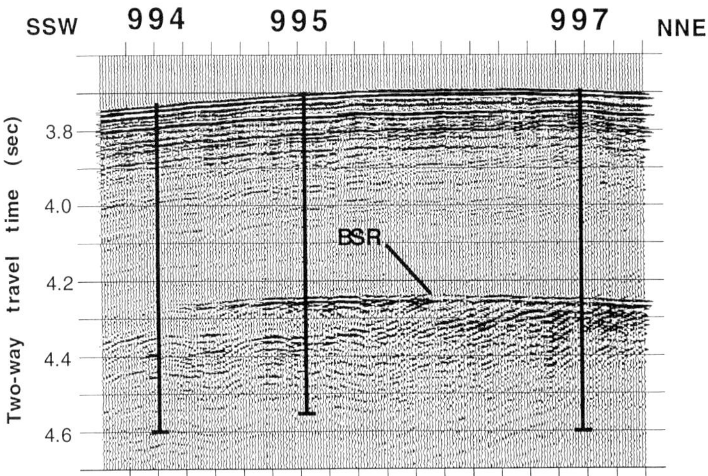

Fig. 2. Locations of ODP sites 994, 995 and 997 with respect to the BSR on a seismic reflection profile across the Blake Ridge (after Paull *et al.* 1996*b*).

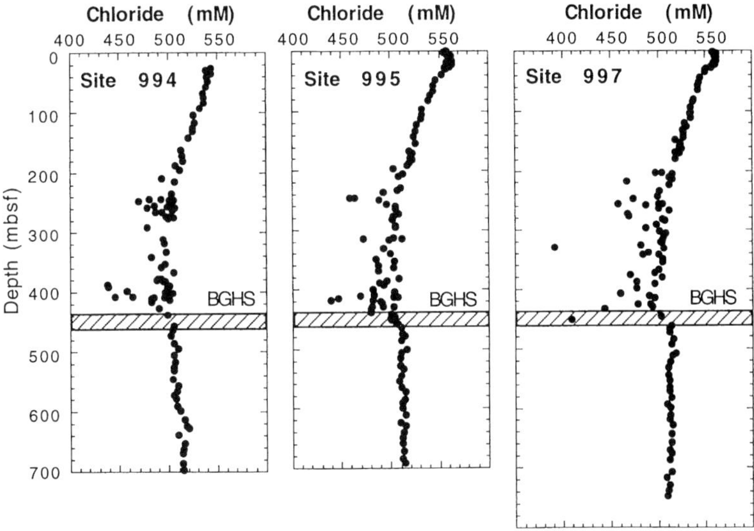

Fig. 3. Interstitial water chloride concentration values vs mbsf at ODP sites 994, 995 and 997 (after Paull *et al.* 1996*b*). The gas hydrate-bearing zone (~190–450 mbsf) is interpreted to occur where there are chloride excursions toward fresher values. Note that this zone is roughly coincident between all three drill sites. BGHS, base of gas hydrate stability.

nearly modern isotopic signatures and suggest a diffusive flux of strontium into the sediments (Fig. 7B). Again, diffusion of strontium into the sediments suggests that upwards advection is too slow to carry significant volumes of methane-bearing fluid from the underlying gas hydrate deposits toward the sea floor.

Point source discharge

Sea-floor venting of microbial gases occurs over the Blake Ridge Diapir (2167 m water depth). Gas-rich plumes were identified acoustically in the water column up to 320 m above a pock-marked sea floor associated with active chemo-

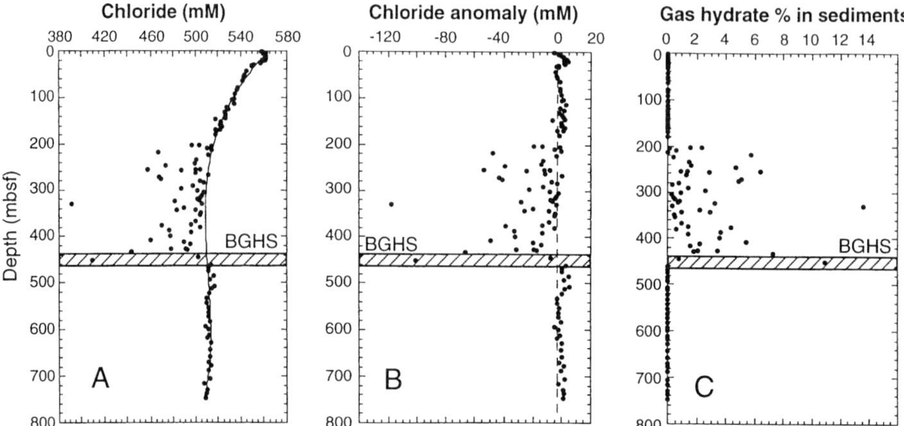

Fig. 4. (**A**) Interstitial water chloride concentration profile at Site 997 indicating a distinct freshening with depth which is highlighted by a zone of low and erratic values between 200 and 455 mbsf. A polynomial fit to the data above and below the 200–455 mbsf zone is shown. (**B**) Chlorinity anomalies calculated with respect to the polynomial. (**C**) Per cent of gas hydrate by volume calculated by assuming that the chloride anomalies in (B) are produced solely by gas hydrate decomposition during core recovery, and corrected by porosity using a linear fit to the porosity data (after Paull *et al.* 1996*b*).

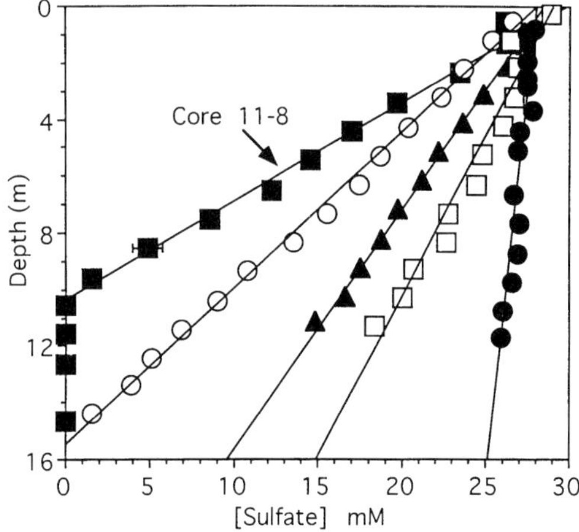

Fig. 5. Pore-water sulphate concentrations vs depth in five piston cores from the Carolina Rise and Blake Ridge. Concentration is expressed in millimolar (mM) units; measurement uncertainties are less than symbol size, unless error bars are present. Sulphate concentration gradients are linear and show a factor of 16 variation with the five illustrated cores representing a spectrum of differing sulphate gradients. Data from core 11-8, highlighted with an arrow (filled squares), show the highest gradient. Cores with the lowest sulphate gradients (e.g. core 31-24, filled circles) represent background gradients (after Borowski *et al.* 1996).

synthetic biological communities. Plumes and venting fluids emanate from sediments near a small fault that extends downward toward a dome in the bottom simulating reflector, suggesting that fluid and/or gas migration is associated with gas hydrate-bearing sediment below. The isotopic compositions of methane captured within the sediments at >50 mbsf at this vent (sampled with piston cores and at ODP Site 996) are consistent with the isotopic composition of methane sampled from the gas hydrate and free gas-bearing zones at more than 150 mbsf at DSDP Site 533 and ODP Leg 164 (Fig. 8). Apparently, faults form conduits for advective degassing

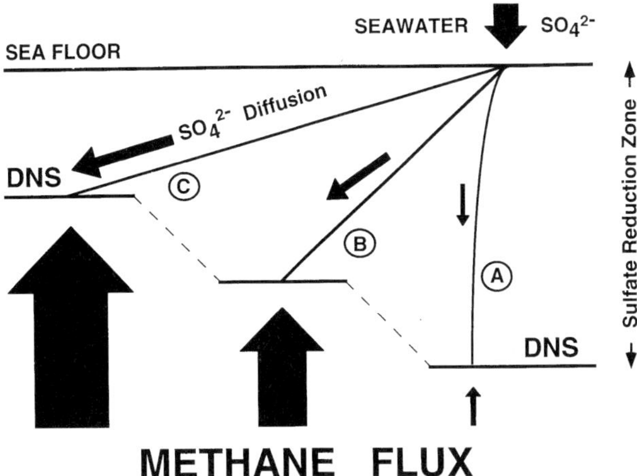

Fig. 6. Schematic diagram showing how upward methane flux may influence sulphate profiles and the depth-of-no-sulphate (DNS). Arrow size is proportional to flux. Typical sulphate profiles display convex-up curvature (Ⓐ) reflecting sulphate reduction of in situ sedimentary organic matter. Linear sulphate profiles (Ⓑand Ⓒ) result when focused sulphate consumption, driven by methane flux from below, occurs at the DNS at rates substantially greater than those for sulphate reduction of *in situ* sedimentary organic matter. In these cases, sulphate diffuses into the sediments and is consumed by reaction with methane at the base of the sulphate reduction zone. The rate of sulphate consumption and the steepness of the sulphate gradients, is thus controlled by the flux of methane from below (after Borowski *et al.* 1996).

of methane associated with the current gas hydrate related reservoirs on the Blake Ridge.

Slumping frequency over the gas hydrate deposits

Climatically controlled breakdown of natural gas hydrates may play a significant role in promoting continental-rise sediment instability (McIver 1977; Summerhayes *et al.* 1979; Carpenter 1981; Bugge *et al.* 1988; Paull *et al.* 1991; Field & Barber 1993; Kayen & Lee 1993; Popenoe *et al.* 1993). Gas hydrate formation will increase the mechanical strength of a sediment column by adding solid material to the interstitial pores. Conversely, when gas hydrates become unstable, they decompose to water plus methane gas, and act to destabilize the sediment column by allowing abnormally high porosity to occur hundreds of metres below the sediment surface. Thus, gravity-induced compression of the sediment column following gas hydrate decomposition may induce sediment failures (Paull *et al.* 1991; Kayen & Lee 1993). Therefore, the prediction has been made that continental-margin slumping related to gas hydrates should be more common during glacial sea-level lowstands (Carpenter 1981; Bugge *et al.* 1988; Paull *et al.* 1991; Field & Barber 1993; Kayen & Lee 1993; Popenoe *et al.* 1993).

During the last major ice age, sea level dropped about 120 m as water was transferred from the ocean to continental glaciers (Fairbanks 1989). Empirical data indicate that a sea-level induced pressure decrease would cause the BGHS to rise about 20 m, decreasing the thickness of the gas hydrate stability zone by several per cent (Dillon & Paull 1983; Paull *et al.* 1991). The chronology of Pleistocene sea-level fluctuations is best delineated by $\delta^{18}O$ values recorded in marine carbonates (Fig. 1). These data (Fig. 9) show that the lowest sea level associated with the last ice age occurred between 14 and 25 ka (Imbrie *et al.* 1984).

Scars left by slumping events are common on the continental rises of the world (Embly & Jacobi 1977; Summerhayes *et al.* 1979; Embly 1980). Many of the major slide scars occur in areas associated with gas hydrates (Summerhayes *et al.* 1979; Bugge *et al.* 1988; Field & Barber 1993; Kayen & Lee 1993; Popenoe *et al.* 1993), although the underlying gas hydrates are not necessarily the cause of the slumping (Booth *et al.* 1994). A significant step toward assessing whether a connection exists is to establish whether or not slumps over gas hydrate

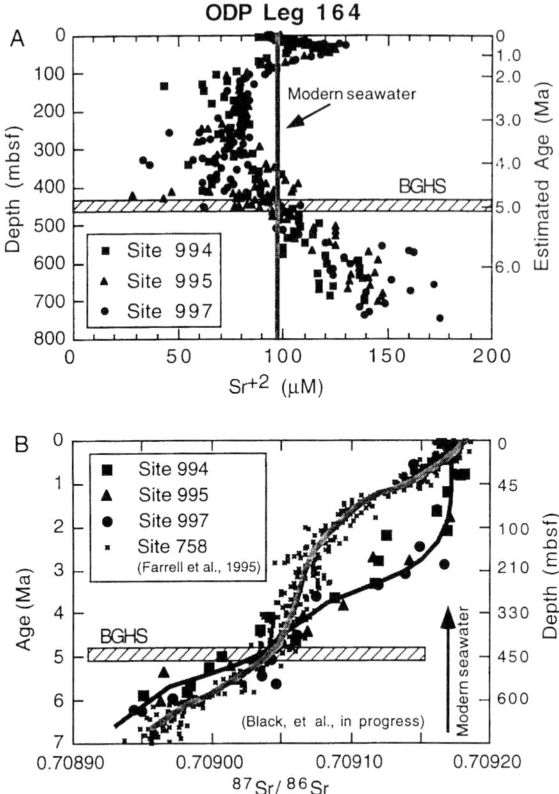

Fig. 7. (**A**) Pore-water strontium concentrations (μM) from ODP sites 994, 995 and 997. Anomalously low Sr^{2+} concentrations (below sea-water values) occur above the base of the gas hydrate stability (BGHS) and correspond to unusually high alkalinity (not shown). This pattern suggests that the likely sink for Sr^{2+} is gas hydrate-related carbonate precipitation. (**B**) Pore-water $^{87}Sr/^{86}Sr$ vs both estimated sediment age (based on shipboard nannofossil ages) and corresponding sediment depth (mbsf) for sites 994, 995 and 997. Leg 164 pore water data are compared to the chronostratigraphic reference curve for $^{87}Sr/^{86}Sr$ presented by Farrell *et al.* (1995). All $^{87}Sr/^{86}Sr$ ratios are normalized to SRM-987 = 0.71025.

deposits occurred more frequently during sea-level lowstands.

Slumping frequency over the Blake Ridge gas hydrate field

^{14}C data on sediment samples from the upper 7 m of the sediment column overlying a major continental-rise gas hydrate field on the southern Carolina Rise and inner Blake Ridge offshore the south-eastern United States show that glacial-aged sediments are under-represented (Fig. 9). The observation is consistent with the predicted association between sea-level lowstands and increased frequency of sea-floor slumping on continental margins containing gas hydrates and is not easily explained by local sedimentological conditions (Paull *et al.* 1996*a*).

Implications

The results of ODP Leg 164 indicate that the Blake Ridge sediments contain significant amounts of methane stored both as gas hydrate and as free gas below the base of gas hydrate stability. Thus, mechanisms of methane venting need to be further considered. Our current understanding of the processes and extent of outgassing from the Blake Ridge gas hydrate deposits suggest that methane only escapes from these gas hydrate reservoirs at point sources today. While methane outgassing at present seems to be relatively quiescent, outgassing activity may increase during sea-level lowstands. Moreover, the occurrence of sedimentary hiatuses over these deposits suggests that the sediments above gas hydrate deposits failed more

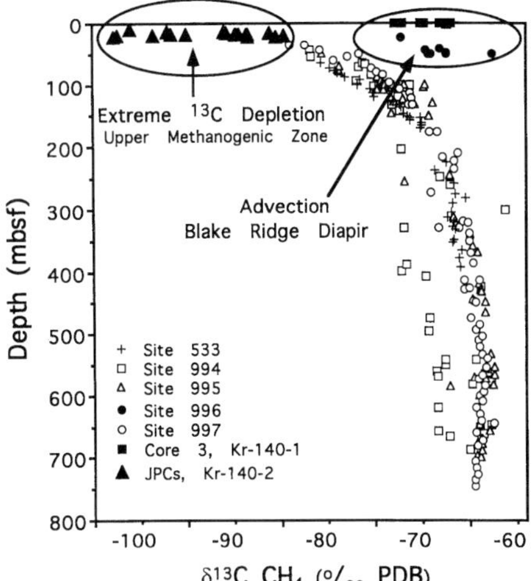

Fig. 8. Depth profiles showing carbon isotopic data for methane (CH_4) gas from DSDP Site 533 (Blake Ridge; Claypool & Threlkeld 1983; Galimov & Kvenvolden 1983), from ODP Leg 164 (sites 994, 995, 996 and 997), from a seep site developed on the Blake Ridge Diapir (Cruise Kr-140-1; Paull *et al.* 1995) and from jumbo piston cores (JPCs) taken at the Carolina Rise and Blake Ridge (Cruise Kr-140-2; Borowski *et al.* 1997) (see map, Fig. 1). Carbon isotopic composition is relative to the Pee Dee Belemnite (PDB) standard. Measurement uncertainties are shown with bars or are less than symbol size.

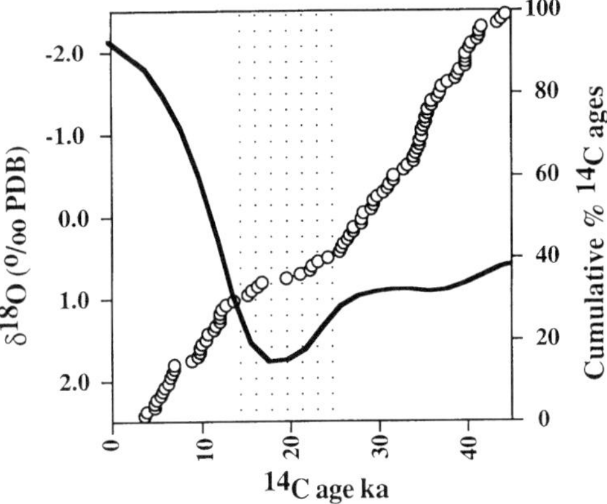

Fig. 9. Chronology of sea-level variations through late Quaternary time is indicated by variation in $\delta^{18}O$ of sea water (solid curve) (Imbrie *et al.* 1984). During last glacial maximum (~18 ka), sea level was about 120 m lower than present (Fairbanks 1989). Dot pattern indicates period between 14 and 25 ka, when eustatic sea level was especially low. Circles indicate cumulative frequency of 95 ^{14}C age measurements (open circles) on sediments from the upper 7 m of the Carolina Rise and Blake Ridge that display no visible slump scars. Inflection between 12 and 25 ka indicates that samples of this age are under-represented. PDB is Pee Dee Belemnite standard. Figure after Paull *et al.* (1996*a*).

frequently during sea-level lowstands, making it likely that more methane from the gas hydrate reservoir reached the sea floor during glacial lowstands.

References

BOOTH, J., WINTERS, W. & DILLON, W. 1994. Circumstantial evidence of gas hydrate and slope failure associations on the United States Atlantic Continental Margin. *In*: SLOAN, E. D., JR. (ed.) *Natural Gas Hydrates*. Annals of the New York Academy of Sciences, **715**, 487–489.

BOROWSKI, W., PAULL, C. & USSLER, W., III. 1996. Marine pore water sulfate profiles indicate methane flux from underlying gas hydrates. *Geology*, **24**, 655–658.

——, PAULL, C. & USSLER, W., III. 1997. Carbon cycling within the upper methanogenic zone of continental rise sediments : An example from the methane-rich sediments overlying the Blake Ridge gas hydrate deposits. *Marine Chemistry*, **57**, 299–311.

BUGGE, T., BELDERSON, R. & KENYON, N. 1988. The Storegga slide. *Philosophical Transactions of the Royal Society of London*, **325**, 357–388.

CARPENTER, G. B. 1981. Coincident sediment slump/clathrate complexes on the U.S. Atlantic slope. *Geo-Marine Letters*, **1**, 29–32.

CLAYPOOL, G. & THRELKELD, C. 1983. Anoxic diagenesis and methane generation in sediments of the Blake Outer Ridge, Deep Sea Drilling Project Site 533, Leg 76. *In*: SHERIDAN, R., GRADSTEIN, F. *et al.* (eds) *Initial Reports of the Deep Sea Drilling Project*, Volume 76. US Government Printing Office, Washington, DC, **76**, 391–402.

DICKENS, G., PAULL, C. & WALLACE, P. 1997. Direct measurement of *in situ* methane quantities in a large gas-hydrate reservoir. *Nature*, **385**, 426–428.

DILLON, W. & PAULL, C. 1983. Marine gas-hydrates: II Geophysical evidence. *In*: COX, J. (ed.) *Natural Gas Hydrates: Properties, Occurrence and Recovery*. Butterworths, Woburn, MA, 73–90.

EMBLY, R. 1980. The role of mass transport in the distribution and character of deep-ocean sediments with special reference to the North Atlantic. *Marine Geology*, **38**, 28–50.

—— & JACOBI, R. 1977. Distribution and morphology of large submarine sediment slides and slumps on Atlantic continental margins. *Marine Geotechnology*, **2**, 205–228.

FAIRBANKS, R. 1989. A 17,000-year glacio-eustatic sealevel record. Influences of glacial melting rates on the Younger Dryas event and deep-ocean circulation. *Nature*, **342**, 637–642.

FARRELL, J., CLEMENS, S. & GROMET, L. 1995. Improved reference curve of late Neogene seawater $^{87}Sr/^{86}Sr$. *Geology*, **23**, 403–406.

FIELD, M. & BARBER, J., JR. 1993. A submarine landslide associated with shallow sea-floor gas and gas hydrates off Northern California. *In*: SCHWAB, W. *et al.* (eds) *Submarine Landslides:*

Selective Studies in the U.S. Exclusive Economic Zone. US Geological Survey Bulletin, **2002**, 151–157.

KAYEN, R. & LEE, H. 1993. Slope stability in regions of sea-floor gas hydrate: Beaufort Sea continental Slope. *In*: SCHWAB, W. *et al.* (eds) *Submarine Landslides: Selective Studies in the U.S. Exclusive Economic Zone*. US Geological Survey Bulletin, **2002**, 97–103.

GALIMOV, E. & KVENVOLDEN, K. 1983. Concentrations and carbon isotopic compositions of CH_4 and CO_2 in gas from sediments of the Blake Outer Ridge, Deep Sea Drilling Project Leg 76. *In*: SHERIDAN, R., GRADSTEIN, F. *et al.* (eds) *Initial Reports of the Deep Sea Drilling Project*, Volume 76. US Government Printing Office, Washington, DC, 403–410.

HOLBROOK, W., HOSKINS, H., WOOD, W., STEPHEN, R., LIZZARRALDE, D. & LEG 164 SCIENCE PARTY. 1996. Methane hydrate and free gas on the Blake Ridge from vertical seismic profiling. *Science*, **273**, 1840–1843.

IMBRIE, J. *et al.* 1984. The orbital theory of Pleistocene climate. Support from a revised chronology of the marine $\delta^{18}O$ record. *In*: BERGER, A. (ed.) *Milankovitch and Climate, Part 1*. D. Reidel, Higham, MA, 269–305.

MCIVER, R. D. 1977. Hydrates of natural gas – Important agent in geologic processes. *Geological Society of America Abstracts with Programs*, **9**, 1089–1090.

PAULL, C., USSLER, W. & DILLON, W. 1991. Is the extent of glaciation limited by marine gas-hydrates? *Geophysical Research Letters*, **18**, 432–434.

——, USSLER, W., III, BOROWSKI, W. & SPIESS, F. 1995. Methane-rich plumes on the Carolina Continental Rise. Associations with gas hydrates. *Geology*, **23**, 89–92.

——, BUELOW, W., USSLER, W., III & BOROWSKI, W. 1996*a*. Increased continental-margin slumping frequency during sea-level lowstands above gas hydrate-bearing sediments. *Geology*, **24**, 143–146.

——, MATSUMOTO, R., WALLACE, P. *et al.* 1996*b*. *Proceedings of the Ocean Drilling Program, Initial Reports*, College Station, TX. Ocean Drilling Program, **164**.

POPENOE, P., SCHMUCK, E. & DILLON, W. 1993. The Cape Fear landslide: Slope failure associated with salt diapirism and gas hydrate decomposition. *In*: SCHWAB, W. *et al.* (eds) *Submarine Landslides: Selective Studies in the U.S. Exclusive Economic Zone*. US Geological Survey Bulletin, **2002**, 40–53.

RODRIGUEZ, N., PAULL, C. & FULLAGAR, P. 1998. Isotopically modern Sr^{2+} in gas hydrate-bearing sediments on the Blake Ridge (ODP Leg 164). In preparation.

SUMMERHAYES, C., BORNHOLD, B. & EMBLEY, R. 1979. Surficial slides and slumps on the continental slope and rise of South West Africa: A reconnaissance study. *Marine Geology*, **31**, 265–277.

Geochemistry of gas hydrates and associated fluids in the sediments of a passive continental margin. Preliminary results of the ODP Leg 164 on the Blake Outer Ridge

R. THIÉRY[1], R. BAKKER[2], C. MONNIN[3] &
THE ODP LEG 164 SHIPBOARD SCIENTIFIC PARTY

[1] *CREGU, BP 23, 54 501 Vandoeuvre-les-Nancy, France*
[2] *Geologisches-Palaeontologisches Institut, Heidelberg, Germany*
[3] *Laboratoire de Gochimie, Université Paul Sabatier, rue des Trente-Six Ponts, Toulouse, France*

Abstract: Marine sediments containing gas hydrates have been drilled up to a depth of 750 m below the sea floor (mbsf) at the Blake Outer Ridge during the ODP Leg 164. Gas hydrates are present between 190 and 450 mbsf. Shipboard analyses of gases and interstitial waters are interpreted with the help of thermodynamic models for gas hydrates and aqueous solutions. Two aspects of the influence of gas hydrates on the chemistry of associated fluids are investigated here. The first one is related to the methane/ethane ratio, which exhibits a sudden change of trend at the base of the gas hydrate zone. This phenomenon could be due to gas hydrates, which act as a concentration barrier for ethane. The second aspect concerns the presence of interstitial waters less saline than sea water above and inside the gas hydrate zone. This could result from the upward expulsion of saline waters during the compaction of sediments in the gas hydrate zone.

New advances on the geochemistry of marine gas hydrates have been obtained during ODP Leg 164 on the Blake Outer Ridge near the southeast coast of North America. This region is characterized by the presence of a strong reflector on seismic profiles, which mimics the topography of the sea floor. This reflector, called the BSR (bottom simulating reflector), is believed to be caused by the high acoustic impedance contrast between a high-velocity gas hydrate-rich layer overlying a low-velocity free gas-rich layer (Minshull & White 1989; Hyndman & Spence 1992; Singh *et al.* 1993; Katzman *et al.* 1994). Three sites (994, 995 and 997) have been drilled along a transect near the crest of the Blake Ridge up to a depth of 750 m below the sea floor (mbsf). Sites 995 and 997 are characterized by the presence of a strong BSR on the seismic profiles, whereas no BSR is distinguishable at Site 994. The drillings have confirmed the presence of gas hydrates in the sediments, as demonstrated by direct sampling on the cores and numerous indirect geochemical and diagraphic evidence. Sediments are homogeneous and mainly composed of greenish nannofossil-rich clays. Thus, the Blake Ridge provides an ideal site to investigate the influence of gas hydrates on the geochemistry of fluids circulating through the sediments. The gas hydrate zone is located between 190 and 450 mbsf. Gas hydrates are finely disseminated throughout this zone. Estimations of the mean gas hydrate content, based on chloride anomalies and logging results, range from 3 to 6% of the volume of porous interstitial space. Two zones, locally richer in gas hydrates, have been observed. The first one is located at depths between 200 and 230 mbsf, whereas the second one is between 420 and 450 mbsf. Two aspects of the influence of gas hydrates on the geochemistry of fluids are described here. The first aspect is related to the methane/ethane ratio, which exhibits a change of its trend near the base of the gas hydrate zone. The second aspect concerns the low salinity of pore waters above and inside the gas hydrate zone. Measured chloride contents of interstitial waters indeed reach values which are about 10% lower than that of sea water. But, before any study of these phenomena, a good understanding of possible phase assemblages involving gas hydrates, aqueous solution and vapour phase is necessary. This will lead us to distinguish different zones in the sedimentary column.

The different zones

Gas hydrates are stable only when certain conditions of low temperature, high pressure and sufficient amount of methane are met. In the Blake Ridge, temperature and pressure are

THIÉRY, R., BAKKER, R., MONNIN, C. & THE SHIPBOARD SCIENTIFIC PARTY OF ODP LEG 164 1998. Geochemistry of gas hydrates and associated fluids in the sediments of a passive continental margin. Preliminary results of the ODP Leg 164 on the Blake Outer Ridge. *In:* HENRIET, J.-P. & MIENERT, J. (eds) *Gas Hydrates: Relevance to World Margin Stability and Climate Change*. Geological Society, London, Special Publications, **137**, 161–165.

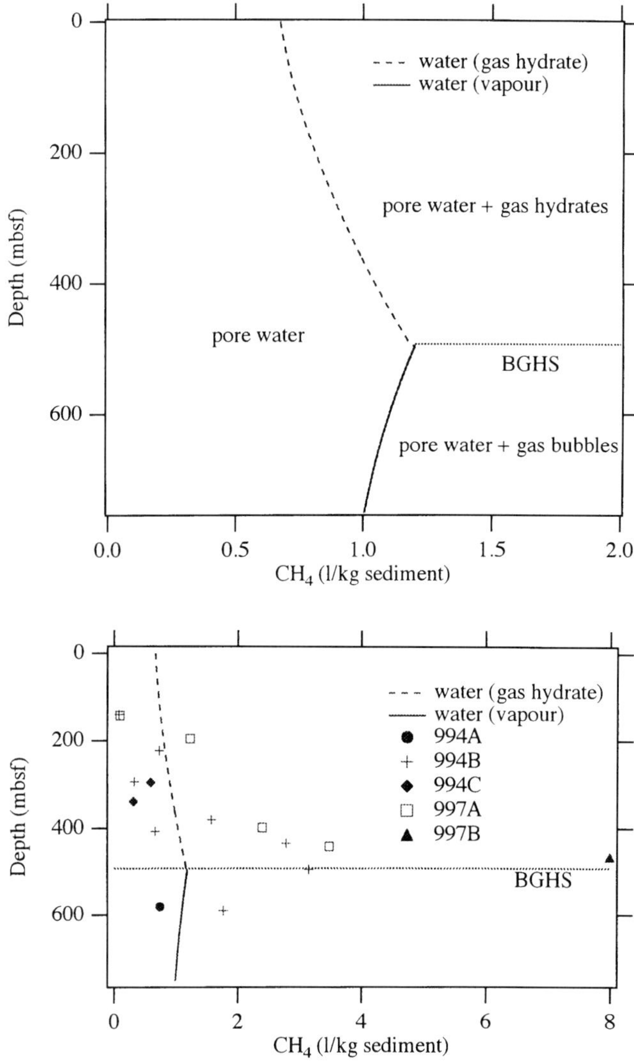

Fig. 1. (**a**) Solubility curves of methane in an aqueous phase in equilibrium either with gas hydrates (labelled as water (gas hydrates)) or with gas phase (labelled as water (vapour)). The BGHS is indicated by the dotted curve. (**b**) Comparison of the phase diagram with methane amounts measured (Dickens, pers. comm.) on PCS samples.

favourable for the formation of gas hydrates between 0 and 450 mbsf. However, no gas hydrates have been sampled or detected by geochemical methods or logging between 0 and 190 mbsf. The absence of gas hydrates in the upper layers of the sedimentary profile may be explained by insufficient quantities of methane. In order to check this hypothesis, it is necessary to quantify the minimum amount of methane for the formation of gas hydrate in a saline aqueous solution. The methane content required for forming gas hydrates depends on the temperature, pressure, water composition and quantity of water in the sediments.

In situ measurements were carried out using downhole tools, such as the WSTP (Westbrook *et al.* 1994) for temperature and the PCS (Petitgrew 1992) for pressure. The water composition was measured by ionic chromatography, and the quantity of water by weighing a piece of sediment before and after being dried in an oven. As there are no experimental data on the methane solubility in a saline aqueous phase at hydrate conditions, thermodynamic models, as

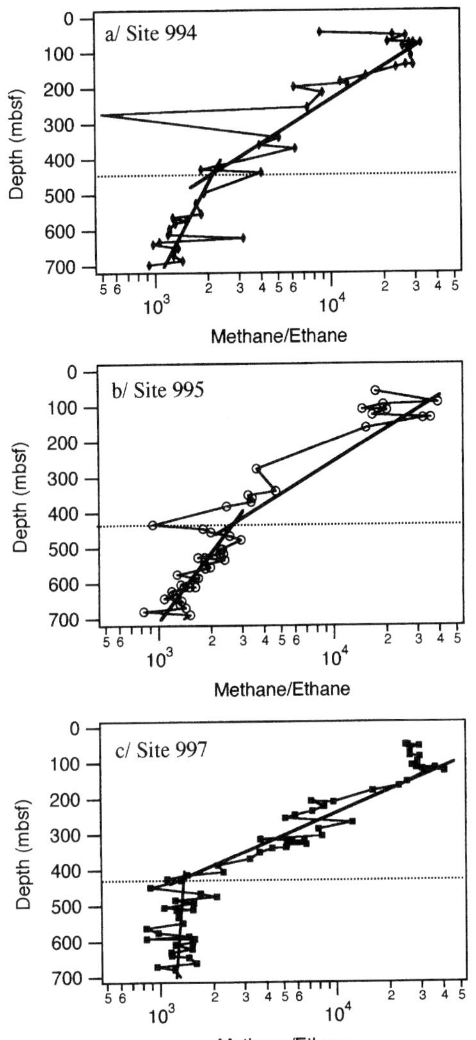

Fig. 2. The methane/ethane ratio vs depth measured in 'free gas': (**a**) Site 994, (**b**) Site 995, (**c**) Site 997.

described in Dubessy *et al.* (1992) and Bakker *et al.* (in press), have been used to estimate this parameter. The model of Duan *et al.* (1992) was applied to model the methane solubility in an aqueous saline solution in equilibrium with a vapour phase.

The result is a phase diagram, which is shown in Fig. 1a and was calculated using the conditions prevailing at Site 997 (geothermal gradient of 36.9°C km^{-1}, hydrostatic regime pressure). The diagram is composed of three fields. The first one, on the left, applies when the amount of methane is low (below 0.8–1.2 l under STP (i.e. 25°C and 1 atm) conditions per kg of sediment). When the methane content is above these values, a CH_4-rich phase appears, which is either a gas hydrate (above the base of gas hydrate stability, abbreviated as the BGHZ, at 490 mbsf) or a gas phase (below the BGHZ). *In situ* measurements of the amount of methane in the sediments were made possible using the PCS (Pettigrew 1992). Comparison of the diagram with values measured by Dickens (pers. comm.) confirms that zones containing no gas hydrates are characterized by pore waters undersaturated in methane (in particular, the zone between 0 and 190 mbsf). This demonstrates that the methane availability is a key for discerning the gas hydrate-rich zones and understanding the methane distribution.

Indeed, methane, produced by the degradation of organic matter by methanogenesis bacteria, migrates upward to the sea floor. Between 0 and 30 mbsf the methane content is negligible, as it is consumed in the reduction of sulphate migrating downward from the sea. A diffusion gradient for the methane probably occur between 30 and 190 mbsf. The top of the gas hydrate zone, located at 190 mbsf, thus corresponds to the depth at which the amount of methane is just sufficient for the formation of gas hydrates (around 0.8 l kg^{-1} of sediment). Between 0 and 190 mbsf, methane is the limiting factor that prevents the formation of gas hydrates. However, abundant quantities of methane are present in the gas hydrate zone and below the gas hydrate zone, as demonstrated by the PCS data (Fig. 1b). The base of the gas hydrate zone, located at 490 mbsf at Site 997, corresponds to the depth at which limiting temperature and pressure values for the stability of gas hydrates are reached. Below 490 mbsf bubbles of free gas appear.

The methane/ethane ratio

Methane/ethane ratios have been measured using the sampling technique, commonly called the 'free gas' sampling technique. A syringe, piercing the plastic around the core of sediments, is used to collect the remaining gases trapped in the sediments. Gases are then analysed on a gas chromatograph. It is important to note that these samples provide only an estimate of the bulk amount of volatile hydrocarbons (methane, ethane, etc.) that were dissolved in pore waters and either trapped in gas hydrates or in gas bubbles.

The methane/ethane ratio is plotted in Fig. 2a–c for sites 994, 995 and 997 as a function of the depth. The ratios are high, well above 1000.

This confirms that most of the methane in these sediments have been produced biogenically. The methane/ethane ratio decreases exponentially with depth. In a logarithmic plot (Fig. 2a–c) the ratio can thus be represented by a straight line. Such a behaviour of the methane/ethane ratio has already been observed at other ODP sites of continental margins. However, at sites 994, 995 and 997, a sudden change of the slope of the methane/ethane line is observed near the base of the gas hydrate zone.

It can be noted that there is a gradation in this effect from Site 994 to Site 997. This observation might indicate that gas hydrates have an influence on the diffusion of ethane. Under the temperature conditions of the upper 750 m of the sedimentary column at the Blake Ridge (3–25°C), ethane is more soluble than methane in aqueous water, as indicated, for example, by comparing the values of the Henry constant for methane (37 600 atm) and ethane (26 300 atm) at 20°C (Perry *et al.* 1984). Gas hydrates, too, tend to preferentially incorporate ethane instead of methane when they are in equilibrium with a gas phase, as indicated by thermodynamic models describing the properties of gas hydrates (Dubessy *et al.* 1992; Bakker *et al.* in press). For example, a gas phase containing 1% of ethane and 99% of methane in mole fraction is predicted to be in equilibrium with a gas hydrate containing 3% of ethane and 97% methane in mole fraction expressed on a water-free basis at 25°C and 210 bar. However, the fractionation of methane and ethane between a gas hydrate and an aqueous solution is not known and remains to be estimated with predictive thermodynamic models. A working hypothesis for explaining the change of the trend of the methane/ethane ratio is that gas hydrates preferentially incorporate ethane, and thus act as a concentration barrier for ethane. It is interesting to note that higher concentrations of other hydrocarbons, such as isobutane, *n*-heptane, etc., are also observed at Site 997 near the BGHS (up to 500 ppm for isobutane in a PCS sample at Site 997).

The salinity of pore waters

Extensive sampling and analysis of interstitial waters have been carried out at sites 994, 995 and 997. The chloride contents vs depth at Site 994 are shown in Fig. 3. The chloride profiles of sites 994, 995 and 997 are very similar. The curve is characterized by irregular and anomalous freshening spikes in the gas hydrate zone. These spikes result from the melting of gas hydrates during the recovery, as gas hydrates do not contain any salts. In addition, the curve

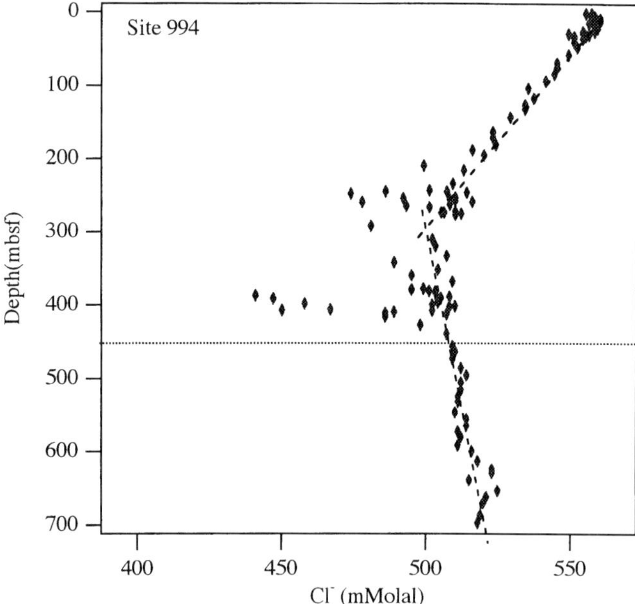

Fig. 3. Chloride content vs depth in interstitial waters at Site 994 (chloride contents are very similar at sites 995 and 997).

exhibits a background trend towards lower salinities in the gas hydrate zone. One possible explanation for this involves gas hydrates. When sediments compact, they expel water. Such a mechanism normally has no incidence on the salinity of pore waters. However, in the gas hydrate zone, salts are excluded from gas hydrates. Thus, waters, that are expelled in the gas hydrate zone as a result of the compaction, are expected to be more saline. This should result in a preferential loss of ions in the gas hydrate zone. Moreover, because of the subsidence, sediments that were in the gas hydrate zone reach the 'free gas' zone below the BGHZ. Gas hydrates melt and release fresh waters that should also contribute to lowering the salinity of pore waters at these depths.

Conclusion

The data gathered during ODP Leg 164 allow us to better understand the influence of gas hydrates on the geochemistry of fluids. Two aspects have been described here. The first one is related to the change of trend of the methane/ethane ratio at the base of the gas hydrate zone. The second aspect concerns the presence of less saline pore waters in the gas hydrate zone. Both phenomena are tentatively explained and would be a consequence of the formation of gas hydrates in the sediments. In the first case, gas hydrates would preferentially concentrate ethane and heavier hydrocarbons with respect to methane, and thus modify the methane/ethane profile. In the second case, crystallization of gas hydrates during the compaction of sediments leads to a preferential migration of dissolved ions to the sea, and thus to a decrease in the salinity of pore waters in the gas hydrate zone. At the present stage, these are only working hypotheses that need to be developed more quantitatively and compared with other studies.

References

BAKKER, R., DUBESSY, J. & CATHELINEAU, M. 1996. Improvements in clathrate modeling. I: the H_2O-CO_2 system with various salts. *Geochimica et Cosmochimica Acta*, **60**, 1657–1681.

DUAN, Z. N., MØLLER, J., GREENBERG, J. & WEARE, J.H. 1992. The prediction of methane solubility natural waters to high ionic strength from 0 to 250°C and from 0 to 1600 bar. *Geochimica et Cosmochimica Acta*, **56**, 1541–1460.

DUBESSY, J., THIERY, R. & CANALS, M. 1992. Modeling of gas phase equilibria involving mixed gas clathrates: Application to the determination of molar volume of the vapour phase and salinity of the aqueous solution in fluid inclusions. *European Journal of Minarology*, **4**, 872–884.

HYNDMAN, R. D. & SPENCE, G. D. 1992. A seismic study of methane hydrate marine bottom simulating reflectors. *Journal of Geophysical Research*, **97**, 6683–6698.

KATZMAN, R., HOLBROOK, W. & PAULL, C. 1994. Combined vertical-incidence and wide-angle seismic study of a gas hydrate zone, Blake Ridge. *Journal of Geophysical Research*, **99**, 17 975–17 995.

MINSHULL, T. & WHITE, R. S. 1989. Sediment compaction and fluid migration in the Makran accretionary prism. *Journal of Geophysical Research*, **94**, 7387–7402.

PERRY, R., GREEN, D. & MALONEY, J. 1984. *Perry's Chemical Engineers's Handbook*, 6th edition. McGraw-Hill, New York.

PETITGREW, T. 1992. *The Design and Preparation of a Wireline Pressure Core Sampler (PCS)*. Ocean Drilling Program Technical Note, College Station, TX. Ocean Drilling Program, **17**.

SINGH, C., MINSHULL, T. & SPENCE, G. 1993. Velocity structure of a gas hydrate reflector. *Science*, **260**, 204–207.

The occurrence of gas hydrates in Eastern Mediterranean mud dome structures as indicated by pore-water composition

G. J. DE LANGE[1] & H.-J. BRUMSACK[2]

[1] *Utrecht University, Institute of Earth Sciences, Department of Geochemistry, Budapestlaan 4, 3584 CD Utrecht, The Netherlands*
[2] *Oldenburg University, Institute of Chemistry and Biology of the Ocean, P.O. Box 2503, D-26111 Oldenburg, Germany*

Abstract: During ODP Leg 160 to the Eastern Mediterranean, two mud dome structures have been sampled. The massive presence of gas hydrates at relatively shallow depths in the sediment of one of these, Milano Dome, contrasts to that of the other, Napoli Dome, where gas hydrates are unlikely to be present at depths greater than 1 m below the sea floor (mbsf). Some observations from very shallow pore-waters at Napoli Dome, however, suggest that some gas hydrates must be present above 1 mbsf. In the case of Milano Dome, it is likely that a massive cap of gas hydrates is overlying natural gas that cannot escape due to this cap and its shape. In this report, some of the mud dome pore-water results of Ocean Drilling Program Leg 160 to the Eastern Mediterranean are discussed in relation to the inferred presence of gas hydrates. In addition, the total quantity of methane, including gas hydrates and 'free gas', has been estimated for Milano Dome ($\sim 5 \times 10^9 m^3 CH_4$) and for the Eastern Mediterranean Ridge mud dome structures ($\sim 1 \times 10^{14} m^3 CH_4$).

The presence of gas hydrates in oceanic sediments was first postulated on the basis of seismic observations (e.g. Tucholke *et al.* 1977). Soon it appeared that vast amounts of gas hydrates must occur in sediments of continental slopes and rises if the bottom simulating reflector (BSR), and the commonly underlying acoustic transparent layer, are to be attributed to gas hydrates and underlying gas-charged sediments (Shipley *et al.* 1979). On the basis of BSR depth and the experimentally determined stability field for gas hydrates, thermal gradients for deep-sea sediments have been estimated (e.g. Shipley *et al.* 1979; Yamano *et al.* 1982; Kvenvolden & McMenamin 1980; De Roo *et al.* 1983; Kvenvolden & Kastner 1990; Hyndman *et al.* 1992). The stability field of gas hydrates appears to depend not only on pressure (water + sediment depth) and temperature, but also on the chemical composition of the gas hydrates, and on the salinity of the pore-water. The methane hydrate dissociation in sea water appears to be at a temperature that is approximately 1°C lower than it is in pure water, whereas that in nearly NaCl-saturated water is 5°C lower (De Roo *et al.* 1983; Dickens & Quinby-Hunt 1994). These experiments appeared to closely follow the stability conditions predicted by thermodynamic calculations (e.g. van der Waals & Platteeuw 1959).

Pure and fully saturated methane hydrate has an average composition of CH_4 $5.75H_2O$. Upon disintegration the volumetric ratio of methane to water is approximately 164 (Davidson *et al.* 1978; Kvenvolden & Kastner 1990). Although some disagreement exists in the literature (Miller 1974; Makogon 1981; Handa 1990), it seems that for the formation of gas hydrates a large amount and a high supply rate of methane are needed as methane concentrations must exceed the solubility level at the *in situ* pore-water conditions, i.e. must exceed a concentration of a few tenths of a mole per litre (e.g. Ginsburg 1996; Kvenvolden 1996). The conditions of a sufficiently high pore-water methane concentration can be met during at least two typical situations: (1) a supply of organic matter that is sufficiently large to generate enhanced methanogenic decomposition of organic matter in the sediment (e.g. the Peru Margin), and (2) a large upward methane flux often related to fault zones or other conduits such as diapirs, mud volcanism (Legs 67, 76, 96, 127, 146, 160 and 164; Black Sea and Eastern Mediterranean). The methane from the first source is biogenic, whereas that from the second source is mainly thermogenic to mainly biogenic. The different origins are reflected in their stable isotope composition and in the methane to ethane + propane (C1/(C1 + C2)) ratio. If $\delta^{13}C-CH_4$ is less than −60‰ and the C1/(C1 + C2) ratio is larger than 1000, the origin is postulated to be biogenic, whereas a ($^{13}C-CH_4$ of larger than −60‰ and a C1/

De Lange, G. J. & Brumsack, H.-J. 1998. The occurrence of gas hydrates in Eastern Mediterranean mud dome structures as indicated by pore-water composition. *In:* Henriet, J.-P. & Mienert, J. (eds) *Gas Hydrates: Relevance to World Margin Stability and Climate Change*. Geological Society, London, Special Publications, **137**, 167–175.

(C1 + C2) ratio of less than 1000, points to a thermogenic origin (Pflaum *et al.* 1985).

One of the first reported observations of gas hydrates in marine sediments was for the Black Sea (Yefremova & Zhizhchenko 1974: references in Brooks *et al.* 1984). The first observations of gas hydrates, and of indirect evidence for the initial presence of gas hydrates, were in sediments recovered by DSDP during Leg 66 and Leg 67 at the Middle America Trench (e.g. Hesse & Harrison 1981; Harrison *et al.* 1982). During several subsequent legs, the initial presence of gas hydrates has been postulated in a similar way (Legs 84, 96, 112, 127, 131, 146, 160 and 164). Leg 164 to the Blake Ridge was devoted to gas hydrates (ODP Leg 164 Shipboard Scientific Party 1996); gas hydrates were found to occur mainly dispersed, and to occupy a few per cent of the pore space.

Gas hydrates

Studies of deep-sea gas hydrates have been seriously hampered by the difficulties in recovering gas hydrates because of their instability at sea-level conditions (e.g. Fig. 1). In addition, gas hydrates mostly occur at depths in the sediment that are beyond the reach of normal gravity- and piston-coring. As a consequence, most reports on the occurrence and composition of gas hydrates have come from ODP research efforts. The presence of gas hydrates in sediments is commonly found (or inferred) to be dispersed in the sediments which, in view of their rapid disintegration upon retrieval from the bottom, makes their recovery extremely difficult. The high-pressure–low-temperature stability field further complicates the proper preservation and analysis of the samples. If corrected for air

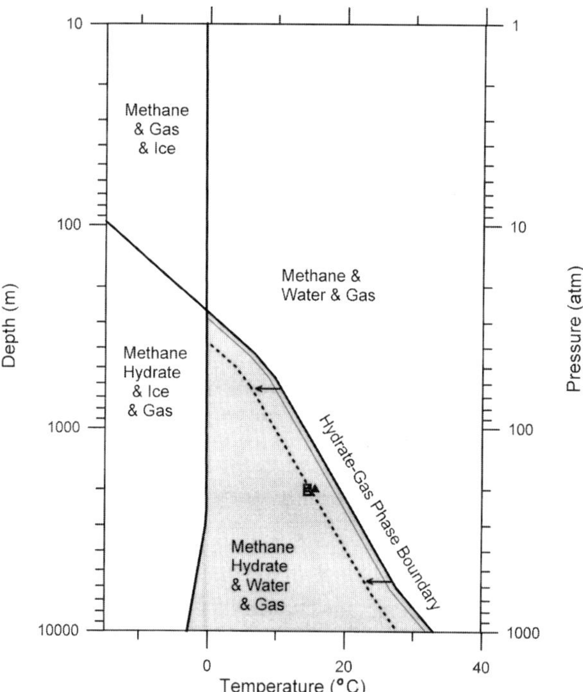

Fig. 1. Phase boundary diagram demonstrating the gas hydrate–freshwater stability field (in grey). Boundaries are given for the pure methane–pure water system. Redrawn after Kvenvolden & McDonald (1984). Directions of hydrate gas phase boundary (HGPB) shift towards stabilities at higher salinities are indicated by arrows (see text also). The thin line and dashed line parallel to the freshwater HGPB indicate the sea-water HGPB and the brine HGPB, respectively (after Hesse & Harrison 1981 and De Roo *et al.* 1983). The brine line is for a 5 M Cl^- solution, comparable to the one assumed for the brine in this study. The open squares and filled triangles indicate the depth–temperature position for crestal sites of, respectively, sediments with inferred presence of gas hydrates at Milano Dome mud volcano and sediments with inferred absence of gas hydrates at Napoli Dome, Eastern Mediterranean. Plotted values are for 10 mbsf depth at all sites; the two ND sites have nearly identical depth and temperature, and consequently coincide.

contamination, all reported analyses indicate that usually more than 99% of the 'encaged' gas is methane, and that the $\delta^{18}O$ of the 'cage' water is approximately +2.5‰.

Pore water derived initial presence of gas hydrates

The pronounced decrease in pore-water salinity has often been interpreted as a good indication for the initial occurrence of gas hydrates in the recovered sediments (Harrison *et al.* 1982). During their formation, gas hydrates not only consume vast amounts of methane, but also extract pure water with a relatively enhanced ^{18}O content from the pore-water (Davidson *et al.* 1983). As a consequence, the remaining pore-water is more saline and is slightly depleted in $\delta^{18}O$. Along the concentration gradient, salinity diffuses to overlying (and underlying) sediment intervals that do not contain gas hydrates. Upon recovery and subsequent disintegration of the gas hydrates, the pore-water in the intervals with initial gas hydrates will demonstrate a decreased chlorinity and enhanced levels of $\delta^{18}O$, whereas in the overlying intervals it should have an increased chlorinity and lower levels of $\delta^{18}O$.

However, not all of the pore-water salinity gradients that decrease with depth can be attributed to the dissolution of gas hydrates during recovery. Decreases may also result from mineral dehydration upon increasing pressure, membrane filtration, dewatering of subducting sediment, input of meteoric waters and from the deposition of a brackish near-coastal sediment 'slab' into the deep sea (Kvenvolden & Kastner 1990; De Lange 1983 and references herein). Therefore, additional confirmation is needed by alternative analyses and observations of pore-water constituents such as methane content and $\delta^{18}O$. Furthermore, the chlorinity in the pore-water overlying the interval of inferred gas hydrate content should be enhanced over that in the bottom water (see above). Examples of this have been reported by Kvenvolden & Kastner (1990, their Fig. 6). In contrast, Harrison *et al.* (1982) were unable to detect such increased salinity, whereas they clearly demonstrated the initial presence of gas hydrates on the basis of their pore-water chlorinity and $\delta^{18}O$ values. It is possible that only enhanced rates of gas hydrate formation may lead to discernible enhanced levels of the pore-water chlorinity in the overlying sediments. Most of the gas hydrate findings have been reported on the basis of pore-water composition. A decrease in the pore-water salinity is one of the more often used diagnostic tools.

Materials and methods

Pore waters for sites 970 and 971 were extracted and analysed using routine ODP methods

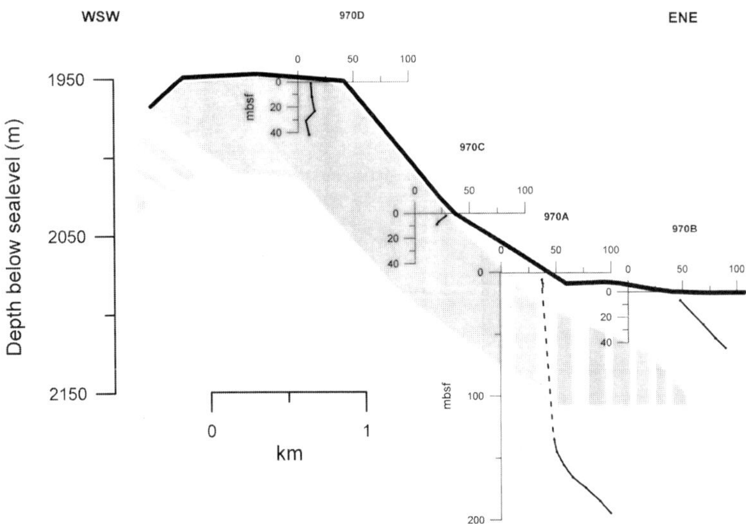

Fig. 2. Cross-section of Milano Dome site, Eastern Mediterranean. Salinity profiles are shown at the approximate positions relative to the dome structure. The grey area indicates gas hydrate occurrence as inferred from pore-water data (see text). Bottom-water salinity is 38‰.

(Gieskes *et al.* 1991), whereas those for cores ND2A, MT3A and KC11 were carried out using shipboard routine and analytical methods described previously (De Lange 1986, 1992; Van Santvoort *et al.* 1996).

Results and discussion

During Leg 160 to the Eastern Mediterranean, two different mud domes (Milano and Napoli) were drilled on the margin of the Mediterranean Ridge. The Milano Dome (MD) seems to be inactive at present, whereas the Napoli Dome (ND) shows evidence of recent gas, fluid and mud flows (e.g. Robertson *et al.* 1996). The salinity of interstitial waters ranges from brackish to brine, and the solid phase is mainly breccia (De Lange *et al.* 1996; Emeis *et al.* 1996). A clear difference exists between the recovered pore-waters of the MD and ND crest sites, the former being brackish and the latter brine-like (Figs 2–4).

The enhanced salinity at ND sites is almost entirely due to an increase in dissolved halite content (sodium chloride). The ND brine is, therefore, likely to be associated with early stage evaporites or 'relict' brines (De Lange *et al.* 1990; Vengosh *et al.* submitted). In fact, halite crystals and crusts have been found in some of the sediments of ND (Robertson *et al.* 1996). Results from pore-waters in previously recovered ND crestal sediments in box cores ND2A and MT3A, and piston core KC11, fit nicely to those of holes 971D and 971E for most elements (not shown in the figure).

The presence of low-salinity waters at MD is unlikely to be related to the expulsion of low-salinity waters from sources deeper downhole because most sediments of the Mediterranean sea floor are underlain by Messinian evaporites. Consequently, at most sites in the Eastern Mediterranean, steep increases in salinity vs depth have been found (Van Santvoort & De Lange 1996). The strong brine influence, i.e. increase in salinity vs depth, can be observed in the non-crestal sites from ND and MD (Figs 2 and 3). The correspondence between the non-crestal sites indicates that the pore-waters in both mud domes are brine-dominated. In the crestal sites, however, the pore-waters appear to be 'fresh-water'-dominated at MD, whereas they are brine-dominated at ND. If the brackish pore-waters at MD were to result from the simple mixing of fresh water and sea water, then the element/chloride ratio would not deviate from that of sea water. A major deviation from sea water occurs for the K/Cl ratio in holes 970C and 970D. Moreover, the element/Cl ratio in the pore-waters at MD indicate a NaCl-brine dominance, in excellent agreement with those at ND (Table 1). It seems, therefore, that the low salinities in the pore-waters at MD are due to the presence of significant amounts of methane hydrates in combination with brine-dominated *in situ* pore-waters. This option is strongly supported by the abundant presence of gas, obvious core expansion features such as numerous gas pockets, the rupture of the core liner of Core 160-970D-2H on deck and the 'soupy' nature of the sediment. The gas appeared to be nearly pure methane (Emeis *et al.* 1996).

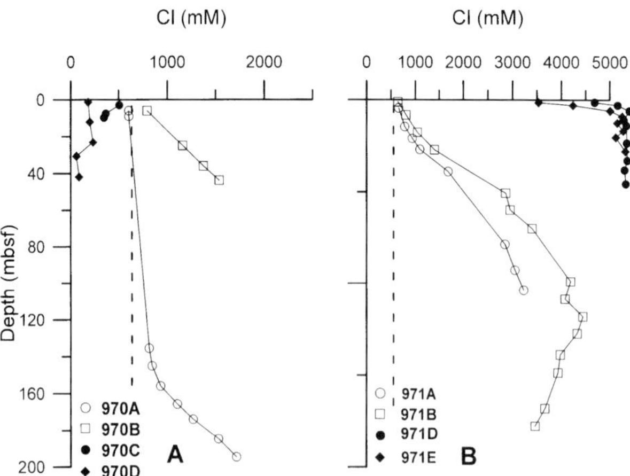

Fig. 3. Pore-water chloride profiles for the four sites at (**A**) Milano Dome and (**B**) Napoli Dome. The bottom-water chloride content is indicated by a dashed line.

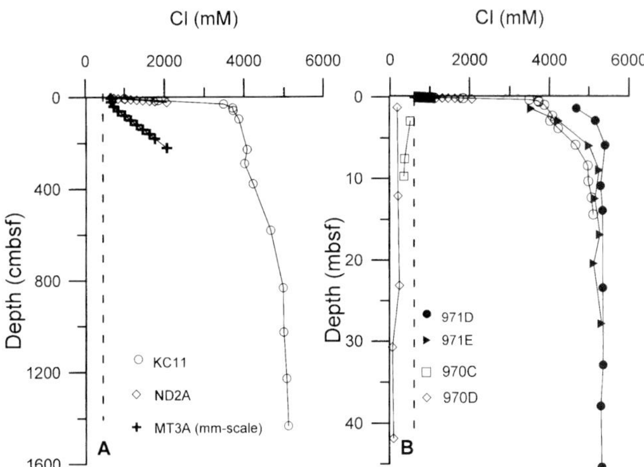

Fig. 4. Pore-water chloride profiles for: (**A**) box cores ND2A and MT3A, and piston core KC11 (both at Napoli Dome crest); and **(B)** ODP holes 970C and 970D (at Milano Dome crest), Box cores ND2A and MT3A, piston core KC11, and ODP Holes 971D and 971E (all at Napoli Dome crest). Note the difference in depth scales: left-hand panel is in cm but MT3A is in mm below sea floor, and the right-hand panel in mbsf. Dashed line indicates bottom-water concentration.

Table 1. *Salinity, chloride and element : chloride ratios for samples from the crestal sites of Milano Dome (970) and Napoli Dome (971) mud volcanoes*

	Depth (mbsf)	Salinity ($g\,kg^{-1}$)	Cl (mM)	Na/Cl (M/M)	Mg/Cl (M/M)	K/Cl (M/M)	Ca/Cl (M/M)	SO_4/Cl (M/M)	Li/Cl (mM/M)
970C									
1H-2, 145–150	2.95	28.7	509	0.99	0.0403	0.0096	0.0035	0.0014	0.090
2H-2, 132–142	7.52	22.0	369	1.13	0.0072	0.0025	0.0018	0.0016	0.089
3H-2, 0–10	9.70	20.3	350	1.11	0.0059	0.0022	0.0015	0.0017	0.095
970D									
1H-1, 127–150	1.27	11.7	183	1.06	0.0085	0.0022	0.0023	0.0043	0.093
2H-5, 130–150	12.1	12.5	202	1.01	0.0059	0.0028	0.0025	0.0011	0.114
3H-6, 130–150	23.1	14.8	236	1.06	0.0072	0.0032	0.0000	0.0013	0.121
4H-5, 130–150	30.71	7.0	61	1.92	0.0115	0.0030	0.0000	0.0018	0.270
5H-6, 130–150	41.8	9.8	88	1.66	0.0068	0.0032	0.0000	0.0051	0.182
971D									
1H-1, 135–150	1.35	258	3827	0.78	0.0007	0.0011	0.0009	0.0027	0.038
1H-2, 135–150	2.85	282	4140	0.82	0.0008	0.0010	0.0023	0.0032	0.038
1H-4, 135–150	5.85	293	4167	0.78	0.0008	0.0010	0.0091	0.0028	0.037
2H-2, 135–150	10.85	292	4129	0.82	0.0004	0.0010	0.0019	0.0025	0.036
2H-4, 135–150	13.85	295	4147	0.77	0.0002	0.0010	0.0012	0.0026	0.034
3H-4, 135–150	23.35	297	4083	0.74	0.0000	0.0010	0.0001	0.0026	0.034
4H-4.135–150	32.85	300	4128	(0.80)	0.0001	0.0009	0.0010	0.0022	0.034
5H-1, 135–150	37.85	297	4145	0.78	0.0007	0.0010	0.0004	0.0028	0.032
5H-6, 135–150	45.35	298	4074	0.80	0.0000	0.0010	0.0000	0.0024	0.027
971E									
1H-1, 140–150	1.4	200	3535	0.87	0.0015	0.0022	0.0004	0.0054	0.047
1H-2, 140–150	2.9	238	4242	0.82	0.0009	0.0019	0.0002	0.0055	0.045
1H-4, 140–150	5.9	278	5010	0.81	0.0005	0.0017	0.0001	0.0057	0.045
1H-6, 140–150	8.9	295	5260	0.78	0.0004	0.0016	0.0000	0.0056	0.044
2H-2, 140–150	12.4	283	5158	0.79	0.0001	0.0022	0.0000	0.0054	0.045
2H-5, 130–140	16.8	293	5281	0.76	0.0001	0.0020	0.0000	0.0031	0.042
3H-1, 135–150	20.35	282	5128	0.77	0.0002	0.0021	0.0000	0.0051	0.042
3H-7, 69–84	27.81	300	5325	0.77	0.0000	0.0023	0.0000	0.0059	0.039
Sea water		35	559	0.86	0.097	0.019	0.019	0.052	0.048

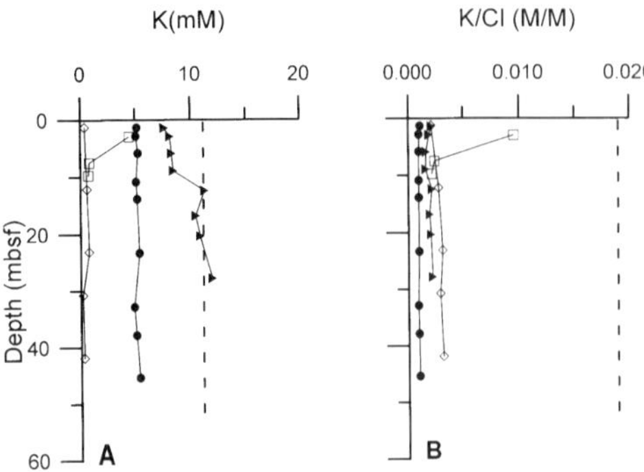

Fig. 5. Pore-water concentrations for the crestal sites of Napoli Dome (971D and 971E) and Milano Dome (970C and 970D) for (**A**) K and (**B**) K/Cl. Dashed line indicates sea-water composition; symbols as indicated in Fig. 4B.

The significant change in element/chloride ratio in the pore-waters of sites 971D (ND) and 970D (MD) can be evaluated with respect to a two end-member mixing of sea water and brine. As Mg, Ca and Sr are involved in carbonate precipitation and recrystallization reactions, and as the Na/Cl ratio of sea water and brine is rather similar (0.86 vs 1), all of these elements are not particularly suitable for such evaluation. Although not entirely free of diagenetic changes, K seems to be the best choice of element for such evaluation at this moment (Fig. 5). In view of the time scales, temperatures, depth ranges and shift in K relative to Cl content involved, a possible diagenetic decrease of K with depth would have only a minor influence on the estimations made below. If a simple two end-member calculation is done for the K/Cl ratio in the pore-waters of ND and MD (holes 971D and 970D), then the average contribution of sea water and brine can be estimated. In this calculation we will assume that the composition of the brine at MD is similar to that at ND. The ND brine being NaCl-dominated, this assumption is likely to lead to an underestimate of the MD brine contribution, if the initial MD brine were to contain a slightly higher K concentration than the ND brine. Consequently, the calculated brine contributions are likely to be underestimated. The K:Cl ratio of end-members considered in this calculation are 0.020, 0.0030 and 0.0010 (M/M) for sea water, MD pore-water and ND brine, respectively. The resulting brine contribution to the interstitial waters of MD is thus estimated to be greater than 85%.

Although gas hydrates have not been recovered at MD, they are likely to occur close to the sediment–water interface in Hole 970D. Salinity values as low as 7‰ can be explained only by a contribution to the pore-water pool of at least 80% hydrate water and of less than 20% pore-water of 'normal' Mediterranean bottom water with a salinity of 38‰. This proportion of gas hydrate-related water is a minimum value, as the remaining *in situ* pore fluid is likely to be brine rather than sea water (see above). In addition, the position of the MD gas hydrate occurrence in the pressure–temperature (p, T) stability plot further supports this view (see above and Fig. 1) (De Roo *et al.* 1983). If the minimum brine contribution of 85% is adopted, it follows that less than 2% of the observed interstitial water may actually have been present *in situ*, whereas more than 98% of the observed pore-water must originate from disintegration of gas hydrates upon recovery.

Sulphate is essentially absent below 1.3 mbsf, and alkalinity values of more than 70 mM reflect the bacterial consumption of methane. The formation of methane hydrates near the sediment–water interface seems to be related to the high supply rate of methane by emanating fluids, the pressure (water depth) and temperature of the bottom water (14.2°C), and the availability of pore space (mud breccia) (Emeis *et al.* 1996). All these prerequisites are found at Hole 970D, and possibly at Hole 970C, and are compatible with the stability field of methane gas hydrates.

Recently, the massive occurrence of gas hydrates at relatively shallow sediment depths (up to a few metres) in mud volcano-related sediments of the Eastern Mediterranean and Black Sea have been confirmed (Woodside & Ivanov pers. comm. 1996; Woodside *et al.* 1996).

Gas hydrates at Milano Dome vs Napoli Dome

In the sediments of MD the presence of gas hydrates is evident, whereas in the deep sediments of the nearby ND they seem to be absent. The sediment temperature at the ND crest is slightly (2°C) higher than the bottom water (Leg 160). At these sites the *in situ* near-bottom conditions for depth, salinity and temperature are close to the stability boundary for gas hydrates (Fig. 1). In addition, the water depth of MD and ND is nearly the same. Consequently, the slightly higher in-sediment temperature at ND relative to that at MD seems to be the factor controlling the presence or absence of gas hydrates at these sites.

The relatively heavy $\delta^{18}O$ of pore-water and carbonate crust, as well as the low $\delta^{13}C$ of the

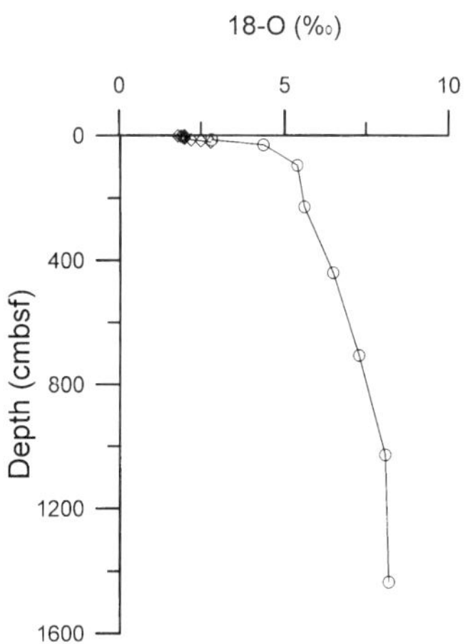

Fig. 6. Oxygen isotopic composition of the pore water at Napoli Dome; symbols as indicated in Fig. 4A.

crust (−24.7‰ relative to approx. −22‰ for organic matter) found at the ND crest, are indications for the shallow presence of gas hydrates (<1 mbsf) and for a possible (deep) regeneration of gas hydrates which may produce 'heavy' water (Fig. 6) and large amounts of methane (Emeis *et al.* 1996). This could be one of the possible ways of initiating mud vulcano formation.

Estimated amounts of methane

By extrapolation of the pore-water results for MD, we can estimate the total quantity of methane present as gas hydrate, and as free gas below the (p, T) stability boundary. In doing so, we will make the same assumption as made for the Blake Outer Ridge, namely that free gas and gas hydrate methane quantities are equal. Assuming a 40 m thick layer of gas hydrates, the total amount of methane at MD can be estimated to be $\sim 5 \times 10^9\,m^3\,CH_4$. At this moment there is no hard evidence to determine what fraction of the Eastern Mediterranean Ridge mud domes actually contain gas hydrates. We will, therefore, make a conservative estimate here, assuming that: (1) only one out of every 20 mud domes contains gas hydrates/methane; (2) this quantity is identical to that of MD; and (3) the mud dome density for the Eastern Mediterranean Ridge is identical to that in Olympi area, i.e. the area around MD and ND. In doing so, the total amount of methane associated with Eastern Mediterranean Ridge mud dome structures is estimated to be $\sim 1 \times 10^{14}\,m^3\,CH_4$.

Owing to the Eastern Mediterranean bottom-water temperature and the water depth of most mud domes, this methane occurs at relatively shallow sub-bottom depths. Slight temperature changes in the bottom water may, therefore, easily disturb the delicate equilibrium situation, and may result in the massive release of methane to the bottom water or to the atmosphere. A recently recorded, possibly anthropogenic, rise in bottom-water temperature could, therefore, result in the release of this extremely strong greenhouse gas in the near future. Clearly, it is essential that further studies are done not only to substantiate the occurrence but also to more accurately determine the amount of gas hydrates and underlying free gas.

Conclusions

Mud dome structures in the Eastern Mediterranean seem to be associated with the presence of a large amount of natural gas, methane. In the now dormant Milano Dome this has led to the massive occurrence of gas hydrates at relatively shallow depth in the sediment (from 1 to over 40 mbsf). In the active Napoli Dome no gas hydrates seem to occur in the sediments recovered during Leg 160, i.e. from 1.4 to 45 mbsf. The pore-water major element and isotopic composition for box core ND2A and MT3A, piston core KC11, and holes 971D and 971E, however, suggest the very shallow (<1 mbsf) occurrence of gas hydrates, and the possible deep regeneration of gas hydrates at this site.

For the Milano Dome the total amount of methane is estimated to be $\sim 5 \times 10^9\,m^3\,CH_4$, whereas that for the total of the Eastern Mediterranean Ridge mud dome structures is thought to be $\sim 1 \times 10^{14}\,m^3\,CH_4$.

All Leg 160 Participants are thanked for their cooperation, but in particular A. Pimel, R. Kemp, J. Rullkoetter and I. Bouloubassi. We acknowledge G. Ginsburg, V. Soloviev, A. Camerlenghi and an anonymous reviewer for their thorough reviews. G. J. De Lange acknowledges the financial support from GAO/NWO by grant No. 750.00.620.7290 for participation in Leg 160, and from MAST programs MAS90-0022C and MAST2-CT93-0051. H. de Waard, D. van der Meent, and A. van Dijk are thanked for their analytical assistance. This is NSG paper No. 970154.

References

BROOKS, J., JEFFREY, A., MCDONALD, T. & PFLAUM, R. 1984. Geochemistry of hydrate gas and water from Site 570, DSDP Leg 84. *In:* VON HUENE, R., AUBOUIN, J. *et al.* (eds) *Initial Reports of the Deep Sea Drilling Project*, Volume 84. US Government Printing Office, Washington, DC, **84**, 699–703.

DAVIDSON, D., LEAIST, D. & HESSE, R. 1983. Oxygen-18 enrichment in the water of a clathrate hydrate. *Geochimica et Cosmochimica Acta*, **47**, 2293–2295.

——, EL-DEFRAWY, M., FUGLEM, M. & JUDGE, A. 1978. Natural gas hydrates in northern Canada. *Proceedings of the 3rd International Conference on Permafrost*, **3**, 937–943.

DE LANGE, G. 1983. Geochemical evidence of a massive slide in the southern Norwegian Sea. *Nature*, 305, 420–422.

—— 1986. Chemical composition of interstitial waters in cores from the Nares Abyssal Plain (Western North Atlantic). *Oceanologica Acta*, **9**, 159–168.

—— 1992. Shipboard routine and pressure-filtration system for pore water extraction from suboxic sediments. *Marine Geol*ogy, **109**, 77–81.

——, BRUMSACK, H.-J. & SHIPBOARD SCIENTIFIC PARTY OF LEG 160. 1996. Gas hydrates in the Eastern Mediterranean: Evidence from interstitial waters recovered during Leg 160 on Milano and Napoli Dome mud volcanoes. *Proceedings of the 1st EuroColloquium, Oldenburg*, 2.

——, MIDDELBURG, J., VAN GAANS, P. *et al.* 1990. Sulphate-related equilibria in the hypersaline brines of the Tyro and Bannock basins, eastern Mediterranean. *Marine Chemistry*, **31**, 89–112.

DE ROO, J., PETERS, C., LICHTENTHALER, R. & DIEPEN, G. 1983. Occurrence of methane hydrate in saturated and unsaturated solutions of sodium chloride and water in dependence of temperature and pressure. *American Institute of Chemical Engineers Journal*, **29**, 651–657.

DICKENS, G. & QUINBY-HUNT, M. 1994. Methane hydrate stability in seawater. *Geophysical Research Letters*, **21**, 2115–2118.

EMEIS, K. C. *et al.* 1996. *Proceedings of the Ocean Drilling Program, Initial Reports*, College Station, TX. Ocean Drilling Program, **160.**

GIESKES, J., GAMO, T. & BRUMSACK, H. 1991. *Chemical Methods for Interstitial Water Analysis Aboard JOIDES Resolution.* Ocean Drilling Program Technical Note, **15**.

GINSBURG, G. 1996. How do gas-hydrates accumulate in deep-water marine sediments: Migration of hydrate-forming reactants. *In*: HENRIET, J.-P. & MIENERT, J. (eds) *First Master Workshop on Gas Hydrates: Relevance to World Margin Stability and Climate Change*, University of Gent, Belgium, 123–147.

HANDA, Y. P. 1990. Effect of hydrostatic pressure and salinity on the stability of gas hydrates. *Journal of Physical Chemistry*, **94**, 2652–2657.

HARRISON, W., HESSE, R. & GIESKES, J. 1982. Relationship between sedimentary facies and interstitial water chemistry of Slope, Trench, and Cocos plate sites from the Middle America Trench transect, active margin off Guatemala, Deep Sea Drilling Project Leg 67 *In*: AUBOUIN, J., VON HUENE, R. *et al.* (eds) *Initial Reports of the Deep Sea Drilling Project*, Volume 67. US Government Printing Office, Washington, DC, 603–614.

HESSE, R. & HARRISON, W. 1981. Gas hydrates (clathrates) causing pore-water freshening and oxygen isotope fractionation in deep-water sedimentary sections of terrigenous continental margins. *Earth and Planetary Science Letters*, **55**, 453–462.

HYNDMAN, R. D., FOUCHER, J. P., YAMANO, M., FISHER, A. & SCIENTIFIC TEAM OF ODP LEG 131 1992. Deepsea bottom-simulating reflectors: calibration of the base of the hydrate stability field as used for heat flow estimates. *Earth and Planetary Science Letters*, **109**, 289–301.

KVENVOLDEN, K. 1996. A primer on the geological occurrence of gas hydrate. *In*: HENRIET, J.-P. & MIENERT, J. (eds) *First Master Workshop on Gas Hydrates: Relevance to World Margin Stability and Climate Change*, University of Gent, Belgium, 39–80.

—— & KASTNER, M. 1990. Gas hydrates of the Peruvian outer Continental Margin. *In*: SUESS, E., VON HUENE, R. *et al.* (eds) *Proceedings of the Ocean Drilling Program, Scientific Results*, College Station, TX. Ocean Drilling Program, **112**, 517–526.

—— & MCDONALD, T. 1984. Gas hydrates of the Middle America Trench – Deep Sea Drilling Project Leg 84. *In:* VON HUENE, R., AUBOUIN, J. *et al.* (eds) *Initial Reports of the Deep Sea Drilling Project*, Volume 84. US Government Printing Office, Washington, DC, **84**, 667–682.

—— & MCMENAMIN, M. 1980. *Hydrates of Natural Gas: A Review of Their Geological Occurrence.* US Geological Survey Circular, **825**.

MAKOGON, Y. F. 1981. *Hydrates of natural gas* (translated from Russian by CIESLEWICS, W. J.). Pennwell, Tulsa, OK.

MILLER, S. L. 1974. The nature and occurrence of clathrate hydrates. *In:* KAPLAN, I. (ed.) *Natural Gases in Marine Sediments.* Plenum, New York, 151–177.

ODP LEG 164 SHIPBOARD SCIENTIFIC PARTY 1996. Methane gas hydrate drilled at Blake Ridge. *EOS, Transactions of the American Geophysical Union*, **77**, 219.

PFLAUM, R., BROOKS, J., COX, H., KENNICUTT, M. & SHEU, D.-D. 1985. Molecular and isotopic analysis of core gases and gas hydrates, Deep Sea Drilling Project Leg 96. *In*: BOUMA A., COLEMAN, J., MEYER, A. *et al.* (eds) *Initial Reports of the Deep Sea Drilling Project*, US Government Printing Office, Washington, DC, 781–784.

ROBERTSON, A. & OCEAN DRILLING LEG 160 SCIENTIFIC PARTY. 1996. Mud volcanism on the Mediterranean Ridge: initial results of Ocean Drilling Program Leg 160. *Geology*, **24**, 239–242.

SHIPLEY, T., HOUSTON, M., BUFFLER, R. *et al.* 1979. Seismic evidence for widespread possible gas hydrate horizons on continental slopes and rises. *AAPG Bulletin*, **63**, 2204–2213.

TUCHOLKE, B., BRYAN, G. & EWING, J. 1977. Gas-hydrate horizons detected in seismic-profiler data from the western North Atlantic. *AAPG Bulletin*, **61**, 698–707.

VAN DER WAALS, J. & PLATTEEUW, J. 1959. Clathrate solutions. *Advances in Chemical Physics*, **2**, 1–57.

VAN SANTVOORT, P. & DE LANGE, G. 1996. Messinian salt fluxes into the present-day Earstern Mediterranean: implications for budget calculations and stagnation. *Marine Geology*, **132**, 241–251.

——, DE LANGE, G., THOMSON, J. *et al.* 1996. Active post-depositional oxidation of the most recent sapropel (S1) in sediments of the eastern Mediterranean Sea. *Geochimica et Cosmochimica Acta*, **60**, 4007–4024.

VENGOSH, A., DE LANGE, G. & STARINSKY, A. Submitted. Boron isotopic evidence for the origin of eastern Mediterranean brines (Bannock and Urania Basins).

WOODSIDE, J., IVANOV, M. & SHIPBOARD SCIENTISTS OF THE ANAXIPROBE-96 EXPEDITION. 1996. Shallow gas and gas hydrates in the Anaximander Mountains region, eastern Mediterranean Sea. (Abstract). *In*: HENRIET, J.-P. & MIENERT, J. (eds) *First Master Workshop on Gas Hydrates: Relevance to World Margin Stability and Climate Change*, University of Gent, Belgium, 48–49.

YAMANO, M., UYEDA, S., AOKI, Y. & SHIPLEY, T. 1982. Estimates of heat flow derived from gas hydrates. *Geology*, **10**, 339–343.

YEFREMOVA, A. & ZHIZHCHENKO, B. 1974. [Occurrence of crystalhydrates of gases in the sediments of modern marine basins.] *Dokl. Akad. Nauk. SSR,* **214**, 1179-1181; *Dokl. Akad. Nauk. SSR Earth. Sci. Sect.*, **214**, 219–220 (in Russian).

Shallow gas and gas hydrates in the Anaximander Mountains region, eastern Mediterranean Sea

J. M. WOODSIDE[1], M. K. IVANOV[2], A. F. LIMONOV[2] & SHIPBOARD SCIENTISTS OF THE ANAXIPROBE EXPEDITIONS

[1] *Free University, De Boelelaan 1085, 1081 HV Amsterdam, The Netherlands*
[2] *Moscow State University, Vorobjevy Gory, Moscow 119899, Russian Federation*

Abstract: Gas hydrates have been sampled from a mud volcano in the eastern Mediterranean for the first time, during a recent expedition to study the nature and origin of the Anaximander Mountains offshore from southwest Turkey. Seven mud volcanoes were sampled as part of a successful strategy to obtain rocks which had been brought up from the deeper parts of the mountains themselves. The mud volcanoes seem to be associated with cross-cutting strike-slip faults defining individual blocks in this mountain complex, and may permit overpressured fluids from thrusts in this compression zone to be expelled at the surface. There is ample evidence in many parts of the area studied for the widespread presence of gas in the sediments, such as pockmarks, acoustic turbidity and acoustic wipe-outs in sub-bottom profiles, gas in sediment samples, and mud volcanoes with carbonate crusts and benthic communities typical of those found elsewhere near gas seeps and fluid vents (e.g. vestimentifera worms, bivalves). It is postulated that the high mobility of some of the sediments is promoted by excess fluids possibly released in part from gas hydrates in the Plio-Pleistocene sediments. A good example of the sediment deformation is shown by a large area of moving sediments.

ANAXIPROBE was a two-phased research programme to determine the origin and evolution of the actively developing Anaximander Mountains in the eastern Mediterranean. Possible scenarios for the origin were that they are a foundered southern part of Turkey (which is one of the conclusions from ANAXIPROBE), a northward-collided part of the African lithospheric plate, or an upthrust block of the neo-Tethyan sea floor. The Anaximander Mountains are comprised of three independent topographic highs rising from sea-floor depths of about 2–3 km to the south of Turkey between Cyprus and Rhodes, at the intersection of the Hellenic and Cyprus arcs (Fig. 1).

A 1995 expedition on the French research vessel *Atalante* carried out multibeam swath mapping and associated underway geophysical measurements to provide basic geophysical data and to determine the best locations for deep-towed side-scan sonar, high-resolution seismic profiling and bottom sampling which took place in the second phase of work on Russian research vessel *Gelendzhik* in 1996. The sampling strategy was to dredge identified outcrops and to core in mud volcanoes, tectonic windows and areas of debris flows or talus from faulted escarpments. The swath mapping using Simrad EM12-D provided bathymetric data at a rate of about 20 $km^2 h^{-1}$, resulting both in a bathymetric map contoured at an interval of 10 m and also a subsidiary map showing the strength of the sea-floor backscattering of the sonar signal. Seismic reflection profiles were made along all *Atalante* lines using the IFREMER 'high-speed seismic' system comprised of two 75 $in.^3$ GI air guns (i.e. each 1.23 l), with a dominant frequency of 40–60 Hz, and a six-channel streamer. The *Gelendzhik* survey made use of a deep-tow acoustic system (MAK-1) with a 30 kHz side-scan sonar and a 4.9 kHz sub-bottom profiler; and sampling was carried out with a wide barrel (14 cm diameter) 6 m long gravity corer and a pipe dredge. This article is a preliminary report on the analysis of these data in the few months since the expedition in 1996.

Geological setting

The three independent mountains in the Anaximander Mountains complex can be identified on the basis of morphology, structural relationships and relief (Fig. 1b). The westernmost mountain, Anaximander, is a broad northward-tilted (at about 4°) block of about 30 × 60 km, elongated in the W–E direction, and rising to a minimum depth of about 1120 m at its steep southern escarpment. Anaximenes, the southernmost mountain, is the shallowest at about 670 m; it resembles a curved ridge almost 50 km in length and concave to the northwest in plan view, with a dip of up to 25° to the northwest. To the east lies Anaxagoras, a rugged range with an area of

WOODSIDE, J. M., IVANOV, M. K., LIMONOV, A. F. & SHIPBOARD SCIENTISTS OF THE ANAXIPROBE EXPEDITIONS 1998. Shallow gas and gas hydrates in the Anaximander Mountains region, eastern Mediterranean Sea. *In:* HENRIET, J.-P. & MIENERT, J. (eds) *Gas Hydrates: Relevance to World Margin Stability and Climate Change*. Geological Society, London, Special Publications, **137**, 177–193.

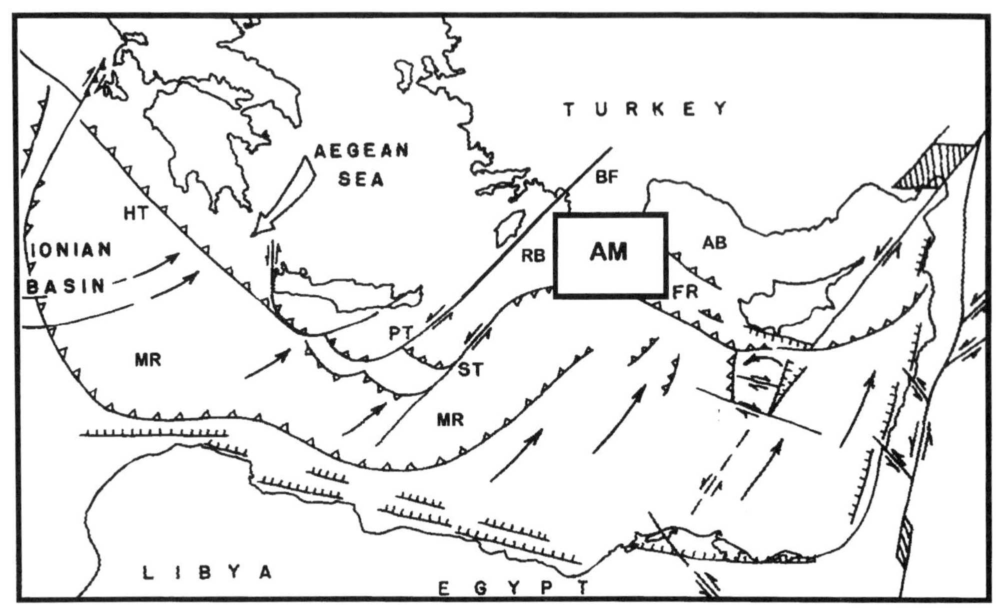
TURKEY
AEGEAN SEA
BF
HT
IONIAN BASIN
RB
AM
AB
FR
PT
MR
ST
MR
LIBYA
EGYPT

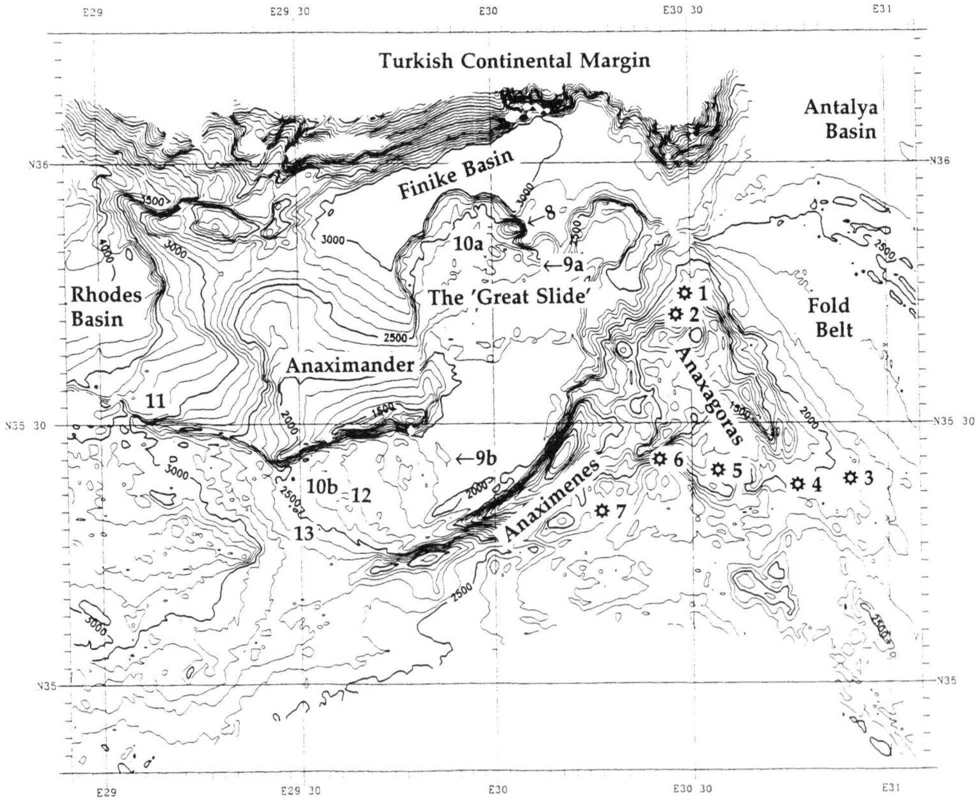
Turkish Continental Margin
Antalya Basin
Finike Basin
Rhodes Basin
The 'Great Slide'
Fold Belt
Anaximander
Anaxagoras
Anaximenes
E29
E29 30
E30
E30 30
E31
N36
N35 30
N35

about 50 × 25 km, elongated northwest–southeast along the top of the Florence Rise, with a minimum depth of about 900 m on a peak near the northwest end. A northwest–southeast ridge along the eastern side of Anaxagoras is offset in the middle by what is inferred to be a fault zone running from the northeast, across the range, and following an escarpment (near location 6 in Fig. 1b) which, if extended beyond Anaxagoras to the southwest, would lie tangential to the southern bend of Anaximenes.

The southern coast of Turkey north of the area is one of the most rapidly subsiding sections of the Mediterranean (Flemming 1972; Flemming *et al.* 1973). Seismic activity associated with the Strabo and (especially) Pliny trenches continues into mainland Turkey as the transpressional plate boundary between Africa and the Aegean–Anatolian microplate (Le Pichon *et al.* 1995). Both shearing and thrusting are observed within a zone associated with the plate boundary (McKenzie 1978; Oral *et al.* 1993; Mascle *et al.* 1986). The margin of Turkey in this region is cut by faults which are observed both in the bathymetry at sea and on land (e.g. Gutnic *et al.* 1979); and a linear ENE–WSW section of the Turkish slope is inferred to be a fault scarp associated with rifting away from Turkey of the blocks now forming the Anaximander Mountains. Thus, the region is tectonically active with evidence of both thrusting and strike-slip faulting in the present and predominantly normal faulting in the past.

The tectonic and structural situation of the mountains is not addressed in this paper but it is an important consideration in interpreting the origin of the mud volcanoes. Anaxagoras lies on the prolongation of the Florence Rise which marks the boundary between the African plate and the Anatolian microplate (McKenzie 1972). The Antalya Basin to the northeast of the Florence Rise (Fig. 1a) is actively subsiding and tilting to the north (e.g. Woodside 1976, 1977) under compression from the African plate, and Anaxagoras is being pushed from the southwest against the sediments in the Antalya Basin creating a fold belt (Fig. 1b). The plate boundary between Africa and the Aegean runs along the Pliny–Strabo Trench system (Fig. 1a) and continues into south-western Turkey as the Burdur fault zone. Some branches of this transpressive fault system may cut through the Anaximander Mountains. Gravity data (Woodside 1976; Ivanov *et al.* 1992; Woodside *et al.* 1993) indicate a major structural discontinuity between the western mountains (Anaximander and Anaximenes) where there is a large positive Bouguer anomaly shared with the Rhodes Basin, and the eastern mountain (Anaxagoras) where the Bouguer anomaly is about 150 mGal less.

The intense tectonic activity since the end of the Miocene is indicated further by the absence throughout the Anaximander Mountains region of the thick Messinian evaporites which are characteristic of the surrounding region. Seismic reflection evidence suggests that the Messinian is represented here by only very thin (a few tens of metres maximum) early evaporites like gypsum. In the Finike Basin, Pliocene–recent sediments appear to lie directly on Bey Daglari basement rocks similar to those sampled by the ANAXIPROBE project as far south as Anaximenes Mountain (paper in preparation); however, in the adjacent Antalya Basin, the thick diapiric Messinian salt layer can be seen. That means that the large relief associated with the Anaximander Mountains is post-Miocene, and that a large part of the mountains were covered with Pliocene sediments before the vertical movements that formed them occurred.

Observations

The 1995 Simrad EM12-D swath mapping survey of the Anaximander Mountain region resulted in two unexpected results: the probable presence of a number of mud volcanoes, and the presence of a remarkable feature resembling a large sediment slide or nappe (Fig. 1b). Both these findings had important implications

Fig. 1. (**a**) Location map showing the location of the Anaximander Mountains (AM) in the eastern Mediterranean Sea, with generalized tectonic setting (after Woodside 1991) showing the Hellenic Trench (HT), the Pliny and Strabo Trenches (PT and ST), Burdur fault zone (BF), the Florence Rise (FR), the Mediterranean Ridge (MR), Rhodes Basin (RB) and the Antalya Basin (AB), relative movement on fault zones (half arrows), relative movement of African plate with respect to Eurasia and the Anatolian–Aegean microplates (full arrows), and rotation of the western Aegean and Greece (full open arrow). (**b**) Bathymetry map of Anaximander Mountains at a 100 m contour interval from ANAXIPROBE Simrad EM12-D survey. Locations of principal features are labelled. Mud volcanoes are numbered 1, San Remo; 2, Kula; 3, St. Ouen l'Aumône; 4, Tuzlukush; 5, Kazan; 6, unnamed fissure eruption; and 7, Amsterdam. Other numbered locations refer to: 8, the 'Anthill' (see text); 9a, 9b, tectonic windows; 10a, 10b, 'gas front' areas; 11, gas seeps and small mounds; 12, area of pockmarks; 13, area of slides. Coordinates are in half-degree intervals (~55 km) for scale.

regarding the likely presence of gas in the sediments, of overpressured formations, and of fluid circulation through the sediments and the sea floor. Both were also important for the second phase of the investigation because bottom sampling could take advantage not only of the rock outcrops on steep slopes but also of clasts erupted from deeper formations at mud volcanoes, and deeper sediments revealed in tectonic windows created by the slide.

The second phase of ANAXIPROBE in 1996 also provided two unexpected results: the first samples of gas hydrates from the Mediterranean sea floor (site 2 in Fig. 1b); and clear evidence of gas in sediments through a large part of the Anaximander Mountains complex. Gas hydrates had previously been inferred in the eastern Mediterranean Sea from the geochemistry of cores from ODP Leg 160 Site 970 on Milano mud volcano (almost 600 km to the WSW of the Anaximander Mountains, and in a different geological setting); and both the gas hydrate stability criteria (e.g. Kvenvolden 1993) and the direct observation of gas emissions from mud volcanoes (e.g. Cita *et al.* 1995), and also of chemosynthetic benthic communities (e.g. Corselli & Basso 1996), suggested that gas hydrates should be present at these depths. The ubiquity of the gas is apparent from the acoustic signature in the deep-tow side-scan images and sub-bottom profiles, and in the seismic data. Acoustic voids, acoustic turbidity and acoustic wipe-outs were observed in a number of locations, as well as what were inferred to be gas fronts and carbonate mounds over gas seeps.

Mud volcanoes

The potential mud volcanoes were inferred from backscatter imagery derived from the Simrad EM12-D swath mapping. All the inferred mud volcanoes that were sampled subsequently proved to be mud volcanoes; thus increasing the probability that the others are also mud volcanoes. Mud volcanoes generally cause higher backscatter because the erupted material, referred to as 'mud breccia' (Cita *et al.* 1981; Staffini *et al.* 1993), contains gas and clasts which form 'volume scatterers' of the acoustic signal (Volgin & Woodside 1996). Often flows from the centre of the mud volcano can also be identified by the high backscatter.

Although the morphology of the mud volcanoes may vary considerably, the mud volcanoes are commonly circular in plan view with a relief of about 100 m and a diameter on the order of 1 km. The largest mud volcano in the Anaxi-

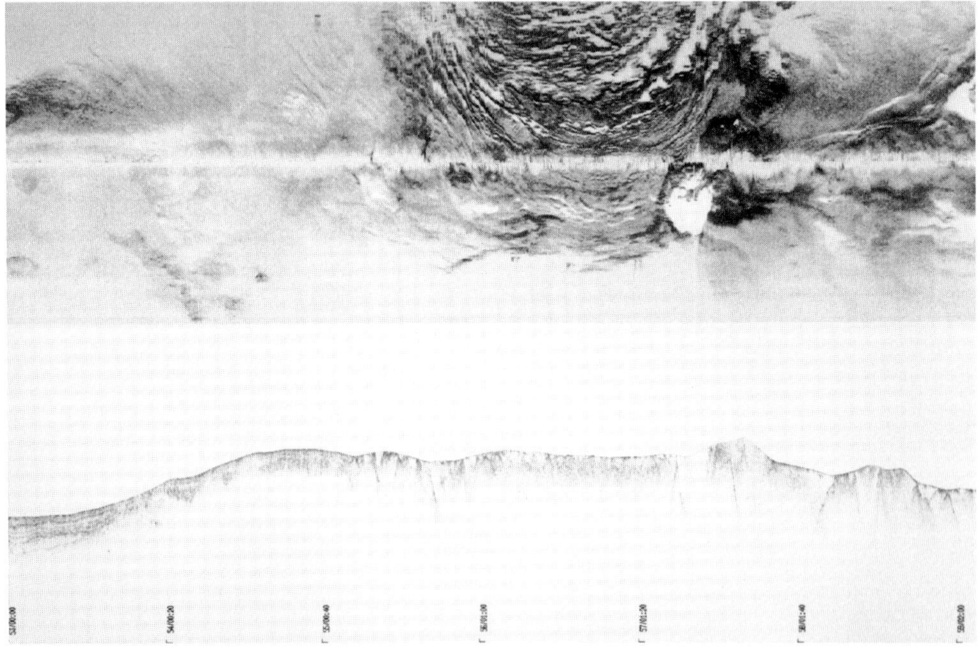

Fig. 2. Deep-tow image and sub-bottom profile across the northwest half of Amsterdam mud volcano. Southwest is to the right. Side-scan sonar image (above) is 2 km across (vertically). High backscatter is in darker tones.

mander field is Amsterdam (location 7 in Fig. 1b; see also Fig. 2), measuring about 2.5 km across with flows extending over an area of about 50 km^2 reaching a maximum thickness of about 360 m (assuming an acoustic velocity of 1800 $m s^{-1}$ for 400 ms of seismic two-way travel time (TTWT) through the flow); and the smallest sampled (San Remo; location 1 in Fig. 1b) is about 700 m across and 60 m high. The shapes vary from conical for San Remo and domed for St. Ouen l'Aumône (location 3 in Fig. 1b) to the 'mud pie' type of the Amsterdam mud volcano (Fig. 2). (The mud volcanoes were named following the tradition in the Mediterranean of using the birthplaces of scientists involved in the research.) Seismic profiles across mud volcanoes commonly show inward-dipping strata generally thought to result from collapse either due to loss of material from below, following eruption, or from flexure under the load of the erupted sediments (Shipboard Scientific Party 1996). Regardless of the diagnostic morphology, high acoustic backscatter and seismic signature, it is usual to verify that such a feature is a mud volcano by sampling the characteristic mud breccia which normally contains clasts of a much older age than the surrounding sea floor.

What made the presence of mud volcanoes interesting in this investigation, besides their important use in providing samples of rocks

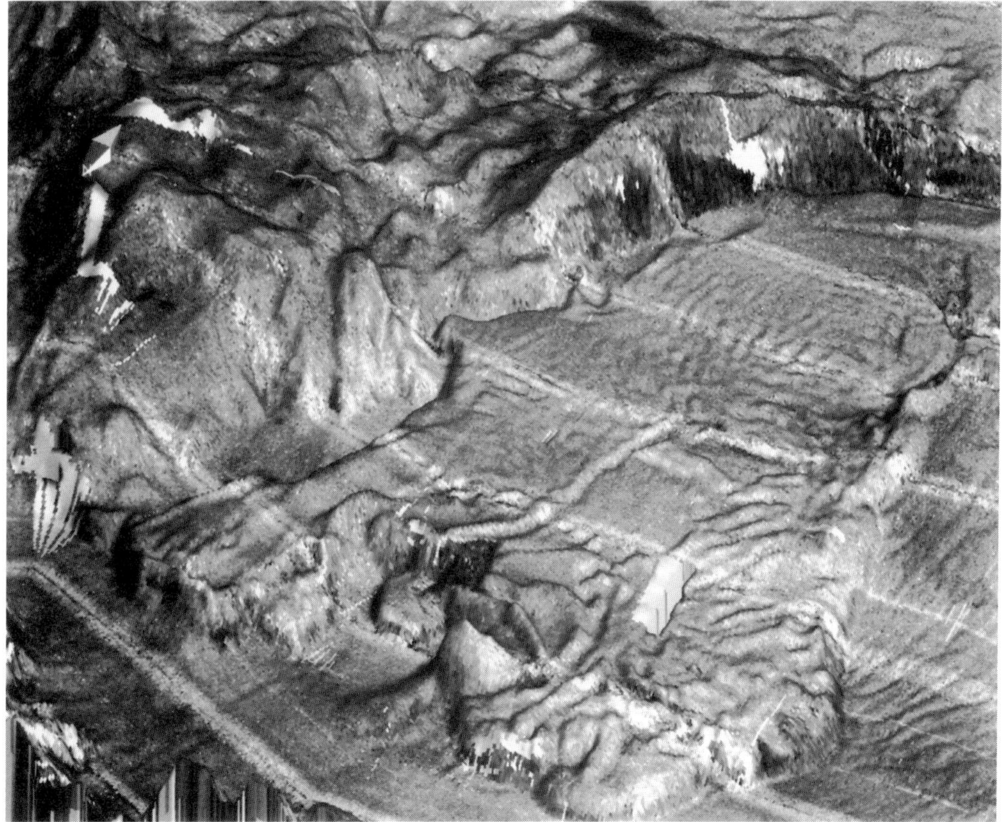

Fig. 3. (**a**) Perspective three-dimensional image of the 'Great Slide' from the northwest (looking to the southeast). Sediment movement is from the south (toward upper right) over a distance of about 60 km to the north. The width of the flow at its narrowest (near the top right) is about 20 km and at its widest (across the entire two main lobes near the bottom of the picture) is about 65 km. The height of the lobes at the front (bottom centre) is about 300 m. The vertical exaggeration is about 5 times. White stripes are artifacts of the data processing. Anaximenes (rising ~1300 m above the flow) is in the upper right corner, Anaxagoras in the upper left quadrant. (**b**) Bathymetric contours of the 'Great Slide' at 50 m interval. The grid lines are geographic coordinates at 10′ intervals (i.e. 10 nautical miles or 18.52 km in the north–south direction). Anaximenes is the curved ridge to the lower right, and Finike Basin and the Turkish continental margin are at the top. The 'Anthill' is visible in the upper central part of the figure, on the right side of the left branch of the lobes.

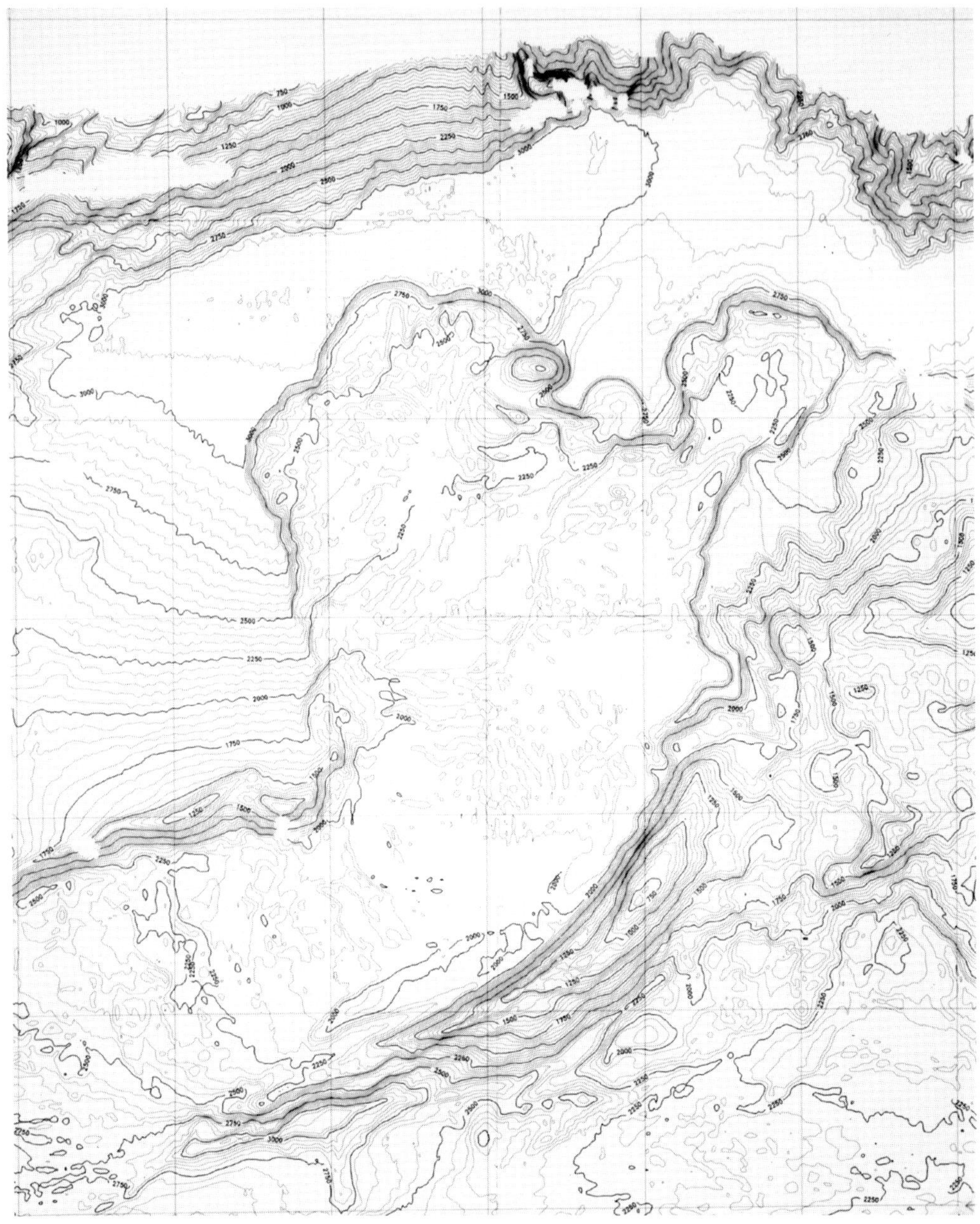

Fig. 3(b).

from deep below the sea floor (from at least several hundreds of metres and possibly from as deep as several kilometres below the sea floor), is the likely presence of overpressured fluids and gas as a driving mechanism. Evidence for gas was present in all samples obtained from mud volcanoes. Not only did the mud breccia smell of gas, but gas escape textures were found in the mud (vesicles and narrow vertical pathways), and samples were obtained of benthic fauna typical of areas of gas and fluid venting through the sea floor (e.g. Suess *et al.* 1985). Furthermore, gas hydrates were sampled in Kula mud volcano (location 2 in Fig. 1b), the first gas hydrates to be sampled in the Mediterranean Sea.

The 'great slide'

Probably the most striking feature in the bathymetry maps is what has been termed the 'Great Slide' (Fig. 1b). It has the appearance of a large glacier-like mass of sediments with great circular lobes at the front (to the north) where it is moving northward out over the Finike Basin, and surface relief in the form of low-amplitude folds or faults perpendicular to the apparent flow direction (Fig. 3). Where the slide turns north, to the east of Anaximander Mountain, it bifurcates, with one branch apparently heading NNW and the other NNE; and there is a complex interaction between surficial fold trends in this area. The slide originates from between the two mountains (Anaximander and Anaximenes) and probably moves not only northward but also to the southwest where it has obscured the continuation of the Strabo Trench. Anaximenes Mountain is partially underthrusting Anaximander and the valley between them, thus narrowing the valley and raising the valley floor.

The area of the slide from the shallowest part between the mountains to the lobes in the north is roughly 1200 km^2 ; and if the area to the southwest is also included this value could rise to about 2200 km^2. The thinnest part is probably between the two mountains where the seismic data indicate Pliocene–recent sediments to be about 250 m thick (i.e. assuming an acoustic velocity in the sediments of 1700 m s^{-1} for a seismic thickness of about 300 ms TWTT). The northern lobes rise around 300 m above the floor of the Finike Basin and the sediments involved in the mass movement may be thicker than this. Thus, a conservative estimate of the volume of sediments involved is about 550 km^3.

Biological observations

The sampling of mud volcanoes for rocks erupted from deep below the sea floor also resulted in a number of samples of benthic fauna (paper in preparation). Vestimentiferan worms (provisionally identified by E. C. Southward (pers. comm.), as lamellibrachia) were sampled at one location (dredge 209D, located on the fissure eruption at location 6 in Fig. 1b), and cold seep molluscs (probably a new species of lucinidae; Salas & Woodside, submitted) were observed in six mud volcanoes with several living specimens present in two of the locations. Molluscs such as these, which survive with the help of chemoautotrophic symbiotic bacteria in extreme environments with sulphide and methane-rich fluid emissions, have been found previously in the Mediterranean only on the Mediterranean Ridge on Napoli mud volcano (Corselli & Basso 1996).

The tube worms, however, are the first to have been found in the Mediterranean and are also known to be indicators of sulphidic environments. The closest known specimens were found in the east Atlantic off northwest Spain (Dando *et al.* 1992; Southward *et al.* 1996). Like the molluscs, these pogonophores contain chemoautotrophs capable of oxidizing sulphur compounds to provide energy for fixation of carbon dioxide. The sampled specimens were attached at one end to the walls of holes running through a crusty material resulting from the oxidation of methane; and they projected out of the holes, presumably within the gas plume (Fig. 4). Fragments of molluscs and tube worms could be seen to be incorporated into the sampled crust.

Acoustic evidence of gas in the sediments

Additional indirect evidence was inferred from an analysis of seismic and 4.9 kHz deep-tow sub-bottom profiles in combination with side-scan data, based on generally accepted acoustic characteristics of the sediments. Gas in sediments can cause sub-bottom profiles of stratified sediment to be disrupted by: (a) areas where high-amplitude scattering obscures any stratification (acoustic turbidity or masking); (b) areas devoid of reflections entirely (acoustic wipe-outs or voids); and (c) areas where the amplitude of reflections from layers of sediment is strengthened (enhanced reflectors). Examples of these phenomena can easily be found in the literature (e.g. Stefanon 1985; Carlson *et al.* 1985; Papatheodorou *et al.* 1993; Baraza & Ercilla 1996), and are discussed in some depth in the book by Hovland & Judd (1988). In particular, Judd & Hovland (1992) use similar sub-bottom acoustic signatures as evidence of the presence of gas in sea-floor sediments.

The acoustic response of the sediment depends on factors related to a given situation. It depends on the amount and concentration of the gas (e.g. whether it is diffusely distributed in the sediments or concentrated in pockets), the type of sediment (which will determine the magnitude of the acoustic impedance contrast with the gas) and the characteristics of the profiling system (primarily the frequency and pulse length). Papatheodorou *et al.* (1993) consider the most important factors to be gas concentration and sediment type, together with the fre-

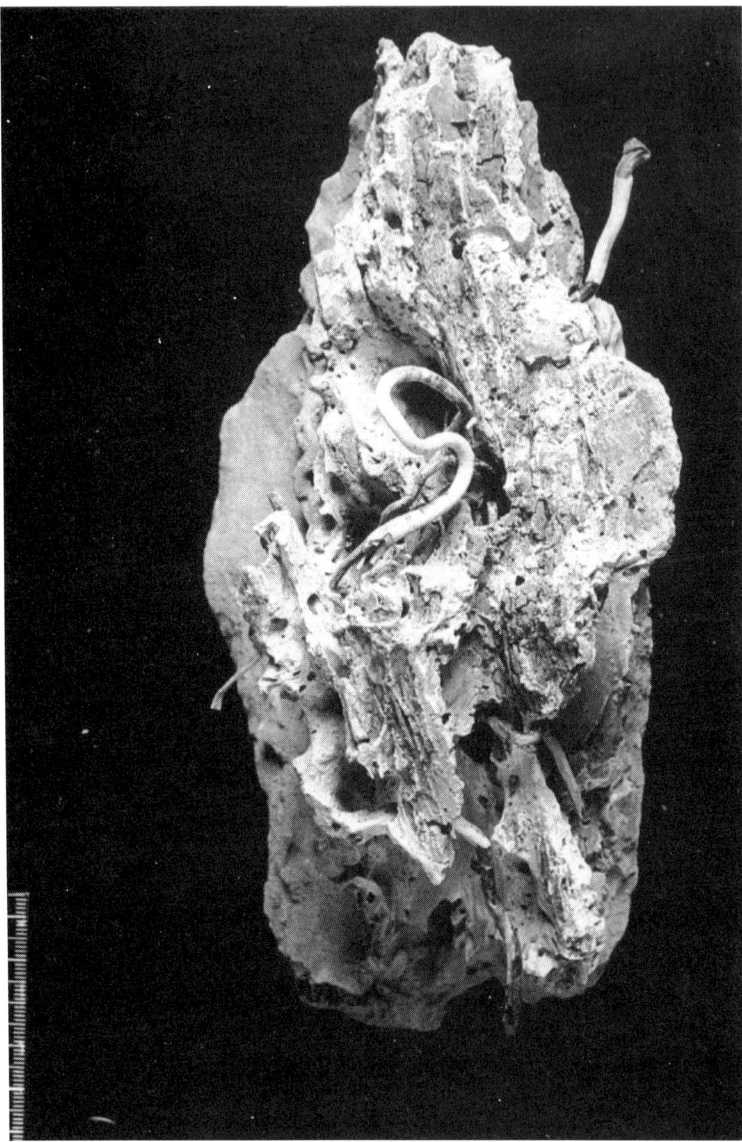

Fig. 4. Vestimentiferan worms protruding from the 'crust' in which they were living in a gas vent. Sampled at location 6 in Fig. 1. Scale is in mm (smallest ticks).

quency of the profiling system used to study the sediments. The normal sedimentary stratification can usually be observed on both sides of the acoustically disrupted section. Often there are other additional indications of gas such as gas plumes in the water column (e.g. observed in the acoustic record as a hyperbolic reflector above the sea floor) or pockmarks on the sea floor (usually observed with side-scan sonar). Furthermore, greater confidence in inferring the presence of gas can derive from the appearance of several of these features together (e.g. pockmarks with acoustic wipe-outs beneath them).

Pockmarks were observed on the Great Slide especially in the region south of Anaximander, but also elsewhere, for example on the northwest lobes. These often appeared to be aligned (Fig. 5). The pockmarks south of Anaximander are up to 60 m across, and one that was crossed by the deep-tow sub-bottom profiler was about

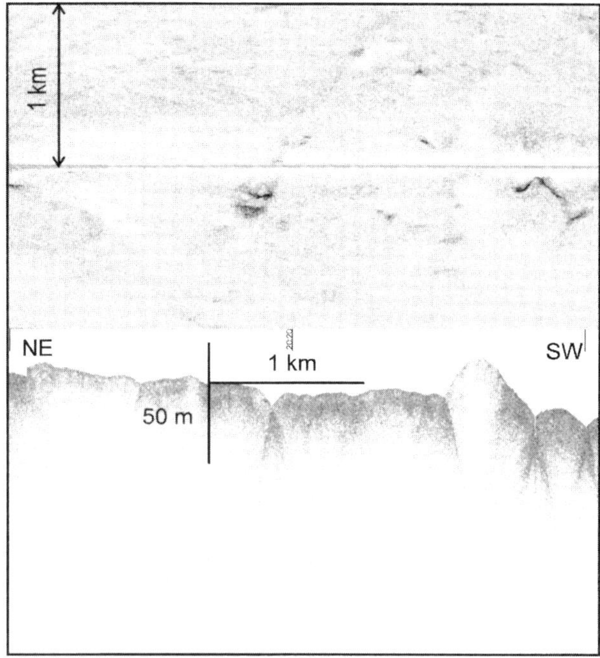

Fig. 5. Pockmarks observed in the deep-tow image and sub-bottom profile near location 12 in Fig. 1. Southwest is to the right. Side-scan sonar image (above) is 2 km across (vertically). High backscatter is in darker tones.

8 m deep (Fig. 5). In some cases enough pockmarks occurred together to create larger pits in the sea floor. They always appear in places with high acoustic turbidity but sometimes have acoustic voids beneath them.

Mass movements down gentle slopes on the sea floor can sometimes be found in the same region as the pockmarks. The small slide in Fig. 6 is accompanied by enhanced reflectors in the underlying sediments, in contrast to the sediments observed away from the slide. Enhanced reflectors are considered to be caused by the accumulation of gas along bedding plains in the sediments. The lower portion of the slide appears to have turned into a debris flow; but the upper part is intact and bears the same shape as the scar from which it moved. The rectangular shaped slide appears from the scar it left to be about 5 m thick and measures about 500–800 m by 400 m. It has moved about 400 m downslope, but the debris flow forming the distal portion of the mass movement extends about 1500 m from the scar.

A good example of acoustic wipe-outs in the sub-bottom profiles can be found on the northwest flank of Anaximander just to the north of the west-trending ridge which, further east, forms the southern faulted escarpment of the mountain. These acoustic wipe-outs show a complete loss of signal beneath sea-floor mounds (Fig. 7) in a region of slightly deeper water. The small hills observed in the deep-tow side-scan images are of the order of a few tens of metres (<50 m) in height and several tens of metres (typically 50–100 m) in diameter. They display high backscatter, sometimes appearing slightly linear. The acoustic wipe-outs or voids are considered to signify the presence of gas and upward-migrating pore fluids, and the hillocks are interpreted to be either carbonate mounds or small mud volcanoes above a region of gas escape. The sub-bottom profiles occasionally show short sections of strongly laminated sediments within the region of the acoustic voids. Similar side-scan sonar records obtained by Traynor & Sladen (1997) in Vietnam were found to be knolls formed of micrite and carbonate concretions inferred to be associated with gas seeps. Judd & Hovland (1992) considered the wipe-outs to be caused by absorption of acoustic energy by gas in the sediments or of reflection of a high proportion of acoustic

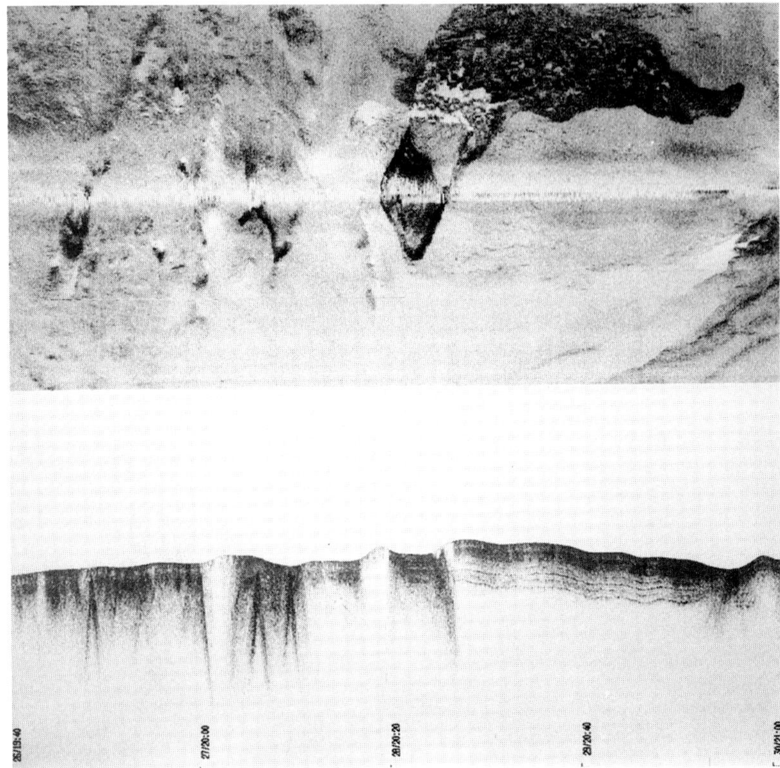

Fig. 6. Deep-tow image and sub-bottom profile across area of small slide (location 13 in Fig. 1). Southwest is to the right. Side-scan sonar image (above) is 2 km across (vertically). High backscatter is in darker tones. Note pockmarks and acoustic turbidity to the left in the sub-bottom profile. About 30 m of well-layered sediments are visible to the right but are poorly imaged elsewhere because of gas.

energy by an overlying hard sediment: both possibilities are likely in this case.

A more puzzling phenomenon is what has been interpreted as gas fronts in the sediments (Fig. 8). They occur in several regions (e.g. south of Anaximander, on the northwest lobes of the slide) and show up in the sub-bottom profiles as very strong irregular (almost jagged in profile) reflectors at a depth of perhaps 10–25 m below sea floor and with a relief of about 6 or 8 m. South of Anaximander the surficial sediments above the gas front appear to have crept slightly down the gentle slope toward the west, creating a crinkled effect in the sea-floor sediments. Because they exhibit a single very strong reflector, often with acoustic turbidity above and loss of signal below, they contrast strongly with the acoustic wipe-outs, where the signal is completely missing, or with the acoustic turbidity, where the backscatter is so strong that any signal is masked. They are often found, as noted above, where sediments have been moving, which in turn may promote the irregular upward movement of gas. Other good examples can be found on the northern lobes of the Great Slide. Similar features were described by Baraza & Ercilla (1996, Fig. 4) as 'plume fronts' indicating local concentrations of gas migrating upward in the Gulf of Cadiz; and both Hovland & Judd (1988) and Yuan *et al.* (1992) show examples of similar gas plumes above zones of acoustic turbidity in the Irish Sea.

Discussion

The mud volcanoes were found only in association with Anaxagoras and to the south and east of Anaximenes. Likewise the only benthic communities and gas hydrates were found in this area also, probably because gas escape in other areas is linked with sediment movement suggesting that benthic communities may not become established and gas hydrates may be unstable. Anaxagoras is considered to be built up of thrust sheets caused by the northeastward

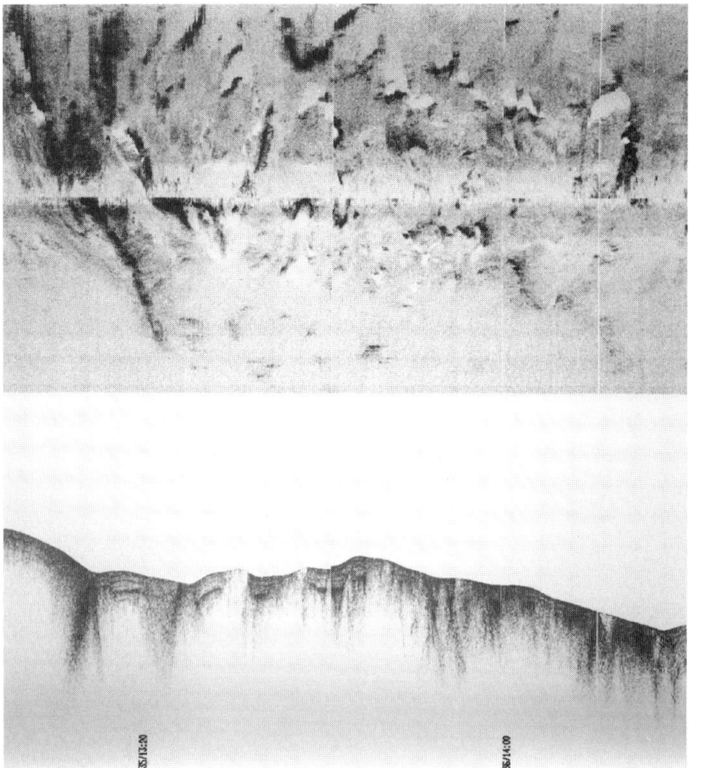

Fig. 7. Deep-tow image and sub-bottom profile across area of gas escape (location 11 in Fig. 1). Northwest is to the right. Side-scan sonar image (above) is 2 km across (vertically). High backscatter is in darker tones. Note the disappearance of sedimentary layering in the sub-bottom profile by acoustic wipe-outs and, to the right, acoustic turbidity. Gas escape is associated with small build-ups on the sea floor as observed in the side-scan image (note acoustic shadows (white) behind them).

advance of the African plate against the Florence Rise. The sampling programme showed that the geological province represented by Anaxagoras is the same as the Antalya Nappes Complex in Turkey (paper in preparation); thus, it is also possible to visualize similar structures in Anaxagoras (including, perhaps, gas venting similar to that known from ancient times in the area of the Antalya Nappes Complex, central to the legend in which Bellerophon slew the fire-breathing Chimera – still 'breathing fire' from where the continuing gas emissions are ignited). Compression is clearly shown by the advancing fold belt in the upper sediments of the Antalya Basin. It may involve Anaximenes also, causing it to bend as it is pushed against Anaxagoras. Continuing compression and the presence of overthrusts may be responsible for the overpressured regions in the sub-surface; and the observed gas (perhaps derived from organic material supplied to the Antalya Basin from nearby land) may be an important source of the overpressuring. Strike-slip faults cutting through the mountains provide potential pathways for the release of the excess pressure and the eruption of water-saturated mud mixtures from formations buried deep below.

Gas in the Anaxagoras area seems to be restricted mainly to the mud volcanoes, although exploration of this mountain with the sub-bottom profiler is still very limited in extent. The distribution of mud volcanoes seems locally to be linear and follow inferred fault zones, both strike-slip and thrust (e.g. mud volcanoes San Remo and Kula, locations 1 and 2 in Fig. 1b, appear to lie along a major NNE running lineation which follows the western side of the Gulf of Antalya to the north, in Fig. 1b). At least one fissure eruption was observed (location 6 in Fig. 1b), on a south-facing escarpment between Anaximenes and Anaxagoras. Fluid expulsion along the northeast–southwest scarp may be

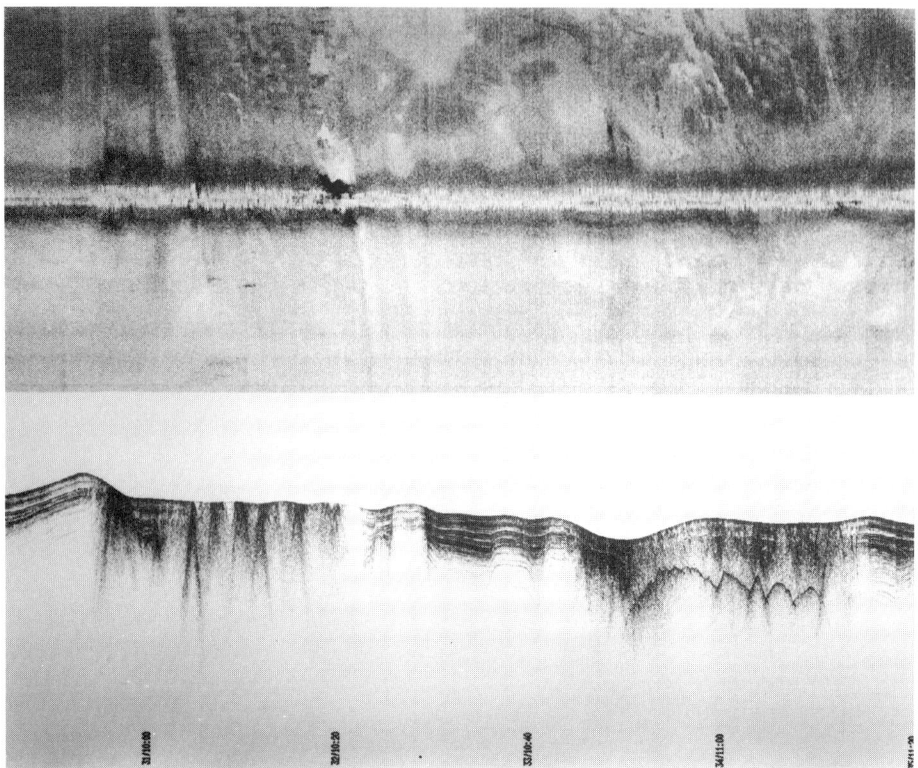

Fig. 8. Deep-tow image and sub-bottom profile (location 10a in Fig. 1) across an area showing 'gas front' (to the right) and gas escape (to the left). Northwest is to the right. Side-scan sonar image (above) is 2 km across (vertically). High backscatter is in darker tones. Note the masking of sedimentary layering in the sub-bottom profile by acoustic turbidity to the left and, to the right, acoustic wipe-outs and turbidity above and below the gas front. The side-scan image shows that the sea-floor sediments are disturbed where there is underlying gas.

associated with a cross-cutting fault which offsets Anaxagoras there in a sinistral sense (35°30′N–30°30′E in Fig. 1b).

Mud volcanoes tend to erupt episodically but fluid seeps are probably a continuous phenomenon. Gas hydrates may build up during dormant stages. This seems to be the case in the Olimpi field where ODP drilling results indicated the presence of a thin layer in which gas hydrates are present on the relatively dormant Milano mud volcano, but none on the more active Napoli mud volcano. On the other hand, the other parts of the Anaximander complex do not appear to have mud volcanoes, but they do display large areas of gas-charged sediments or areas where gas is seeping out of numerous vents.

Gassy sediments appear to be concentrated in two areas besides the mud volcanoes: the sediments involved in the moving mass of sediments between the east and west mountains; and where the inferred carbonate mounds occur, in the west-central part of the area just to the north of a faulted east–west ridge (Fig. 9). Gas is not observed elsewhere such as the Finike or Antalya basins, or on Anaximander Mountain. The connection between gas and sediment movement suggests two possibilities: the gas was mobilized because of sediment movement and deformation; or the sediments were mobilized because of the presence of gas. The interpretation favoured in this paper is that the gas which was mobilized by decomposition of gas hydrates in the sediments (accompanied by water) facilitated the movement of the sediments under gravity.

It is not obvious either whether the pockmarks that coalesce do so as a response to or as a cause of the extension of sediment during sliding. The transverse tectonic windows which appear to have developed above a décollement at the Miocene–Pliocene boundary between Anaximander and Anaximenes could result where excessive gas escape has disrupted the continuity

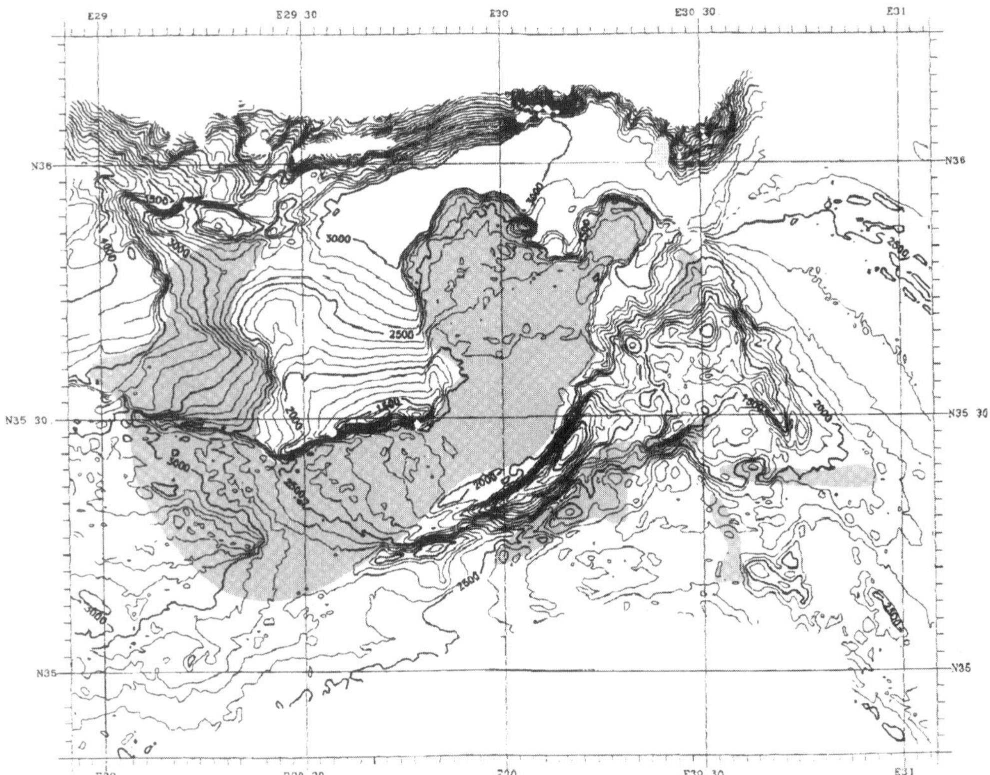

Fig. 9. Bathymetry of the Anaximander Mountains regions with shading to indicate the general known distribution of gassy sediments.

of the upper sediments enough that they separate horizontally; however, this implies the unlikely removal by gas venting of large volumes of sediment from the growing and coalescing pockmarks (e.g. at least 600 m^3 of sediment at location 9b in Fig. 1b) in the absence of space provision for the stretching. Moreover, the orientation of the transverse gaps in the sediment cover argues more for a tectonic origin, as the sediment moves out from between the Anaximander and Anaximenes mountains, forming linear basins of the order of 2 km wide parallel to the direction of shortening between the mountains. That could in turn facilitate gas escape. In the example given (Fig. 5) the pockmarks have developed along a lineation, probably a line of weakness like a cross-cutting fault, thus supporting this contention.

Where the small slide occurred, the underlying sediments which were revealed by the slide showed higher reflectivity than the same strata in the surrounding region (Fig. 6). The amplitude enhancement of the reflectors is probably caused by elevated concentrations of gas in those sediments. The gas may find it easier to escape at the slide scar than elsewhere, with a consequent build-up of the gas in this area. As a result, more detail in the sedimentary layering is seen below the slide scar than in the same sediments to the side. The manner in which the lower part of the slide block has disintegrated suggests that it liquidized as it moved. This could be a result perhaps of incorporation of sea water during rapid sediment movement, a normal occurrence in such deep-sea slides, or possibly through the decomposition of gas hydrates in the sediments during or just prior to the slide, which in turn would create a surplus of water. In any case, the slide was facilitated by a break-up of the sheet of sediments and further disintegration into a distal debris flow carrying with it larger blocks of sediments.

The source of the sediments involved in the great slide is not yet determined; however it seems most likely that the slide is formed from mobilized Pliocene–Recent sediments moving on a décollement at the Miocene–Pliocene interface. 'Stretching' or extension of the nappe of sediments between Anaximander and Anaximenes mountains, as it moves both to the north towards the lobes and to the southwest, has produced several tectonic windows oriented perpendicular to the flow and extending in depth only as far as the top of the base of the Pliocene. The Pliocene–Miocene interface is inferred from a strong and characteristic seismic reflector, and the presence of gypsum fragments of assumed Messinian age in samples cored from sediments in one of the tectonic windows (at core location 217G, at location 9a in Fig. 1b).

It might be presumed that the sediments were dumped from Anaximenes when it was tilted to the northwest; however, this is unlikely for many reasons. For example, the sediments between Anaximander and Anaximenes are fairly uniform in thickness and show no evidence of having slid off Anaximenes; moreover, if the thickness of sediments on Anaximenes had been roughly 250 m, then the volume of sediments which could have slid off would be no more than 10 km^3 (assuming a maximum surface area of 40 km^2 for Anaximenes before it tilted to the northwest), in comparison to the estimated 550 km^3 of sediments which are thought to be involved in the entire slide.

The lobes of the slide are smoothly semicircular in plan view and give the impression of a sediment mass spreading out over the Pliocene–recent sediments of the Finike Basin (e.g. Figs 1b and 3a). Their height (about 250–300 m) and the steepness of their frontal slopes (>10°) suggests that they are still moving or relatively recently stopped. Finike Basin sediments can be followed southward also, across the north flank of Anaximander Mountain to their abrupt termination at the escarpment marking the southern boundary of the mountain (Fig. 10), 1800 m shallower. The differential vertical movement implied by the tilted sediments on Anaximander Mountain indicate either elevation

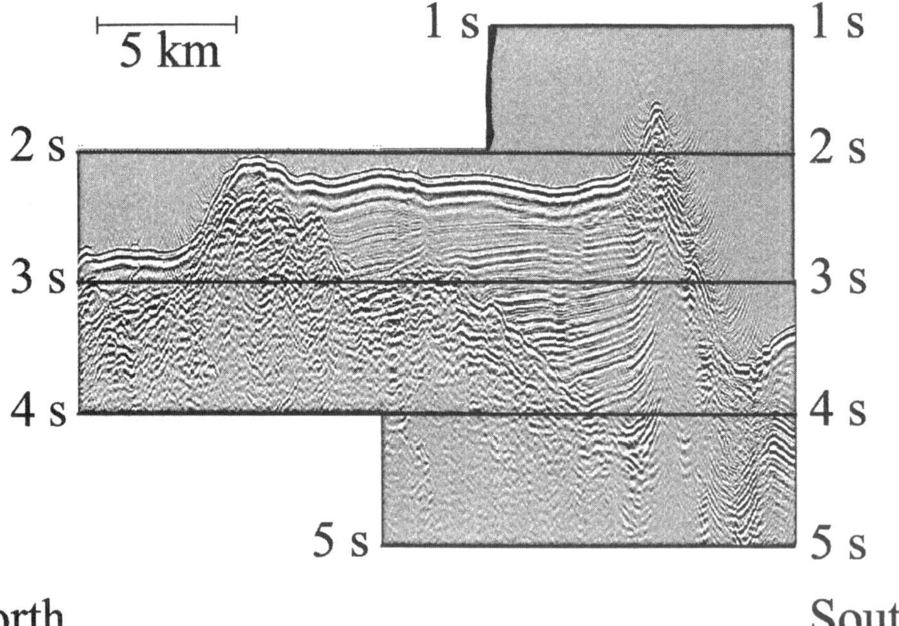

Fig. 10. Seismic reflection profile from north to south (left to right) across the top and southern escarpment of Anaximander Mountain at 29°30′E (Fig. 1). Vertical scale is in seconds TWTT. Note the absence of Messinian evaporites (no M-reflector) at the unconformity between the pre-Pliocene Bey Daglari basement rocks and the overlying Pliocene–recent sediments. Note also the presence of faulting and folding with the sediments and the absence of evidence for gas in the post-Miocene sediments on the mountain.

of Anaximander Mountain in Recent time or continuing subsidence of Finike Basin (or a combination of both). Elevation of Anaximander Mountain is favoured over subsidence of Finike Basin because of the low degree of deformation in the ponded sedimentary deformation there in comparison with the sediments on Anaximander Mountain (e.g. Fig. 10).

It is conjectured that the former, now missing, continuation of the Pliocene–Recent sediments southward of Anaximander Mountain is the principal source of sediments in the slide. Rapid elevation of the sediments out of the gas hydrate stability field during the elevation of Anaximander Mountain could have resulted in the decomposition of the gas hydrates and the collapse of the sediments into a large-scale debris flow toward the north over Recent Finike Basin sediments, and southwestward over a trench visible in the southwest corner of the area (Fig. 1). An estimate of the volume of sediments which might have extended south of the mountain (Fig. 10), and to the east of it, is similar in magnitude to the total estimated volume of sediments in the slide (i.e. about $550\,km^3$).

Mobilization of the sediments into a flow out over the Finike Basin would have created high pore water pressures in the fresh wet sediments forming that basin, beneath the flow. Release of the overpressuring in the Finike Basin sediments beneath the slide would have occurred through the lobes, possibly contributing to their shape. At the northern edge of the central region of the lobes is an elliptical hill about 800 m high and roughly 6–7 km long by 4–5 km across, which is called the 'Anthill' (location 8 in Fig. 1b). Preliminary investigations of a core from the top of the Anthill indicate a higher sedimentation rate than elsewhere in the region (i.e. about $12\,cm\,ka^{-1}$ compared to values of about $2–5\,cm\,ka^{-1}$ elsewhere). The only realistic source of sediments to produce such a high and symmetrical sedimentary build-up seems to be from below (as is the case for real anthills). The inferred build-up of the Anthill by a sediment slurry brought up internally through the centre of the structure, and deposited over the top, is considered to be evidence of dewatering of the underlying Finike Basin sediments.

One of the frontal northern lobes of the Slide, immediately to the east of the Anthill and between the major western and eastern branches of the flow, may be slightly different to the others. To the south of it is a small basin with high acoustic backscatter (location 9a in Fig. 1b), which, like the tectonic windows to the south appears to have developed by removal of the upper Pliocene–recent sediments. In this case the sediments washed out to form a small delta fan to the north. For such a washout to occur requires the liquidizing of the sediments. This is perhaps also a clue to the origin of the great slide.

The presence of gas throughout the slide area (Fig. 9) suggests that the gas is widespread below, which in turn suggests the presence of gas hydrates in the upper sediments. The stability requirements are met with bottom water temperatures of the order of 14°C and depths around 1500–2000 m (e.g. Kvenvolden 1993), and there have been gas hydrates sampled from the same region, 50 km to the east (location 2 in Fig. 1b). Sediments with gas hydrates are known to have a higher water content than overlying non-hydrated sediments (Soloviev & Ginsburg 1997), and the host sediments are underconsolidated (McIver 1982). In a tectonically active area such as the Anaximander Mountains, unconsolidated sediments with high water content would be mobilized without difficulty. Additionally, there are several mechanisms by which the gas hydrates could be at least partially broken down to produce gas and water: for example, by circulation of warm briny water from several hundreds of metres below the underlying sediments or, as suggested earlier, by tectonics (such as uplift of sediments with gas hydrates to depths less than 1400 m, thereby lowering their ambient pressure). With enough of the sediment in the mass becoming weak and fluid-saturated, the entire mass would easily move under even the low slopes present. The Pliocene–Miocene boundary is a convenient décollement, in part because the level of consolidation increases in the Miocene section below; and the top of the Miocene is probably an erosional surface with a thin layer of early stage evaporites like gypsum (or alternatively shallow water gypsum from a later stage), as indicated in the seismic profiles.

Conclusions

Continuous emission of gas from mud volcanoes is indicated by carbonate crusts and by the presence of well-established chemoautotrophic benthic communities. Mud volcanoes seem to be restricted to the eastern part of the Anaximander complex where compression is occurring and there are overthrusts and nappes. They are probably formed by the release of overpressured gas and water-saturated sediments from deep below the overthrusts along conduits created by cross-cutting strike-slip faults or which are present along the thrust planes. In less active mud volcanoes the gas has been trapped as gas hydrates.

Elsewhere (west of Anaxagoras) in the Anaximander Mountains complex, gas is commonly observed in sediments in an extensive area associated with mass movement. The sediments of Anaximander or Anaximenes do not appear to be gas-saturated, but the region of the Great Slide does.

The Great Slide began movement during the late Pliocene or Pleistocene. Gas hydrates formed in the sediments as a result of continuous gas seepage, resulting in poorly consolidated sediments with higher than normal water content. This alone could be the source of mass sediment movement in a tectonically active region. However, it is further suggested that sediment undergoing deformation within the compression zone between Anaximander and Anaximenes began to release gas from decomposing gas hydrates as it was elevated above the gas hydrate stability zone. Deformation in this region comprises underthrusting of Anaximander by Anaximenes and shortening between them, but also with strike-slip components, and some large vertical movements, as Anaximander was raised by underthrusting of Anaximenes, and Anaximenes was tilted up toward the northwest.

This work was made possible through a grant from the Netherlands Foundation for Geosciences (Stichting GOA Project No. 750.195.02) to J. M. Woodside. Among the shipboard scientists, deep-tow side-scan images and sub-bottom profiles were prepared mainly by R. Almendinguer, and seismic processing and display was done by A. Volkonskaya, S. Bouriak and V. Gainanov. Identification of clasts from the mud breccia was made by J.-F. Dumont, G. Akhmanov, J. Henderiks and H. van der Bosch. All shipboard scientists contributed in some way to this work. Preliminary identification of biological samples was made by E. Southward and C. Salas. Furthermore, we are indebted to the officers and crews of the R/V *Atalante* and the R/V *Gelendzhik* for facilitating the research, and to the superb technical crews of both ships for ensuring the success of the expedition through the excellent results obtained.

References

BARAZA, J. & ERCILLA, G. 1996. Gas-charged sediments and large pockmark-like features in the Gulf of Cadiz slope (SW Spain). *Marine and Petroleum Geology*, **13**(2), 253–261.

CARLSON, P., GOLAN-BAC, M., KARL, H. & KVENVOLDEN, K. 1985. Seismic and geochemical evidence for shallow gas in sediment on Navarin Continental margin, Bering Sea. *AAPG Bulletin*, **69**, 422–436.

CITA, M., RYAN, W. & PAGGI, L. 1981. Prometheus mud breccia. An example of shale diapirism in the western Mediterranean Ridge. *Annales Geologiques des Pays Helleniques*, **30**, 543–570.

——, WOODSIDE, J., IVANOV, M., KIDD, R., LIMONOV, A. *et al.* 1995. Fluid venting from a mud volcano in the Mediterranean Ridge Diapiric Belt. *Terra Nova*, **7**(4), 453–458.

CORSELLI, C. & BASSO, D. 1996. First evidence of benthic communities based on chemosynthesis on the Napoli mud volcano (Eastern Mediterranean). *Marine Geology*, **132**(1/4), 227–239.

DANDO, P., SOUTHWARD A., SOUTHWARD, E. *et al.* 1992. Shipwrecked tube worms. *Nature*, **356**, 667.

FLEMMING, N. 1972. Eustatic and tectonic factors in the relative vertical displacement of the Aegean coast. *In:* STANLEY, D. (ed.) *The Mediterranean Sea*. Dowden, Hutchinson & Ross, Stroudsberg, 189–201.

——, CZART'RYSKA, N. & HUNTERE, P.1973. Archeological evidence for vertical earthmovements in the region of the Aegean island arc. *In:* FLEMMING, N. (ed.) *Science Diving International. Proceedings of the Confédération Mondiale des Activités Subaquatiques*, London, 47–65.

GUTNIC, M., MONOD, O., POISSON, A. & DUMONT, J.-F. 1979. Géologie des Taurides Occidentales (Turquie). *Mémoires de la Société Géologique de France*, Tome LVIII, **137.**

HOVLAND, M. & JUDD, A. 1988. *Seabed Pockmarks and Seepages: Impact on Geology, Biology and the Marine Environment*. Graham and Trotman, London.

IVANOV, M., LIMONOV, A. & WOODSIDE, J. 1992. Geological and geophysical investigations in the Mediterranean and Black Seas. *In: UNESCO Reports in Marine Science*, Volume 56. UNESCO, Paris.

JUDD, A. & HOVLAND, M. 1992. The evidence of shallow gas in marine sediments. *Continental Shelf Research*, **12** 1081–1095.

KVENVOLDEN, K. 1993. Gas hydrates – geological perspective and global change. *Reviews of Geophysics*, **31**(2), 173–187.

LE PICHON, X., CHAMOT-ROOKE, N. & LALLEMANT, S. 1995. Geodetic determination of the kinematics of central Greece with respect to Europe: Implications for eastern Mediterranean tectonics. *Journal of Geophysical Research*, **100**, 12 675–12 690.

MASCLE, J., LE CLEAC'H, A. & JONGSMA, D. 1986. The eastern Hellenic margin from Crete to Rhodes: example of progressive collision. *Marine Geology*, **73**, 145–168.

MCIVER, R. D. 1982. Role of naturally occurring gas hydrates in sediment transport. *AAPG Bulletin*, **66**, 789–792.

MCKENZIE, D. 1972. Active tectonics of the Mediterranean region. *Geophysical Journal of the Royal Astronomical Society*, **30**, 109–185.

—— 1978. Active tectonics of the Alpine–Himalayan belt: the Aegean Sea and surrounding regions. *Geophysical Journal of the Royal Astronomical Society*, **55**, 217–254.

ORAL, M., REILINGER, R., TOKSÖZ, BARKA, A. & KINIK, I. 1993. Preliminary results of 1988 and 1990 GPS measurements in western Turkey and their tectonic implications. *In:* SMITH, D. & TURCOTTE, D. (eds) *Contributions of Space Geodesy to Geodynamics: Crustal Dynamics.* American Geophysical Union, Geodynamics Series, **23**, 407–416.

PAPATHEODOROU, G., HASIOTIS, T. & FERENTINOS, G. 1993. Gas-charged sediments in the Aegean and Ionian Seas, Greece. *Marine Geology*, **112**, 171–184.

SALAS, C. & WOODSIDE, J. 1998. Lucinoma Kazami n. sp. (Mollusca: Bivalvia): First evidence of a living cold seep community in the Eastern Mediterranean Sea. Submitted.

SHIPBOARD SCIENTIFIC PARTY. 1996. Site 970. *In*: EMEIS, K.-C., ROBERTSON, A., RICHTER, C. *et al.* (eds) *Proceedings of the Ocean Drilling Project, Initial Reports*, College Station, TX Ocean Drilling Program, **160**, 377–413.

SOLOVIEV, V. & GINSBURG, G. 1997. Water segregation in the course of gas hydrate formation and accumulation in submarine gas-seepage fields. *Marine Geology*, **137**, 59–68.

SOUTHWARD, E., TUNNICLIFFE, V., BLACK, M., DIXON, D. & DIXON, L. 1996. Ocean-ridge segmentation and vent tube worms (Vestimentifera) in the NE Pacific. *In:* MACLEOD, C., TYLER, P. & WALKER, C. (eds) *Tectonic, Magmatic, Hydrothermal and Biological Segmentation of Mid-ocean Ridges.* Geological Society, London, Special Publications, **118**, 211–224.

STAFFINI, F., SPEZZAFERRI, S. & AGHIB, F. 1993. Mud diapirs of the Mediterranean Ridge: sedimentological and micropaleontological study of the mud breccia. *Rivista Italiana Paleontologia et Stratigrafia*, **99**(2), 225–254.

STEFANON, A. 1985. Marine sedimentology through modern acoustical methods: II. Uniboom. *Bolletino di Oceanologia Teorica ed Applicata*, **III/2**, 113–144.

SUESS, E., CARSON, B., RITGER, S. *et al.* 1985. Biological communities at vent sites along the subduction zone off Oregon. *Biological Society of Washington Bulletin*, **6**, 475–484.

TRAYNOR, J. & SLADEN, C. 1997. Seepage in Vietnam – onshore and offshore examples. *Marine and Petroleum Geology*, **14**(4), 345–362.

VOLGIN, A. & WOODSIDE, J. 1996. Sidescan sonar images of mud volcanoes from the Mediterranean Ridge: possible causes of variations in backscatter intensity. *Marine Geology*, **132**(1/4), 39–53.

WOODSIDE, J. 1976. Regional vertical tectonics in the eastern Mediterranean. *Geophysical Journal of the Royal Astronomical Society*, **47**, 493–514.

—— 1977. Tectonic elements and crust of the eastern Mediterranean Sea. *Marine Geophysical Researches*, **3**, 317–354.

—— 1991. Disruption of the African Plate margin in the eastern Mediterranean. *In*: SALEM, M., SBETA, A. & BAKBAK, M. (eds) *The Geology of Libya*, Volume 6. Elsevier, Amsterdam, 2319–2329.

——, IVANOV, M. & SHIPBOARD SCIENTISTS R/V GELENDZHIK. 1993. Training Through Research cruise 1993: Is the African Plate tearing between the Hellenic and Cypriot Arcs? (Abstract) *Terra Nova*, **5**(1), 272.

YUAN, F., BENNELL, J. & DAVIS, A. 1992. Acoustic and physical characteristics of gassy sediments in the western Irish Sea. *Continental Shelf Research*, **12**, 1121–1134.

Extensive deep fluid flux through the sea floor on the Crimean continental margin (Black Sea)

M. K. IVANOV[1], A. F. LIMONOV[1] & J. M. WOODSIDE[2]

[1] *UNESCO-MSU Centre for Marine Geosciences, Faculty of Geology, Moscow State University, Vorobjevy gory, 119899 Moscow, Russia*
[2] *Free University, Faculty of Earth Sciences, De Boelelaan 1085, 1081 HV Amsterdam, The Netherlands*

Abstract: The ANAXIPROBE/TTR-6 Cruise of the Russian R/V *Gelendzhik* in the summer of 1996 resulted in the discovery of a new area of extensive deep fluid flux through the sea floor on the lower Crimean continental margin, within the Sorokin Trough in the Black Sea. The study area is noted for the widespread development of acoustic anomalies in bottom and sub-bottom sediments indicating the presence of gas. Bottom sampling confirmed a high content of gas in the sediments, the presence of gas hydrates and other features normally associated with gas escape. These features include authigenic carbonates with a light carbon isotope composition and bacterial mats. The geochemical data indicate that the gas has in part a thermogenic origin.

Numerous investigations world-wide have shown that active submarine fluid discharge (particularly of gases) produces specific structures on the sea floor (Hovland & Judd 1988; Ivanov *et al.* 1996*a*, *b*). Primarily, such structures are represented by mud volcanoes and pockmarks. Mud volcanoes are expressed in the sea-floor relief either as mounds or as negative collapse features. Pockmarks generally are depressions developed in unconsolidated bottom sediments. Through feeder channels of mud volcanoes, a large volume of clastic material, with gas and other fluids, is transported in a clay–silt fluidized matrix and deposited on the sea floor. This material issues from a mud volcano crater, forms a mounded body of a mud volcano and, depending on the degree of fluidization, is able to form extensive mud flows spreading from mud volcano slopes sometimes over distances of several kilometres.

Intensive fluxes of fluids, especially of hydrocarbon gases, hydrogen sulphide, carbon dioxide and nitrogen, result in specific geochemical conditions on the sea bottom and in sub-bottom sediments which significantly differ from 'normal' conditions of a marine basin. Such local changes of geochemical environment and the supply of material from the depth provoke a blossom of biological activity manifested in the appearance of bacterial colonies (bacterial mats) and chemosynthetic communities of benthic organisms (Cita *et al.* 1989, 1995; Corselli & Basso 1996). At the same time, these areas are noted for fast mineral transformations under the influence of fluid venting and bacterial activity. Carbonate and sulphide build-ups grow in sediments and on the sea floor, and carbonate crust and concretions develop (Hovland & Judd 1988).

Migrating hydrocarbon gases, and methane in particular, often form under favourable *P–T* conditions lenses and layers of gas hydrates lying directly on the sea floor or in sediments at a depth ranging from a few tens of centimetres to a few metres (Ginsburg & Soloviev 1997; Soloviev & Ginsburg 1997).

In fluid escape zones, fluidized sediments have an enhanced instability and are capable of flowing down very gentle slopes (Limonov *et al.* 1997), leading to complex deformation in the uppermost sediments. The sediment sliding and slumping is also favoured by the presence of gas hydrate lenses and layers, along whose smooth surface such sliding of fluidized sediments takes place.

All the above listed phenomena produce a remarkable inhomogeneity of physical properties of bottom sediments, which, in turn, is expressed not only in bottom relief but also in sea-floor acoustic reflectivity and acoustic anomalies in sub-bottom sediments.

The Black Sea provides an excellent opportunity to study gas venting through the sea floor. It is a unique restricted deep-water basin marked by a natural hydrogen sulphide production starting from a depth of 150–200 mbsl. The basin is underlain by oceanic crust, and the total thickness of its Mesozoic–Cenozoic sedimentary infill reaches 13–15 km.

The high gas content in the Black Sea bottom and sub-bottom sediments has long been known (Hunt & Whelan 1978), and recent investigations

IVANOV, M. K., LIMONOV, A. F. & WOODSIDE, J. M. 1998. Extensive deep fluid flux through the sea floor on the Crimean continental margin (Black Sea). *In:* HENRIET, J.-P. & MIENERT, J. (eds) *Gas Hydrates: Relevance to World Margin Stability and Climate Change*. Geological Society, London, Special Publications, **137**, 195–213.

demonstrate that widespread fluid venting takes place in many Black Sea areas, both in the deep-water region and on the shelves (Ivanov *et al.* 1992; Dimitrov 1994; Shnyukov *et al.* 1995). Gas hydrates were repeatedly recovered from the central, deep-water Black Sea region and also on the Crimean continental margin during several marine expeditions, including those carried out by the Geological Faculty of Moscow State University. In the central Black Sea, gas hydrates were found to occur in mud breccia of large mud volcanoes (Ivanov *et al.* 1989, 1992; Konyukhov *et al.* 1990; Kruglyakova *et al.* 1993; Limonov *et al.* 1994). Ginsburg *et al.* (1990) also believed that gas hydrates from the Crimean margin originated from a mud volcano.

Geological setting

During the second leg of the ANAXIPROBE/TTR-6 cruise of the Russian R/V *Gelendzhik* in 1996 a large number of geological–geophysical investigations were carried out in the area of the Sorokin Trough, south of the Crimean Peninsula (Fig. 1). The Sorokin Trough is located on the lower continental slope and rise SSE of the Crimean Peninsula, at water depths of 800–2000 m. The southwest–northeast oriented trough, which is 150 km long and 50 km wide, is considered as the Crimean Alpine foredeep founded in the Oligocene–Early Miocene (Maikop) time. The thickness of the Maikop Formation, largely clay deposits, exceeds 5 km in the study area. The total thickness of the overlying Middle Miocene–Pliocene sediments is rarely over 1 km, but Quaternary sediments attain up to 2–3 km in thickness, mainly due to thick accumulations in the palaeo-Don/palaeoKuban Pleistocene composite fan which covers the whole north-eastern Black Sea Basin (Tugolesov *et al.* 1985).

Diapiric folds consisting of Maikop clay were first discovered in the Sorokin Trough in the mid-1970s based on single- and multi-channel seismic reflection profiles oriented roughly south-north (Andreev 1976; Morgunov *et al.* 1976; Peklo *et al.* 1976). The structure of the diapirs was afterwards elaborated by Andreev *et al.* (1981) and Shnyukov *et al.* (1981), and the final results were summarized by Tugolesov *et al.* (1985). Since then, it has been widely accepted that the diapiric folds in the Sorokin Trough form several elongated zones and have a trend strictly parallel to the general trend of the trough. The shallow sub-bottom structure of the sediments in the trough area, however, has remained unclear.

Data acquisition

The following geological–geophysical methods were applied during the cruise: (1) single-

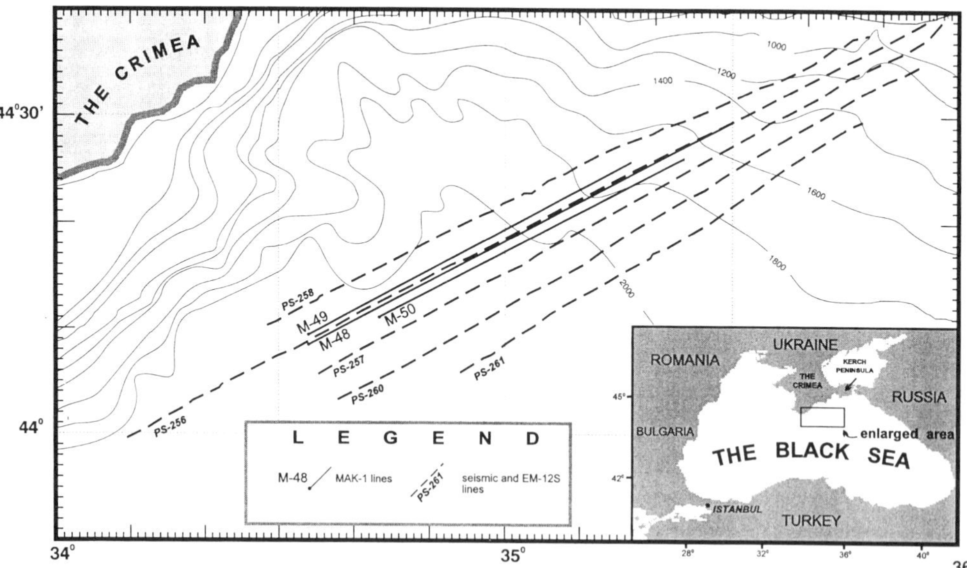

Fig. 1. Location map of seismic, multi-beam echosounder and MAK-1 lines. Bathymetry after IOC–UNESCO (1981). Inset shows the general location of the study area.

channel high-resolution seismic reflection profiling with a 3 litre airgun; (2) bathymetric survey with the Simrad EM-12S multi-beam echosounder; (3) MAK-1 deep-towed side-scan sonar survey accompanied by sub-bottom profiling; and (4) bottom sampling with a gravity corer.

The seismic profiling provided seismic time sections to a depth of about 1.5 s (TWTT) bsf with a vertical resolution of about 20 ms. Five seismic lines with a total length of 630 km were shot (Fig. 1). Line spacing was such to ensure overlapping of the bathymetric images accompanying the seismic survey (swath range ~3.5 times the water depth). The bathymetric survey resulted in the first detailed bathymetric and bottom reflectivity maps for this area (1 : 150 000 scale, 10 m contour interval).

The MAK-1 side-scan sonar, towed approximately 100 m above the sea floor, was operated at 30 kHz and had a swath range of 2 km. The sub-bottom profiler was operated at 4.9 kHz, and had a penetration varying between a few metres and 50 m, depending on bottom and sub-bottom acoustic conditions. Three MAK profiles were run (Fig. 1) narrowly spaced to ensure overlapping of the sonographs. These images were used for the characterization of the sea-floor pattern and for choosing sampling sites on specific objects.

The gravity corer used for bottom sampling had a length of 660 cm, an external diameter of 168 mm and weighed about 1.5 tons. It was supplied with a plastic liner facilitating the extraction of cores and preventing any disturbance. The cores were extracted on deck immediately after recovery of the corer. Each core was cut lengthwise in two, and, first of all, gas hydrates (if they were present) and gas were sampled at specific lithological intervals. The cores were then photographed, described and additionally sampled for different analyses.

The gas hydrate samples were stored in plastic bags in a laboratory at −18°C, and the gas

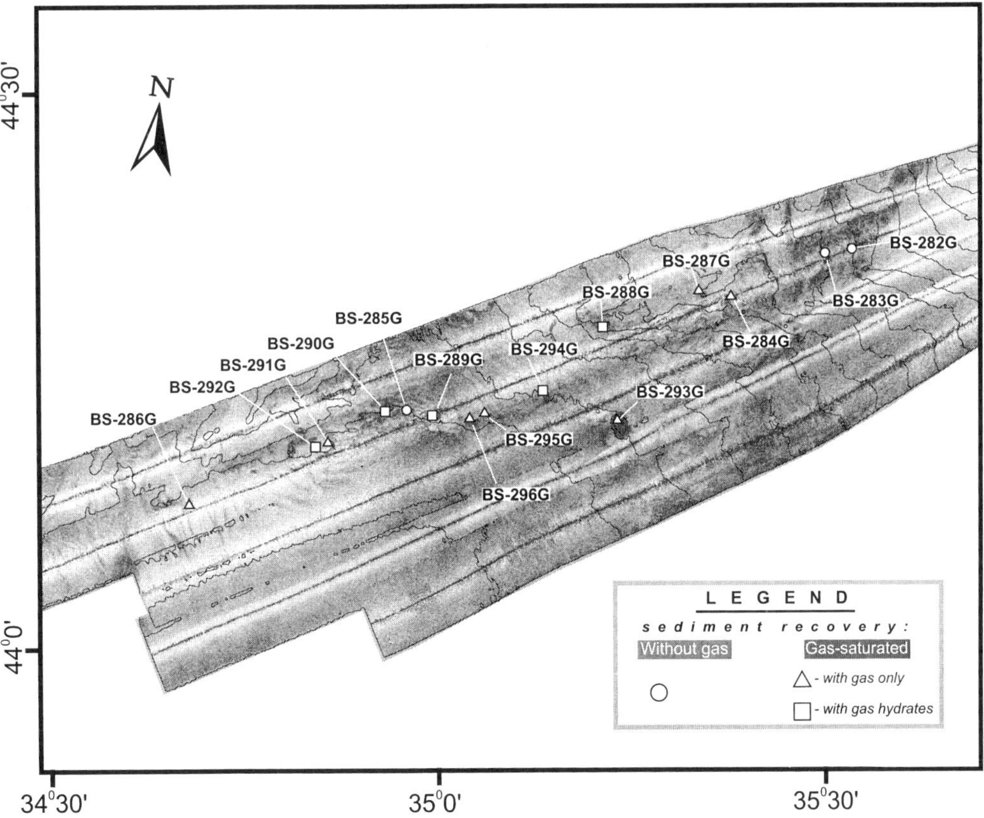

Fig. 2. Sea-bottom reflectivity map constructed on the basis of the Simrad EM-12S data, with the location of sampling sites.

samples were stored in glass jars with salt seals at +15–+20°C. In total 15 cores were taken in the area (Fig. 2), varying in length from a few cm to 602 cm.

Seismic and acoustic evidence for fluid fluxes, shallow gas and gas hydrate occurrence

Seismic reflection data

The seismic lines were run along the seaward zone of the diapiric folds known from previous investigations. A typical example of the structure of the diapiric folds is illustrated in Fig. 3. The diapirs have a complex pattern and consist of structures of several orders of magnitude. The first-order structures are represented by diapiric zones 12–40 km long and 2–10 km wide. In turn, these zones are sub-divided into diapiric ridges, varying from one to four for each zone (second-order structures). Finally, the third-order structures are represented by individual diapirs or mud volcanoes rising to the sea floor from the summits and flanks of the diapiric ridges.

The feeder channels of the observed mud volcanoes have a visible width of about 300–400 m, and their roots are masked by the acoustically homogeneous Maikop clay from where the volcanoes rise. Most mud volcanoes are situated on the flanks of the diapiric ridges.

It was found that the trend of the diapiric zones in the trough differs significantly from that assumed earlier. The zones have variable trends, controlled by the underlying structures in the Cretaceous–Eocene sequence.

Simrad EM-12S data

The recorded bathymetric images show the presence of isometric, positive structures, a few hundred metres to 2.5 km across and up to 120 m high. On the sea-floor reflectivity map these features, as a rule, have an enhanced level of reflectivity (dark colour). A number of large, irregular patches of high reflectivity can also be

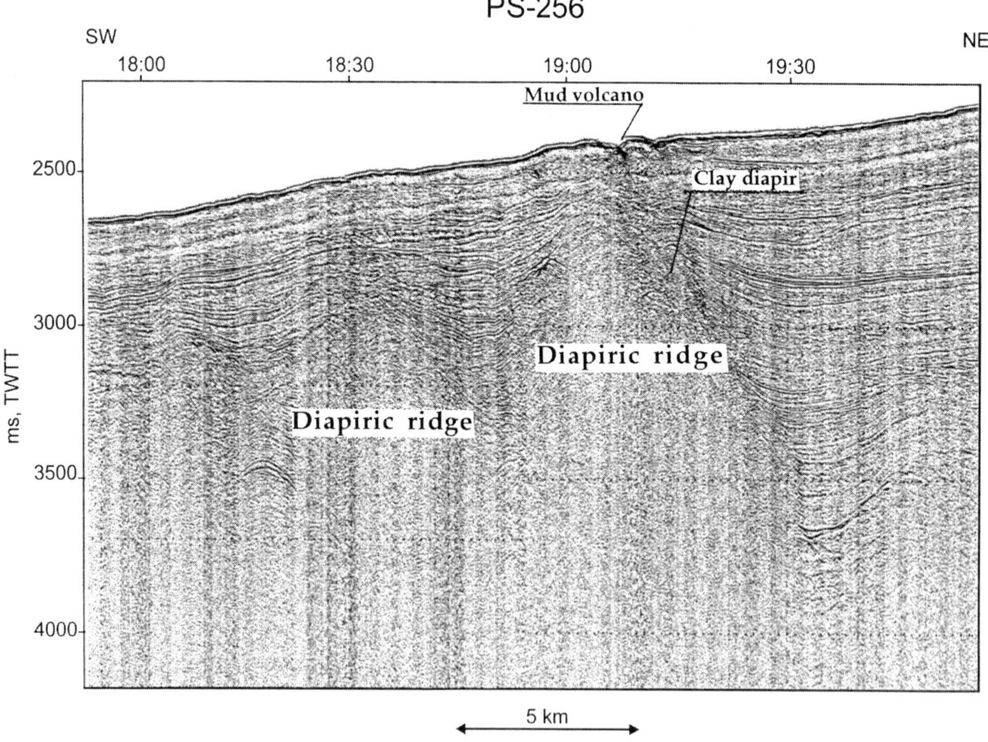

Fig. 3. Detail from seismic profile PS-256 showing a large diapiric zone with diapiric ridges. One of the ridges is crowned by a mud volcano as proved by coring.

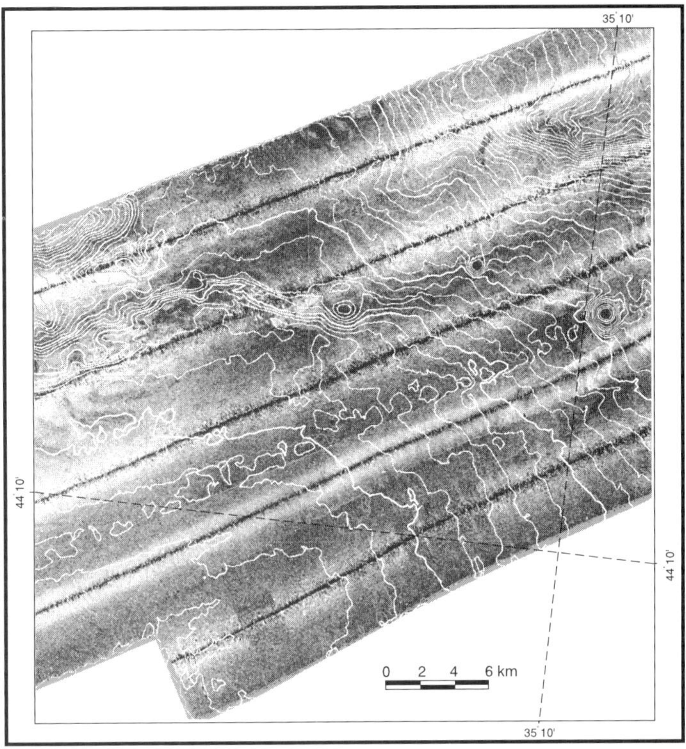

Fig. 4. Detail from the sea-bottom reflectivity map with bathymetric contours (every 10 m). Stronger reflectivity is imaged as darker tones. Small, circular, high-reflective patches indicate mud volcanoes. The largest (Kazakov) mud volcano, about 120 m high and 2.5 km across, is seen in the right centre. Large irregular areas with strong reflectivity normally mark sediment slumping and sliding.

observed (Fig. 4). These may be related to the escape of deep material through small channels and faults in the sea-floor relief, as well as to the extensive downslope mass movement in the form of slumps, debris flows, mud flows and turbidity currents. A close-up view of these strong reflectivity zones is given by the MAK-1 side-scan sonar.

Side-scan and sub-bottom profiler data.

Both positive and negative circular structures are seen in much more detail on the MAK-1 digital mosaic (Fig. 5). The central part of the mosaic differs clearly from the rest. This area, with a roughly amphitheatric shape, is marked by extremely high backscattering zigzag-shaped bends, sub-parallel to the continental slope, and individual irregular patches with high backscatter which alternate with areas of intermediate and weak backscatter, forming a contrasting acoustic pattern. According to the seismic and sub-bottom profiler data, this area is located on the southwest side of a large diapiric zone.

The observed high contrast in sea-floor reflectivity seems not only to be related to the complex small-scale seabed relief, but also to significant changes in physical properties of the shallow sediments due to different degrees of gas-charging. For example, it is clearly seen from the M-48 sonograph and its sub-bottom profile (Fig. 6) that the area with more contrasting backscatter (on the left) is characterized by a flatter sea-floor relief.

It is evident that the dark zigzag-shaped lines on the sonographs are related to sediment movement. The stronger backscatter of these lines may be conditioned by: (1) outcrops of older and denser sediment (e.g. the Novoeuxinian (the upper Pleistocene) clay which is rich in hydrotroilite) exposed as a result of sliding of the overlying fluidized layer; (2) formation of slump breccia in the frontal parts of slump bodies; and (3) intensive discharge of fluids through initial tensional fissures in sub-bottom sedi-

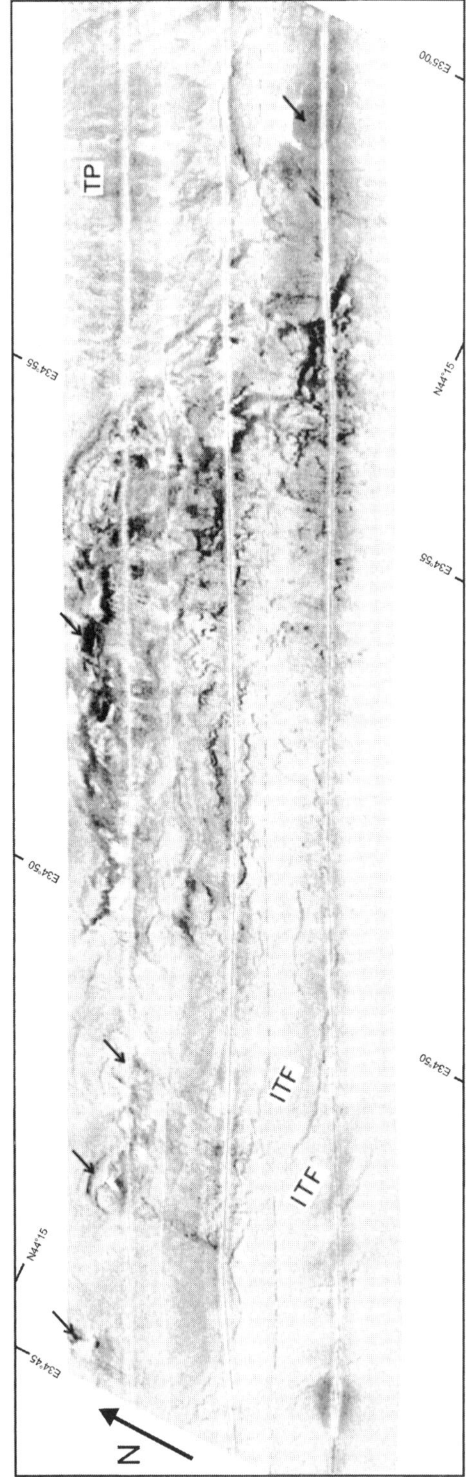

Fig. 5. Detail from the MAK-1 mosaic showing a large, amphitheatre-shaped slump area. The slumping is associated with the southwest slope of a large diapiric zone expressed in the sea-floor relief. Mud volcanoes are indicated by arrows. Also shown are assumed initial tensional fractures in surficial sediments (ITF) and possible turbidite pathways (TP).

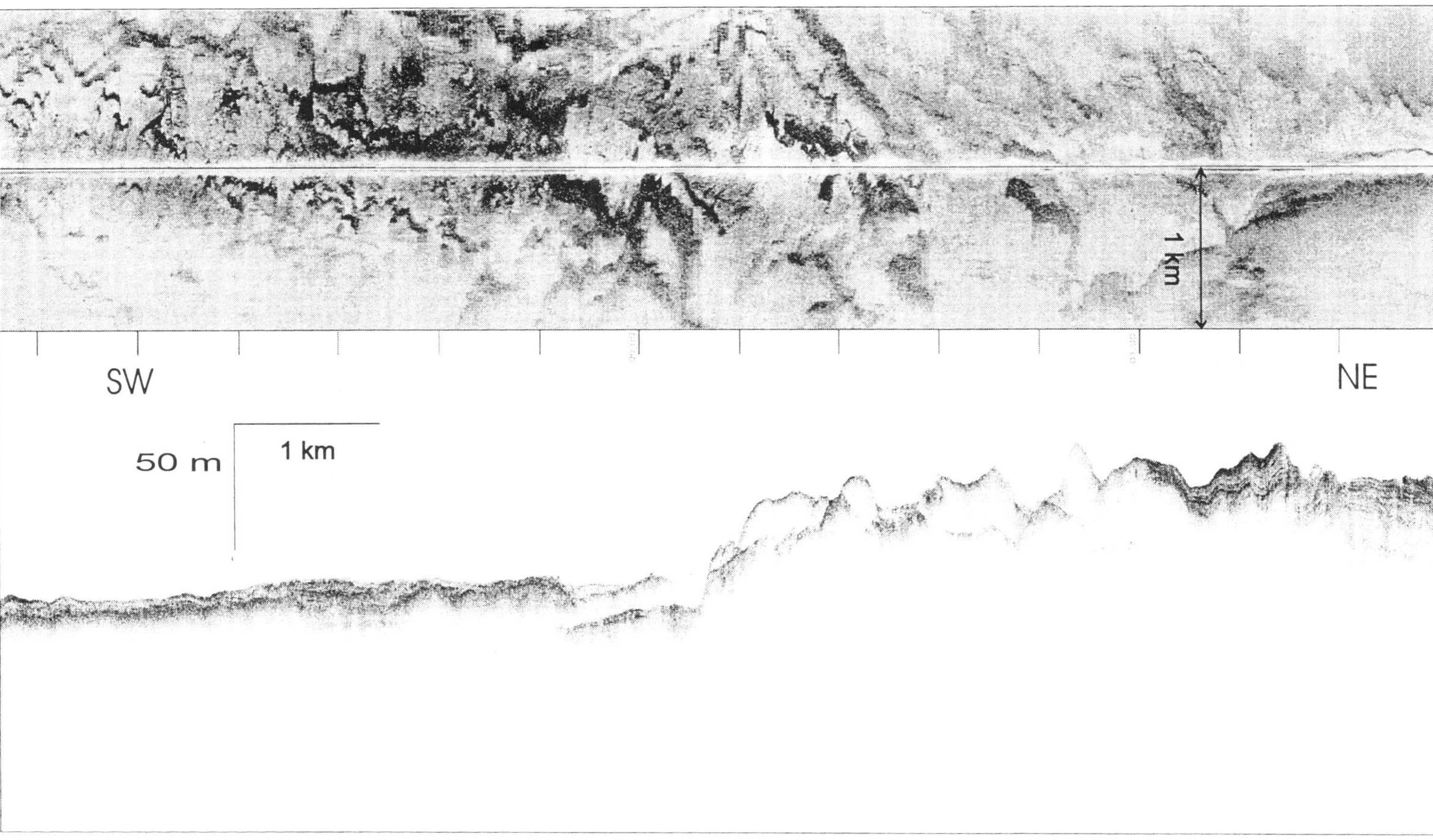

Fig. 6. MAK-1 sonograph and sub-bottom profile showing a close-up view of the same slump area of Fig. 5. The profile shows an irregular sea-floor relief with several enhanced reflectors and a possible gas front. Despite the rougher sea-floor relief in the right part of the record, the backscatter is less contrasting then on the left. Note that the irregular strips of strong backscatter on the sonograph correspond to those places on the profile where the enhanced reflectors or gas front rise to or approach the sea floor. This indicates a close relation between the slumping and gas saturation.

ments, which might result in the concentration of gas hydrates, carbonate nodules and sulphide mineralization in these places.

The areal distribution of gas content in the sub-bottom sediments can be illustrated by two fragments of profiles M-48 and M-49 (Figs 7 and 8). On both records one can see that the layered sub-bottom sequence changes to a more acoustical turbid zone, which is normally interpreted as an increase in gas content. Because the most acoustical turbid sediments are observed on the uprisen areas of the sea floor, there is an impression that the lateral gas migration occurs up-dip. Figure 7 also shows a charac-

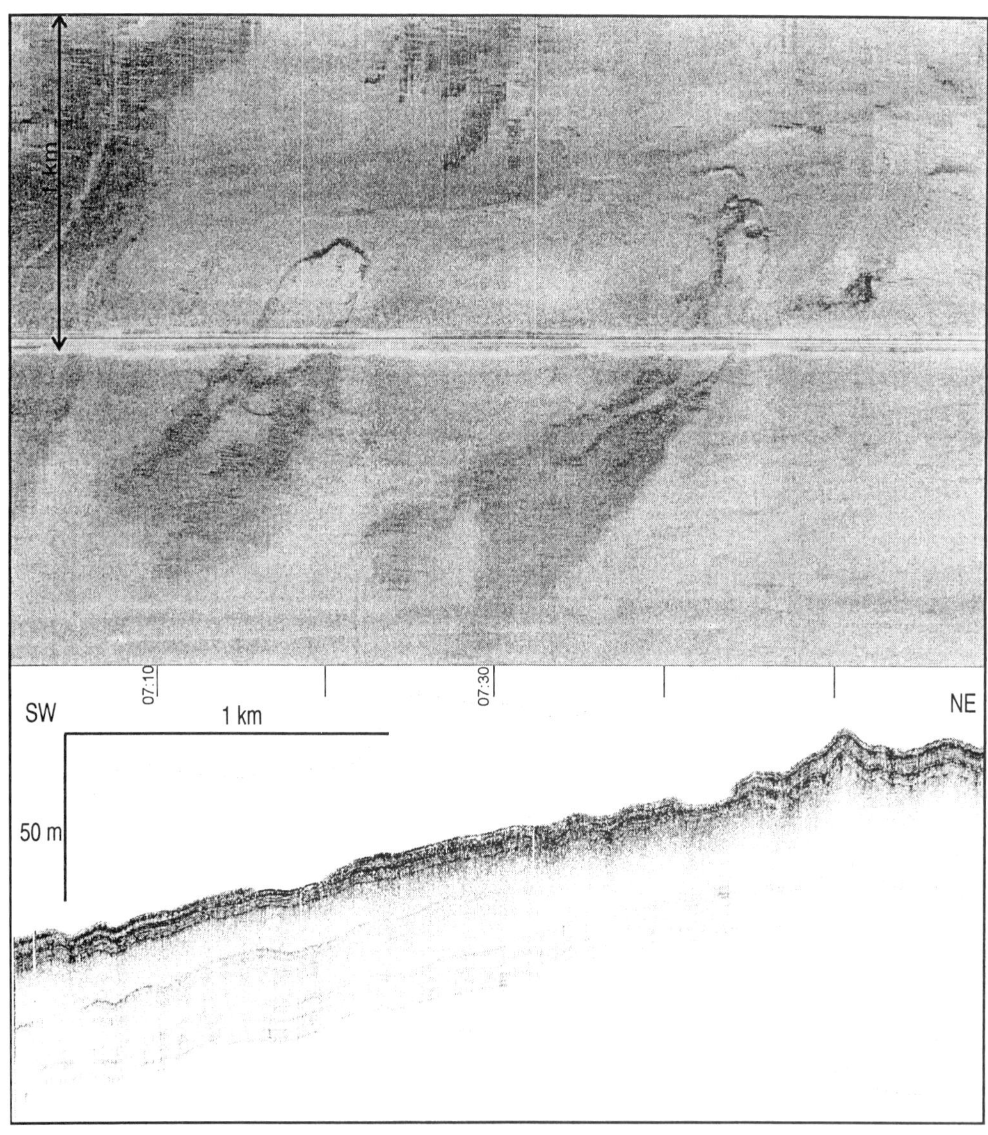

Fig. 7. MAK-1 sonograph and sub-bottom profile showing mud breccia eruptions. Although they may be related to faults as in Fig. 8, the fault traces are not so evident here. The eruption of mud breccia is accompanied by sliding of a thin layer of the overlying sediments. The traces of slides are seen as shallow sea-floor depressions on the profile. Note that the acoustic turbidity in sub-bottom sediments increases upslope.

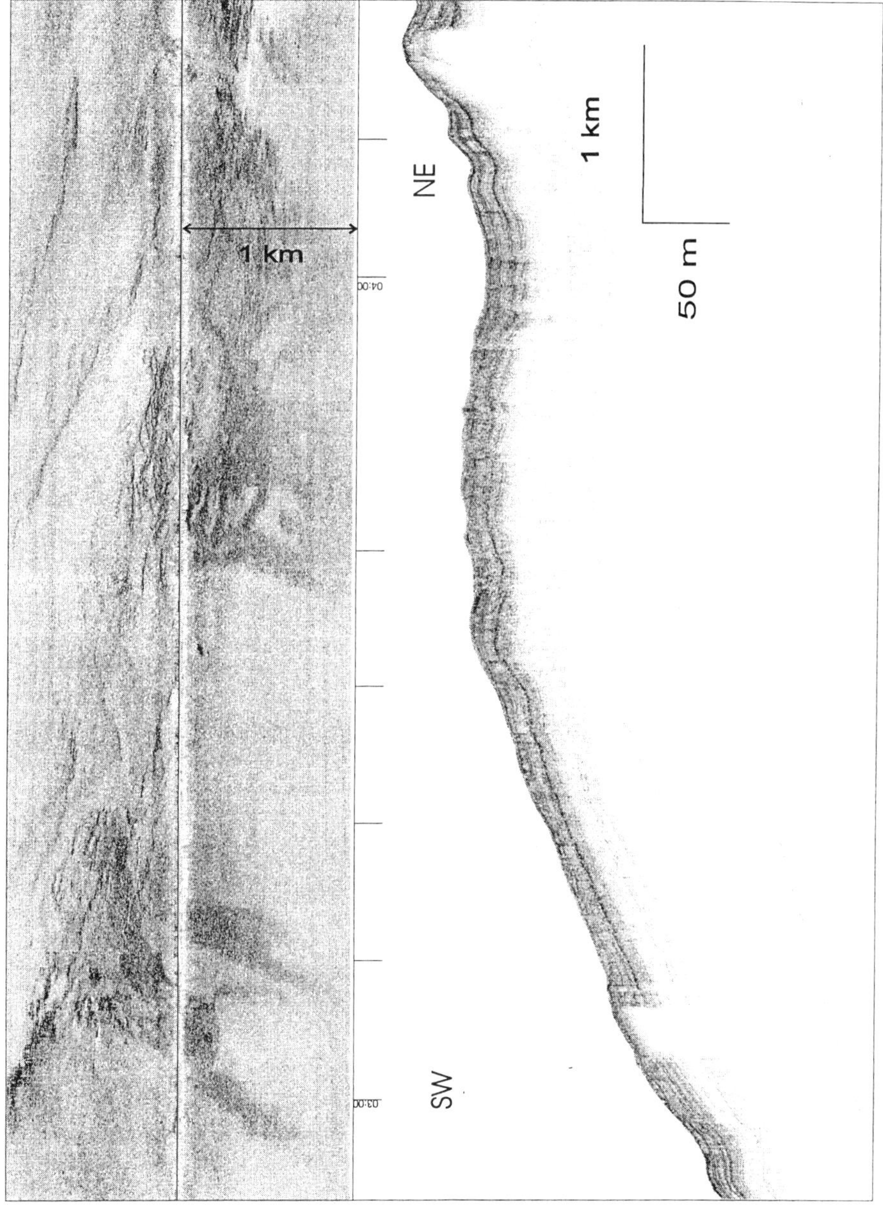

Fig. 8. MAK-1 sonograph and sub-bottom profile showing mud breccia eruptions. Mud flows spread down the regional slope from a system of parallel faults, and a prominent acoustic wipe-out is observed on the sub-bottom profile (left). Abundant gas hydrates were found in the core from this site. The sub-surface sediments are characterized by acoustic turbidity and the presence of an enhanced reflector at about 10 mbsf. The turbidity generally increases upslope and reaches its maximum below the uppermost part of a gentle swell on the profile.

teristic example of fluid venting from negative topographic features. Extensive flows of fluidized sediments apparently spread downslope from these pockmark-like depressions. Small ravines are seen on the profile where these flows erode the uppermost sediments of the sea floor.

Figure 8 shows a similar pattern, but the percolation of gas-saturated sediments takes place via linear structures which are interpreted to be faults oriented roughly along the continental slope. From one of these mud flows, which is marked by a classical acoustic void, a core was raised with gas-saturated sediment and a large amount of gas hydrates. Similar acoustic anomalies in the profiler records are observed throughout the study area (Fig. 9).

A very unusual sub-bottom reflector has been observed in the study area. The reflector is much stronger than the bottom reflection, and has a relief which does not follow the sea-floor topography and which is discordant to the layering of the overlying sediments (Fig. 10). It has a depth of no more than 10 metres below the sea floor (mbsf), and sometimes approaches the sea floor. Below the reflector the sediments are either acoustically transparent or opaque. In areas where the reflector weakens or crops out on the sea floor, the entire sedimentary sequence becomes acoustically transparent. On the sonographs, these places correspond to areas with an enhanced backscatter, e.g. in the northeastern part of sonograph M-48 (Fig. 10). In the southwest, this reflector is exposed on the sea floor at the margin of a collapse structure whose origin is evidently related to an intensive fluid flux. The core taken from this structure (BS-284G) contained gas-saturated mud breccia referred to as 'mousse-like' after Staffini *et al.* (1993).

The characteristics of this strong reflector indicate its direct tie with the underlying gas-charged sequence, and hence it was tentatively interpreted as a gas front. However, gas fronts were repeatedly observed in other areas of the Black Sea and even in other profiles from this area, but never this strong. This allows us to suggest that this reflector could be the upper surface of a gas hydrate layer or a 'gas hydrate front'.

Geological evidence for deep fluid flux

The sampling strategy was directed towards the search for intensive fluid flows through the sea floor and traces of these flows in the sedimentary record in the form of specific structures. Therefore the sampling sites, as a rule, were chosen within the corresponding acoustic anomalies on the sonographs and profile records (mud volcano craters, pockmarks, mud flows, acoustic wipe-outs, etc.).

A total of 15 cores were taken (Fig. 2). At 12 sampling sites, cores were found to have an anomalously high gas content, and two cores were characterized by a background gas content typical of the basin as a whole. At one site we did not manage to take a core because the corer could not penetrate through a superficial sand layer (turbidites?), the traces of which remained in the core catcher. Ten of the 12 gas-saturated cores contained variable amounts of mud breccia (Fig. 11). In cores BS-288G, BS-289G, BS-290G, BS-292G and BS-294G, in addition to the high gas content, gas hydrates were observed. In cores BS-284G, BS-287G, BS-288G, BS-292G and BS-294G carbonate crust and concretions were found, which are unusual for normal basinal sediments of the deep Black Sea. The carbonate crust was often associated with an increased concentration of iron sulphide.

Core BS-284G included small mussel(?) shells, weakly cemented by carbonate material. Bacterial mats were found in cores BS-288G, BS-292G and BS-294G.

Mud breccia

The mud breccia recovered in the Sorokin Trough vary in consistency, lithological compo-

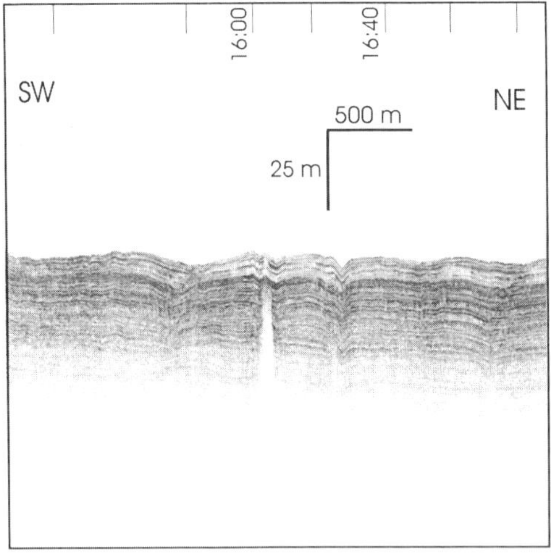

Fig. 9. Sub-bottom profile showing characteristic acoustic anomalies. The acoustic void (centre) is accompanied by sea-floor doming. The columnar disturbance 500 m northeast of this is related to a pockmark.

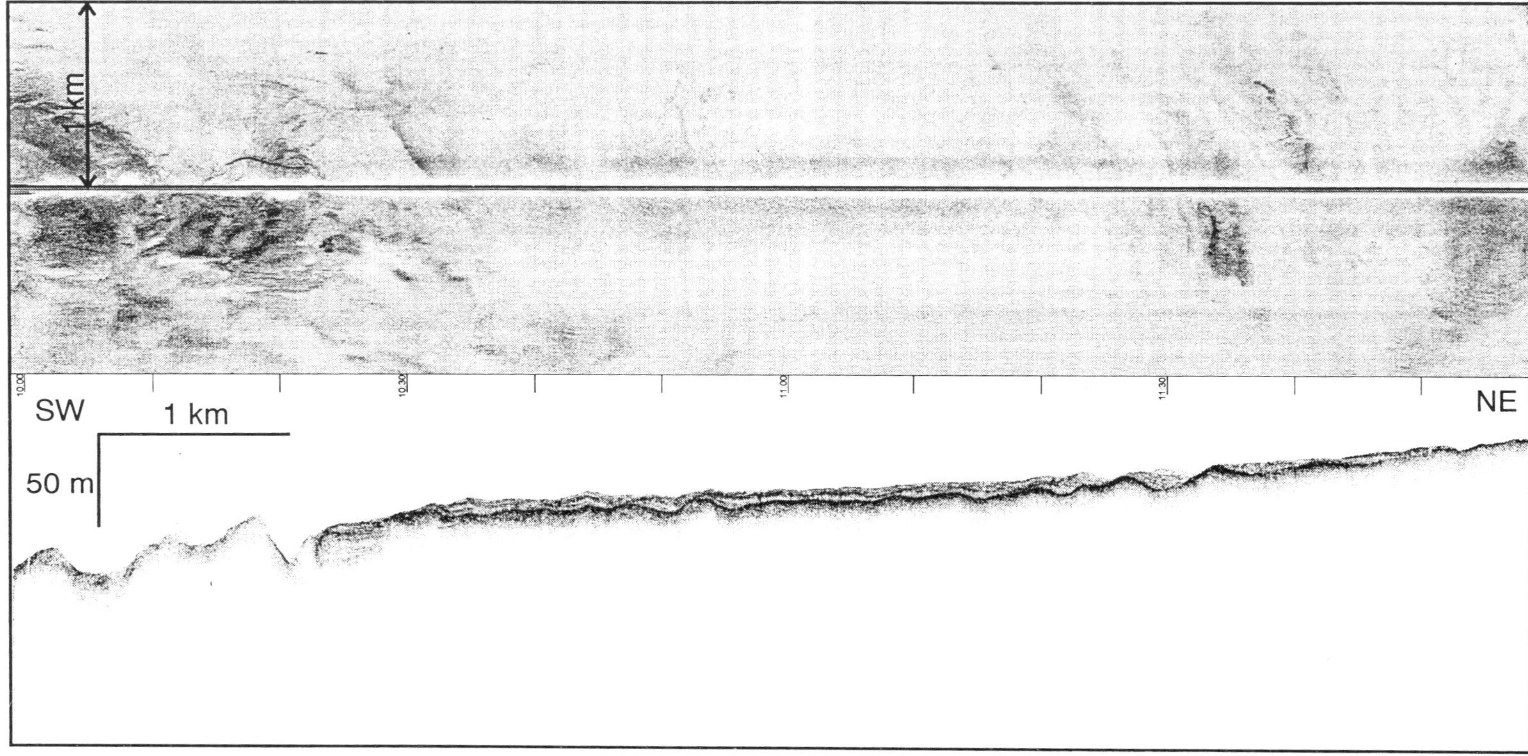

Fig. 10. MAK-1 sonograph and sub-bottom profile showing a very irregular collapsed mud volcano (left) and an enhanced reflector (probable gas front) in sub-surface sediments. This reflector crops out at a margin of the volcano and gives a strong backscatter in the northeastern part of the sonograph where it approaches the sea floor.

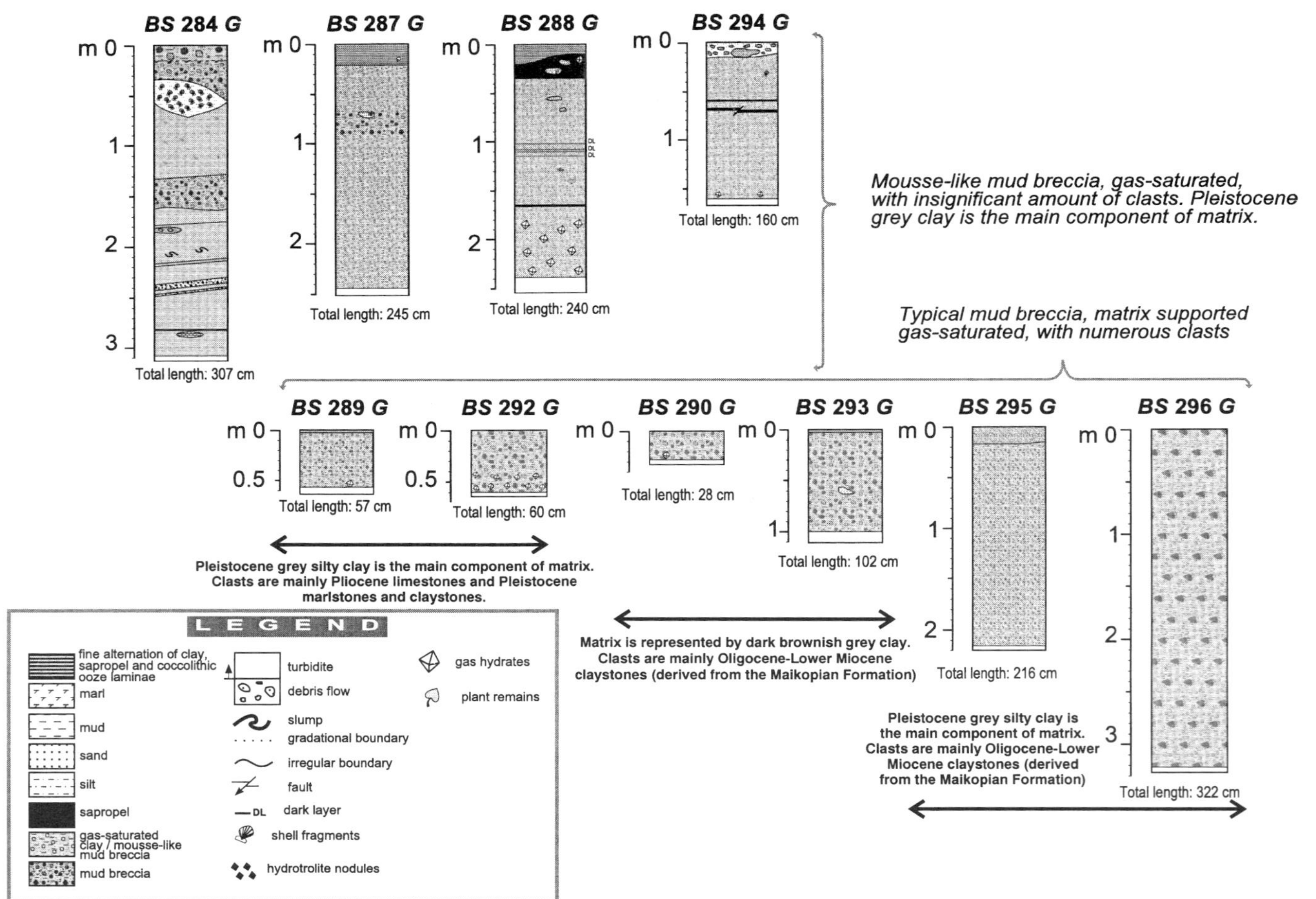

Fig. 11. Columnar sections of the TTR6 cores showing different types of mud breccia.

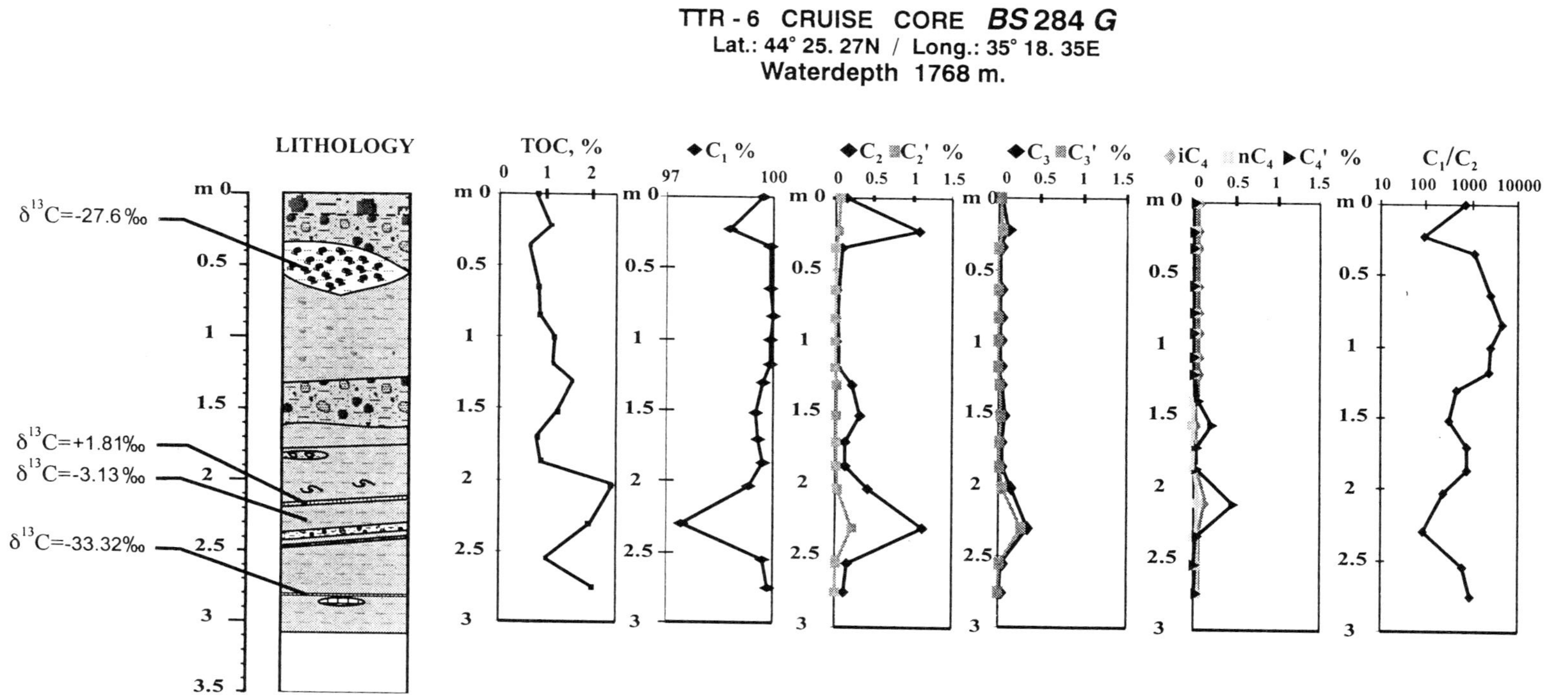

Fig. 12. Distribution of TOC, methane and its homologues (% of the total content of hydrocarbon gases) in Core BS-284G. The methane/ethane ratio is given in a logarithmic scale. The results of carbon isotope analysis from carbonate crust and concretions are shown for the different sampling intervals.

sition and age of rock clasts, which could be evidence for different source formations for fluids. Three types of mud breccia can be distinguished (Fig. 11): (1) fluidized homogeneous clay–silt sediment (mousse-like breccia) containing millimetre-sized rock clasts which are uniformly dispersed in the matrix. The composition and age of the clasts are difficult to define owing to their small size; (2) typical mud breccia containing fragments of different rocks (mudstone, siltstone, sandstone, marlstone, limestone, etc.) largely of Oligocene–Early Miocene age. Such a breccia is characteristic of mud volcanoes from the Kerch Peninsula and the central Black Sea and is well studied and described in many papers; and (3) mud breccia containing only Pliocene and Pleistocene rock clasts. The wide development of the mud breccia and the associated high gas content strongly suggest a stable deep fluid flow towards the sea floor.

Gas

An enhanced gas content was found not only in cores with mud breccia, but also in two of four hemipelagic cores, probably testifying to the general high gas concentration in sediments from the Sorokin Trough. The main gas components, according to the chromatographic analysis data, are hydrocarbons, nitrogen and carbon dioxide. A hydrogen sulphide content was not defined due to technical reasons, although its presence is beyond question because many samples had a strong smell of hydrogen sulphide. Among the hydrocarbons, methane strongly prevails. Its content in most samples was 97.7–99.9%. A small increase in methane homologue content in the composition of the hydrocarbon gases and a decrease in the methane/ethane ratio is typical of mud breccia interlayers (Fig. 12), indicating a constant supply of gas from below. However, in some cores the content of heavy methane homologues sharply increases. For example, in Core BS-293G taken from the Kazakov mud volcano the methane content decreases from 92.2 to 1.5% down the core, and the total heavy methane homologue content correspondingly increases up to 98.5% (Fig. 13).

Because the carbon isotope composition study is still in progress, it is not absolutely clear yet whether the gas is thermogenic or biogenic in origin. However, the presence of heavy methane homologues, up to iso-hexane in some cores, and a methane/ethane ratio of more than 100 (Fig. 13), suggest a deep thermogenic origin of the

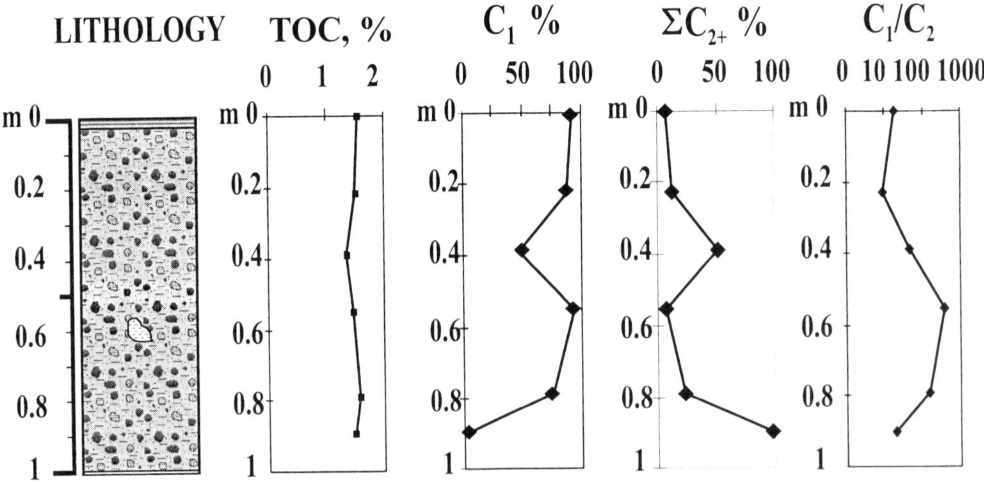

Fig. 13. Distribution of TOC, methane and the sum of its homologues (% of the total content of hydrocarbon gases) in Core BS-293G. The methane/ethane ratio is given in a logarithmic scale.

Fig. 14. Photograph of gas hydrates from Core BS-288G.

gas (Kaplan 1974). From the diagrams in Fig. 13 it is clearly seen that the quantity and composition of the gas are not related to the distribution of total organic carbon content (TOC) over the core length, and the gas seems to be allochthonous rather than being the product of *in situ* organic matter transformation. The ratio of methane to ethane varies from 9 to 379, evidently indicating the presence of thermogenic gas in samples.

Gas hydrates

Gas hydrates are also likely to be very widespread in the study area. The location map (Fig. 2) shows that gas hydrates were discovered at sites separated by a distance of some kilometres and even some tens of kilometres. The presence of gas hydrates was acknowledged only if their macrocrystals were observed in cores. However, some gas-charged cores demonstrated indirect evidence for the presence of gas hydrate microcrystals (e.g. liquefication of discrete core intervals, slow gas escape as a core warmed up in laboratory, etc.).

The bottom of the gas hydrate layer was penetrated at none of the five sites. At all but one site gas hydrates occurred in mud breccia in the lower part of the cores or in the core catcher. Only in Core BS-288G were small accumulations additionally found in a sapropel layer in the upper part of the core.

The maximum recovered thickness of the gas hydrate-containing layer (in BS-288G) is about 0.5 m. At some sites (BS-289G, BS-290G and BS-292G) the corer, with a speed of 2.2 m s^{-1}, was stopped by a layer of massive gas hydrates lying almost on the sea floor. While hoisting of the corer on board large pieces of gas hydrates, stuck in the core catcher, quickly dissociated in the warm surface water layer, producing large gas bubbles.

The morphology of the gas hydrate accumulations was described schematically, on deck under normal P–T conditions, due to their instability and fast dissociation. Three principal varieties of gas hydrate accumulations were distinguished: (1) layers of massive hydrates (with unknown thickness). Large fragments of such hydrates were seen in the core catcher; (2) massive hydrates with irregular shape, chaotically distributed in mud breccia between rock clasts; and (3) individual thin interlayers and lenses of gas hydrates occurring sub-parallel to sediment stratification.

Fragments of gas hydrates sampled from Core BS-288G (Fig. 14) were described in a cold laboratory, and the results are summarized in Fig. 15. Hydrocarbon gases from the gas hydrates are represented mainly by methane (99.8–99.9%), and its homologues are virtually absent. This implies that either the source for this gas produced almost pure methane (e.g. microbial origin of the gas), or that of the whole mixture of hydrocarbon gases present in the sediments only methane was fixed in the crystalline structure of gas hydrates.

Carbonate crust and concretions

These were observed in the gas-saturated cores (BS-284G, BS-287G, BS-288G, BS-292G and BS-294G) in different intervals of the core sections (Fig. 11). The crust has a thickness of a few millimetres to 1–2 cm; and its fragments reach a size of 10 × 10 cm. The upper and lower surfaces are irregular. The concretions are often characterized by an elongated or irregular shape with a length of up to 5–6 cm. The preliminary results of microscopic study and X-ray analysis show an unusual (for this water depth in the Black Sea) association of carbonate minerals, in which high magnesium calcite, dolomite and aragonite often predominate over normal calcite. Similar mineralogical associations from carbonate knolls and crust related to methane escape have been previously reported from several marine basins, including the Black Sea (e.g. Ritger *et al.* 1987; Hovland & Judd 1988; Ferrell & Aharon 1994; Ivanov *et al.* 1992; Shnyukov *et al.* 1995). Normally the

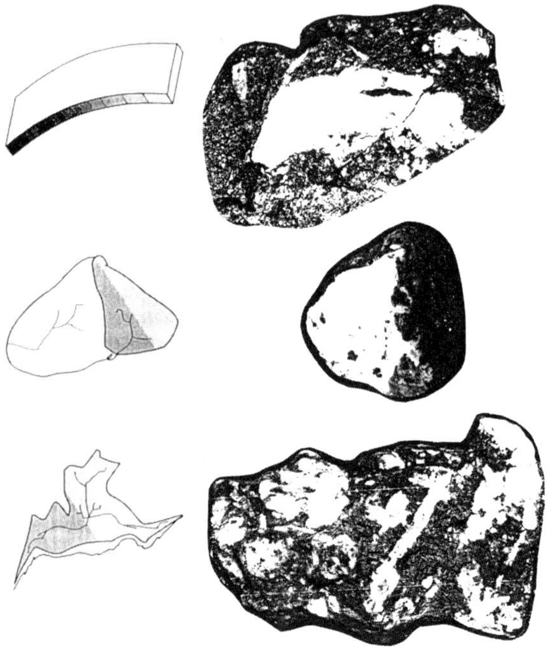

Fig. 15. Morphology of gas hydrate inclusions from different cores gathered during the cruise.

$\delta^{13}C$ value in such knolls is 20–30‰ less than in sea water and bottom sediments, which allowed the authors to infer that the light carbon isotope from methane participates in the formation of these carbonates.

The authigenic carbonates investigated here have $\delta^{13}C$ values ranging from +1.81 to −33.32‰ suggesting that they, at least partially, were formed by methane oxidation. In some cores several intervals containing authigenic carbonates with either 'normal' or lighter carbon isotope composition were observed (Fig. 12). This may indicate that the methane flux had an episodic character.

Bacterial mats

Bacterial mats were discovered for the first time in the deep Black Sea (water depth 2000 m) in the hydrogen sulphide zone. In cores BS-288G and BS-292G they were found on fragments of carbonate crust and had a similar appearance. They are spherical aggregates, 2–3 mm in size, of jelly-like material with light yellow, green, pink and violet colours, filling pores and cavities inside the crust and on its surface (Fig. 16). The study of the smear-slide of these aggregates at a ×1000 magnification made it possible to distinguish several characteristic forms of these bacteria, which are shown in Fig. 16. Also, in these cores, some small bivalve shells were noted.

The bacterial mat from Core BS-294G formed a continuous coating on the surface of a carbonate crust fragment (Fig. 16). It was colourless on the interior and pink–yellow on the exterior. Its texture was not revealed in thin-section, but a coloured smear-slide showed, at a high magnification, the presence of variably shaped and sized bacteria. Detailed definitions of the bacteria are not yet available, but one may suppose that they largely consist of methanogenic organisms capable of anaerobic methane oxidation to CO_2 in the reaction opposite to that of methane formation (Zehnder & Brock 1979).

Conclusions

- The investigations carried out during the *Gelendzhik* 1996 cruise on the Crimean deep-water margin in the Black Sea resulted in the discovery of an area with extremely extensive fluid venting through the sea floor.

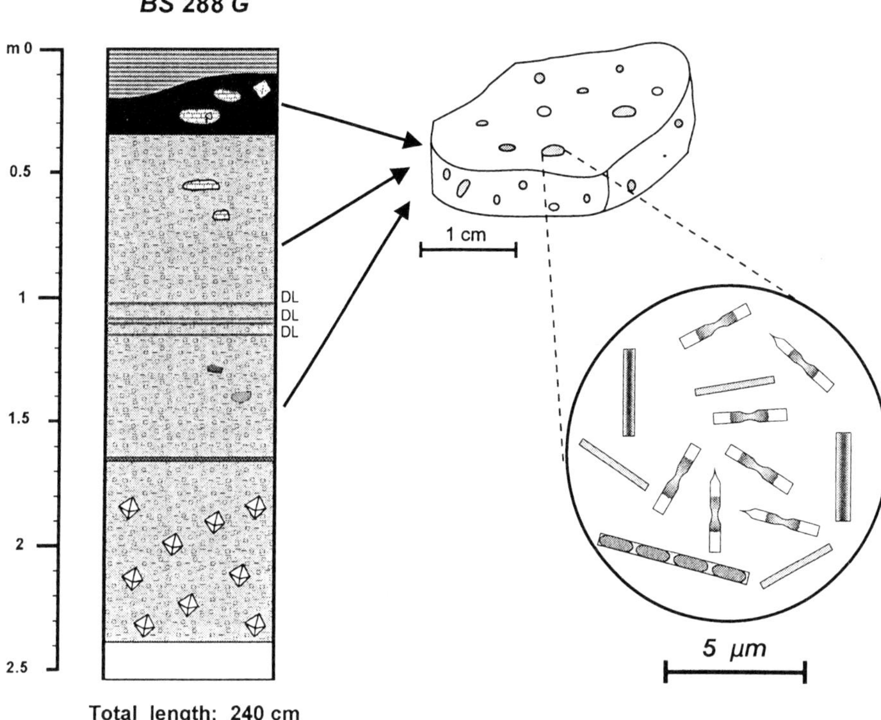

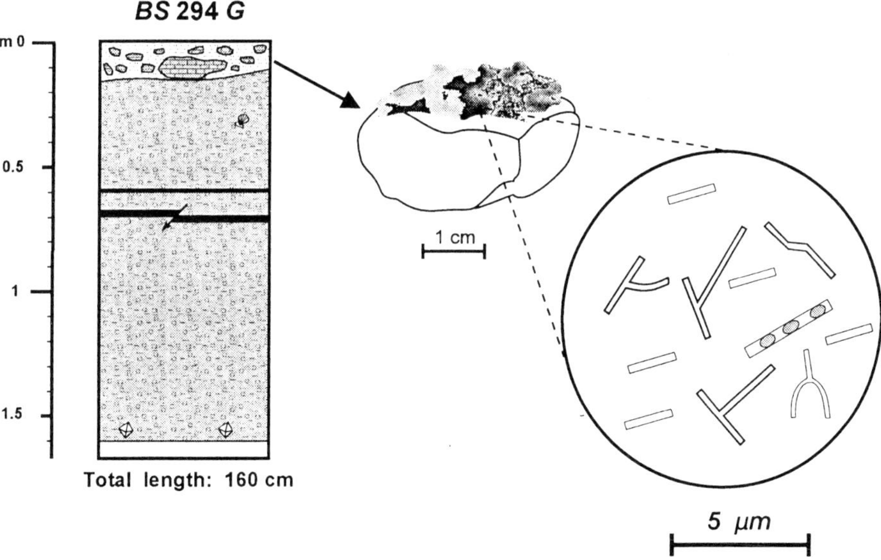

Fig. 16. Location of the bacterial mats on fragments of carbonate crust and the appearance of bacteria from these colonies (magnification ×1000).

- The acoustic methods applied during the cruise allowed us to distinguish the areas with active extrusion of material from depth onto the sea floor, manifested in numerous mud volcanoes, fissure mud breccia eruptions, pockmarks, acoustic anomalies and other associated phenomena.
- A very high gas content is noted in the uppermost part of the sedimentary sequence in the form of both free gas and gas hydrate accumulations. The composition of the hydrocarbon gases and the distribution over the sequence give evidence for the important contribution of deep thermogenic gases.
- The saturation of sub-bottom sediments with gas leads to large-scale slope instability in this area and to sliding and slumping of surficial sediments.
- The unusual association of carbonate minerals in carbonate crust and nodules found in gas-saturated cores, as well as the light carbon isotope composition of some of them, suggest a perceptible influence of methane flux on their formation. This flux seems to have had an episodic character.
- Bacterial mats were discovered for the first time in the Black Sea at a depth of about 2000 m, in a hydrogen sulphide zone.

The authors thank the Ministry of Natural Resources and the Ministry of Science and Technical Policy of the Russian Federation, the Ministry of Science of the Netherlands, and IOC–UNESCO for the financial support of the expedition. We are also grateful to the crew and technical staff of the R/V *Gelendzhik* for valuable help in carrying out the investigations. Special thanks is addressed to M. Hovland (Statoil, Norway) for his fruitful revision of the manuscript.

References

Andreev, V. 1976. [The Crimean and Caucasian foredeeps in the Sea.] *Izvestiya Akademil Nauk SSSR*. Seriya Geologicheskaya, **11**, 130–133 (in Russian).

——, Kazantsev, R. & Panaev, V. 1981. [Tectonics of the junction area of the Crimea and the Caucasus.] *Tektonika I Stratigrafiya*, **19**, 17–28 (in Russian).

Cita, M., Aghib, F., Arosio, S. *et al.* 1989. Bacterial colonies and manganese micronodules related to fluid escape on the crest of the Mediterranean Ridge. *Rivista Italiana Paleontologia et Stratigrafia*, **95**(3), 315–336.

——, Woodside, J., Ivanov, M. *et al.* 1995. Fluid venting from a mud volcano in the Mediterranean Ridge Diapiric Belt. *Terra Nova*, **7**(4), 453–458.

Corselli, C. & Basso, D. 1996. First evidence of benthic communities based on chemosynthesis on the Napoli mud volcano. *Marine Geology*, **132**, 227–240.

Dimitrov, L. 1994. Seabed pockmarks in the southern Bulgarian Black Sea zone. *Bulletin of Geological Society of Denmark*, **41**(1), 24–33.

Ferrell, R. & Aharon, P. 1994. Mineral assemblages occurring around hydrocarbon vents in the northern Gulf of Mexico. *Geo-Marine Letters*, **14**, 74–80.

Ginsburg, G. & Soloviev, V. 1997. Methane migration within the submarine gas-hydrate stability zone under deep-water conditions. *Marine Geology*, **137**, 49–57.

——, Kremlev, A., Grigoriev, M. *et al.* 1990. [Filtrogenic gas hydrates in the Black Sea. 2lst Cruise of R/V Evpatoria.] *Geolagiya I Geofizika*, **3**, 10–19 (in Russian).

Hovland, M. & Judd, M. 1988. *Seabed Pockmarks and Seepages. Impact on Geology, Biology and the Marine Environment*. Graham and Trotman, London.

Hunt, J. & Whelan, J. 1978. Dissolved gases in Black Sea sediments. *Initial Reports of the Deep Sea Drilling Project*, Volume 42. US Government Printing Office, Washington, D.C., **42**, 661–665.

IOC–UNESCO. 1981. *International Bathymetric Chart of the Mediterranean. Scale 1 : 1 000 000*. Ministry of Defence, Leningrad.

Ivanov, M. K., Limonov, A. & Cronin, B. 1996*a*. Mud volcanism and fluid venting in the eastern part of the Mediterranean Ridge. *In: UNESCO Reports in Marine Science*. UNESCO, Paris.

——, Limonov, A. & van Weering, T. 1996*b*. Comparative characteristics of the Black Sea and Mediterranean Ridge mud volcanoes. *Marine Geology*, **132**, 253–271.

——, Limonov, A. & Woodside, J. 1992. Geological and geophysical investigations in the Mediterranean and Black Sea. *In: UNESCO Reports in Marine Science*. UNESCO, Paris.

——, Konyukhov, A., Kulnitskii, L. & Musatov, A. 1989. [Mud volcanoes in the deep part of the Black Sea.] *Vestnik Moskovskogo Universiteta*. Seriya Geologiya, **3**, 21–31 (in Russian).

Ivanov, M. V., Polikarpov, G. & Lein, A. 1991. [Biochemistry of carbon in an area of methane venting in the Black Sea.] *Dokiady Akademil Nauk SSSR*, **320**(5), 1235–1240 (in Russian).

Kaplan, I. 1974. *Natural Gases in Marine Sediments*. Plenum, New York.

Konyukhov, A., Ivanov, M. & Kulnitskii, L. 1990. [On mud volcanoes and gas hydrates in the deep Black Sea Basin.] *Litologiya I Poleznye Iskopaemye*, **3**, 12–23 (in Russian).

Kruglyakova, R., Prokoptsev, G. & Berlizeva, N. 1993. [Gas hydrates in the Black Sea as a potential source of hydrocarbons]. *Razvedka I Okhrana Nedr*, **12**, 7–10 (in Russian).

Limonov, A., Ivanov, M & Woodside, J. 1994. Mud volcanism in the Mediterranean and Black Seas and shallow structure of Eratosthenes Seamount. *In: UNESCO Reports in Marine Science*. UNESCO, Paris.

——, van Weering, T., Kenyon, N., Ivanov, M. & Meisner, L. 1997. Seabed morphology and gas venting in the Black Sea mud volcano area: observations with the MAK-1 deep-tow side-scan sonar and bottom profiler. *Marine Geology*, **137**, 121–136.

Morgunov, Y., Kalinin, V., Gainanov, V. *et al.* 1976. [Diapiric folds in the Black Sea of the Mountainous Crimea.] *Doklady Akademil Nauk SSSR*, **228**(5), 1159–1162 (in Russian).

Peklo, V. Malovitskii, Y., D'yakonov, A. & Sidorenko, S. 1976. [Tectonics of the junction area of Taman, the western Caucasus, and the adjacent part of the Black Sea.] *In: Kompleksnye Issledovaniya Chernomorskoi. [Comprehensive Investigations of the Black Sea Basin.]* (Nauka, Moscow (in Russian).

Ritger, S., Carson, B. & Suess, E. 1987. Methane-derived authigenic carbonates formed by subduction-induced pore water expulsion along the Oregon/Washington margin. *Geological Society of America Bulletin*, **98**, 147–156.

Shnyukov, E., Alenkin, V., Put, A. *et al.* 1981. [*The Kerch Strait*]. Naukova Dumka, Kiev (in Russian).

——, Sobolevskii, Y. & Kutnii, V. 1995. [Unusual carbonate build-ups on the continental slope in the northwestern Black Sea – a probable consequence of gas inflow from the depth.] *Litologiya I Poleznye Iskopaemye*, **5**, 451–461 (in Russian).

Soloviev, V. & Ginsburg, G. 1997. Water segregation in the course of gas hydrate formation and accumulation in submarine gas-seepage fields. *Marine Geology*, **137**, 59–68.

Staffini, F., Spezzaferri, S. & Aghib, F. 1993. Mud diapirs of the Mediterranean Ridge: sedimentological and micropaleontological study of the mud breccia. *Rivista Italiana Paleontologia et Stratigrafia*, **99**(2), 225–254.

Tugolesov, D., Gorshkov, A., Meisner, L., Soloviev, V. & Khakhalev, E. 1985. [*Tectonics of Mesozoic–Cenozoic Deposits of the Black Sea Basin*.] Nedra, Moscow (in Russian).

Zehnder, A. & Brock, T. 1979. Methane formation and methane oxidation by methanogenic bacteria. *Journal of Bacteriology*, **137**, 420–432.

Origin of gas hydrate accumulations on the continental slope of the Crimea from geophysical studies

S. V. BOURIAK & A. M. AKHMETJANOV

Moscow State University, Vorobjevy Gory 119899 Moscow, Russia

Abstract: The Sorokin Trough on the south-eastern Crimean margin (northern part of the Black Sea) is known for mud diapirism. Gas hydrates in the sea-bed sediments have previously been recovered from this area in1988. During the TTR-6 cruise of the R/V *Gelendzhik* in 1996 gas hydrates were observed in five cores containing mud breccia. A comprehensive geophysical survey was carried out, which allowed more insight to be gained into the local distribution of gas hydrates and the linkage of hydrate accumulations to fluid vents. It is therefore suggested that gas hydrates in the study area have most probably been formed from allochthonous gas. Preliminary results from a lithological study of the gas hydrate bearing cores suggested a complex mechanism for the mud volcanoes from which the gas hydrates were sampled, and a complicated history of the gas which was a source for the hydrates.

Five cores containing gas hydrates were sampled in an area known for mud diapirism on the south-eastern margin of the Crimea Peninsula, during the 6th cruise of the UNESCO–IOC Training-through-Research programme (TTR-6 Cruise) with the R/V *Gelendzhik* in the summer of 1996. All coring sites were covered using a deep-towed side-scan sonar survey with a sub-bottom profiler, and high-resolution seismic records were obtained near the sites (Fig. 1). Gas hydrates have previously been recovered from this area by Kremlev & Ginsburg (1989). However, the new data collected during the TTR-6 cruise give more regional information which allows for a better understanding of the formation mechanism of gas hydrates in the study area, and their relation to the mass and fluid transport through the sea floor.

Geological setting and previous studies of gas hydrates

The study area is located within the Sorokin Trough, one of the large depressions in the deep part of the Black Sea (Figs 1 and 2). The

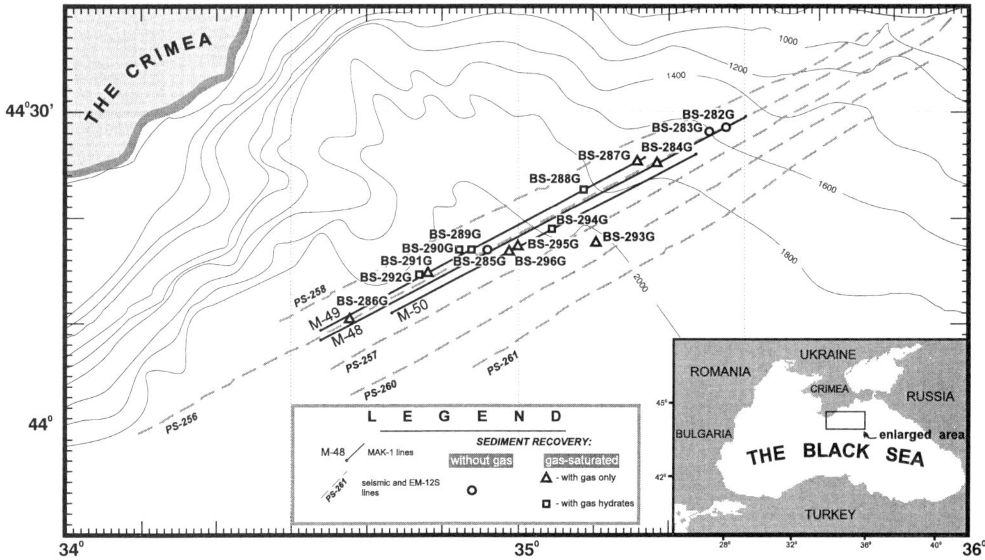

Fig. 1. Schematic overview of the study area showing the location of the seismic, deep-tow (MAK-1) and sampling sites.

BOURIAK, S. V. & AKHMETJANOV, A. M. 1998. Origin of gas hydrate accumulations on the continental slope of the Crimea from geophysical studies. *In:* HENRIET, J.-P. & MIENERT, J. (eds) *Gas Hydrates: Relevance to World Margin Stability and Climate Change*. Geological Society, London, Special Publications, **137**, 215–222.

trough lies along the south-eastern margin of the Crimean Peninsula, with a length of 150 km and a width of 45–50 km, and water depths of 800–2000 m (Tugolesov *et al.* 1985). To the southeast it is bounded by the Cretaceous–Eocene Shatskii and Tetyaev rises, as recognized by seismic investigations. The trough is considered to be a foredeep of the Crimean mountains, and its formation began in the Oligocene (Andreev 1976). Two main units have been recognized in the sedimentary cover of the trough from seismic profiling conducted during the TTR-6 cruise. The thickness of the units varies from 5 to 6 km. The lower unit is likely to represent the upper part of the Maikopian Formation (Oligocene–Lower Miocene) and Pliocene deposits. It is intensively folded and disturbed by numerous faults which also can be traced into the upper unit (Limonov *et al.* 1997).

The Quaternary deposits representing the upper unit are characterized by sub-parallel bedding, and form a blanket above the lower unit. The thickness of this unit is largely controlled by the underlying diapirs, but it generally increases toward the northeast (Limonov *et al.* 1997). The diapiric folds are of particular interest as they originated from the protrusion of plastic, fluid-saturated Maikopian clays. Previous works established the elongated pattern of the diapirs coaxial with the general trend of the Sorokin Trough (Tugolesov *et al.* 1985). This pattern has been revised, using the data obtained on the TTR-6 cruise (Fig. 2). It was discovered that the trend of the diapirs is basically controlled by ledges of Cretaceous–Eocene rises, and that the formation of the diapirs is likely to have been connected with a lateral compression. The ledges of rises act as stamps pressing the Maikopian clays and causing the diapirs to grow (Limonov *et al.* 1997). The diapirs can also be complicated by the presence of mud volcanoes.

The existence of mud volcanoes in the study area has been suggested since the late 1980s, but was documented for the first time with data from the TTR-6 cruise. The morphology of the mud volcanoes is highly variable and both depressed and dome-shaped structures are observed (Limonov *et al.* 1997; Ivanov *et al.* 1998). Their activity is usually accompanied by eruptions of large volumes of mud breccia onto the sea floor. Our acoustic data gave much evidence for gas saturation of the surrounding sub-bottom sediments.

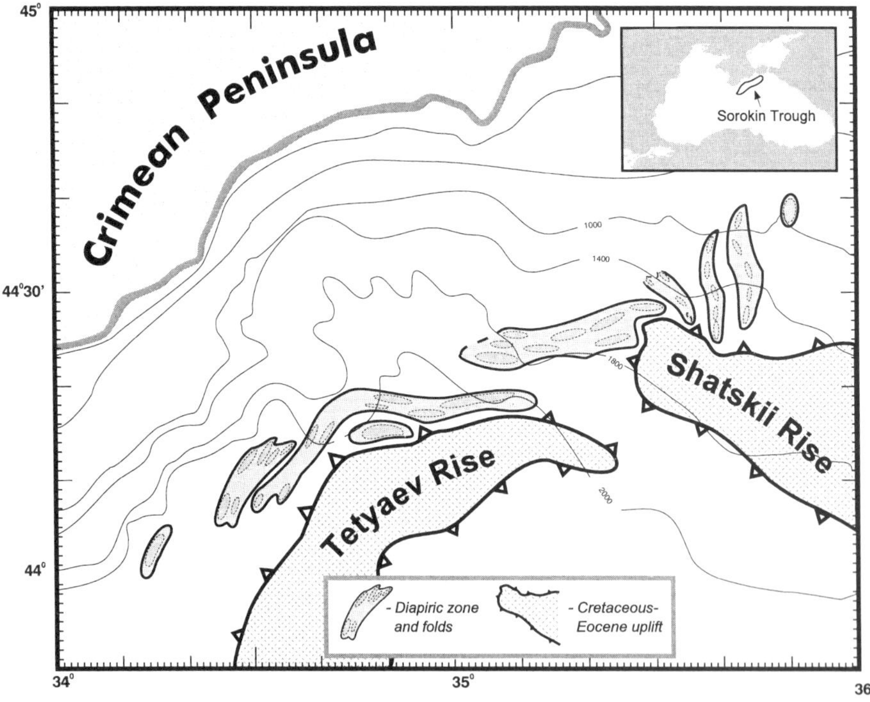

Fig. 2. Distribution of mud diapirs within the Sorokin Trough (after Limonov *et al.* 1997).

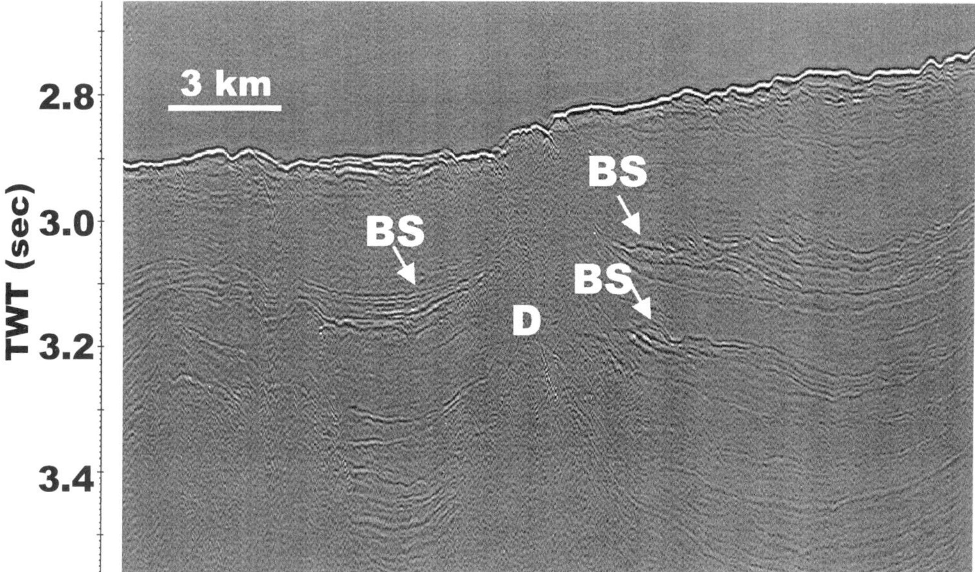

Fig. 3. Seismic section of profile 256 showing bright spots (BS) which indicate accumulations of gas at the flanks of a diapiric structure (D).

As the gas hydrate stability zone in the Black Sea is situated at water depths greater than 700 m (Ginsburg *et al.* 1990), the study area completely falls within the boundaries where gas hydrate occurrence is theoretically possible. Gas hydrates were recovered in the study area by Kremlev & Ginsburg (1989), who investigated the Sorokin Trough for gas hydrate occurrence in 1988 and believed that these have been formed as a result of fluid filtration through the hydrate stability zone on its way upwards in the sedimentary pile. The gas hydrates were recovered at the top of diapiric structures, and a mud volcanic control over their formation was assumed. The visually estimated content of gas hydrate crystals in sediments was approximately 10% (Ginsburg *et al.* 1990).

Data and methods

In this work we present the results of high-resolution seismic profiling, side-scan sonar investigations accompanied by subbottom profiling, and bottom sampling carried out in the study area in 1996.

The high-resolution seismic system included a 3 l airgun with a frequency range of 20–250 Hz and a six-channel streamer with all channels connected in parallel. The data were recorded digitally on magneto-optical discs. The towing speed was 7–7.5 knots.

For the side-scan survey a MAK-1 deep-towed side-scan sonar system was used, accompanied by a sub-bottom profiler. The working frequency of the sonar was 30 kHz with a swath width of 2 km. The sub-bottom profiler was operated at a frequency of 4.9 kHz. The MAK-1 system was towed at a speed of 2–2.5 knots at a depth of about 100 m above the sea floor. Fish positioning was carried out using a short-base underwater navigation system.

Bottom sampling was carried out with a gravity corer (6.6 m length, 168 mm diameter, 1500 kg weight) with an inner plastic liner, which allowed the collection of up to 6 m of undisturbed sediments.

A total of five seismic lines and three lines with the MAK-1 deep-towed system were shot, and 15 sites were sampled with the gravity corer. Twelve of the cores sampied were gas-saturated, of which five contained fragments of gas hydrates.

Indications of gas on seismic profiles

Analysis of the seismic data does not show any bottom simulating reflector (BSR), which is often associated with submarine gas hydrates, and which is interpreted as the base of the hydrate stability zone. However, the seismic records do contain numerous signs of gas presence in the sediments. Figure 3 shows part of a time-section along line 256 where bright spots

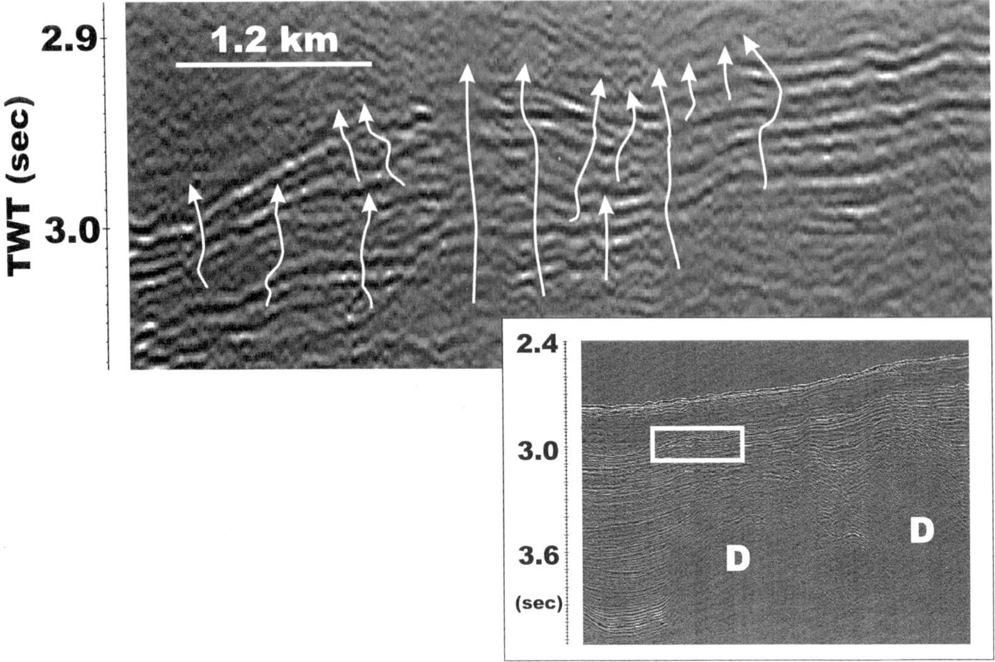

Fig. 4. Enlarged seismic section of profile 256 showing acoustically transparent disturbances which probably indicate paths (marked by arrows) of upward fluid migration through faults above a diapiric structure (D). The inset on the bottom right shows the location of the enlarged section on the profile.

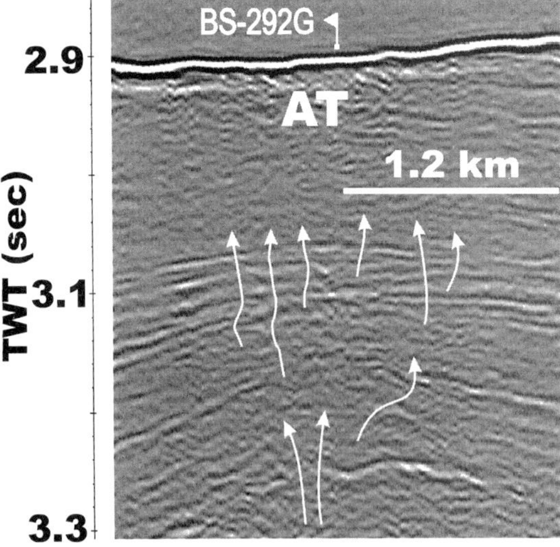

Fig. 5. Seismic section of profile 256 near Site BS-292G. The paths of fluid migration through faults and fracture zones, reflected by acoustically transparent disturbances, are indicated by arrows. Also note the sub-bottom zone of intensive diffraction forming some sort of acoustic turbidity (AT).

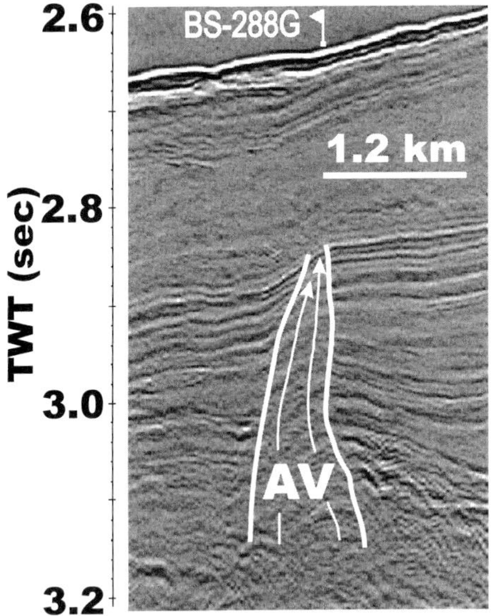

Fig. 6. Seismic section of profile 256 near Site BS-288G showing a cone-like acoustic void associated with flexure. Interpreted boundaries are indicated by solid lines. Arrows indicate the interpreted paths of fluid migration.

indicate accumulations of gas at the flanks of a diapiric structure. On a smaller scale, one can observe acoustically transparent disturbances above the diapiric structures which may indicate the paths of the upward migration of fluid through faults (Fig. 4).

Similar features were also noted near the sites where gas hydrates were sampled. For example, acoustically transparent disturbances can be observed at coring Site BS-292G on line 256 at a depth of more than 160 ms TWT bsf. In addition to this we can see a sub-bottom zone of intensive diffraction forming some sort of acoustic turbidity (Fig. 5). As the sea-floor topography at this site is rather flat, one can assume that the diffraction is caused by gas plumes.

At Site BS-288G on seismic line 256 a cone-like acoustic void can be observed at a depth of up to 200 ms TWT bsf (Fig. 6). This zone, which is over 500 m wide, is associated with a flexure and seems to indicate gas migration through the deformed and faulted sedimentary sequence. It is interesting to note that a similar acoustic void, about 200 m wide, is also observed on the sub-bottom profile over this coring site (Fig. 7). Taking into consideration the high(er) frequency involved, this seems to reflect either: (1) a thin uppermost layer of gas-saturated sediments attenuating the acoustic energy; or (2) an intensive gas flux. In

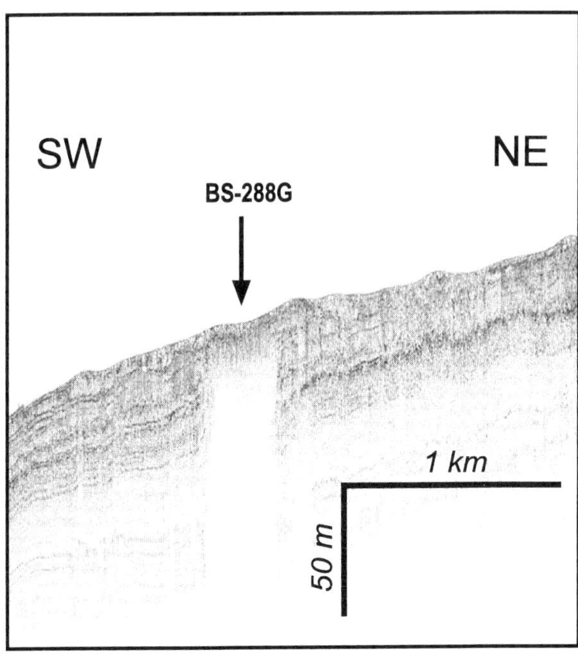

Fig. 7. Section of sub-bottom profile M-49 near Site BS-288G showing an acoustic void of approximately 200 m wide.

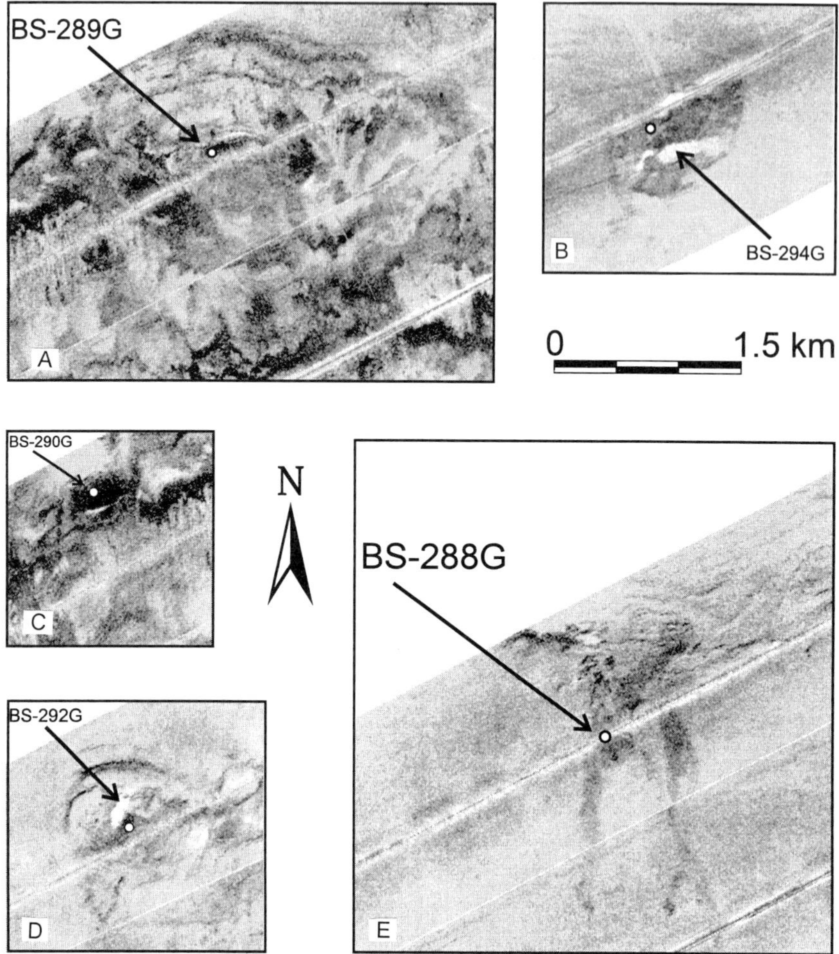

Fig. 8. Sonographs of the MAK-1 deep-towed side-scan sonar over five different sites where gas hydrates have been sampled.

either case the acoustic data are indirect evidence that the coring site is located within a local zone of abnormal gas content contrasting with the adjacent areas of the sea floor.

Gas hydrate-bearing cores

All cores containing gas hydrates were sampled from areas with high backscatter on the deep-towed side-scan sonographs. Cores BS-289G, BS-290G and BS-294G were taken from dark isometric spots interpreted as mud volcanoes (Fig. 8a–c). Core BS-292 is located within an irregularly-shaped dark zone which occurs in the centre of a circular depression (Fig. 8d) that was interpreted as a sea floor collapsed structure of a mud volcano. Core BS-288G was taken from an elongated dark zone (Fig. 8e) which was assumed to be a mud flow from an open fault. All of these cores recovered mud breccia, confirming the preliminary interpretation of these structures. In all cases, gas hydrates were present in the mud breccia layers in the lower part of the cores. This suggests that hydrates in the study area are related to zones of intensive fluid exhaust on the sea floor. Recovery of the gas hydrates in the sampled cores and the morphology of gas hydrate fragments is described in detail by Ivanov *et al.* (1998).

Table 1 shows the dominant type of sediments for each core sampled in the study area during the cruise. It is based on the core descriptions and preliminary age definitions of mud breccia

Table 1. *Lithology and gas/gas hydrate content of the cores sampled in the study area during the TTR-6 cruise*

Core number	Type of sediments recovered	Gas/gas hydrate content
BS-282G	Hemipelagic	No gas
BS-284G	Mousse-like mud breccia	Gas-saturated
BS-285G	Hemipelagic	No gas
BS-286G	Hemipelagic	Gas-saturated
BS-287G	Mousse-like mud breccia	Gas-saturated
BS-288G	Mousse-like mud breccia	Gas hydrates
BS-289G	Mud breccia with Pliocene–Pleistocene clasts	Gas hydrates
BS-290G	Mud breccia with Oligocene–Lower Miocene clasts	Gas hydrates
BS-291G	Hemipelagic	Gas-saturated
BS-292G	Mud breccia with Pliocene–Pleistocene clasts	Gas hydrates
BS-293G	Mud breccia with Oligocene–Lower Miocene clasts	Gas-saturated
BS-294G	Mousse-like mud breccia	Gas hydrates
BS-295G	Mud breccia with Oligocene–Lower Miocene clasts	Gas-saturated
BS-296G	Mud breccia with Oligocene–Lower Miocene clasts	Gas-saturated

The ages of clasts are from the preliminary field definition. Gas content was visually estimated. All cores with gas hydrates were gas-saturated as well. Core BS-283G had no recovery and thus is omitted here.

clasts made onboard (Woodside *et al.* 1997). As the diapirs in the Sorokin consist of the Oligocene–Lower Miocene Maikopian Formation (Tugolesov *et al.* 1985) we expected an abundance of Oligocene–Lower Miocene clasts in mud volcanic products in the study area. It can, however, be seen from Table 1 that only one of the cores containing gas hydrates, core BS-290G, consisted of mud breccia with Oligocene–Lower Miocene clasts. Cores BS-289G and BS-294G contained mud breccia with Pliocene–Pleistocene clasts, and cores BS-288G and BS-294G included the so-called 'mousse-like' mud breccia (liquid and gassy type of mud breccia with a negligible amount of clasts), which in our view may indicate a relatively shallow location of the roots of the mud volcanoes. In addition, the results of a total organic carbon analysis (Stadnitskaia 1997) also imply that the hydrate-bearing mud breccia probably originates from a different source than the breccia with Maikopian clasts.

Discussion

The absence of BSR on the seismic records is likely to suggest that there is no continuous layer of hydrated sediments in the study area. Gas hydrates seem to form local accumulations, which laterally are too small to cause a continuous seismic event or to affect the reflection pattern. This, together with the numerous signs indicating the presence of free gas below the sites, seems to suggest that the hydrate accumulations in the area are linked to local zones of active fluid exhaust through the sea floor. The sampling data show these zones to be mud volcanoes but one can also anticipate local gas hydrate occurrence in the area associated with faults, pockmarks, etc. Hence, the allochthonous nature of gas comprising the gas hydrates can be supposed.

The results of bottom sampling suggest that the activity of the mud volcanoes is controlled by relatively shallow gas accumulations. A possible mechanism could be that gas escapes from the body of the diapirs through faults and/or fracture zones and first gets trapped in some shallower layers. Then the shallow gas breaks up the overlying sequence for some reason (movements caused by diapir growing, change of sea level, etc.) and escapes to the sea floor, catching recent unconsolidated sediments and forming a mud volcano. This rather complex scenario remains quite speculative and needs to be verified in the future; for example, if this hypothesis is true it should probably be reflected somehow in the chemical composition of the gas hydrates.

The lack of gas hydrates in other mud volcanic cores (see Table 1) requires an additional explanation. Two contradictory hypotheses might be considered: (1) too small a gas inflow; and (2) too strong a gas inflow. In the former case, there would not be enough gas to form a saturated water solution which is believed to be necessary for the formation of gas hydrates (Ginsburg & Soloviev 1994). In the latter case, a strong inflow of fluids from greater depths could result in a local temperature increase in the sub-bottom sediments, making gas hydrate formation impossible.

Conclusions

Analysis of the geophysical data indicates that the gas hydrates on the continental slope off the Crimea Peninsula most probably form local accumulations associated with zones of intensive fluid transport through the sea floor, such as active gas vents and mud volcanoes, rather than a continuous layer of hydrated sediments. Therefore, the gas hydrates in the study area seem to be made up of allochthonous gas which was not formed *in situ* but has migrated from depth.

The results of the onboard core analyses and preliminary age definitions of mud breccia clasts seem to suggest that the mud volcanoes containing gas hydrates are controlled by some shallow accumulations of fluids. In this case, the gas that formed the hydrates is likely to have escaped from the body of a diapir, re-deposited and, only afterwards, entered the hydrate stability zone and formed hydrated sediments around a gas vent.

The absence of gas hydrates in other mud volcanic cores in different parts of the study area could be explained either by an insufficient amount of gas, or by a too intensive inflow of warm deep fluids resulting in a local temperature increase.

We wish to acknowledge UNESCO, IOC, Uzhmorgeologia Inc., Russia, the Russian Committee for Geology, the Ministry of Science and Technological Policy, Russia, the Dutch Ministry of Science (NWO) and all other institutions supporting the TTR-6 Cruise and this investigation. We are sincerely grateful to M. K. Ivanov and A. F. Limonov (UNESCO-MSU Centre for Marine Geosciences, Moscow State University) for their valuable assistance, insight and helpful comments.

References

ANDREEV, V. 1976. [Foredeeps of the Crimea and the Caucasus in the Black Sea.] *Izvestia Akademii Nauk SSSR. Seria Geologicheskaia,* **11**, 130–133 (in Russian).

GINSBURG, G. & SOLOVIEV, V. 1994. [*Submarine Gas Hydrates*]. VNII Okeangeologia, St. Petersburg (in Russian).

GINSBURG, G., KREMLEV, A., GRIGORIEV, M. *et al.* 1990. [Filtrogenic gas hydrates in the Black Sea (21st voyage of the R/V *Evpatoria*).] *Geologia i Geofizika*, **3**, 10–19 (in Russian).

IVANOV, M., LIMONOV, A. & WOODSIDE, J. 1998. Extensive deep fluid flux through the sea floor on the Crimean continental margin (Black Sea). *This volume*.

KREMLEV, A. & GINSBURG, G. 1989. [The first results of the search for submarine gas hydrates in the Black Sea (the 21st trip of the R/V *Evpatoria*).] *Geologia i Geofizika*, **4**, 110–111 (in Russian).

LIMONOV, A., IVANOV, M., MEISNER, L., GLUMOV, I., KRYLOV, O. & KOZLOVA, E. 1997. [New data about the structure of sedimentary cover in the Sorokin Trough (Black Sea).] *Vestnik Moskovskogo Universiteta. Seria Geologicheskaia*, **3**, 36–43. (in Russian).

STADNITSKAIA, A. 1997. Distribution and composition of hydrocarbon gas in the seabed sediments of the Sorokin Trough (south-eastern part of the Crimea margin). *In: Gas and Fluids in Marine Sediments: Gas Hydrates, Mud Volcanoes, Tectonics, Sedimentology and Geochemistry in Mediterranean and Black Seas*. IOC Workshop Report No 129.

TUGOLESOV, D., GORSHKOV, A., MEISNER, L. *et al.* 1985. [*Tectonics of Mezo-Cainozoic Deposits of the Black Sea Basin*.] Nedra, Moscow (in Russian).

WOODSIDE, J., IVANOV, M. & LIMONOV, A. 1997. *Neotectonics and Fluid Flow Through Sea Floor Sediments in the Eastern Mediterranean and Black Seas*. Intergovernmental Oceanographic Commission Technical Series, in press.

Possible hydrate mounds within large sea-floor craters in the Barents Sea

D. LONG[1], S. LAMMERS[2] & P. LINKE[2]

[1] *British Geological Survey, West Mains Road, Edinburgh, UK*
[2] *Geomar, Wischhofstrasse 1–3, D-24148 Kiel, Germany*

Abstract: Interpretation of several surveys across a 'crater field' in the Barents Sea provide further evidence that the craters (large depressions, 300–500 m diameter, 10–30 m deep) are related to gas escape after deglaciation some 15 000 years BP. The disposition of the craters suggests that the flow of gas was controlled by fractures within the Triassic siltstone bedrock. Topographic highs within several craters, comprising angular blocks of rock locally rising above the level of surrounding crater walls, are interpreted as hydrate mounds indicating that gas flow continued after the formation of the craters. This may be the first reported occurrence of hydrate mounds in lithified sediments. Assuming the gas was methane and sea-bed temperature was similar to that at present then the hydrate mounds were formed at a time when the sea-bed was between 280 and 340 metres below sea level (mbsl) (i.e. 10–80 m lower than at present). Geochemical studies provide evidence that gas hydrates in the sub-bottom adjacent to the crater field are presently decomposing in accordance with seasonal temperature variations.

Solheim & Elverhøi (1993) reported a cluster of crater-like depressions in the sea floor of the Barents Sea centred at 74°55′N–27°36′E, in about 340 m water depth. Initial surveys (1987) using a 3.5 kHz echo sounder (PDR), 50 kHz side-scan sonar and 3.6 kJ single-channel sparker system (Solheim *et al.* 1988) indicated 15–20 depressions in the sea floor within approximately 30 km^2. A subsequent detailed multi-beam echo sounder survey of 7 km^2 suggested that the density of depressions was even greater (Solheim & Elverhøi 1993). Another (1991) multi-beam echo sounder survey (Suess & Altenbach 1992) of a wider area suggested an overall slightly lower density. These surveys thereby suggested that the depressions were concentrated about a centre. As part of the latter survey CTD casts revealed anomalously high methane concentrations in the bottom water of this area (Suess & Altenbach 1992; Lammers *et al.* 1995). This led support to the suggestion of Solheim & Elverhøi (1993) that these craters were related to the release of methane gas. While today none of the craters is an active gas source (Lammers *et al.* 1995), repeated surveys with R/V *Meteor* in October 1993 confirmed that the methane plume is a persistent feature in this area and the near-bottom water column is affected by seasonal temperature variations in the range of 1.4°C. As will be discussed later, hydrate reservoirs in the shallower vicinity are inferred to respond to these temperature changes and release the observed large quantities of methane on a seasonal amplitude.

A recent (1993) multi-beam echo sounder (hydrosweep) survey of 50 km^2 extending to 75°00′N and 27°20′E has suggested a western limit of 27°25′E to the crater field, with the most northerly crater at 74°59′N. Parasound, a swept-frequency pinger-type profiler, was run simultaneously with the multi-beam echo sounder. For part of this survey a 2 kHz deep-tow Huntec boomer was run across several of the sea-floor depressions to obtain deeper penetration. Penetration in soft sediments may be more than 300 ms with resolution less than 1 m (e.g. Evans *et al.* 1996) but this is greatly reduced in hard rock. Sea-floor visual inspection was made with the aid of video and stills cameras deployed within an adapted benthic barrel (Linke *et al.* 1994). Sea-bed sampling was by means of a 0.25 m^2 box corer. Analysis of the overlying water mass was by means of CTD and multiple corer casts (Suess *et al.* 1994).

The aim of the 1993 survey was to examine possible sources of methane in the crater area of the Barents Sea. The identification of mounds within several craters initiated this examination of their form and consideration of their potential origin.

Morphology

The sea floor is generally smooth, sloping at about 2.5 m km^{-1} to the southeast, varying from 310 mbsl in the northwest of the study

LONG, D. LAMMERS, S. & LINKE, P. 1998. Possible hydrate mounds within large sea-floor craters in the Barents Sea. *In:* HENRIET, J.-P. & MIENERT, J. (eds) *Gas Hydrates: Relevance to World Margin Stability and Climate Change.* Geological Society, London, Special Publications, **137**, 223–237.

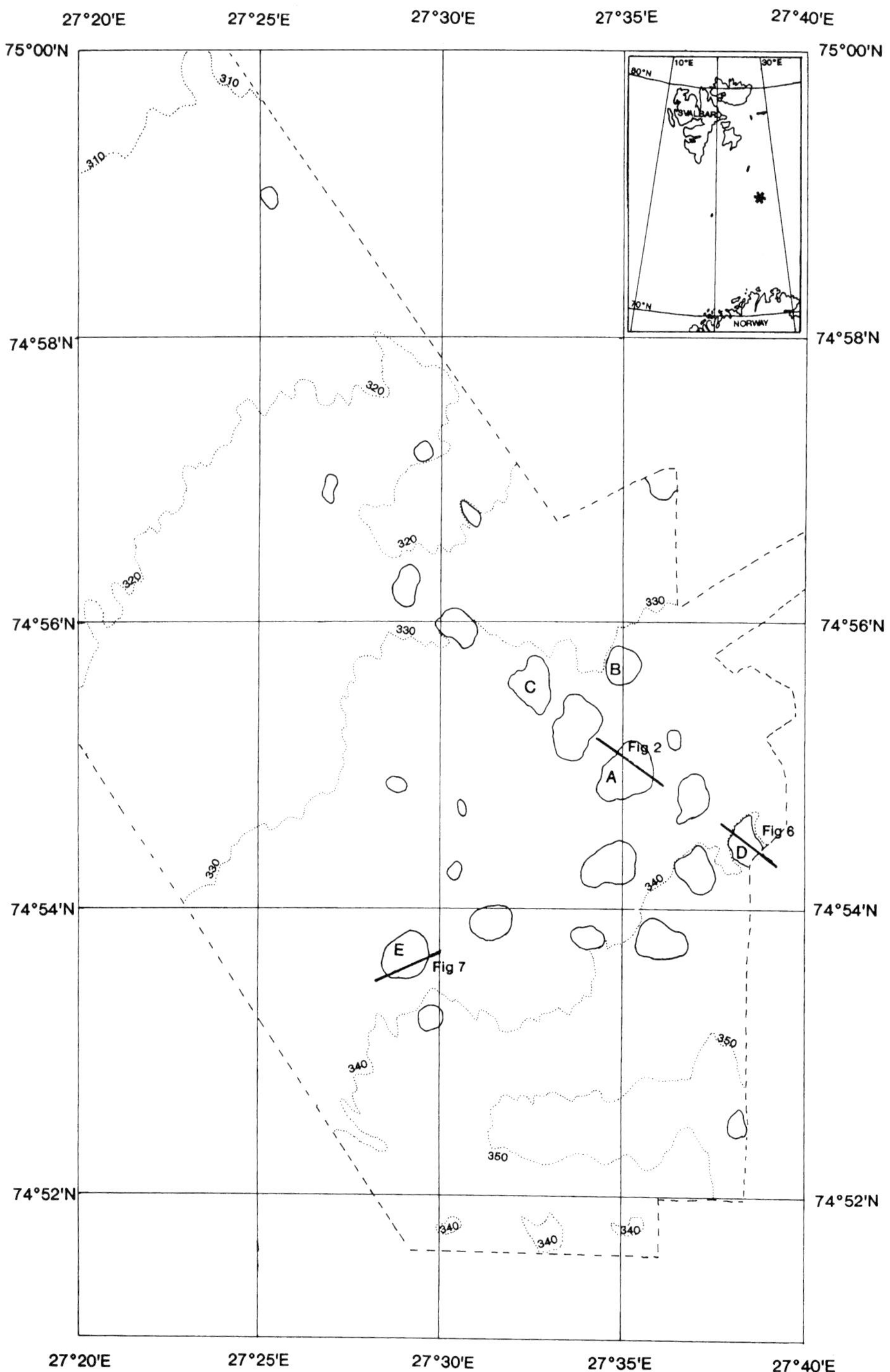

Fig. 1. Location of Barents Sea craters, illustrating the apparent lineation in their distribution. Dotted lines indicate bathymetric contours outside the craters (depths in metres on the upside of contour lines). Dashed line indicates limit of swath bathymetric data from the 1991 and 1993 hydrosweep surveys. Craters are lettered to provide locational information for discussion in text. Positions of Figs. 2, 6 and 7 are given. Star (*) on inset map indicates the location of the crater field.

area to 350 m in the southeast corner (Fig. 1). Side-scan sonar images of the sea bed display glacial flutes cut by iceberg plough marks, the latter typically 20 m across (Solheim & Elverhøi 1993). Within this area are many large depressions, typically semi-circular in profile. They vary from 10 to 30 m in depth and from 200 to 600 m in diameter. The walls of the depressions are steep with gradients of up to 50° (Solheim & Elverhøi 1993).

The 1991 hydrosweep data suggested the presence of several mounds (5–10 m high) on the sea bed outside the craters (Suess & Altenbach 1992). These features could not be recognized in the 1993 survey nor in the data presented by Solheim & Elverhøi (1993) and are thought to have been an artefact in data acquisition due to 'cross-talk' (Suess *et al.* 1994).

Strong hyperbola on the boomer records from the floors of several of the craters (Fig. 2), together with no acoustic penetration by the Parasound, suggested a hard irregular surface. This was subsequently proved by sampling and video to comprise broken fragments of rock. The sea floor surrounding the craters is primarily a flat surface, attributed to planar erosion such as that by glaciers, and dips gently to the south. This flat surface has a very strong acoustic response. Locally, adjacent to the craters (Fig. 2) small mounds occur above this reflector giving an uneven sea floor suggesting post-glacial deposition. Solheim & Elverhøi (1993) suggest that these small mounds comprise ejecta from the craters.

Video surveying and photographs of the base of several of the crater depressions revealed a muddy sea floor with occasional angular slabs of mudstone–siltstone lying on the bottom or slightly protruding through the sediment cover (Fig. 3D). Locally these slabs project up to 1.5 m above the surrounding crater floor (Fig. 4) (estimated from the video image). Suspension-feeding animals like sponges, soft corals (octocorallia) and anemones (actinians) were common on top of the rocks. Shrimps were occasionally visible, as were dense swarms of Euphausids. In the deepest part of the depressions numerous fish, particularly cod, were observed. Crustacean burrows were not common suggesting only a thin soft sediment cover (Fig. 3D). The frequency of rock slabs increased towards the centre of the depression (Fig. 4A). The sides of the depression were steep with abundant mudstone–siltstone slabs (Fig. 3C), which also occurred on the sea floor around the rim of the depression (Fig. 3A and B).

Internal features

The swath surveys show that some of the depressions contain internal mounds and ridges. These

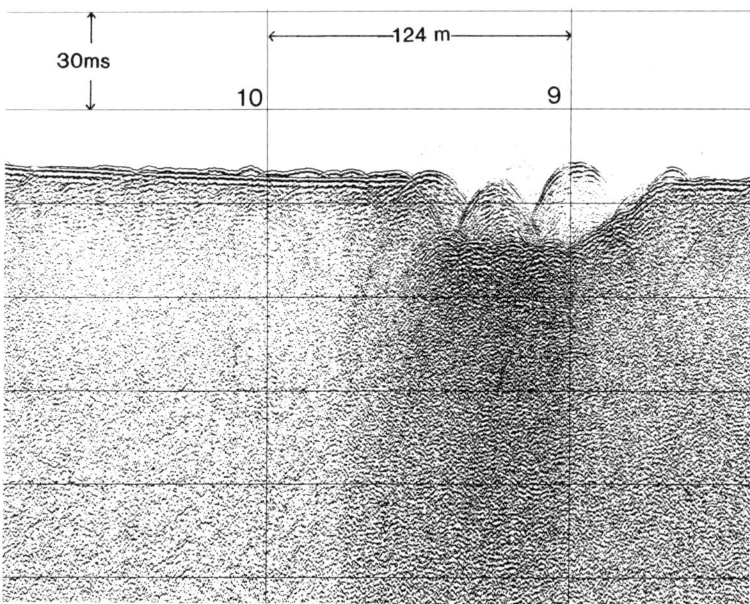

Fig. 2. Boomer profile (93/03/488/2) illustrating mounds within Crater A rising above the rims of the crater. The surrounding sea bed is smooth with a thin uneven cover of sediment close to the crater.

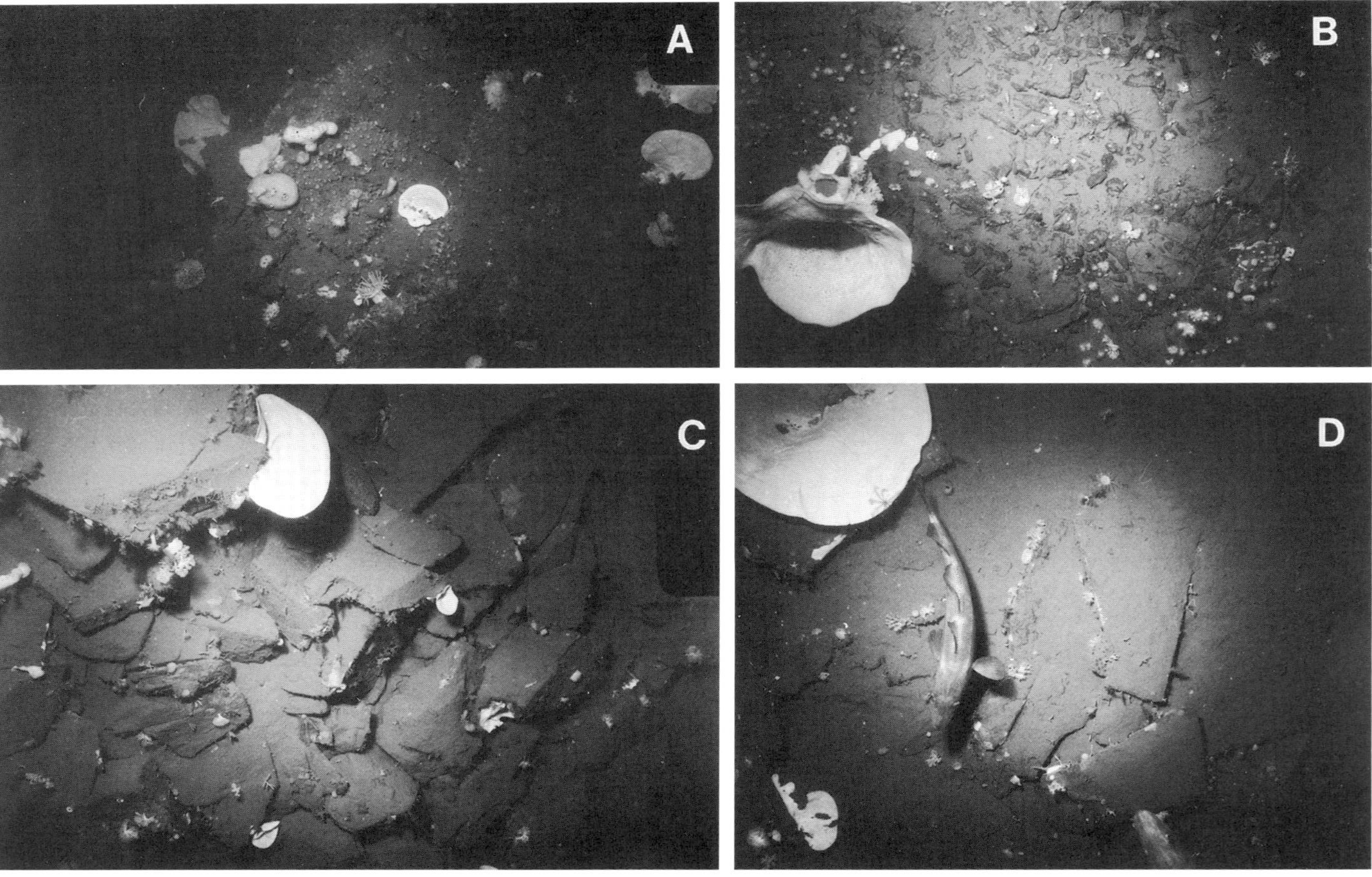

Fig. 3. (**A**) Rim of Crater A. The mudstone–siltstone slabs are densely colonized by a suspension feeding community consisting of small hydrozoans, actinians, soft corals and sponges: *Phakettia* sp. (big earshaped type), *Polymastia* sp.(small with papillae), *Geodia* sp. (spherical type with few openings), *Cladorhiza* sp. (branched type). (**B**) Sea floor around the rim of crater A with a mixture of attached species (bryozoans, soft corals and a big hexactinellid sponge *Chonelasma* sp.), and sediment-colonizing species (*Cerantharia* and a smaller branched demosponge *Cladorhiza* sp.). (**C**) Steep side of Crater A with abundant mudstone–siltstone slabs. (**D**) Base of Crater C with occasional angular slabs of mudstone–siltstone lying on the bottom or slightly protruding through the sediment cover. Fish, mainly cod, are present in large numbers.

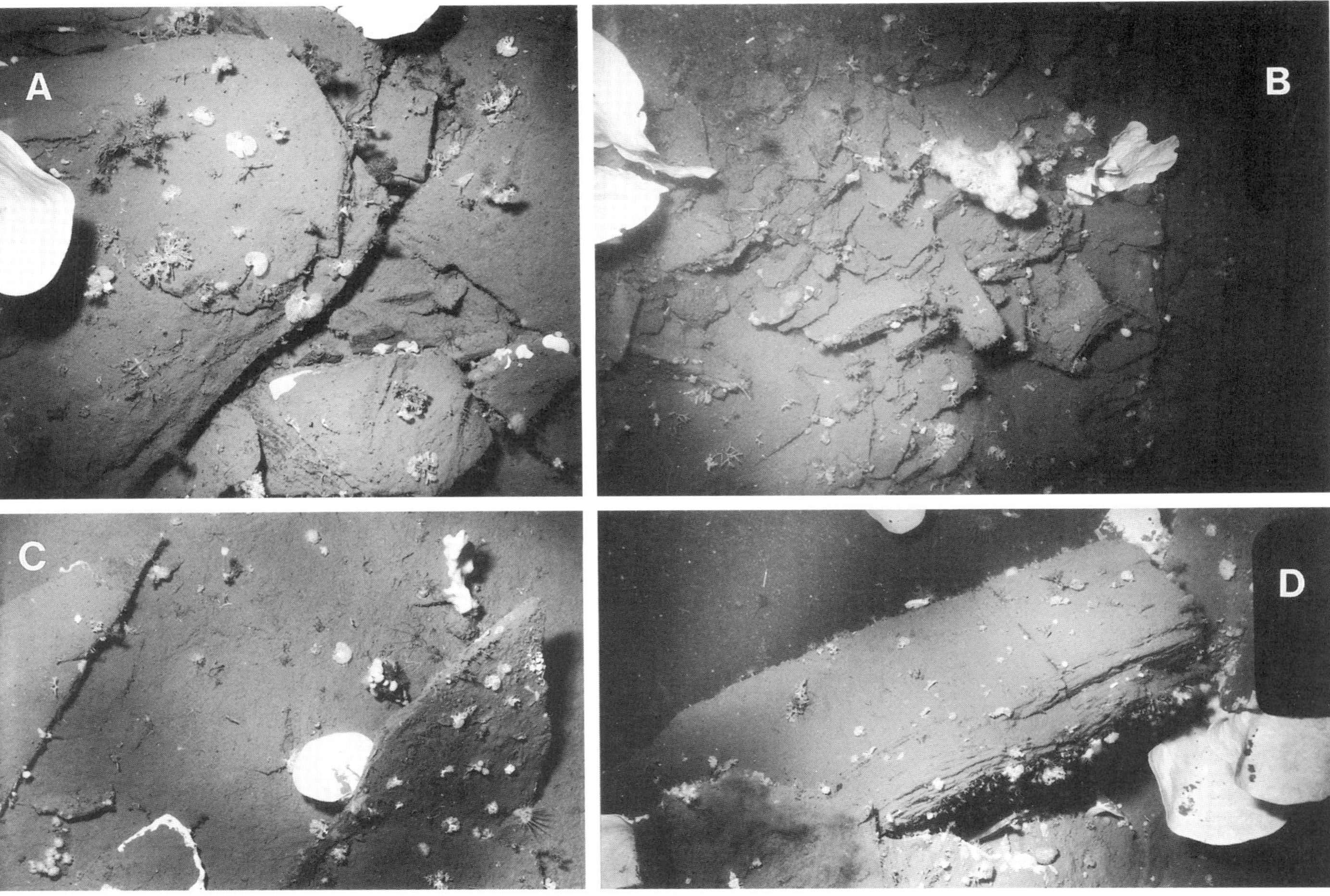

Fig. 4. (**A**) The frequency of rock slabs increase towards the centre of Crater C. (**B**) Mound/pinnacle within Crater A comprising many slabs with angular protuberances. (**C**) and (**D**) Slabs standing upright protruding through the thin sediment cover in Crater C are densely colonized.

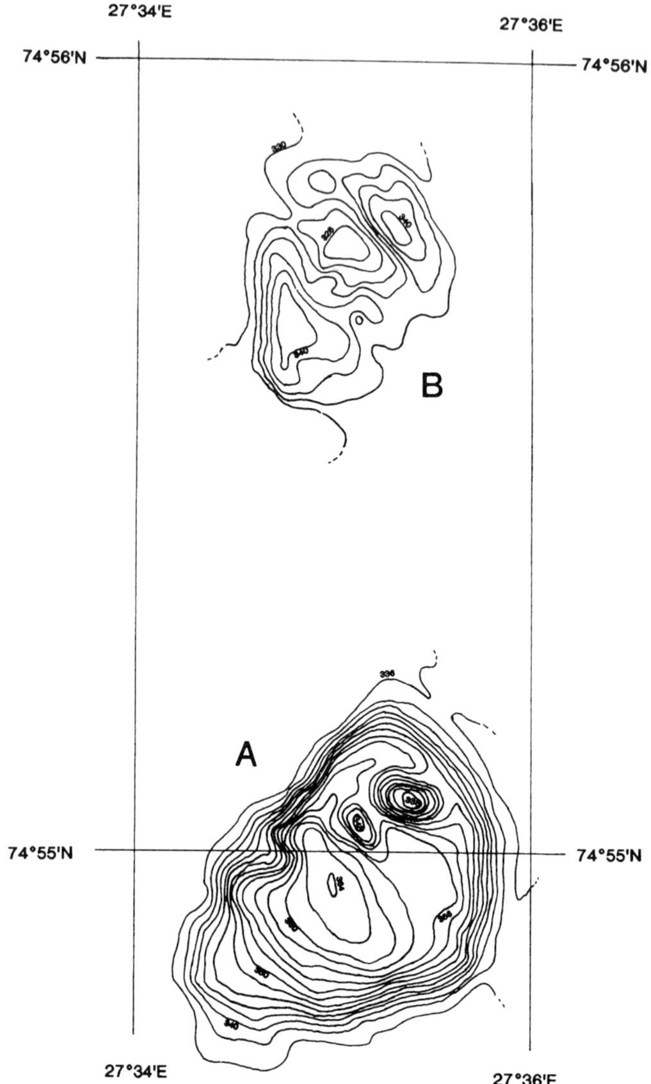

Fig. 5. Detail bathymetry of craters A and B (see Fig. 1 for location) with contours at 2 m intervals. Topographic mounds extend above the rim of the craters. (Contour values in metres positioned on upside of contour line.)

mounds and ridges are very steep sided and occasionally are as high as, or even greater than, the crater walls (Fig. 2). The most notable internal features occur in the larger craters near the centre of the crater field (Fig. 5).

In Crater A, centred on 74°54.8′N–27°35.1′E, there are two mounds/pinnacles rising from the floor of the crater, one rising nearly 20 m from the crater floor (Fig. 5) to the level of the crater rim (Fig. 2). A third mound is located on the western side wall of the crater (Fig. 5). These features can be equally well seen on the multi-beam echo sounder bathymetry map of Solheim & Elverhøi (1993, fig 3). These three features occur on a line with a bearing of 67°.

Crater B, centred on 74°55.7′N–27°35.1′E, has an elongate mound orientated at 123° forming a ridge dividing the crater floor into two halves (Fig. 5). This ridge rises a few metres above the height of the crater rim (Fig. 5).

Within the craters the boomer records frequently resolve both the crater floor and the crest of a mound within the footprint of the boomer (e.g. Fig 2). This means that for the tallest mound in Crater A (Fig. 5) the side wall has a minimum slope of 39°. The hydrosweep data indicate that the mounds and ridges locally maintain slope angles in excess of 25°. Locally the small area resolved on the video images appeared near vertical, certainly suggesting slope angles in excess of those above. Video examination of the mound/pinnacles showed they comprised rock slabs giving it rather angular protuberances (Fig. 4B). The slabs were easily dislodged by collision with the video deployment vehicle. The top of the pinnacle was flat with a thin sediment cover and abundant anemones growing on it. Sponges were observed growing on the rock slabs comprising the sides of the mounds (Fig. 4B).

Distribution of the observed features

Combining data from all the surveys known from this area, the majority of depressions occur within 3 km of 74°55′N–27°34′E, although there is little hydrosweep information east of 27°39′E from the 1991 and 1993 Meteor surveys. Solheim & Elverhøi (1993, fig. 2) show craters at 27°40′E. Although the largest depressions occur close to the centre of the crater field there is no apparent direct correlation between location and size of the depression, with several small depressions also located in the centre of the crater field. However, based on the existing bathymetry survey, depressions containing internal mounds and ridges are restricted to an area near the centre of the crater field.

The hydrosweep data suggest that the majority of depressions are located along lineations with bearings of approximately 60° and 120°. Where there are sufficient mounds or pinnacles within individual craters to define a lineation, they too occur along similar alignments, (e.g. 67° in Crater A and 123° in Crater B).

Bedrock

Box cores (0.25 m^2) taken from within the craters are dominated by large angular blocks (up to 30 cm across) of dark to light grey siltstone and mudstone, frequently with abundant reworked plant or coal fragments. The clasts are very angular, sometimes with conchoidal fractures, and occur within a matrix of soft mud. These samples confirmed the video evidence.

Acoustic reflectors were noted to be downwarped beneath several of the craters indicative of significant acoustic velocity differences

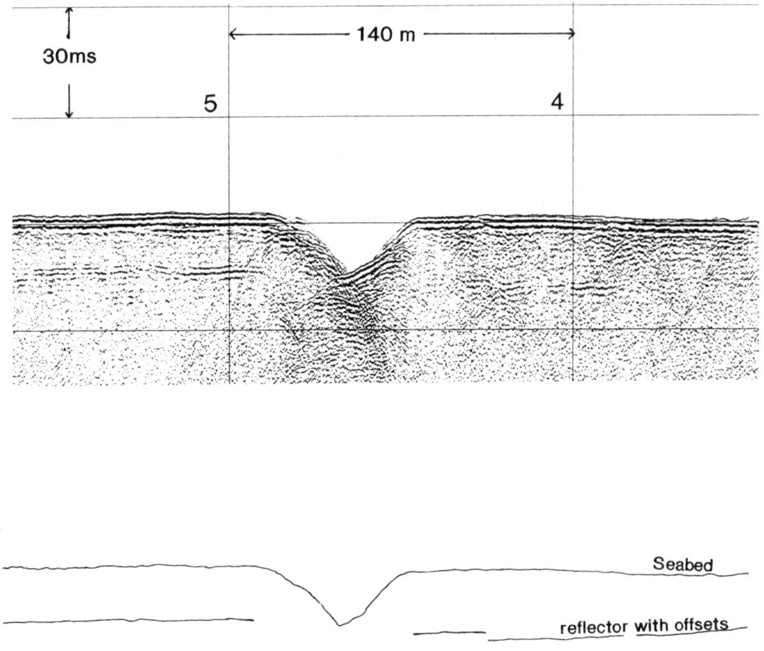

Fig. 6. Boomer profile (93/03/488/2) illustrating an offset in depth to internal reflector suggesting bedrock faulting in the vicinity of Crater D.

between the sediment of the crater walls and the water within the craters. Basic calculations assuming a velocity for the water within the crater to be 1480 m s^{-1} indicated that, to explain all the down-warping of the reflectors, the crater walls have acoustic velocities of 2200–2400 m s^{-1}. These values are considerably greater than values from Quaternary sediments in a nearby borehole (7425/09-U-01) which ranged from 1700 to 1900 m s^{-1} (Sættem *et al.* 1991). The figures are low for solid rock but might be explained if the rock is weathered or fractured, thereby reducing its velocity. Therefore, it can be assumed that the craters are cut into bedrock, as indicated by Solheim & Elverhøi (1993), and these acoustic velocities are considered minimum values for the bedrock.

Direct velocity measurements on recovered samples proved unsuccessful due to the friable nature of the sediments. Acoustic velocity measurements of Triassic siltstone on Björnöya have yielded V_p values of about 3100 m s^{-1} (Grønlie *et al.* 1980). This is a little higher than the 2200–2400 m s^{-1} estimated above using the down-warping of reflectors beneath the craters. The latter value is considered to be a minimum value. Other physical property measurements on the recovered samples of rock gave values for dry density and effective porosity of 2.62 g cm^{-3} and 3% respectively for mudstone, and similarly 2.22 g/cm^3 and 18% for a laminated siltstone.

Based on the intersection of boomer profiles and an acoustic velocity of 3000 m s^{-1} the internal acoustic reflectors of the bedrock dip at about 3° with a bearing of 274°. Some of the internal reflectors appear to be offset on either side of some of the craters (Fig. 6), which may be interpreted as minor fault movement within the vicinity of the crater.

Acoustic features

Strong hyperbola on the boomer records from the floors of several of the craters, together

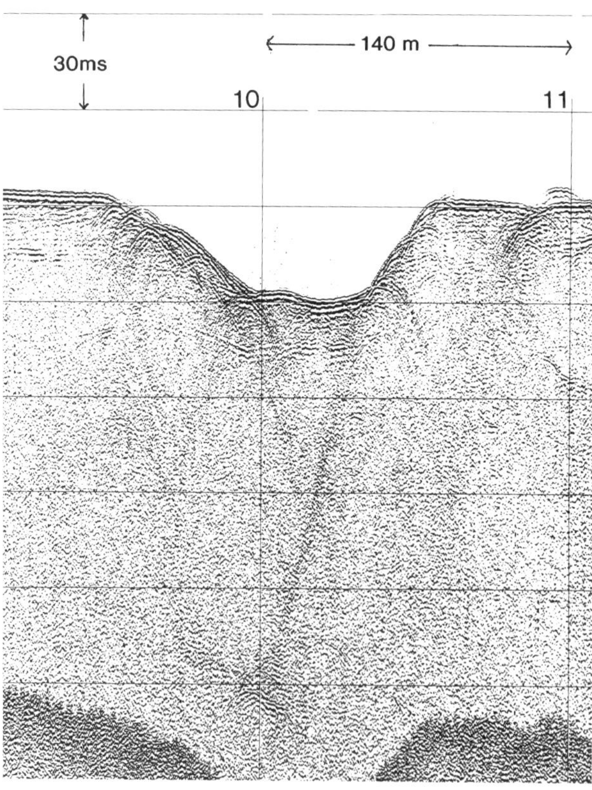

Fig. 7. Boomer profile (93/03/488/1) illustrating enhanced reflector located about 125 ms beneath the floor of Crater E.

with no acoustic penetration by the Parasound, suggested a hard irregular surface. The internal mounds have an amorphous acoustic character and no internal structure could be observed. Deeper reflectors were only noted on the boomer profiles.

Below several of the larger craters a single, laterally restricted, enhanced reflector was observed (Fig. 7). Its depth ranged between 30 and 130 ms below the floor of the craters. Assuming an acoustic velocity of $3000 \mathrm{m\,s^{-1}}$ for the rock, these reflectors are 45–185 m below the floor of the craters (65–210 m below the sea bed outside the craters). Such features have been attributed elsewhere to gas accumulations, including gas beneath hydrate zones (Dillon & Paull 1983). No similar reflectors were noted beneath areas devoid of craters. Although Solheim & Elverhøi (1993) reported that no BSR could be identified in the seismic data from the area, the laterally restricted reflectors evident on these boomer profiles may be too short to be detected on sparker profiles used by Solheim & Elverhøi (1993).

Assuming that the sea floor is in thermal equilibrium with the overlying bottom-water temperatures as measured on the cruise (about 0.2 °C), and as there is approximately 350 m of hydrostatic pressure on the floor of the craters, any methane and water mixture present is likely to occur within the solid phase as methane hydrate (Fig. 8). However, as the temperature of the sediment increases with depth below sea bed there will be position at which free methane gas becomes stable and may accumulate beneath the base of the hydrate stability zone, creating the acoustic properties required to form a reflector such as seen on Fig 7. The restricted lateral continuity of these enhanced reflectors (Fig. 7) may imply that if they represent an hydrate–free gas boundary it has limited lateral extent, such as may occur where gas migrates up through a fault within a generally tight formation.

Shallow gas is present locally in the Barents Sea. Sættem *et al.* (1991) report methane concentrations up to 40 ml of gas per kg of dry sediment at borehole 7425/09-U-01, located approximately 75 km to the southwest. Pockmarks have been noted in the area to the northeast of the crater field (Solheim & Elverhøi 1985). Gas hydrates have been reported extending over $55 \mathrm{km^2}$ of the southern Barents Sea (Laberg & Andreassen 1996).

Methane hydrate stability

Using the hydrate stability formulae from JOIDES Pollution Prevention and Safety Panel (1992) and applying an offset of 1.1°C for saline waters (Dickens & Quinby-Hunt 1994) it is possible to calculate the depth at which free gas (assuming it is pure methane) would be stable. A geothermal gradient of $31 ^\circ\mathrm{C\,km^{-1}}$ has been determined in the western Barents Sea

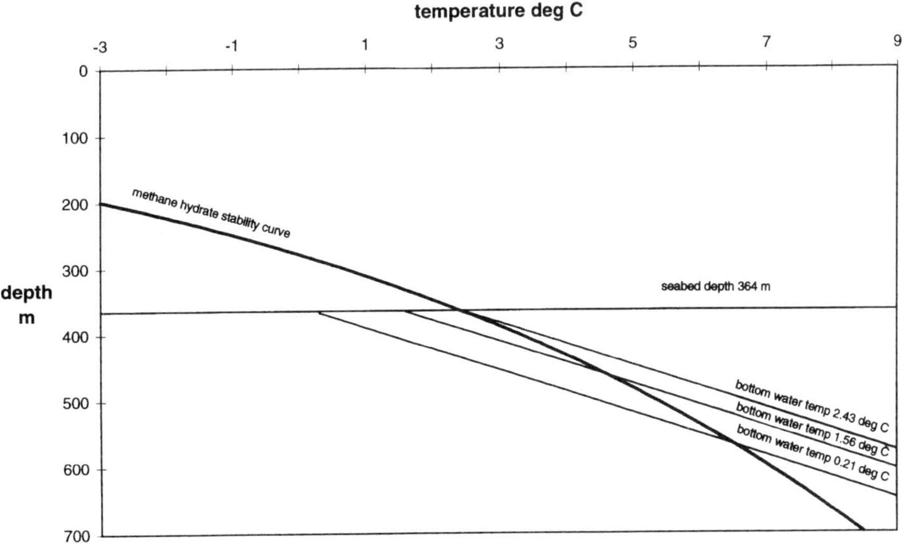

Fig. 8. Methane hydrate stability curve and suggested depth–temperature profiles for the floor of Crater A (364 mbsl). The sub-sea-bed profiles are for bottom-water temperatures of 0.21, 1.56 and 2.43°C with a geothermal gradient of $31 ^\circ\mathrm{C\,km^{-1}}$.

(Sundvor & Eldholm 1992), and values between 30 and 35°C km^{-1} in the southern Barents Sea (Andreassen *et al.* 1990). Using the conditions at Crater A (the deepest), where the base of the crater floor is 364 m (Fig. 5), a bottom-water temperature of 0.21°C (Suess *et al.* 1994) and a geothermal gradient of 31°C km^{-1}, the base of the hydrate stability zone would be 200 m below the floor of the crater (Fig. 8). This is comparable with the maximum depth of the enhanced reflectors. No consideration for effective gas pressure has been made in the above calculation.

Variations in the depth of the enhanced reflectors noted on the boomer records may be due to variations in bottom-water or sea-bed temperatures, gas composition, salinity of pore waters, geothermal gradient and the depth of the sea bed. These would create changes in the depth below sea bed to free gas and explain the range of levels for the enhanced reflectors. The last three features might be expected to vary most within the small area of the crater field and the variation would be linked to the presence of faults and fractures.

It should be noted that bottom-water temperatures vary seasonally as values 1.4°C higher were recorded in August 1991 (Lammers *et al.* 1995). Comparison of the records from both surveys also indicates that methane concentrations co-vary seasonally with the bottom-water temperatures by roughly 40% between 40 nM at 1.6°C in August and 25 nM at 0.2°C in October (Fig. 9). If the sea bed were in equilibrium with these higher bottom-water temperatures and the same geothermal gradient applied, the base of the hydrate stability zone would be 100 m below the crater floor. If bottom-water temperatures rose even higher to 2.4°C and this warming was transmitted throughout the underlying sea bed, then the base of the hydrate stability zone would rise to the crater floor and any hydrate would sublimate causing methane gas to escape at the sea bed. Owing to the lower hydrostatic pressure in other craters (they are shallower) and at the top of the mounds, a sea-bed temperature less than 2.4°C would be needed for any hydrate to sublimate. At bottom-water temperatures of 0.2°C, as observed in October of 1993, sublimation would occur at any water depth less than 300 m. The implication is that sub-surface gas hydrate reservoirs decompose in response to these seasonal temperature variations and release gaseous methane. According to the pressure–temperature conditions required for this process, it is assumed that the hydrates are exposed to a hydrostatic pressure of less than 300 dbar.

Age of craters

The Late Weichselian Barents Sea Ice Sheet was grounded in this area forming sub-glacial geomorphological features (Solheim *et al.* 1990). On the basis of the morphology of the craters and their relationship to iceberg plough marks and glacial flutes, Solheim & Elverhøi (1993) suggested that these crater-like depressions formed after the retreat of the ice. The samples recovered by the box corer would support this. The friable nature, angularity and near-monolithic composition of the rock recovered in the box cores suggests that the craters are post-glacial in age as the clasts are likely to have been quickly rounded by ice action. There were only a few sub-rounded ice-rafted pebbles, including a large clast with glacial striae. The topography of the depressions could be expected to have been modified by ice action if the craters were formed prior to the removal of the Barents Sea Ice Sheet.

Recent modelling of the Barents Sea Ice Sheet (Elverhøi *et al.* 1993) suggests an almost fully deglaciated Barents Sea as early as 15 000 years BP. This together with dates of 23 000 years BP for marine conditions prior to the expansion of the Barents Sea Ice Sheet implies that the ice sheet existed for only a short time and applied minimal load in this part of the Barents Sea (Elverhøi *et al.* 1993; Lambeck 1995).

The video evidence of sponges growing on the crater wall and pinnacles within the craters may provide minimum ages for these features and therefore minimum ages for the formation of the craters. Although it is not possible to give an exact taxonomic description of the sponge species using black and white photography without sample material, the overall impression is that of a vivid sponge community (J. Reitner pers. comm. 1995). Dead specimens would appear grey not bright white, covered by sediment and tipped over. The sponges observed on the sea-bed photographs (Figs 3 and 4) are thought to be *Phakettia* sp. (big earshaped type), *Chonelasma* sp. (big hexactinellid sponge), *Polymastia* sp. (small with papillae), *Geodia* sp. (spherical type with few openings) and *Cladorhiza* sp., (branched demosponge). Except for *Cladorhiza* sp. all sponges are typical colonists of hard substrate in deeper waters (J. Reitner written comm. 1995). At present no actual growth rates of deep water sponges are available, but the apparent size of the sponges (20–50 cm in diameter), their vivid appearance and the environmental conditions present at the site imply that the pinnacles have been in place for a minimum of 100 years.

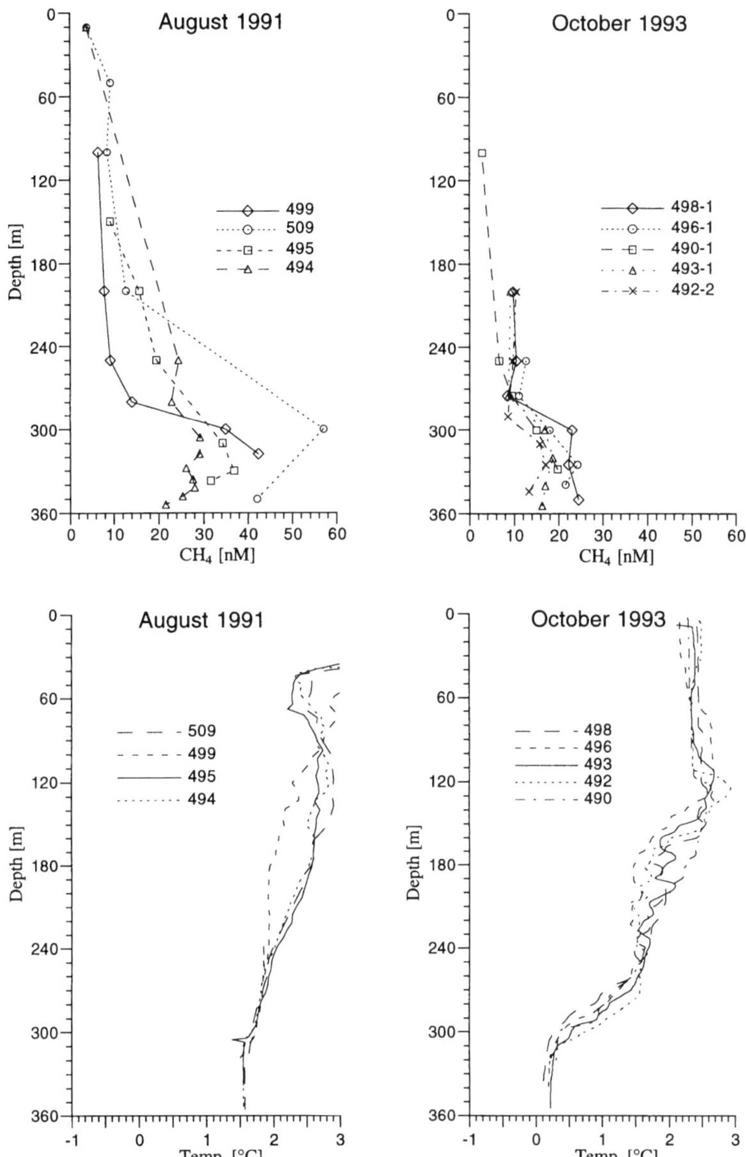

Fig. 9. Compiled records of dissolved methane and corresponding water temperatures in the Barents Sea crater area, observed in October 1993 and in August 1991.

Discussion

The conclusion reached by Solheim & Elverhøi (1993) that these crater-like depressions are most probably formed by gas eruption is supported by this recent data set. As some of the mounds rise above the level of sea bed surrounding the craters they must post-date the glacial erosion of the sea floor. Owing to the fragile nature of the sides of the mounds it is difficult to consider a mechanism for mound formation prior to the violent formation of the craters. Therefore, it is presumed that the mounds formed after the craters erupted.

The site of these craters (74°55′N–27°34′E) is located on the southern edge of the Gardarbanken High structure as mapped by Gabrielsen *et al.* (1990). According to the sub-crop map produced by Dowdeswell (1988) the location of these depressions is on an anticlinal axis which has a core of Late Triassic–Early Jurassic sediments, possibly plunging to the southeast. Sub-cropping rocks of the same age but differing structure have been mapped by Elverhøi *et al.* (1988), and more recent maps (Sigmond 1992) indicate that the crater locality is near the western limit of an anticlinal core of early–middle Trias sediments orientated approximately eastwest. It is suggested that the reduction in pressure due to the removal of ice loading would permit the opening of fissures and/or faults in the Mesozoic rocks, and thereby provide pathways for the migration of shallow gas to the sea bed during periods of lower sea level (lower hydrostatic pressure). The possible presence of faults close to the position of the craters has been noted by the vertical displacement of internal reflectors within the bedrock. Sea-bed mounds associated with gas migration up through faults have been reported elsewhere (e.g. Hovland *et al.* 1994) but not with the formation of hydrate.

Mounds containing hydrate have been reported from the Gulf of Mexico (Campbell *et al.* 1986; MacDonald *et al.* 1994). The growth of gas hydrates in near-surface sediments has been attributed as a possible mechanism for the formation of these mounds in the Gulf of Mexico (Prior *et al.* 1989). This occurs by local expansion of sediment, water and hydrate mixtures, and is analogous to the development of pingos (Prior *et al.* 1989). The model suggests that the mounds form when the hydrate–free gas boundary is close to the sea bed. If bottom-water temperatures remained constant at 0.2°C during the Holocene then a water depth of 285 m would cause the hydrate–free gas boundary to be close to the crater floor (Fig. 10A) implying a significant change in water depth (approximately 80 m). A higher sea-bed temperature (e.g. 1.56°C as recorded for bottom waters in August 1991 (Lammers *et al.* 1995)) would require a water depth of 330 m for the hydrate–free gas boundary to be at the floor of Crater A (Fig. 10B). The latter scenario would imply only a small change (10 m for the floor of Crater B) from current sea levels for shallower craters.

Assuming there was gas seepage at the craters following eruption, gas would have continued to flow until pressure and temperature conditions were such that hydrate formed at the seep sites in the floor of the crater, thereby creating a cap to the gas (MacDonald *et al.* 1994). If gas seepage to the surface was via faults, then sites of seepage within an individual crater would have been restricted. The internal pinnacles/mounds may have been formed when pressure and temperature were such that hydrate formed at the sea floor within the rubble-strewn floor of the craters above a seepage point. As more gas seeped up, the weakly frozen rubble may have been pushed upwards before the free gas went into a solid phase, growing in a similar manner to a pingo. Such a mechanism has been suggested for hydrate mounds reported in the Gulf of Mexico (e.g. Prior *et al.* 1989) where significant changes in size and shape can occur in less than 1 year (MacDonald *et al.* 1994). Therefore, it is suggested that this process formerly permitted the development of the large topographically positive features seen within the craters, with a cover of broken rock. With rising sea level causing increased pressure the hydrate–free gas front moved to deeper depths below sea bed thereby freezing the mounds. The hydrate–free gas front may now relate to the reflector noted beneath several of the craters.

Unlike the hydrate mounds reported from the Gulf of Mexico and the Florida Escarpment, the mounds within the Barents Sea craters are not covered with seep-associated fauna or chemosynthetic communities. This difference may be explained by the inactive state of the mounds now that the hydrate–free gas boundary is currently located well below the sea bed, and methane is not diffusing into the overlying water mass or at least not on a regular enough basis to sustain a seep fauna. By contrast, those sites reported in the Gulf of Mexico located close to the hydrate–free gas boundary are extremely active (MacDonald *et al.* 1994) and support chemosynthetic communities (Brooks *et al.* 1987).

It is suggested that seasonal temperature variations during times of shallower water allowed sublimation of the hydrate on the surface of the mounds leaving an outer covering just comprising rock slabs. If bottom-water temperatures increased in the future, and this input of heat was sustained to raise the temperature of the underlying rocks, additional hydrate sublimation could occur in the sub-bottom of the surveyed craters leading to more intense methane expulsion or even gas blow-outs and the formation of new craters.

The suggestion that the mounds formed when water depths were about 80 m less than at present may seem anomalous due to the significant Holocene uplift evident on Svalbard (e.g. Salvigsen & Österholm 1982). However, recent modelling (Lambeck 1995) indicates that the sea bed was *c.* 150 m below present at the crater field locality at 16 000 years BP due to eustatic and isostatic effects. These changes in sea level would provide

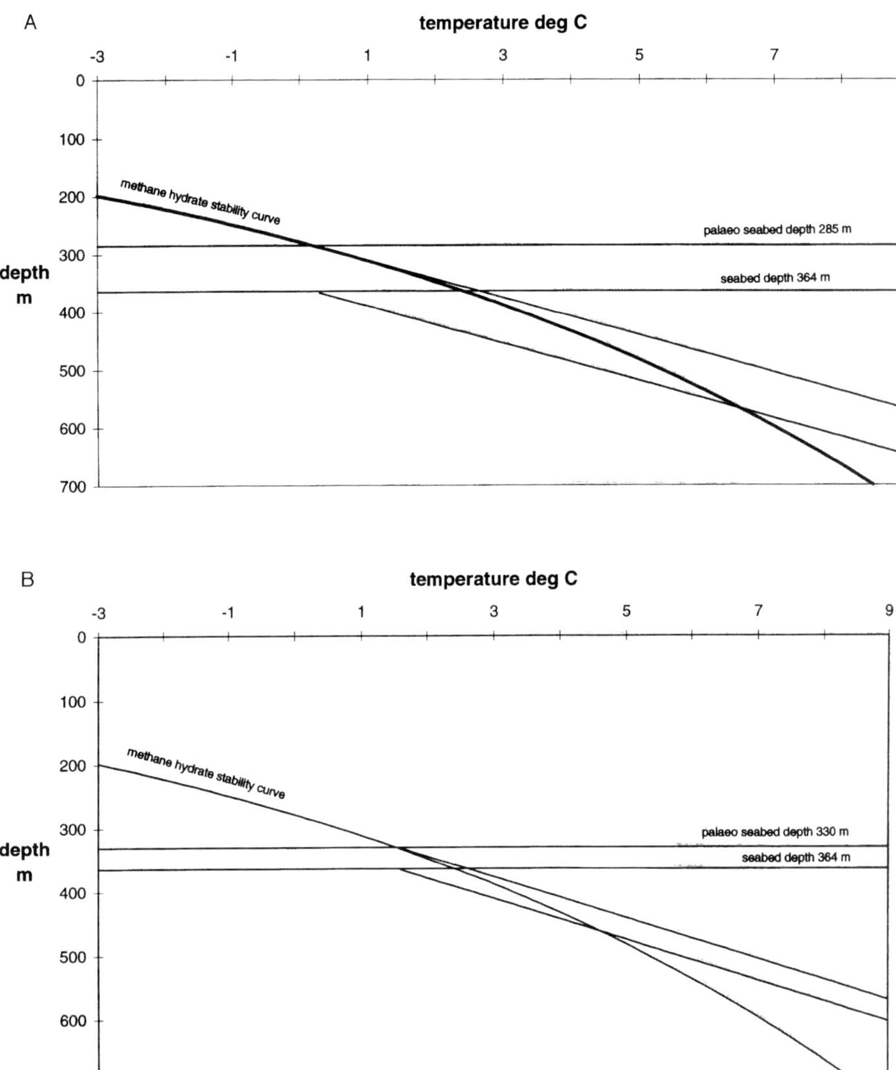

Fig. 10. (**A**) Methane hydrate stability curve and suggested depth–temperature profiles for the floor of Crater A, currently 364 mbsl and at a palaeosea level (80 m less than today) when the base of the hydrate stability zone cropped out on the crater floor. The sub-sea-bed profiles are for bottom water temperatures of 0.21°C with a geothermal gradient of 31°C km^{-1}. (**B**) Methane hydrate stability curve and suggested depth–temperature profiles for the floor of Crater A, currently 364 mbsl and at a palaeosea level (34 m less than today) when the base of the hydrate stability zone cropped out on the crater floor. The sub-sea-bed profiles are for bottom-water temperatures of 1.56°C with a geothermal gradient of 31°C km^{-1}.

the necessary lower water depth to permit the free flow of methane at the crater floor following crater formation then, as sea level rose, the pressure and temperature conditions at the seepage site would have changed such that hydrate formed, creating the mounds reported here.

Conclusions

- Following deglaciation, fractures within the Triassic siltstone opened due to the reduction in pressure.
- Large volumes of free gas escaped explo-

sively via these fractures to the sea bed causing the formation of the large sea-bed depressions (craters).

- A reduced volume of gas continued to seep from individual sites on the floor of a few craters, principally at the centre of the crater field. When pressure and temperature conditions permitted the gas formed hydrate within the rubble at the base of the crater which built up as mounds or ridges.
- Depending on the sea-bed temperature the hydrate mounds could have formed when water depths were between 10 m and 80 m less than at present.
- The mounds were formed some time between deglaciation of the Barents Sea and 100 years ago.
- Today, decomposing hydrates in nearby areas at less than 300 m water depth supply the observed widespread methane plume in response to seasonal warming of the bottom water.

The authors would like to thank officers and crew of R/V *Meteor* and fellow scientists on cruise M26/2 for assisting in the collection of this data set. J. Reitner (Georg-August-University, Göettingen) is thanked for identification of the sponges and supply of information on them. H. E. Lie (Norsk Polar Institutt) is thanked for identifying the bedrock samples. The comments of K. Andreassen and F. Theilen on an earlier version of this paper and the comments of two anonymous referees of this paper are gratefully acknowledged. This work was undertaken as part of CEC MAST funded project SEEPS (Gas and Water Seepage on the Continental Margin) project number MAS2-CT92-0040. D. Long publishes with permission of the Director, British Geological Survey (NERC).

References

ANDREASSEN, K., HOGSTAD, K. & BERTEUSSEN, K. 1990. Gas hydrate in the southern Barents Sea, indicated by a shallow seismic anomaly. *First Break*, **8**, 235–245.

BROOKS, J. , KENNICUTT, M. , BIDIGARE, R. R. *et al.* 1987. Hydrates, oil seepage and chemosynthetic ecosystems on the Gulf of Mexico slope: an update. *EOS, Transactions of the American Geophysical Union*, **68**, 498–499.

CAMPBELL, K. , HOOPER, J. & PRIOR, D. 1986. Engineering implications of deepwater geologic and soil conditions, Texas–Louisiana slope. *Proceedings of the 18th Offshore Technology Conference*, **5105**, 225–232.

DICKENS, G. & QUINBY-HUNT, M. 1994. Methane hydrate stability in seawater. *Geophysical Research Letters*, **21**, 2115–2118.

DILLON, W. & PAULL, C. 1983. Marine gas hydrates–II. Geophysical evidence. *In*: COX, J. (ed.) *Natural Gas Hydrates: Properties, Occurrence and Recovery*. Butterworth, Boston, MA, 73–90.

DOWDESWELL, E. 1988. The Cenozoic stratigraphy and tectonic development of the Barents Shelf. *In*: HARLAND, W. & DOWDESWELL, E. (eds) *Geological Evolution of the Barents Shelf Region*. Graham and Trotman, London, 131–155.

ELVERHØI, A., ANTONSEN, P., FLOOD, S. B., SOLHEIM, A. & VULLSTAD, A. 1988. *The Physical Environment. Western Barents Sea, 1:1 500 000. Shallow Bedrock Geology – Structure, Litho- and Biostratigraphy. Norsk Polarinstitutt, Skrift*, **179D**.

——, FJELDSKAAR, W., SOLHEIM, A., NYLAND-BERG, M. & RUSSWURM, L. 1993. The Barents Sea Ice Sheet – a model of its growth and decay during the last ice maximum. *Quaternary Science Reviews*, **12**, 863–873.

EVANS, D., KING, E., KENYON, N., BRETT, C. & WALLIS, D. 1996. Evidence for long-term instability in the Storegga Slide region off western Norway. *Marine Geology*, **130**, 281–292.

GABRIELSEN, R., FAERSETH, R., JENSEN, L., KALHEIM, J. & RIIS, F. 1990. *Structural Elements of the Norwegian Continental Shelf, Part I: The Barents Sea Region*. Norwegian Petroleum Directorate Bulletin, **6**.

GRØNLIE , G., ELVERHØI, A. & KRISTOFFERSEN, Y. 1980. A seismic velocity inversion on Bjornoya – the western Barents Shelf. *Marine Geology*, **35**, 17–26.

HOVLAND, M., CROKER, P. & MARTIN, M. 1994. Fault-associated seabed mounds (carbonate knolls?) off western Ireland and north-west Australia. *Marine and Petroleum Geology*, **11**, 232–246.

JOIDES POLLUTION PREVENTION AND SAFETY PANEL. 1992. Ocean Drilling Program Guidelines for Pollution Prevention and Safety. *JOIDES Journal*, **18**(7).

LABERG, J. & ANDREASSEN, K. 1996. Gas hydrate and free gas indications within the Cenozoic succession of the Bøjrnøya Basin, western Barents Sea. *Marine and Petroleum Geology*, **13**, 921–940.

LAMBECK, K. 1995. Constraints on the Late Weichselian ice sheet over the Barents Sea from observations of raised shorelines. *Quaternary Science Reviews*, **14**, 1–16.

LAMMERS, S., SUESS, E. & HOVLAND, M. 1995. A large methane plume east of Bear Island (Barents Sea): implications for the marine methane cycle. *Geologische Rundschau*, **84**, 59–66.

LINKE, P., SUESS, E. M., TORRES, M. *et al.* 1994. In situ measurement of fluid flow from cold seeps at active continental margins. *Deep-Sea Research*, **41**, 721–739.

MACDONALD, I., GUINASSO, N., JR, SASSEN, R., BROOKS, J., LEE, L. & SCOTT, K. 1994. Gas hydrate that breaches the sea floor on the continental slope of the Gulf of Mexico. *Geology*, **22**, 699–702.

PRIOR, D., DOYLE, E. & KALUZA, M. 1989. Evidence for sediment eruption on deep sea floor, Gulf of Mexico. *Science*, **243**, 517–519.

SÆTTEM, J., RISE, L. & WESTGAARD, D. 1991. Composition and properties of glacigenic sediments in the southwestern Barents Sea. *Marine Geotechnology*, **10**, 229–255.

SALVIGSEN, O. & ÖSTERHOLM, H. 1982. Radiocarbon dated raised beaches and glacial history of the northern coast of Spitsbergen, Svalbard. *Polar Research*, **1**, 97–115.

SIGMOND, E. M. 1992. Bedrock Map of Norway and Adjacent Ocean Areas. Scale 1:3 million. Geological Survey of Norway.

SOLHEIM, A. & ELVERHØI, A. 1985. A pockmark field in the central Barents Sea; gas from petrogenic source? *Polar Research*, **3**, 11–19.

—— & ——. 1993. Gas-related sea floor craters in the Barents Sea. *Geo-Marine Letters*, **13**, 235–243.

——, —— & FINNEKASA, O. 1988. *Marine Geophysical/Geological Cruise in the Northern Barents Sea 1987 – Cruise Report. Norsk Polarinstitutt Rapportserie*, **43**.

——, RUSSWURM, L., ELVERHØI, A. & NYLAND BERG, M. 1990. Glacial geomorphic features in the northern Barents Sea: direct evidence for grounded ice and implications for the pattern of deglaciation and late glacial sedimentation. *In*: DOWDESWELL, J. & SCOURSE, J. (eds) *Glaciomarine Environments: Processes and Sediments*. Geological Society, London, Special Publications, **53**, 253–268.

SUESS, E. & ALTENBACH, A. 1992. *Europaisches Nordmeer*, Reise Nr. 17, University of Hamburg.

——, KREMLING, K. & MIENERT, J. 1994. *Nordatalantic 1993, Cruise No. 26*, Meteor-Berichte, University of Hamburg, **94–4**.

SUNDVOR, E. & ELDHOLM, O. 1992. Norway: Off-shore and north-east Atlantic. *In*: HURTIG, E., GERMÁK, V., HAENEL, R. & ZUI, V. (eds) *Geothermal Atlas of Europe*. Hormone Hack, Gotha, 63–66.

Detection of gas-charged sediments and gas hydrate horizons along the western continental margin of India

M. VEERAYYA, S. M. KARISIDDAIAH, K. H. VORA, B. G. WAGLE & F. ALMEIDA

National Institute of Oceanography, Dona Paula, Goa-403004, India

Abstract: High-resolution seismic reflection and sub-bottom profiling on the continental margin off western India has revealed the presence of characteristic acoustic maskings in the form of wipe-outs, reflector terminations and seep-associated features in the inner shelf. These maskings suggest the presence of gas-charged sediments. Further seaward on the outer shelf–middle slope, pockmarks and prominent plumes in the overlying water column indicate a significant seepage of gas from the slope sediments, and it is the seepage which may demonstrate the existence of source rocks. Seismic profiles also revealed the presence of bottom simulating reflectors (BSRs) in the mid-lower slope–rise regions, presumably suggesting the presence of gas hydrates. The BSRs occur at roughly 300–600 ms (TWTT) beneath the sea floor at water depths between 525 and 2200 m; they occasionally show discontinuities. Distinct blanking zones as well as acoustic voids have also been observed above the BSR. In contrast, chaotic and/or scattered hyperbolic reflectors occur in places below the BSR, which may suggest the presence of gas-charged sediments. Folds, diapiric features and faults present in the slope–rise areas may probably serve as traps and conduits for upward migration of fluids and methane gas from the deep.

Knowledge regarding the presence and extent of shallow gas horizons and gas hydrate compounds within sediments on continental margins, which appear to be the largest reservoirs of natural gas on Earth, has greatly improved during the last few decades (Kaplan 1974; Claypool & Kvenvolden 1983; Kvenvolden 1993). Shallow hydrocarbon gas is considered to have a number of geophysical and geomorphological effects, of which the most well known are absorption of acoustic energy in the seismic record and the formation of acoustic maskings and pockmarks (Anderson & Bryant 1990). Data on shallow gas-charged horizons of sediments displaying wipe-outs, reflector terminations, seeps, acoustic masking and acoustic turbidity, have been variously reported (Kaplan 1974; Schubel 1974; Watkins & Worzel 1978; Carlson *et al.* 1985; Behrens 1988; Hovland & Judd 1988; Bouma *et al.* 1990; Aharon *et al.* 1992, and others.). A thorough account on the origin, composition and distribution of surface and sub-surface methane has been dealt with by Schoell (1988) and Davis (1992).

Natural gas hydrates, or clathrates, are ice-like crystalline substances composed of cages of water molecules that host low molecular weight gases, mainly methane, and form in marine sediments when gas concentrations are adequate, temperature is low and pressure is high (Sloan 1990). Theoretical studies together with geological and geophysical investigations have demonstrated that gas hydrates are suspected to be spread over the permafrost regions and in most of the upper few hundred metres of ocean floor on many continental margins (e.g. Kvenvolden & Barnard 1983; Kvenvolden 1988, 1993; Brooks *et al.* 1991; Miller *et al.* 1991; Hyndman & Davis 1992; Hyndman & Spence 1992; Dillon *et al.* 1993, 1994; MacDonald *et al.* 1994; MacKay *et al.* 1994; Lee *et al.* 1994, 1996; Paull *et al.* 1995; Trehu *et al.* 1995; Brown *et al.* 1996; Pecher *et al.* 1996; Sloan 1996; Yuan *et al.* 1996; Zwart *et al.* 1996; Dickens *et al.* 1997).

Most occurrences of gas hydrates in the sediments of continental margins have been inferred mainly from the presence of an anomalous seismic reflector on seismic profiles. This reflector usually coincides with the predicted transition boundary at the base of the stability field for methane hydrate (Kvenvolden 1988) and mimics the sea floor, and is commonly referred to as the bottom simulating reflector (or BSR). BSRs have been shown prominently in seismic records from the Blake Outer Ridge (Markl *et al.* 1970; Paull & Dillon 1981), the western north Atlantic (Tucholke *et al.* 1977) and in the references cited earlier. The only reported occurrences of BSRs in the Indian Ocean are from the Gulf of Oman (White & Klitgard 1976; White 1977, 1979; Minshull *et al.* 1992) and offshore Andaman (Chopra 1985). A detailed account of the sampling areas of gas hydrates during DSDP and ODP drilling in the Atlantic and the Pacific oceans has been given by Brooks *et al.*

VEERAYYA, M., KARISIDDAIAH, S. M., VORA, K. H., WAGLE, B. G. & ALMEIDA, F. 1998. Detection of gas-charged sediments and gas hydrate horizons along the western continental margin of India. *In:* HENRIET, J.-P. & MIENERT, J. (eds) *Gas Hydrates: Relevance to World Margin Stability and Climate Change*. Geological Society, London, Special Publications, **137**, 239–253.

(1991), Miller *et al.* (1991), Kvenvolden *et al.* (1993) and Henriet & Mienert (1996).

The possible occurrence of BSRs along the western continental margin of India has been reported recently (Veerayya *et al.* 1996). This paper describes the detailed occurrence of: (1) characteristic acoustic maskings in the inner continental shelf; (2) plumes and pockmarks along the upper-middle slope, and (3) BSRs on seismic profiles as possible evidence of gas hydrates along the lower slope and rise of the western continental margin of India.

Data acquisition

Approximately 30 000 km of high-resolution, single-channel seismic reflection profiles were collected using a 500–16 000 J sparker (40–2000 Hz), together with an ORE sub-bottom profiler (3.5 kHz) and echo sounders (MS-45 and EK-12). The data were collected along track lines spaced at 20 km intervals across the continental margin during various cruises of R/V *Gaveshani*. Positions were obtained using satellite navigation. An assumed seismic velocity of 1500 m s^{-1} was used to derive the water depth.

Tectonic and geological setting

The continental margin off western India is a divergent and passive margin. It is characterized by structural elements such as horst- and graben-like features, northwest–southeast trending regional faults and ridge complexes like the Laxmi and Laccadive–Chagos Ridge systems, the latter forming the boundary between the western and the eastern basin of the eastern Arabian Sea (Fig. 1) (Naini & Talwani 1982; Biswas 1987). The regional MCS sections off Ratnagiri–Kerala show that on the slope, the thickness of the sediments gradually increases to 1500–2000 m on the Miocene shelf edge – a few kilometres landward of the present-day shelf edge. This zone is marked by a sudden thickening of sediments up to 3500–4000 m and shows progradational depositional features. The shelf edge, which forms a prominent tectonic zone, appears faulted in places with possible deformation of sedimentary cover (Raju *et al.* 1981).

The margin is characterized by its distinct physiographic provinces – continental shelf, slope (shelf margin basin, marginal high) and rise. The shelf is about 345 km wide off Tarapur in the north and narrows down to 60 km off Cochin in the south. It slopes gently to the west (1 : 400–1 : 3000). The shelf break occurs between 80 and 145 m water depth. Based on topographic variations and sedimentological characteristics, the shelf is divided into two sub-provinces – an inner shelf and outer shelf. The inner shelf is marked by an even, gently seaward-sloping topography and is covered by 15–35 m thick, weakly to well stratified or acoustically transparent, Holocene muds. This even topography extends up to 50–60 m water depth. Further seaward (deeper than 60–65 m), the outer shelf is characterized by an uneven or rugged topography with amplitude variations of up to 20 m (Veerayya *et al.* 1991; Rao *et al.* 1994). Prominent reefs are a common feature at the shelf edge (Vora *et al.* 1996).

Shallow seismic data of the continental slope show a typical sub-bottom penetration of 200–500 m with seaward-dipping clinoforms, outbuilding and upbuilding. Sediment slumps, subsurface faults and gullies are also discernible along some sectors of the slope. The slope is dominated in places by prominent topographic highs of considerable length (30–40 km) and width (13 km) – e.g. the Prathap Ridge Complex (not shown in Fig. 1) – often protruding several hundreds of metres above the surrounding sea floor. Such highs fringe a basinal structure (shelf margin basin), characterized by about 2.5–4 km thick sediments (Raju *et al.* 1981; Basu *et al.* 1982). The deep sea floor adjacent to the slope is relatively smooth, with occasional basement highs bounding basin structures characterized by 2–3 km thick sediments.

Results and Discussion

Gas-charged sediments

High-resolution seismic profiles of the inner continental shelf showed ~5–35 m thick, weakly stratified to acoustically transparent clays which overlie the Pleistocene–Holocene unconformity. Gas-charged sediments occur extensively on the inner shelf and are confined to water depths shallower than 60 m (Figs 2 and 3). Within these clayey sediments, the sub-bottom profiles are characterized by anomalous seismic signatures in the form of acoustic maskings with different shapes ranging from inverted U or mushroom types, dome or conical shapes, cone-shaped features surfacing above the sea floor, seeps and wipe-outs (Karisiddaiah *et al.* 1993; Karisiddaiah & Veerayya 1994, 1996), indicating the presence of gas in the sediments and gas escape features in the study area.

Shallow seismic profiles clearly indicate that

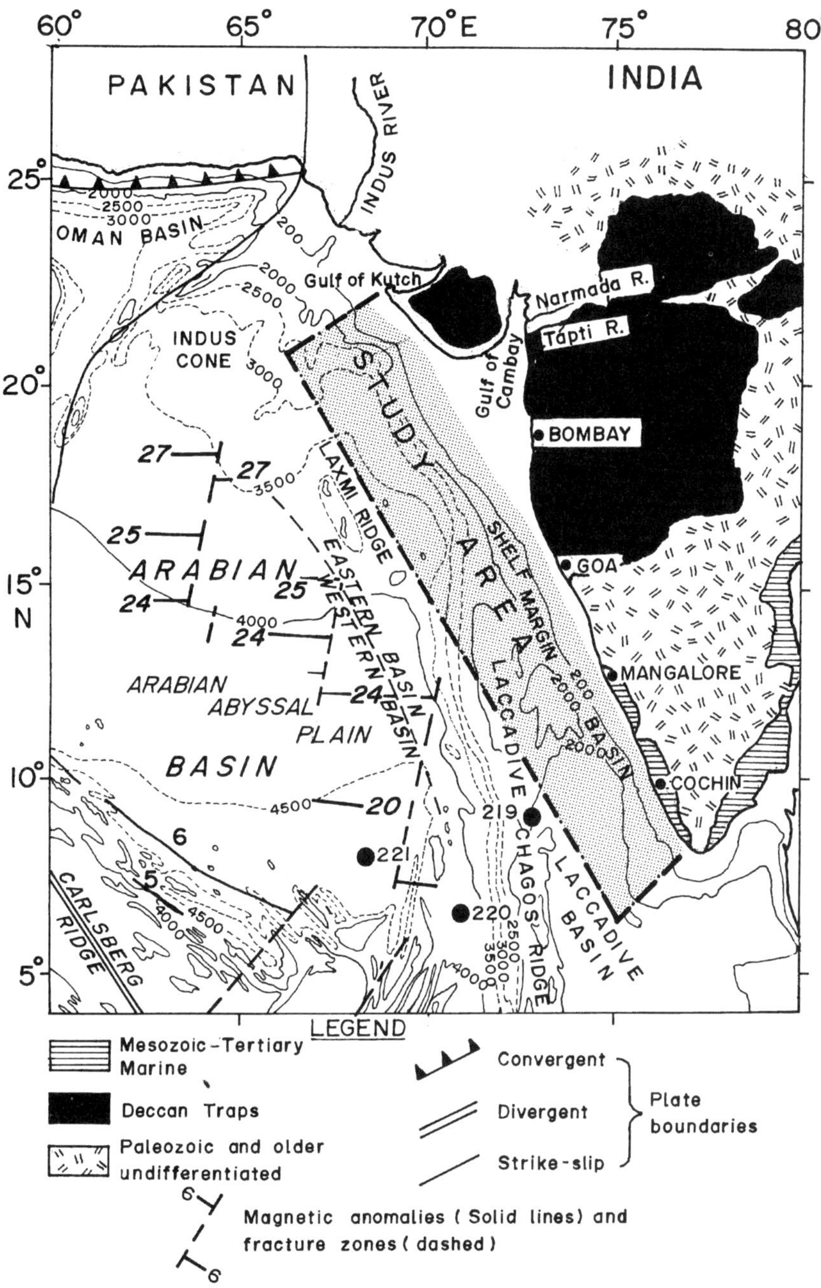

Fig. 1. Regional bathymetric and tectonic map of the Arabian Sea (modified after Whitmarsh *et al.* 1974; Biswas 1987; Whiting *et al.* 1994) showing the location of the study area (in grey). Depth contours are given in metres.

below the sea floor the clays are often characterized by anomalous seismic zones (Fig. 3) extending from the sea floor down to 5–10 m. The sides of these acoustic maskings are sharp, their tops being flat with rounded to sub-rounded edges. Cone-shaped acoustic maskings occur 3–4 m below the sea floor and well above the Late Pleistocene–Holocene unconformable compact sandy

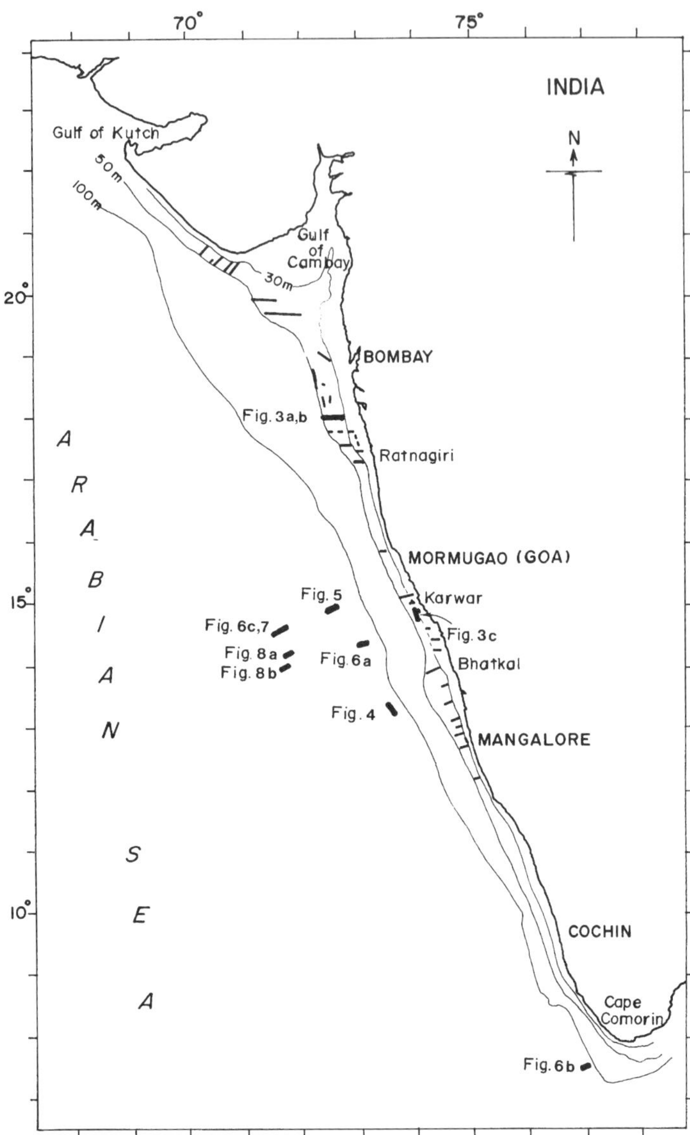

Fig. 2. Locations of 3.5 kHz and sparker profiles along the western continental margin of India. Black bars within the 60 m water depth zone indicate zones of acoustic masking, while those at greater depths mark the occurrence of pockmarks, seeps and gas hydrate horizons. Thick bars indicate profiles presented in Figs 3–8.

layer (Karisiddaiah & Veerayya 1996). Such acoustic voids are assumed to represent zones of possibly over-pressured sediments with pore water containing high concentrations of dissolved gas/or micro-bubbles.

Wipe-outs or reflector terminations ranging from 35 to 800 m wide are a common feature (Fig. 3). Acoustic maskings with uniformly rounded edges are often conspicuous. Some maskings are confined to bedding planes. Vast stretches of inverted U-shaped slab-like features are mostly 1.5–2 km wide and often stretch to more than 40 km off Bombay, 5–10 m below weakly stratified clays in the inner shelf. Numerous individual conical-shaped and inverted U-shaped acoustic maskings are ubiquitous throughout the inner shelf. These are 75–150 m wide and occur below 6 m thick clays (Karisiddaiah & Veerayya 1994).

The anomalous seismic signatures appear just

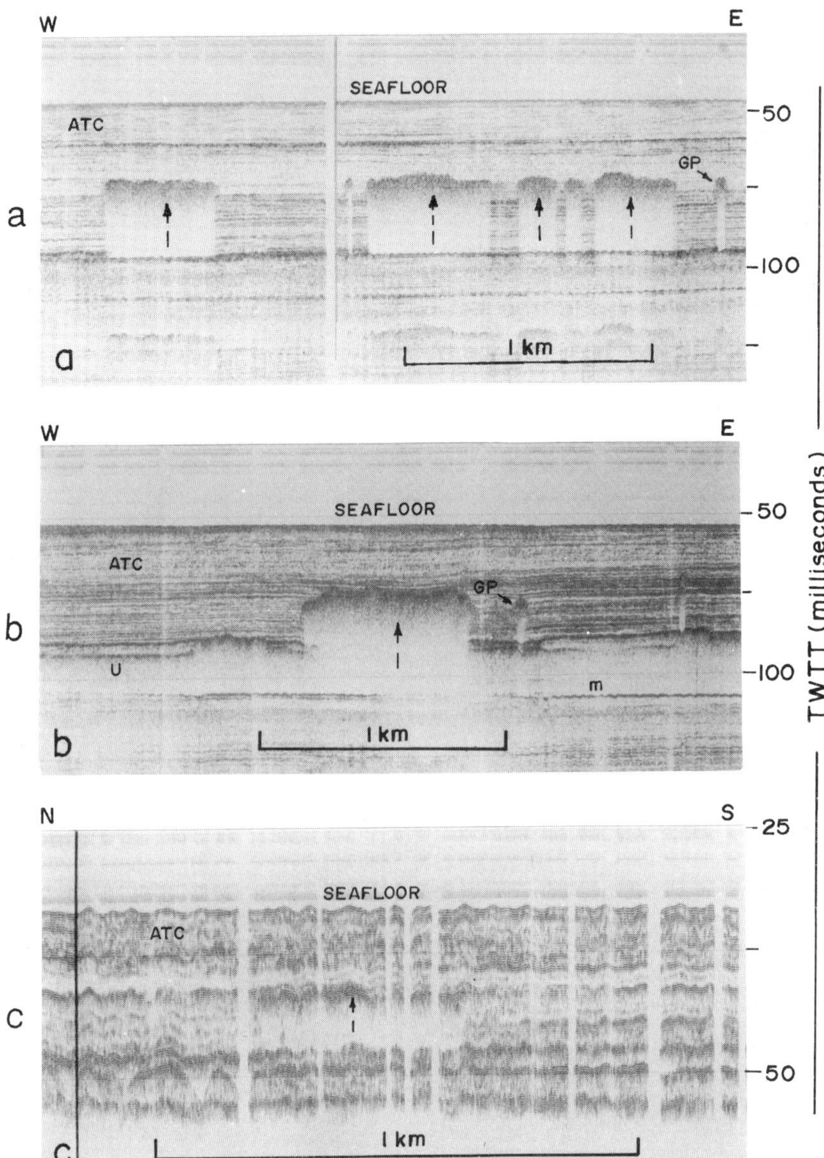

Fig. 3. A 3.5 kHz sub-bottom profile of the inner continental shelf showing prominent acoustic maskings due to gas-charged sediments as: (**a**) inverted U-shaped features overlain by 25 ms thick well-stratified to acoustically transparent clayey sediments (ATC); seismic reflectors between 30 and 50 ms bsf terminate where acoustic maskings are present, (**b**) mushroom-shaped features with well-defined reflector terminations, and (**c**) distinct wipe-out zone confined to bedding planes. Broken arrow denotes acoustic masking/wipe-outs. U, Late Pleistocene–Holocene unconformity; m, multiple. See Fig. 2 for location.

a few metres below the sea floor. However, upward migration of hydrocarbon gases by either molecular diffusion or bubble formation or ebullition (Martens & Klump 1980; Sweeney 1988) from these sediments could be possible. In some places, the anomalous features are seen on the sea floor indicating the existence of sufficient pressure in the gas for its escape. Side-scan sonar imaging has revealed bubble vents (Karisiddaiah *et al.* 1993). Often well-stratified sedimentary horizons that contain wipe-outs act as traps for the accumulation of gas, whereas the acoustic maskings present at various levels in the inner shelf sediments represent stages in the vertical migration of gas from underlying sediments.

The formation of gas-charged sediments confined to the inner shelf might be due to relatively large fluxes of labile organic carbon to the sea bed. These sediments are characterized by a high (2–4%) content of organic carbon which is mainly terrigenous in nature, as indicated by high C/N ratios (10–30) (Paropkari *et al.* 1987). This probably reflects higher accumulation rates leading to an increased decomposition rate (Whiticar 1990). It is known that in more rapidly accumulating marine sediments with a higher organic matter content bacterial methane formation is common once the dissolved sulphate has been utilized (Claypool & Kaplan 1974). In some instances methane concentrations can exceed the saturation level relative to the interstitial fluids and free gas may result (Whiticar 1990). Hence, high-accumulation rates (0.44–0.88 mm $year^{-1}$; Nambiar *et al.* 1991) trapped terrestrial organic matter within the inner shelf sediments; turning it into the main biogenic source.

Seeps and pockmarks

Shallow seismic profiles on the mid-outer shelf, mostly at about 65 m water depth, showed in some places 2 m deep and 150 m wide V-shaped depressions resembling pockmarks. These depressions could have formed due to the removal of sea-floor material by escaping hydrocarbon gas, mainly methane.

Pockmarks are also observed on seismic profiles of the upper slope, and are markedly different from those occurring on the mid-outer shelf as described above. On a typical profile, these slope-bound pockmarks (both active and relict) are confined to a prominent synclinal–anticlinal structure between 170 and 260 m water depths (Fig. 4). Out of 30 pockmarks demarcated along this section, six are buried. In general, they are 80–130 m in diameter and 0.75 to 2.5 m deep (Fig. 4). Similar pockmark features were also observed in Belfast Bay, Maine, USA (Kelley *et al.* 1994), in the Patras Gulf, Greece (Hasiotis *et al.* 1996), in the Persian Gulf (Uchupi *et al.* 1996), and have been thoroughly dealt with by Hovland & Judd (1988) and Fader (1991). Their formation has been ascribed to the escape of biogenic natural gas and pore water (Kelley *et al.* 1994), to gas seepage triggered by earthquakes (Hasiotis *et al.* 1996), and

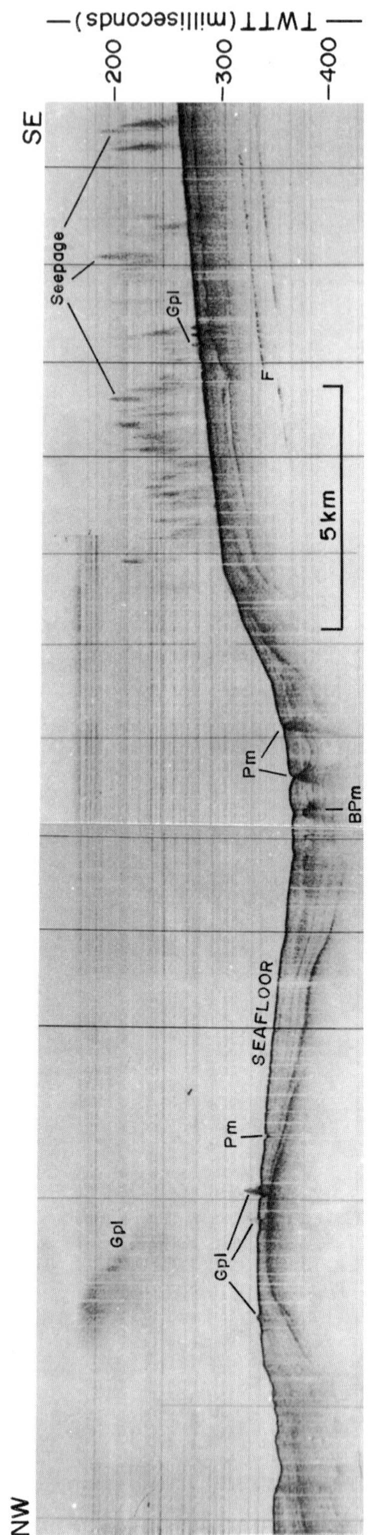

Fig. 4. A 3.5 kHz sub-bottom profile crossing the upper continental slope north of Mangalore showing distinct gas plumes (Gpl) emanating from the sea floor into the water column, Pm, pockmark; BPm, buried pockmark; F, fault. See Fig. 2 for location.

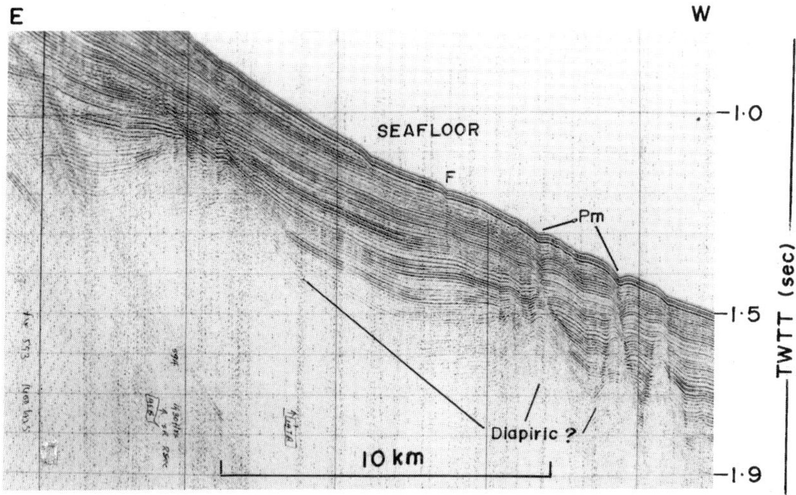

Fig. 5. Single-channel seismic reflection profile from the central part of the study area off Goa–Bhatkal showing diapiric-like folded features on the middle slope. Note the pockmarks and deep-seated gas/fluid escaping features emanating from 300–350 ms bsf on the seaward side. Pm, pockmark; F, fault. See Fig. 2 for location.

to ground water and gases of thermogenic and volcanic origin (Uchupi *et al.* 1996). In addition, water column reflections in the form of plumes observed in the study area are usually associated with pockmarks and probably represent seepage of gas. This upward migration of shallow gas through the sediment and the subsequent seepage into the overlying water partially changes the water's velocity–density property. Thus, an acoustic impedance boundary is formed with the adjacent homogeneous water resulting in the appearance of a plume-shaped gassy reflector (Li & Jin 1989) (Fig. 4).

Further seaward on the middle slope, at about 1100 m water depth, pockmarks are quite distinct (Fig. 5). These pockmarks are underlain by contorted, folded and faulted strata which in turn are underlain by diapiric-like folded structures. The diapiric-like structures and pockmarks in the upper–middle slope region also suggest the

Table 1. *Distribution of BSRs along the western continental margin of India*

Destination on Fig. 9	Line No.	Corresponding Figure	Water depth (m)	Overburden (ms, TWTT)	BSR distribution
1	191-6		525	441	Upper to middle slope
2	191-6		975	335	Middle slope, shelf margin basin
3	191-7		562	367	Slope
3A	191-7		590	369	Slope
4	191-7	6a	950	320	Shelf margin basin
5	191-11	7	2025	312	Rise (?)
6	191-11	6c	1937	458	Rise (?)
7	191-9	8b	1689	306	Rise (?)
8	150-9		875	377	Middle to lower slope
9	150-5		2195	408	Rise
16	167-4		1826	293	Shelf margin basin
19	167-7	6b	1350	240	Slope
22	167-6		1626	338	Slope
23	150-5		840	392	Slope
24	191-10		525	300	Upper slope
25	191-8	8a	1650	375	Rise
26	191-8		675	237	Slope
27	167-8		1160	600	Slope

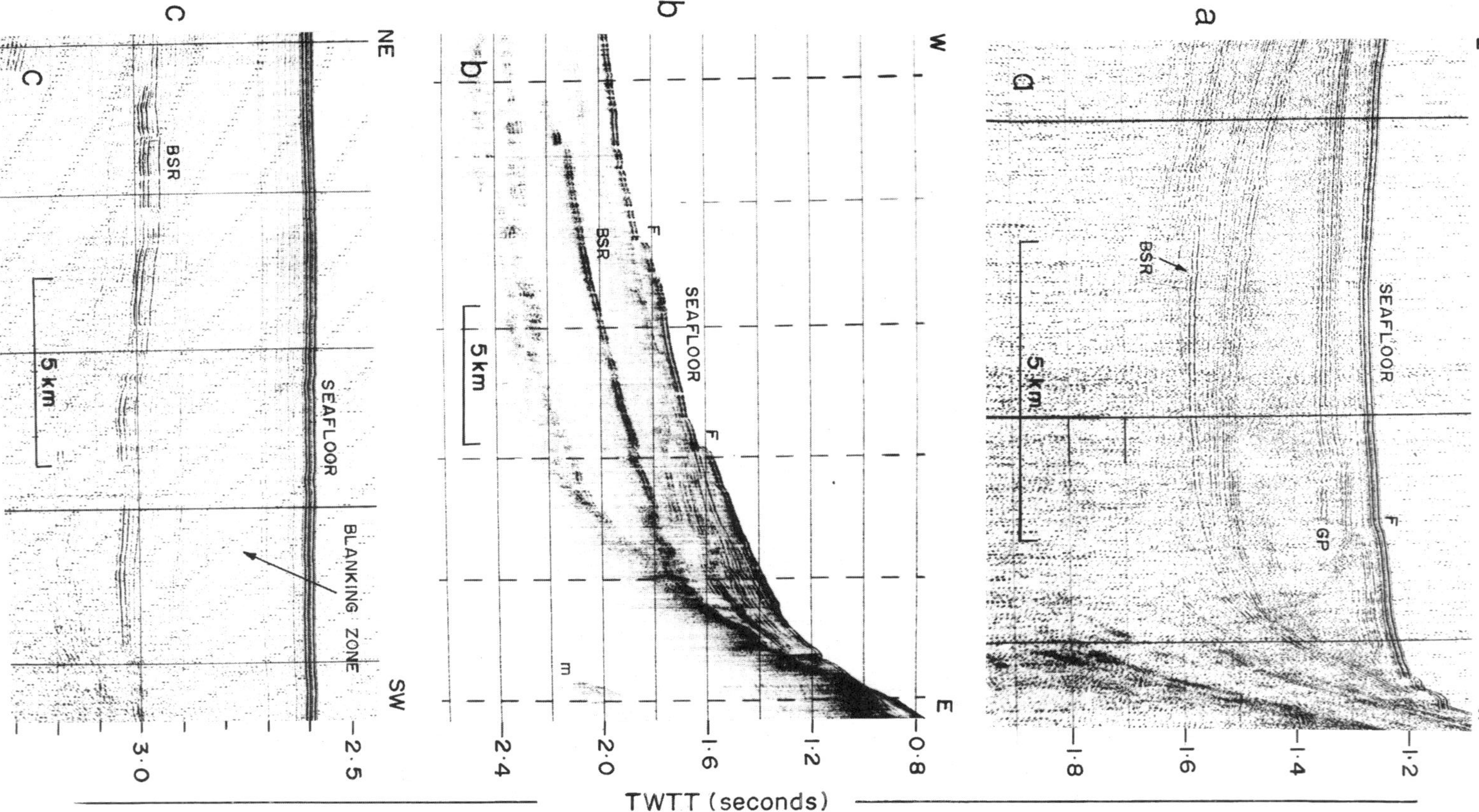

Fig. 6. Single-channel seismic reflection profile showing a (possible) BSR: (**a**) off Goa–Bhatkal in the shelf margin basin; the BSR cuts across the strata towards the east; also note the sea-floor doming and an acoustic void at 1.3 s TWTT, presumably due to gas; (**b**) off the southwest coast of India, on the middle–lower continental slope between 1350 and 1650 m water depth; and (**c**) off Goa–Bhatkal, with a distinct blanking zone and local breaching of the BSR. GP, gas pocket; F, fault; m, multiple. See Fig. 2 for location.

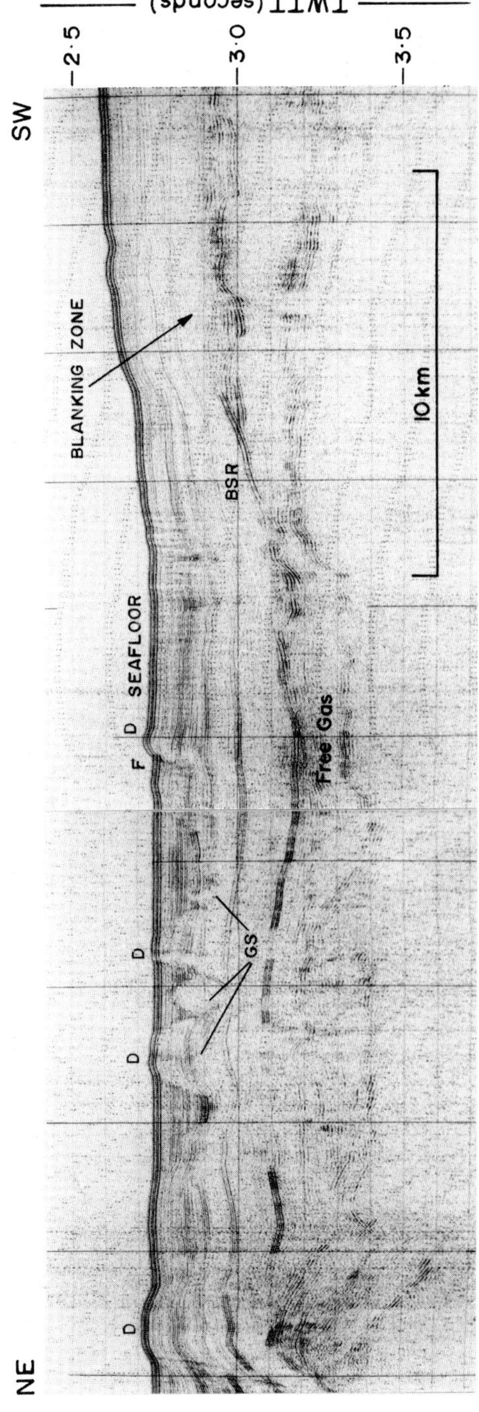

Fig. 7. Single-channel seismic reflection profile from the central part of the study area off Goa–Bhatkal showing an anomalous reflector presumably indicating a BSR, breached at several places. Note the upward tilted BSR in the southwest, overlain by a distinct blanking zone. Sea-floor doming (D), sometimes faulted (F), is also observed, with prominent acoustic voids characterized by reflector terminations in the upper 200 ms record. Chaotic reflectors below the BSR may indicate the presence of free gas (?). GS, gassy sediments (fluid expulsion features). The seaward continuation of this profile (towards the southwest) is shown in Fig. 6c (separated by a 4 km gap). See Fig. 2 for location.

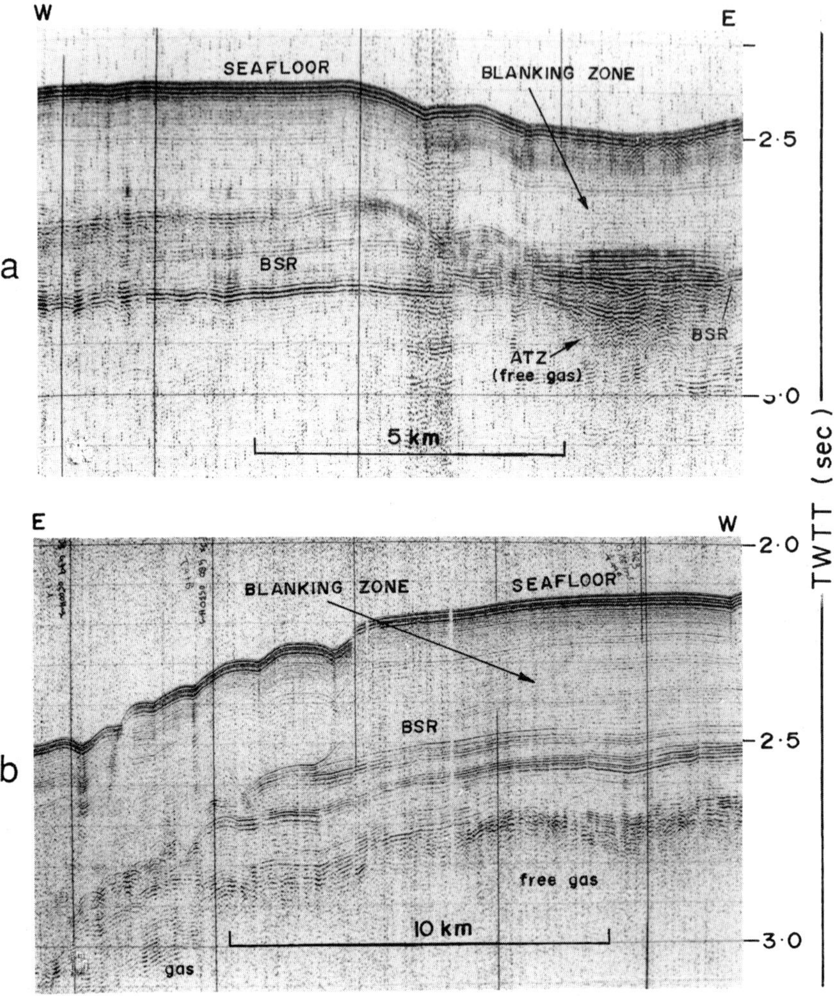

Fig. 8. Single-channel seismic reflection profile from the central part of the study area off Goa–Bhatkal showing: (**a**) BSR overlain by a distinct blanking zone and an acoustically turbid zone (ATZ) of nearly 3 km wide, presumably due to free gas below the BSR; and (**b**) BSR overlain by a prominent blanking zone, showing an upward tilt near the middle of the profile presumably reflecting a change in geothermal gradient (?); towards the east the profile is marked by numerous folds and faults. See Fig. 2 for location.

existence gas-bearing horizons. The acoustically transparent columnar disturbances below the pockmarks could indicate the presence of vertical drainage routes for gas or pore fluids.

Bottom simulating reflectors (BSRs)

Several seismic reflection profiles in the mid-lower continental slope and rise areas show, in places, a prominent reflector displaying a clear affinity with a BSR. The reflector is located at sub-bottom depths of 300–600 ms TWTT, between 525 and 2200 m water depth (Figs 6–8 and Table 1). For example, a relatively flat and well-defined (possible) BSR can be observed in the shelf margin basin at about 320 ms below the sea floor (bsf) at 950 m water depth, cutting across the strata (Fig. 6a). The sedimentary column immediately above the BSR is apparently seismically transparent. The profile also shows a distinct acoustic void zone 2 km wide at 50 ms bsf in the west, presumably indicating a gas pocket, which is characterized by reflector

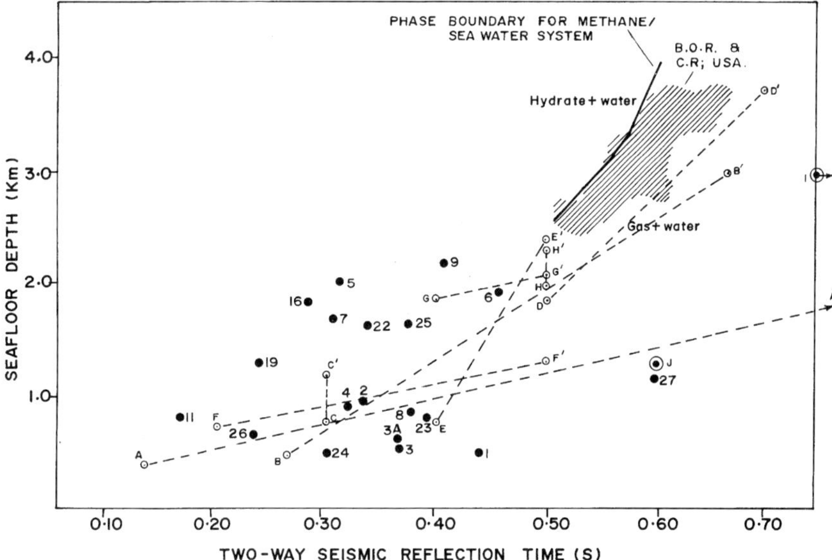

Fig. 9. BSR burial depth (TWTT) vs water depth for present data (numbered dots) and reported data from the Blake Outer Ridge (B.O.R.) and the continental rise (C.R.) off the south-eastern USA (striped area; modified after Tucholke *et al.* (1977)). Details of the studied BSRs are given in Table 1 and those of the reported BSRs designated by AA′, BB′, etc. (after Kvenvolden & Barnard 1983; Kvenvolden 1993) are given in Table 2.

terminations. Although the sea-floor topography of the upper slope is mostly smooth in the east, faulting and doming of strata above the acoustic void in the west suggest upward migration of gas through the fault planes into the overlying water column, corroborating the seismic evidence of seeps, pockmarks and water column anomalies on the upper slope described earlier.

The seismic reflection data in the mid-lower slope along the south-western margin of India show a remarkably prominent sub-bottom reflector at ~240 ms bsf between 1350 and 1650 m water depths, which closely mimics the sea-floor topography (Fig. 6b). At shallower depths in the east, however, this (supposed) BSR shows slight updoming and faulting. The high impedance contrast may be due to low-density gas horizons, but also to carbonate cementation or high-density sulphides (Behrens 1988).

Further seaward, in the rise area north of the Laccadive ridge complex, the BSR lies at 460 ms bsf at about 1950 m water depth, and is overlain by a prominent and widespread blanking zone, still characterized by weak reflections Fig. 6c). Towards the east it shows some discontinuity and also an upward tilt. The faintly visible sub-bottom reflectors above the BSR exhibit a marked change in their configuration. The blanking zone is occasionally occupied by numerous domal structures and faults. The domal structures are characterized by prominent acoustic voids, weak reflections and/or reflector terminations in the upper 250 ms of the profile. Chaotic and/or scattered hyperbolic reflectors occur below the BSR, which may suggest the presence of gas-charged sediments or free gas (Fig. 7).

Approximately 50–60 km south of this profile and at about 1650–1875 m water depth, the seismic profiles also show a distinct blanking zone underlain by a prominent reflector, most probably representing a BSR (Fig. 8a and b). The latter is again marked by some discontinuity and an upward tilt (Fig. 8b). The shallower reflector at about 260 ms bsf could possibly indicate a second BSR (Fig. 8a). The acoustically turbid zone (ATZ) below the deeper BSR may indicate the presence of gas-bearing horizons (Fig. 8a). Sea-floor folding, faulting, contorted sedimentary layers and pull-down structures mostly above the BSR seem to suggest the existence of probable pathways for upward migration of fluids/gas from the deep. Thus, a network of tectonic features may provide numerous channelways that feed gas to shallow levels where hydrate is formed from the deep (Dillon *et al.* 1993).

A plot of the sub-bottom depth of BSR horizons vs water depth (Fig. 9) reveals that there is some affinity between the currently studied BSRs and the ones reported off the Gulf of

Table 2. *Reported BSRs referred to in Fig. 9 (after Kvenvolden & Barnard 1983; Kvenvolden 1993)*

Designation on Fig. 9	Location	Water depth (s TWTT)	Sub-bottom depth of inferred base of gas hydrate (s TWTT)	Reference
AA′	Beafort Sea off Alaska	400–2500 m	100–800 m	Grantz *et al.* (1976)
BB′	Beringian Margin off Alaska (Continental slope)	500–2000 m	200–500 m	Marlow *et al.* (1981)
CC′	Pacific Ocean-Eel River Basin off California	800–1200 m	0.3 s	Field & Kvenvolden (1985)
DD′	Middle America Trench off Mexico	2.5–5.0 s	0.5–0.7 s	Shipley *et al.* (1979)
EE′	Middle America Trench off Nicaragua	800–2400 m	0.4–0.5 s	Shipley *et al.* (1979)
FF′	Middle America Trench off Costa Rica	1.0–1.8 s	0.2–0.5 s	Shipley *et al.* (1979)
GG′	Pacific Ocean off Panama	2.5–2.8 s	0.4–0.5 s	Shipley *et al.* (1979)
HH′	Western Gulf of Mexico off Mexico	1200–2000 m	0.5 s	Shipley *et al.* (1979) Hedberg (1980)
I	Northwest Indian Ocean, Makran margin, Gulf of Oman	3000 m	600–700 m	White (1979)
J	Andamans, Indian Ocean	1300 m	0.6 s	Chopra (1985)

Mexico, Panama, Costa Rica and the Blake Outer Ridge area (see also Table 2).

From the three main seismic characteristic features marking gas hydrate sediments, i.e. (a) a clear BSR; (b) a blanking zone above the BSR; and (c) a velocity inversion (Dillon *et al.* 1993; Lee *et al.* 1994), the first two typify the presently studied gas hydrate horizons. The BSR is thought to be caused by a reduction in sound velocity at the base of the methane gas hydrate stability zone (Shipley *et al.* 1979; Field & Kvenvolden 1985; Hyndman & Davis 1992; Hyndman & Spence 1992). The marked impedance contrast is believed to be the contact between sediments cemented with gas hydrates and underlying sediments having a lower velocity due to the absence of hydrate and the possible presence of free gas (Scholl *et al.* 1970; Ewing & Holister 1972; Hein *et al.* 1978; Shipley *et al.* 1979; Field & Kvenvolden 1985). The blanking is due to the reduction of the amplitude of seismic reflections caused by hydrate cementation (Dillon *et al.* 1993).

Source of hydrate methane

The western continental margin of India is biologically a highly productive margin. Organic matter deposited in such environments is diagenetically transformed by bacteria and/or heat and breaks down to methane and other organic compounds (von Rad *et al.* 1996). Organic matter which is considered to be the main source for hydrate methane (Kvenvolden, 1993), shows an average of 6–7% on the continental slope and 2% in the deep Arabian Sea sediments (Paropkari *et al.* 1987; Rao *et al.* pers. comm.). Stable carbon isotope data, $C_{organic}/N_{total}$ ratios and I/C organic ratios suggest that the organic matter in the sediments of the western margin appear to be of marine origin (Calvert *et al.* 1995). Therefore, methane might probably have formed at a sub-bottom depth of a few metres to hundreds of metres within the carbonate reduction–methane generation zone, below the bacterial sulphate reduction zone (von Rad *et al.* 1996). In addition to organic matter content, the production of biogenic gas is enhanced by rapid sedimentation. The slope and the sedimentary basins in the adjoining deep sea floor, characterized by fairly thick, organic carbon-rich sedimentary strata, serve as congenial sites for the formation and accumulation of hydrocarbon gases. However, the presence of acoustic voids, gas emission features and faults suggests that pathways prevail for the migration of gas to the upper horizons from the deep.

Summary

The present study reveals that the western continental margin of India is characterized by gas-charged sediments on the inner shelf, pockmarks and seeps on the mid-outer shelf upper slope, whereas distinct blanking zones and (supposed) BSRs occur in the mid-lower slope (shelf margin basin) and rise regions. The presence of pockmarks and seeps suggest that they arise from the venting of gas. The slope and rise areas of the margin, characterized by rapid sedimentation and organic carbon-rich sediments coupled with optimal lithostatic and hydrostatic pressures, favoured the formation of gas hydrates in the deeper horizons. However, structural elements such as folds, diapir-like features and faults present in the area may act as channelways and traps for upward-migrating methane gas from of thermogenic nature. Gas hydrate sample recovery from the study area and their carbon isotope analysis would help in ascertaining their origin.

We are grateful to E. Desa, Director of the National Institute of Oceanography, and R. R. Nair for their help and permission to publish this work. We also sincerely acknowledge J.-P. Henriet (RCMG, University of Gent, Belgium) for his constant help during M. Veerayya's visit to Gent to attend the First Master Workshop on Gas Hydrates. This is NIO contribution 2562.

References

AHARON, P., GRABER, E. & ROBERTS, H. 1992. Dissolved carbon and δ^{13}anomalies in the water column caused by hydrocarbon seaps on the northern Gulf of Mexico slope. *Geo-Marine Letters*, **12**, 33–40.

ANDERSON, A. & BRYANT, W. 1990. Gassy sediment occurrence and properties: northern Gulf of Mexico. *Geo-Marine Letters*, **10**, 209–220.

BASU, D., BANERJEE, A. & TAMHANE, D. 1982. Facies distribution and petroleum geology of the Bombay Offshore basin, India. *Journal of Petroleum Geology*, **5**, 51–75.

BEHRENS, E. W. 1988. Geology of a continental slope oil seep, northern gulf of Mexico. *AAPG Bulletin*, **72**, 105–114.

BISWAS, S. K. 1987. Regional tectonic framework, structure and evolution of the western marginal basins of India, *Tectonophysics*, **135**, 320–327.

BOUMA, A. H., ROBERTS, H. H. & COLEMAN, J. M. 1990. Acoustic and geological characteristics of near surface sediments, upper continental slope of northern gulf of Mexico. *Geo-Marine Letters*, **10**, 200–208.

BROOKS, J. M., FIELD, J. M. & KENNICUTT, M. C. II, 1991. Observations of gas hydrates in marine sediments, offshore northern California. *Marine Geology*, **96**, 103–109.

BROWN, K. M., BANGS, N. L., FROELICH, P. L. & KVENVOLDEN, K. A. 1996. The nature, distribution, and origin of gas hydrate in the Chile Triple junction region. *Earth and Planetary Science Letters*, **139**, 471–483.

CALVERT, S. E., PEDERSON, T. F., NAIDU, P. D. & VON STACKELBERG, U. 1995. On the organic carbon maximum on the continental slope of the eastern Arabian Sea. *Journal of Marine Research*, **53**, 269–296.

CARLSON, P. R., GOLAN-BAC, M., KARL, H. A. & KVENVOLDEN, K. A. 1985. Seismic and geochemical evidence for shallow gas in sediments on Navarin continental margin, Bering Sea. *AAPG Bulletin*, **69**, 422–436.

CHOPRA, N. N. 1985. Gas hydrate-an unconventional trap in forearc region of Andaman offshore. *Bulletin of Oil and Natural Gas Commission, India*, **22**, 41–54.

CLAYPOOL, G. W. & KAPLAN, I. R. 1974. The origin and distribution of methane in marine sediments, *In:* KAPLAN, I. R. (ed.) *Natural Gases in Marine Sediments*. Plenum, New York, 99–139.

—— & KVENVOLDEN, K. A. 1983. Methane and other hydrocarbon gases in marine sediments. *Annual Review of Earth & Planetary Sciences*, **11**, 299–327.

DAVIS, A. M. 1992. Shallow gas: an overview. *Continental Shelf Research*, **12**, 1077–1079.

DICKENS, G. R., PAULL, C. K., WALLACE, P. & THE ODP LEG 164 SCIENTIFIC PARTY, 1997. Direct measurement of in situ methane quantities in a large gas-hydrate reservoir. *Nature*, **385**, 426–429.

DILLON, W. P., LEE, M. W. & COLEMAN, D. F. 1994. Identification of marine hydrates in situ and their distribution off the Atlantic coast of the United States. *Annals of the New York Academy of Sciences*, **715**, 364–380.

DILLON, W., LEE, M., FEHLHABER, K. & COLEMAN, D. F. 1993. Gas hydrates on the Atlantic continental margin of the United States – Controls on Concentration. *US Geological Survey Professional Paper*, **1570**, 313–330.

EWING, J. & HOLLISTER, C. 1972. Regional aspects of deep sea drilling in the western North Atlantic. *In*: *Initial Reports of the Deep Sea Drilling Project*, Volume 11. US Government Printing Office, Washington, DC, 951–973.

FADER, G. 1991. Gas-related sedimentary features from the eastern Canadian shelf. *Continental Shelf Research*, **11**, 1123–1153.

FIELD, M. & KVENVOLDEN, K. 1985. Gas hydrates on the northern California continental margin. *Geology*, **13**, 517–520.

GRANTZ, A., BOUCHER, G. & WHITNEY, O. T. 1976. Possible solid gas hydrate and natural gas deposits beneath the continental slope of the Beaufort Sea. U. S. Geological Survey Circular, 733.

HASIOTIS, T., PAPATHEODOROU, G., KASTANOS, N. & FERENTINOS, G. 1996. A pockmark field in the

Patras gulf (Greece) and its activation during the 14/7/93 seismic event. *Marine Geology*, **130**, 333–344.

HEDBERG, H. D. 1980. Methane generation and petroleum migration. *In*: ROBERTS, W.H. III & CORDELL, R.J. (eds) *Problems of Petroleum Migration*. American Association of Petroleum Geologists, Studies in Geology, **10**, 179–206.

HEIN, J., SCHOLL, D., BARRON, J., JONES, M. & MILLER, J. 1978. Diagenesis of late Cenozoic diatomaceous deposits and formation of bottom simulating reflector in the southern Bering Sea. *Sedimentology*, **25**, 155–181.

HENRIET, J.-P. & MIENERT, J. 1996. *First Master Workshop on Gas hydrates: Relevance to World Margin Stability & Climatic Change*, University of Gent, Belgium.

HOVLAND, M. & JUDD, A. 1988. *Seabed Pockmarks and Seepages: Impact on Geology, Biology and the Marine Environment*. Graham & Trotman, London, 180–203.

HYNDMAN, R. & DAVIS, E. E. 1992. A mechanism for the formation of methane hydrate and seafloor bottom-simulating reflectors by vertical fluid expulsion. *Journal of Geophysical Research*, **97**, 7025–7041.

—— & SPENCE, G. 1992. A seismic study of methane hydrate and seafloor bottom simulation reflectors by vertical fluid expulsion. *Journal of Geophysical Research*, **97**, 6683–6698.

KAPLAN, I. 1974. *Natural Gases in Marine Sediments*. Plenum, New York, 1–10.

KARISIDDAIAH, S. & VEERAYYA, M. 1994. Methane-bearing sediments in the eastern Arabian Sea: a probable source for greenhouse gas. *Continental Shelf Research*, **14**, 1361–1370.

—— & —— 1996. Potential distribution of subsurface methane in the sediments of the eastern Arabian Sea and its possible implications. *Journal of Geophysical Research*, **101**, 25, 887–25, 895.

——, ——, VORA, K. & WAGLE, B. 1993. Gas-charged sediments in the inner continental shelf off western India. *Marine Geology*, **110**, 143–152.

KELLEY, J., DICKSON, S., BELKNAP, D., BARNHARDT, W. & HENDERSON, M. 1994. Giant sea-bed pockmarks: evidence for gas escape from Belfast Bay, Maine. *Geology*, **22**, 59–62.

KVENVOLDEN, K. 1988. Methane hydrate: A major reservoir of carbon in the shallow geosphere? *Chemical Geology*, **71**, 41–51.

—— 1993. Gas hydrates – geological perspective and global change. *Reviews of Geophysics*, **31**, 173–187.

—— & BARNARD, L. 1983. Hydrates of natural gas in continental margins *Studies in Continental Margin Geology*. American Association of Petroleum Geologists Memoir, **34**, 631–640.

LEE, M., HUTCHINSON, D., AGENA, W. *et al.* 1994. Seismic character of gas hydrates on the south-eastern U.S. continental margin. *Marine Geophysical Research*, **16**, 163–184.

——, ——, COLLETT, T. & DILLON, W. 1996. Seismic velocities for hydrate-bearing sediments using weighted equation. *Journal of Geophysical Research*, **101**, 20,347–20,358.

LI, T. & JIN, B. 1989. Seismic facies analysis of the seafloor instabilities in the Pearl River Mouth region. *Marine Geotechnology*, **8**, 19–31.

MACDONALD, I. , GUINASSO, JR, N., SASSEN, , BROOKS, J., LEE, L. & SCOTT, K. 1994. Gas hydrate that breaches the seafloor on the continental slope of the gulf of Mexico. *Geology*, **22**, 699–702.

MACKAY, M., JARRARD, R., WESTBROOK, G., HYNDMAN, R. & SHIPBOARD SCIENTIFIC PARTY OF ODP LEG 146. 1994. Origin of bottom-simulating reflectors: Geophysical evidence from the Cascadia accretionary prism. *Geology*, **22**, 459–462.

MARKL, R., BRYAN, G. & EWING, J. 1970. Structure of the Blake–Bahama Outer Ridge. *Journal of Geophysical Research*, **75**, 4539–4555.

MARLOW, M., CARLSON, P., COOPER, A. *et al.* 1981. *Hydrocarbon Resource Report for Proposed OCS Sale No. 83, Navarin Basin, Alaska*. US. Geological Survey Open File Report, **81-252**, 83.

MARTENS, C. & KLUMP, J. 1980. Biogeochemical cycling in an organic rich coastal marine basins. 1. Methane sediment-water exchange processes. *Geochimica Cosmochimica Acta*, **44**, 471–490.

MILLER, J., LEE, M. & VON HUENE, R. 1991. An analysis of a seismic reflection from the base of a gas hydrate zone, offshore Peru. *AAPG Bulletin*, **75**, 910–924.

MINSHULL, T., WHITE, R., BARTON, P. & COLLIER, J. 1992. Deformation at plate boundaries around the Gulf of Oman. *Marine Geology*, **104**, 265–277.

NAINI, B. & TALWANI, M. 1982. Structural framework and evolutionary history of the continental margin of western India. American Association of Petroleum Geologists Memoir, **34**, 167–191.

NAMBIAR, A., RAJAGOPALAN, G. & RAO, B. 1991. Radiocarbon dates of sediment cores from inner continental shelf off Karwar, west coast of India. *Current Science*, **61**, 353–354.

PAROPKARI, A., RAO, C. & MURTHY, P. 1987. Environmental controls on the distribution of organic matter in recent sediments of the western continental margin of India. *In*: KUMAR, R. *et al.* (eds) *Petroleum Geochemistry and Exploration in the Afro-Asian Regions*. Balkema, Rotterdam, 347–361.

PAULL, C. & DILLON, W. 1981. Appearance and Distribution of the Gas Hydrate Reflection in the Blake Ridge Region, Offshore Southeastern United States. *US Geological Survey Miscellaneous Field Studies Map*, **MF-1252**.

——, USSLER, W., III, BOROWSKI, W. & SPIESS, F. 1995. Methane-rich plumes on the Carolina continental rise: associations with gas hydrates. *Geology*, **23**, 89–92.

PECHER, L., MINSHULL, T., SINGH, S. & VON HUENE, R. 1996. Velocity structure of a bottom simulating reflector offshore Peru: Results from full waveform inversion. *Earth and Planetary Science Letters*, **139**, 459–469.

RAJU, A., SINHA, R., RAMAKRISHNA, M., BISHT, H. & NASHIPUDI, V. 1981. Structure, tectonics and hydrocarbon prospects of Kerala–Laccadive basin. *In:* PRASADA RAO, R. (ed.) *Workshop on*

Geological Interpretation of Geophysical Data. Oil and Natural Gas Commission, Debradun, India, 123–127.

RAO, B. , VEERAYYA, M. & RAO, C. Submitted. Influence of marginal highs on the accumulation of high organic carbon along the continental slope off western India. *Deep-Sea Research*.

RAO, V., VEERAYYA, M., NAIR, R., DUPEUBLE, P. & LAMBOY, M. 1994. Late Quaternary Halirneda bioherms and aragonitic fecal pellet-dominated sediments on the carbonate platform of the western continental shelf of India. *Marine Geology*, **121**, 293–315.

SCHOLL, D., CHRISTENSEN, M., VON HUENE, R. & MARLOW, M., 1970. Peru–Chile trench sediments and seafloor spreading. *Geological Society of America Bulletin*, **81**, 1339–1360.

SCHOELL, M. 1988. Multiple origins of methane in the Earth. *Chemical Geology*, **71**, 1–11.

SCHUBEL, J. 1974. Gas bubbles and the acoustically impenetrable, or turbid character of some estuarine sediments. *In:* KAPLAN, I. (ed.) *Natural Gases in Marine Sediments*. Plenum, New York, 275–298.

SHIPLEY, T., HOUSTON, M., BUFLER, R. *et al.* 1979. Seismic evidence for widespread possible gas hydrate horizons on continental slopes and rises. *AAPG Bulletin*, **63**, 2204–2213.

SLOAN, E. D., JR. 1990. *Clathrate Hydrates of Natural Gases*. Marcel Dekker, New York.

——. 1996. Chemical properties of gas hydrates and application to world margin stability and climatic change. *In*: HENRIET, J.-P. & MIENERT, J. (eds) *First Master Workshop on Gas Hydrates: Relevance to World Margin Stability and Climatic Change*. University of Gent, Belgium, 1–38.

SWEENEY, R. E. 1988. Petroleum-related hydrocarbon seepage in a recent North Sea sediment. *Chemical Geology*, **71**, 53–64.

TREHU, A., LIN, G., MAXWELL, E. & GOLDFINGER, C. 1995. A seismic reflection profile across the Cascadia subduction zone offshore central Oregon: New constraints on methane distribution and crustal structure. *Journal of Geophysical Research*, **100**, 15, 101–15, 116.

TUCHOLKE, B., BRYAN, G. & EWING, L. 1977. Gas hydrate horizons detected in seismic-profiler data from the western North Atlantic. *AAPG Bulletin*, **61**, 698–707.

UCHUPI, E., SWILL, S. & ROSS, D. 1996. Gas venting and Late Quaternary sedimentation in the Persian (Arabian) gulf. *Marine Geology*, **129**, 237–269.

VEERAYYA, M., WAGLE, B., VORA, K., KARISIDDAIAH, S. & ALMEIDA, F. 1991. Geomorphology and shallow structure of the western continental margin of India. (Abstract). *In: International Symposium on Oceanography of Indian Ocean*. National Institute of Oceanography, Goa, India, 51–52.

——, KARISIDDAIAH, S., VORA, K., WAGLE, B. & AIMEIDA, F. 1996. Detection of gas-charged sediments and gas hydrate horizons in high-resolution seismic profiles from the western continental shelf of India. *In:* HENRIET, J.-P. & MIENERT, J. (eds) *First Master Workshop Gas hydrates: Relevance to World Margins Stability and Climatic Change*, University of Gent, Belgium, 44.

VON RAD, U., ROSCH, H., BERNER, U. *et al.* 1996. Authigenic carbonates derived from oxidized methane vented from the Makran accretionary prism off Pakistan. *Marine Geology*, **136**, 55–77.

VORA, K., WAGLE, B., VEERAYYA, M., ALMEIDA, F. & KARISIDDAIAH, S. M. 1996. 1300 km long Pleistocene–Holocene shelf edge barrier reef system along the western continental shelf of India: occurrence and significance. *Marine Geology*, **134**, 145–162.

WATKINS, J. S. & WORZEL, J. L. 1978. Serendipity gas seep area, south Texas offshore. *AAPG Bulletin*, 62, 1067–1074.

WHITE, R. S. 1977. Seismic bright spots in the Gulf of Oman. *Earth and Planetary Science Letters*, **37**, 29–37.

—— 1979. Gas hydrate layers trapping free gas in the Gulf of Oman. *Earth and Planetary Science Letters*, **42**, 114–120.

—— & KLITGARD, K. 1976. Sediment deformation and plate tectonics in the Gulf of Oman. *Earth and Planetary Science Letters*, **32**, 199–209.

WHITICAR, M. J. 1990. A geochemical perspective of natural gas and atmospheric methane. *Organic Geochemistry*, **16**, 531–547.

WHITING, B., KAMER, G. & DRISCOLL, N. 1994. Flexural and stratigraphic development of the western Indian continental margin. *Journal of Geophysical Research*, **99**, 13, 791–13, 811.

WHITMARSH, R., WESER, O. & ROSS, D. *et al.* 1974. *Initial Reports of the Deep Sea Drilling Project*, Volume 23. US Government Printing Office, Washington, DC, 1180.

YUAN, T., HYNDMAN, R., SPENCE, G. & DESMONS, B. 1996. Seismic velocity increase and deep-sea gas hydrate concentration above a bottom-simulating reflector on the northern Cascadia continental slope. *Journal of Geophysical Research*, **101**, 13, 655–13, 671.

ZWART, G., MOORE, J. & COCHRANE, G. 1996. Variations in temperature gradients identify active faults in the Oregan accretionary prism. *Earth and Planetary Science Letters*, **139**, 485–495.

Reflection characteristics, depth and geographical distribution of bottom simulating reflectors within the accretionary wedge of Sulawesi

S. NEBEN, K. HINZ & H. BEIERSDORF

Bundesanstalt für Geowissenschaften und Rohstoffe (BGR), Stilleweg 2, D-30655 Hannover, Germany

Abstract: Several reflection seismic profiles across the subduction zone north of Sulawesi show strong bottom simulating reflectors (BSRs) below the lower and upper slopes of the accretionary wedge. The geographical distribution, depth and the reflection characteristics of the BSRs have been described in detail. Along one reflection seismic profile the depth of BSR was correlated with heat flow data. These correlations indicate that at locations, where the heat flow is high and the BSR interrupted, active venting of methane may occur at the sea floor. The occurrence of BSRs is limited to the central part of the North Sulawesi subduction zone (between 121°30′E and 123°30′E). In the westernmost part of the surveyed area only short BSR segments are found, which may be a result of slope instability and slumping of sediments. On the easternmost profile, no bottom simulating reflectors are found at all.

The German–Indonesian project GIGICS (German–Indonesian Geoscientific Investigations in the Celebes Sea) aimed to investigate of the sub-bottom of the Celebes Sea with regard to its age, internal structure and tectonic evolution. Based on data of the former R/V *Sonne* cruise SO49 (Hinz *et al.* 1991), additional geophysical and geological data were collected during the R/V *Sonne* cruise SO98 (Fig. 1). On the first two legs of the cruise over 3000 km of multi-channel reflection seismic data, together with over 6700 km of gravity, magnetics and swath bathymetry data, were acquired.

On the southernmost profiles across the subduction zone north of Sulawesi strong bottom simulating reflectors (BSRs) in the lower and upper slopes of the accretionary wedge were found. They gave rise to heat flow measurements and a special geochemical sampling programme during the third and the fourth legs of the cruise because the BSRs were thought to represent the lower boundary of frozen gas hydrates.

Acquisition and seismic data processing

The seismic data were acquired using BGRs reflection seismic equipment consisting of a 48-channel streamer with a total active length of 2400 m and two tuned airgun arrays of 10 guns each. The total volume of the arrays was 51.21 and the operating pressure was 13.5 MPa. For recording of the seismic data a SYNTRAK 480 reflection seismic acquisition system was used.

To process over 3000 km of multi-channel reflection seismic data, a standard data processing sequence was applied with special emphasis on the velocity determination in the area of the subduction zone off North Sulawesi.

After computation of the recording geometry and correlation between shotpoints and navigation data, the data were sorted into CMP (Common Mid Point) gathers. This resulted in a 2400% coverage for each reflection element. The shotpoint interval was 50 m resulting in a CMP spacing of approximately 25 m. A very dense spacing for the velocity analysis was applied. Instead of performing velocity determinations equidistant along the seismic lines, as usual in standard seismic processing, the locations for the analysis were adapted to the topography of the ocean bottom and the igneous basement, respectively – the stronger the undulations the denser the spacing of the velocity analysis. The optimum stacking velocity was determined interactively using velocity spectra based on the semblance method (Taner & Köhler 1969). From the stacking velocities the interval velocities were determined using the Dix algorithm (Dix 1955). Deconvolution processes were applied to the data before and after stacking in order to reduce reverberations and to suppress peg-leg multiples.

After several tests with different migration methods, a (ω–x) migration (Claerbout 1985; Lee & Suh 1985) was used for time migration of the seismic data.

Tectonic setting

The Celebes Sea is situated at the convergence centre of three major crustal plates: the Indo-Australian plate, the Eurasian plate and the Paci-

Neben, S., Hinz, K. & Beiersdorf, H. 1998. Reflection characteristics, depth and geographical distribution of bottom simulating reflectors within the accretionary wedge of Sulawesi. *In:* Henriet, J.-P. & Mienert, J. (eds) *Gas Hydrates: Relevance to World Margin Stability and Climate Change*. Geological Society, London, Special Publications, **137**, 255–265.

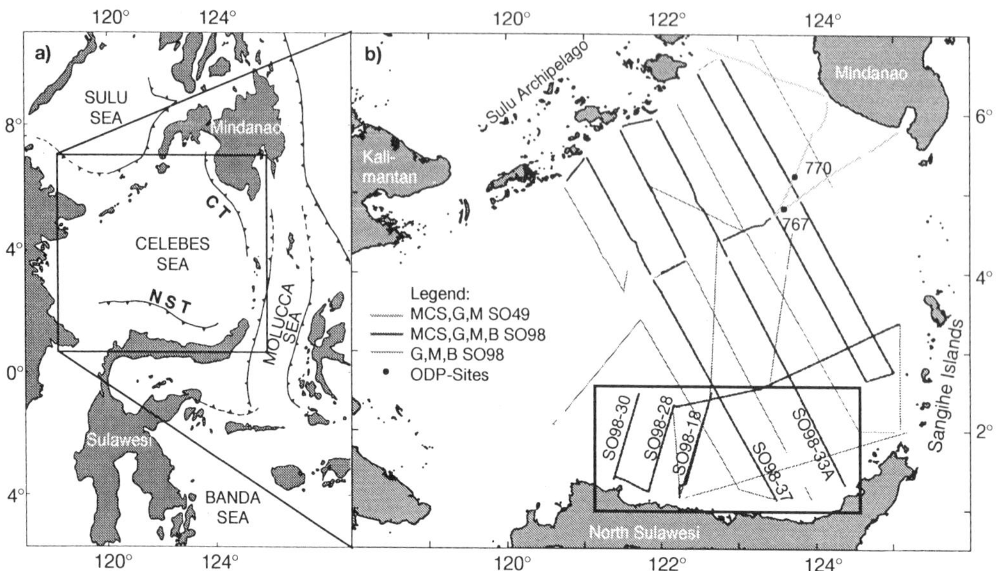

Fig. 1. (**a**) The Celebes Sea, the surrounding oceanic micro-plates and the thrust fronts of the different subduction zones after Hamilton (1979). CT, Cotabato Trench; NST, North Sulawesi Trench. (**b**) Location of the combined geophysical profiles of cruises SO49 and SO98, and position of ODP sites 767 and 770 in the Celebes Sea. MCS, multi-channel reflection seismic; G, gravity; M, magnetics; B, bathymetry. The box marks the study area over the accretionary wedge of North Sulawesi.

fic plate (including the Philippine plate) moving towards each other. This in turn results in a complex, but mainly compressional, stress regime (Hamilton 1976).

The Celebes Sea covers an area of approximately 270 000 km^2 (Fig. 1a). It is surrounded by Kalimantan in the west, the Sulu Archipelago in the north, the island of Mindanao in the northeast, the Sangihe Islands in the east and Sulawesi in the south (Fig. 1a and b). From early refraction seismic measurements (Murauchi *et al.* 1973) and the identification of magnetic lineaments (Weissel 1980; Lee & McCabe 1986) it was suggested that the Celebes Sea is underlain by oceanic crust. This was confirmed by the results of the ODP sites 767 and 770 (see Fig. 1b) (Rangin *et al.* 1990). From the drilling results, a middle–late Eocene age was determined for the basement of the Celebes Sea. Ar–Ar datings by the Bundesanstalt für Geowissenschaften und Rohstoffe, Hannover (BGR) gave an age 40.8 ± 0.3 Ma for a basalt dredged from an abyssal hill in the southern Celebes Sea (Beiersdorf *et al.* 1997).

The compressional tectonic regime is demonstrated by the presence of several subduction zones bordering the Celebes Sea and the adjacent oceanic basins (e.g. Sulu Sea, Molucca Sea and Banda Sea, Fig. 1a). The oceanic crust of the Celebes Sea is subducted under Mindanao at the Cotabato Trench (CT in Fig. 1a) and under Sulawesi at the North Sulawesi Trench (NST in Fig. 1a). Here, the deepest water depths of the Celebes Sea, of more than 5500 m, are observed.

Reflection characteristics of BSRs off North Sulawesi

Our reflection seismic data from the subduction zone off North Sulawesi show a number of reflections parallel to the sea floor, which cross-cut bedding interfaces and are characterized by strong negative amplitudes. These properties define BSRs as described by numerous authors (Shipley *et al.* 1979; Biju-Duval *et al.* 1982; Katzman *et al.* 1994; MacKay *et al.* 1994; Wood *et al.* 1994; Andreassen *et al.* 1995). Typical examples of reflection seismic profiles with BSR are shown and discussed.

In contrast to sea-floor reflectors, BSRs show reflections with a negative polarity (Fig. 2). In this figure a BSR lies approximately 0.4 s TWTT below the sea floor (bsf). The phase shift of 180° is caused by a decrease of the interval velocity below the BSR forming a strong negative contrast in the acoustic impedance (MacLeod 1982). This impedance contrast can

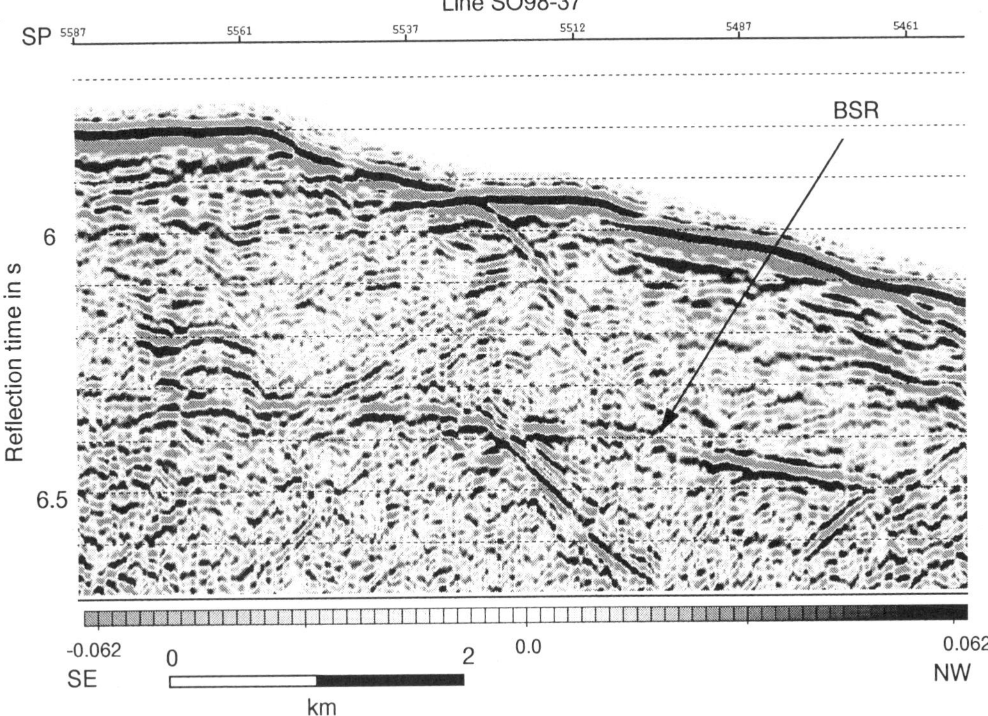

Fig. 2. Seismogram section (shots 5452–5587) of reflection seismic profile SO98-37. The strong negatively reflective horizon parallel to and approximately 0.4 s TWTT bsf is identified as BSR (SP, shotpoint).

be caused by the presence of free gas in the pore space of the sediments. Only a few per cent of free gas in the pore space of the sediments can cause the P-wave velocity to decrease well below the sea-water velocity (Gregory 1978; Gimpel 1987). For data from the Cascadia subduction zone, Yuan *et al.* (1996) show that the strong impedance contrast is not only caused by free gas below the BSR, but that the gradual velocity increase in the gas hydrate stability zone has also to be taken into account. A very dense grid of velocity analysis was applied to the multi-channel reflection seismic lines over the subduction zone north of Sulawesi. We derived average velocities of approximately 1900 m s^{-1} for in the sediments between the ocean bottom and the BSR. In the undeformed sedimentary sequence immediately seaward of the deformation front the average velocity for a similar depth below the sea floor is between 1650 and 1700 m s^{-1}. Such an increase in interval velocities was also found by Yuan *et al.* (1994, 1996) for the layer above the BSR of the Cascadia subduction zone.

In our analyses we found only rare indications of decreases in the interval velocities below the BSR. This most probably is due to two reasons:

- the target depths were too large and, hence, the move-out of our measuring configuration was too small (minimum water depth with BSR on line SO98–28: 1500 m (Fig. 5a), maximum offset on this line: 1400 m); and/or
- the thickness of the layer containing the free gas was too small to be seismically resolved.

Andreassen *et al.* (1995) report thicknesses of between 11 and 30 m for the free gas layer in the Beaufort Sea, and Yuan *et al.* (1996) report thicknesses of approximately 20 m for the Cascadia continental slope.

The most distinctive reflection characteristic which allows the recognition of BSRs in reflection seismograms is the cross-cutting of bedding or stratigraphical interfaces. A seismic example is shown in Fig. 3. Here, shots 1009–1142 of line SO98–18 are displayed. Again, the BSR is shown as a strong negative reflection paralleling the sea-floor reflector approximately 0.45 s TWTT below it. The sediments in this part of the subduction zone were strongly folded and upthrusted by the accretionary processes (Fig. 3). The BSR cross-cuts these deformed sediment sequences. This behaviour shows that the BSR

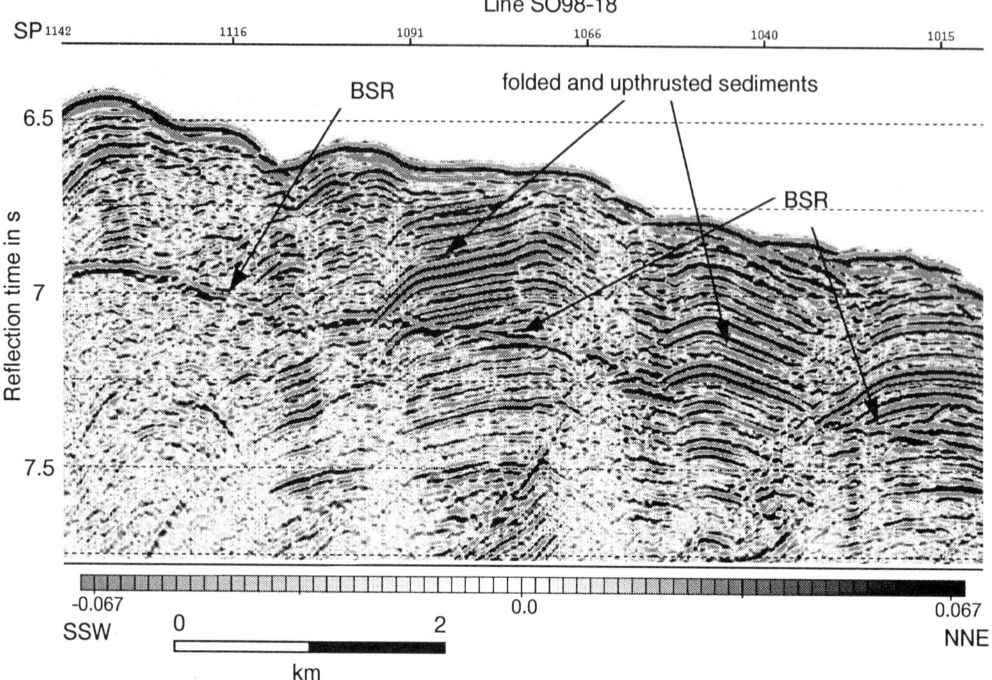

Fig. 3. Seismogram section (shots 1009–1142) of reflection seismic profile SO98-18. Here, the BSR parallels the sea floor and cuts through upthrusted and folded reflection sequences revealing its nature as a physical but not a stratigraphic boundary (SP, shotpoint).

resembles a physical boundary whose depth depends solely on pressure and temperature, and that it is neither a lithological nor stratigraphical interface (MacLeod 1982).

The BSRs north of Sulawesi are often interrupted by landward-dipping reflectors. These reflectors are interpreted as thrust faults in the accreted sediments. A seismic example from line SO98–28 is shown in Fig. 4. Here, the BSR is interrupted at shotpoints 3792 and 3814 by two thrusts in depths of about 5.1 s and 5.3 s TWTT. Heat flow measurements, bottom-water sampling and sediment coring were carried out along this profile. At some locations, where thrusts intersect with the sea floor especially at the lower slope and the toe of the accretionary prism, the measured heat flow values in the uppermost sediments of 75–97 mW m^{-2} were very high in comparison to those on the upper slope of between 26 and 42 mW m^{-2} (Delisle *et al.*, 1998). Additionally, the sediment samples taken from the lower slope and in the vicinity of the deformation front show very high methane concentrations of between 34 000 and 49 000 ppb (Beiersdorf *et al.* 1997). We interpret the absence of BSRs at the thrust faults of the lower slope as a consequence of a loss of free methane through venting along these thrusts. Permeability along the thrust planes may be enhanced, and methane can be mobilized and migrate along these planes towards the ocean bottom.

Distribution of BSRs within the accretionary prism north of Sulawesi

An overview on the internal structure of the accretionary prism north of Sulawesi and the distribution of BSR based on seismic profiles is shown in Fig. 5a and b. Here, the line drawings from the migrated seismic sections of the five lines crossing the subduction zone are presented as fence diagrams. The lateral extent of the BSRs is indicated by grey bars and the position of the seismic sections described in the text are outlined with boxes. Two structural units can be seen at the continental slope:

- an accretionary complex with numerous thrust sheets;
- a sedimentary forearc basin (Fig. 5a and b).

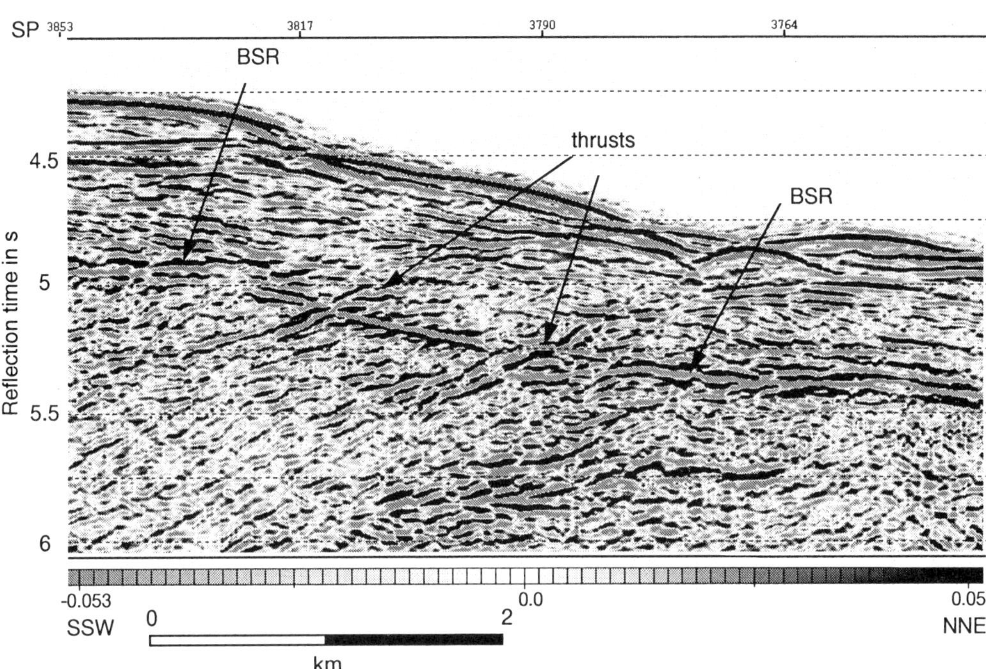

Fig. 4. Seismogram section (shots 3743–3853) of reflection seismic profile SO98-28. At two locations (shotpoints 3792 and 3814) the continuity of the BSR is interrupted by upgoing reflections which are interpreted as thrust folds originating from the accretionary processes (SP, shotpoint).

Based on slope angles the accretionary complex can be further sub-divided into a lower slope unit and an upper slope unit. The angles of the lower slope decrease from east ($\sim 12^\circ$) to west ($\sim 4^\circ$), those of the upper slope from $\sim 4^\circ$ in the east to $\sim 2^\circ$ in the west. The boundary between lower and upper slope unit is between 3000 and 4000 m water depth, and the boundary between upper slope and forearc basin lies between 2000 and 2500 m water depth.

The total width of the accretionary prism is approximately 25 km in the eastern part (line SO98-33A, Fig. 5b), and increases westward to about 85 km (line SO98-30, Fig. 5a). Profile SO98-37 (see Figs 1b and 5b) crosses the subduction zone obliquely, hence giving the impression of a broadening accretionary complex. The compressional style of the accretionary complex varies. The number of thrust sheets increases westward.

The areal distribution and the depth of the BSR below the sea floor is shown in Fig. 6. The BSR sub-bottom depth contours were derived by freehand interpolation between the seismic lines assuming constant pressure–temperature conditions along strike.

Like the width and compressional style of the accretionary prism the areal distribution of the BSR varies from east to west. The sea-floor area underlain by BSR increases to the west. On the easternmost profile SO98-33A (Figs 5b and 6) no BSRs were detected. On profile SO98-37 the water depth interval underlain by BSR is between 5200 and 4200 m. On line SO98-18 (Figs 5b and 6) the water depth interval is between 5200 and 3200 m. The maximum lateral extent was observed on line SO98-28 (Figs 5a and 6) where the interval underlain by BSR covers water depths between 5200 and 1500 m.

This westward increasing water depth interval underlain by BSRs correlates with a decreasing inclination of the lower and upper slope (Fig. 5a and b). From this we suggest that a critical angle of inclination exists above which there is no accumulation of free gas under the gas hydrate stability zone. One reason could be that thrusting is intense and free gas escapes along thrust faults to the sea floor. An alternative explanation is that there is gas migration to higher parts of the accretionary wedge. The existence of gas hydrates with no BSRs was proven by ODP drilling at the Blake Ridge (Leg 164,

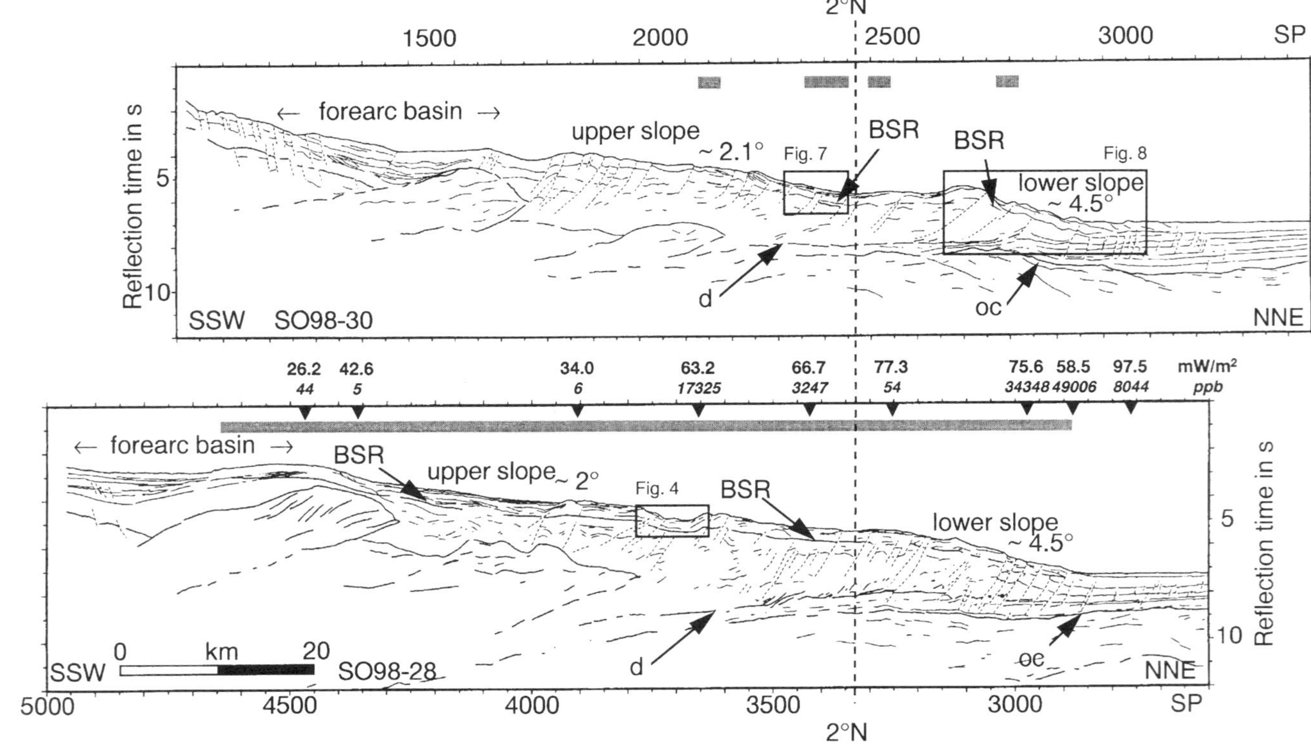

Fig. 5. (**a**) Line drawings of multi-channel reflection seismic profiles SO98-30 and SO98-28 across the subduction north of Sulawesi. The lateral extent of the BSRs is shown by the grey bars and the angles of the lower and upper slopes are indicated; oc, top of igneous oceanic crust of the Celebes Sea; d, décollement. Black triangles denote the positions of heat flow and methane concentrations measurements along profile SO98-28, the values are shown over the stations (heat flow in mW m^{-2}, methane concentrations in ppb). For a better orientation, the position of the 2°N latitude is indicated by the stippled line. For locations of the profiles see Fig. 1. (**b**) Line drawings of multi-channel reflection seismic profiles SO98-18, SO98-37 and SO98-33A across the subduction zone of North Sulawesi. The lateral extent of the BSRs (BSR) is shown by the grey bars in the diagrams, the angles of the lower and upper slopes are indicated; oc, top of igneous oceanic crust of the Celebes Sea; d, décollement. For a better orientation, the position of the 2°N latitude is indicated by the stippled line. For locations of the profiles see Fig. 1.

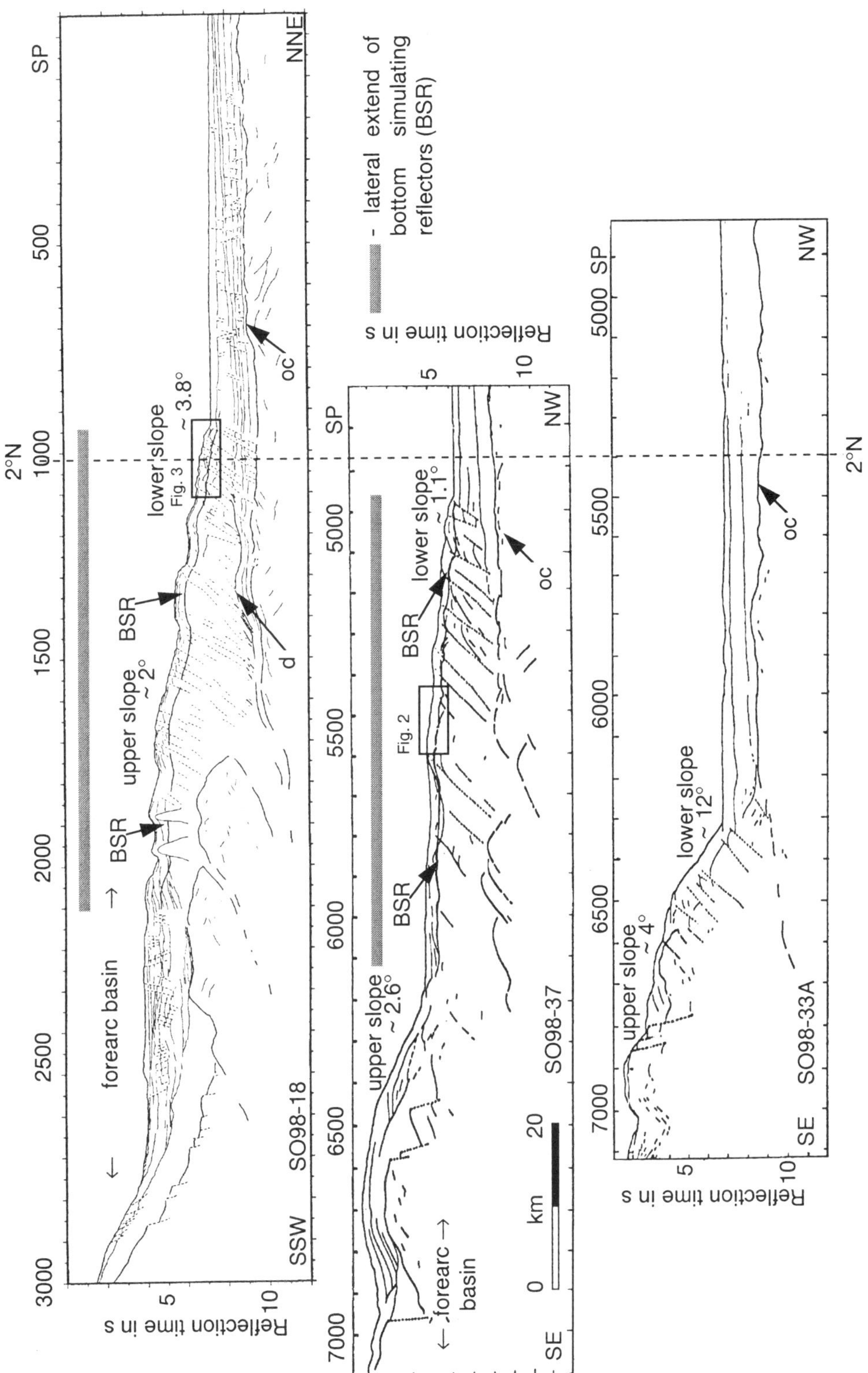
SP
500
1000
2°N
1500
2000
2500
3000
5
10
Reflection time in s
NNE
SSW
SO98-18
lower slope
~ 3.8°
Fig. 3
oc
BSR
d
upper slope
~ 2°
forearc basin
- lateral extend of bottom simulating reflectors (BSR)
SO98-37
5000
5500
6000
6500
7000
lower slope
~ 1.1°
Fig. 2
upper slope
~ 2.6°
forearc basin
0
km
20
NW
SE
SO98-33A
lower slope
~ 12°
upper slope
~ 4°

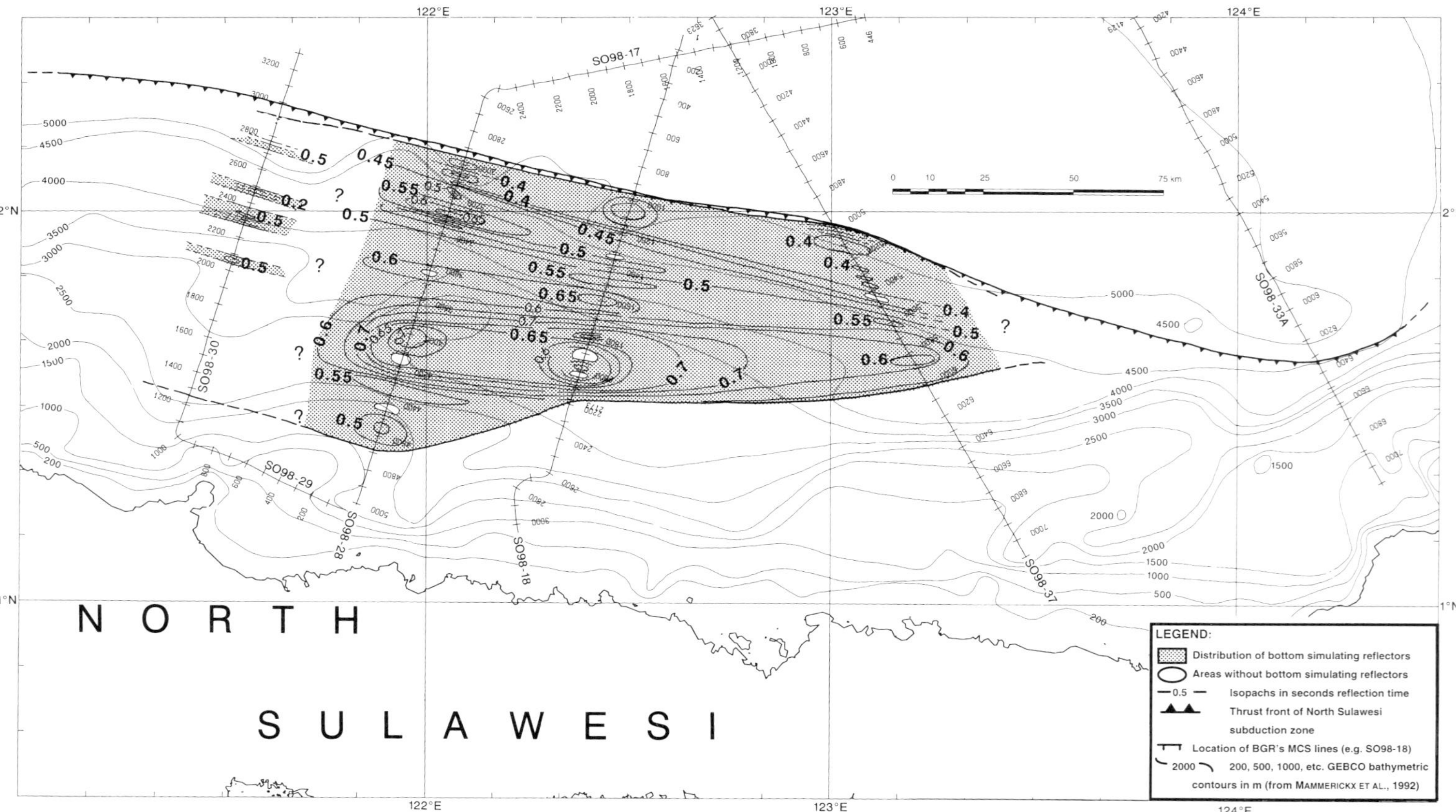

Fig. 6. Distribution of BSRs (shaded area) and depth below sea floor (in s TWTT) derived from migrated multi-channel reflection seismic data of the R/V *Sonne* cruise SO98 across the subduction zone north of Sulawesi.

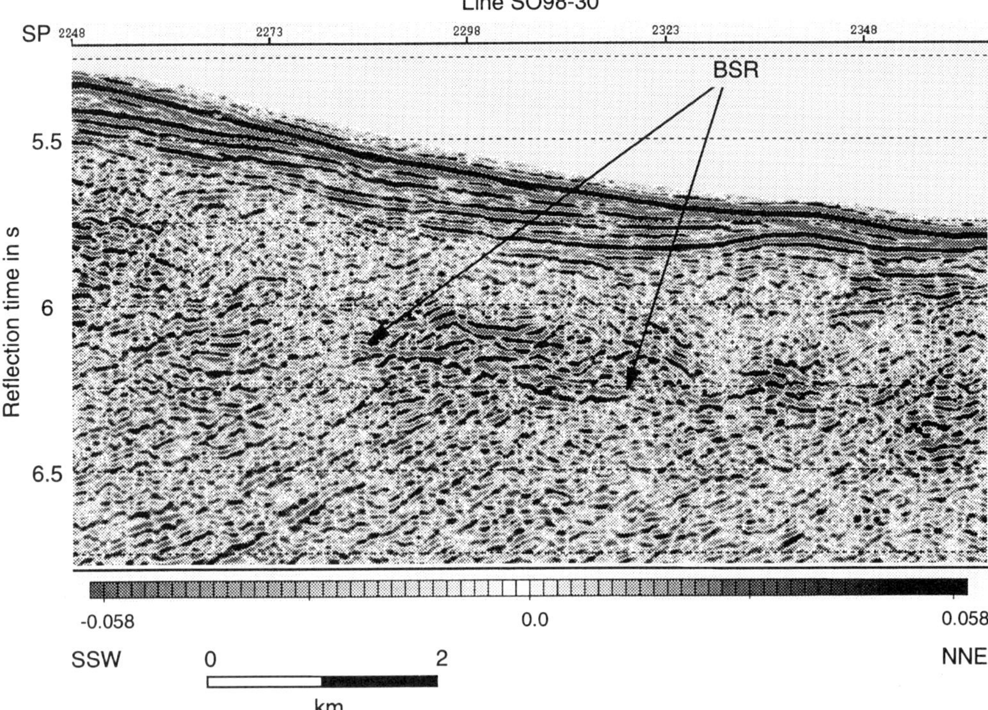

Fig. 7. Seismogram section (shots 2248–2364) of reflection seismic profile SO98-30. A strong negative reflective event parallel to the sea floor and at approximately 0.6 s TWTT bsf could resemble a BSR. The lateral extent (shotpoints 2285–2317) of this reflector is only 1.6 km (SP, shotpoint).

sites 994, 995 and 997, Paull *et al.* 1996). There it was found that the free gas migrated from the flanks of the Blake Ridge to the summit and accumulated there.

Although morphology and slope inclinations resemble those of profiles SO98–28 and SO98–18, the westernmost profile SO98–30 (Figs 1b and 5a) shows a peculiarity regarding the presence of BSRs (Fig. 6). On line SO98–30 indications for BSRs are found for only very short distances of 1–2 km (Fig. 7). The section (shots 2248–2364) shows a negative reflection in a sub-bottom depth of 0.6 s TWTT. The interval velocity as derived from the rms-velocities using the Dix formula (Dix 1955) for the interval between the sea floor and this reflector results to 1850 m s^{-1} but this horizon can be recognized only for 1.6 km.

Another reflection seismic example is shown in Fig. 8. Here, a section over 25 km long covers the lower slope of the accretionary prism. BSRs were found on this section only at two small places (Fig. 8). In front of the deformation front and on the slope the uppermost sediments show no or only chaotic internal reflection patterns. The sediment masses in front of the lower slope make it difficult to locate the position of the deformation front. The reflection pattern is interpreted as slump scars and slump deposits. The initiation of the slumps could be caused by slope instabilities, which would affect the gas hydrate layer first. Such slope instabilities can be caused by slope oversteepening as a consequence of lateral tectonic compression. McIver (1982) suggested earthquakes as one triggering mechanism. The southern Celebes Sea is an earthquake-prone region (Hamilton 1979). A possible explanation for the discontinuous BSR is that the slumping activity does not provide enough time with a stable sediment structure for a BSR to develop.

The minimum area underlain by gas hydrates as indicated by the BSRs is about 8200 km^2 if only the area between profiles SO98-37 and SO98-28 is considered. Yuan *et al.* (1996) estimated that the total volume of methane per m^2 of sea floor for the gas hydrate field at the Cascadia continental slope amounts to 800 m^3. They show that a similar volume per m^2 can be deduced from the data of the Blake–Bahama

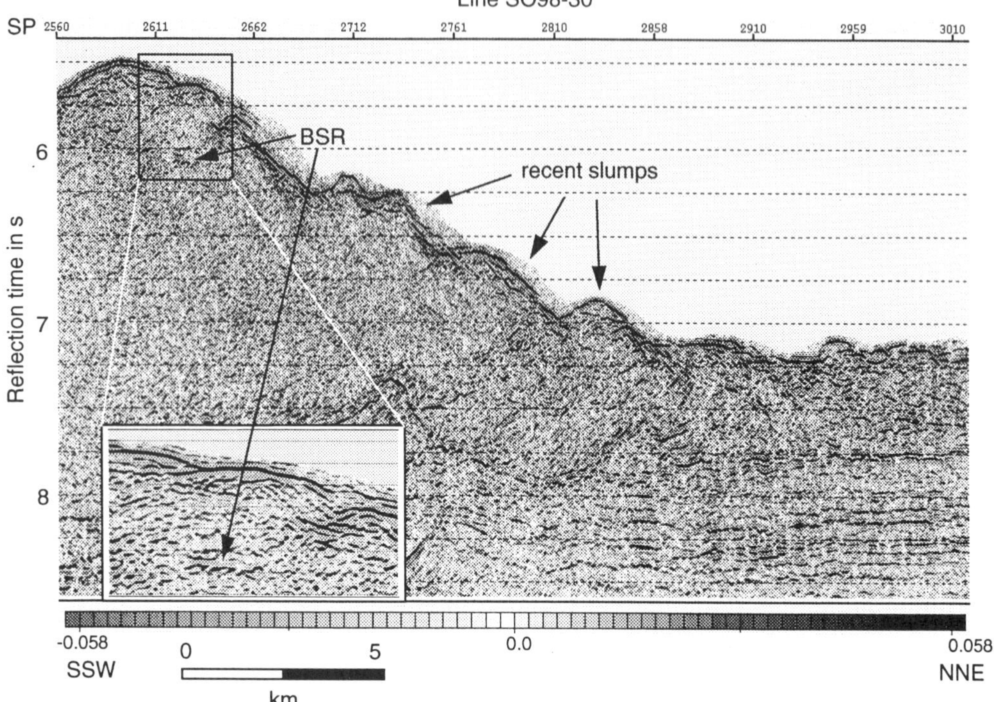

Fig. 8. Seismogram section (shots 2560–3015) of reflection seismic profile SO98-30 showing the lower slope and the toe of the subduction zone north of Sulawesi. Slumped sediment masses extend over the whole lower slope and obscure the deformation front in the North Sulawesi Trench. Inlay shows a close-up of the area between shots 2605 and 2650 where a very short negative reflection occurs which could resemble a BSR (SP, shotpoint).

region. If we use these estimates the minimum amount of methane in the gas hydrate layer off North Sulawesi is approximately 6.6×10^{12} m^3.

Conclusions

The multi-channel reflection seismic data across the subduction zone north of Sulawesi show strong bottom simulating reflections (BSRs) in the lower and upper slopes of the accretionary wedge. This indicates the presence of gas hydrates in the accreted sediments of the wedge. This is the first time that the presence and areal distribution of bottom simulating reflectors and, hence, the presence of gas hydrates are reported for the Celebes Sea east of longitude 121°E. The areal distribution of BSRs mapped by us show that at least 8200 km^2 of the accretionary prism north of Sulawesi are underlain by gas hydrates.

BSRs are interrupted at thrust faults. Heat flow and geochemical data there suggest that active methane venting occurs at the sea floor. This venting is highest at the toe of the subduction zone and in the lower slope region.

The occurrence of BSRs is restricted to the central part of the northern slope of the subduction zone. On the westernmost reflection seismic profile (SO98-30) the seismograms show reflections below the lower slope of the accretionary wedge which are interpreted as sediment slumps. In these regions no or only very few and short BSRs are found, which suggests that the slope has been very unstable. It is not clear whether this state still prevails. On the easternmost profiles, no BSRs are found at all. Here, the slopes of the accretionary wedge are 2–3 times steeper than in the middle part of the northern slope of Sulawesi between 121°30′E and 123°30′E and probably too steep for the establishment of stable conditions for the formation of BSRs.

The R/V *Sonne* cruise SO98 was funded by Bundesministerium für Bildung, Wissenschaft, Forschung und Technolgie (BMBF), Bonn (Germany) project No. 03G0098A. We thankfully acknowledge this support.

We are particularly indebted to the Agency for the Assessment and Application of Technology (BPP Teknologi), Jakarta (Indonesia) for the assistance in planning the cruise and in providing the assistance of M. T. Zen and his colleagues B. Ganie, Y. Djajadihardja, R. Trimanadi, R. Anantasena, H. Manato, as well as Letkol S. Poospoo, during the cruise. We also thank all shipboard and non-shipboard colleagues who have spent time and energy to make R/V *Sonne* cruise SO98 successful.

References

ANDREASSEN, K., HART, P. & GRANTZ, A. 1995. Seismic studies of a bottom simulating reflection related to gas hydrate beneath the continental margin of the Beaufort Sea. *Journal of Geophysical Research*, **100**, 12, 659–12, 673.

BEIERSDORF, H., BACH, W., DELISLE, G. *et al.* 1997. Age and possible modes of the formation of the Celebes Sea basement and thermal regimes within the accretionary complexes off SW Mindanao and N Sulawesi. *In: Proceedings of the International Conference on Stratigraphy and Tectonic Evolution of SE Asia and the SW Pacific*, Bangkok, 369–397.

BIJU-DUVAL, B., LE QUELLEC, P., MASCLE, A., RENARD, V. & VALERY, P. 1982. Multibeam bathymetric and high-resolution seismic investigations of the Barbados Ridge complex (Eastern Carribean): A key to the knowledge and interpretation of an accretionary wedge. *Tectonophysics*, **86**, 275–304.

CLAERBOUT, J. F. 1985. *Imaging the Earth's Interior*. Blackwell, Oxford.

DELISLE, G., BEIERSDORF, NEBEN, S. & STEINMANN, D. 1998. The geothermal field of the North Sulawesi accretionary wedge and a model on BSR migration in unstable depositional environments. *This volume*.

DIX, C. H. 1955. Seismic velocities from surface measurements. *Geophysics*, **20**, 68–86.

GIMPEL, P. 1987. Marine flachseismische Untersuchungen in der Kieler Bucht unter besonderer Berücksichtigung von Scherwellenmessungen. PhD thesis, University of Kiel.

GREGORY, A. 1977. Aspects of rock physics from laboratory and log data that are important to interpretation. *In*: PAYTON, D. (ed.) *Seismic Stratigraphy – Application to Hydrocarbon Exploration*. American Association of Petroleum Geologists Memoir, **26**, 15–46.

HAMILTON, W. 1979. Tectonics of the Indonesian Region. *US Geological Survey* Professional Paper, **1078**.

HINZ, K., BLOCK, M., KUDRASS, H. R., & MEYER, H. 1991. Structural elements of the Sulu Sea, Philipines. *Geologisches Jahrbuch*, **A127**, 483–506.

KATZMAN, R., HOLBROOK, W. & PAULL, C. 1994. Combined vertical-incidence and wide-angle seismic study of a gas hydrate zone, Blake Ridge. *Journal of Geophysical Research*, **99**, 17, 975–17, 995.

LEE, C. & MCCABE, R. 1986. The Banda Celebes Sulu Basins: A trapped piece of Cretaceous–Eocene oceanic crust? *Nature*, **322**, 51–54.

LEE, M. & SUH, S. 1985. Optimization of one-way wave equations. *Geophysics*, **50**, 1634–1637.

MACKAY, M., JARRARD, R., WESTBROOK, G., HYNDMAN, R. & THE SHIPBOARD SCIENTIFIC PARTY OF LEG 146. 1994. Origin of bottom-simulating reflectors: Geophysical evidence from the Cascadia accretionary prism. *Geology*, **22**, 459–462.

MACLEOD, M. K. 1982. Gas hydrates in ocean bottom Sediments. *AAPG Bulletin*, **66**, 2649–2662.

McIver, R. D. 1982. Role of naturally occurring gas hydrates in sediment transport. *AAPG Bulletin*, **66**, 789–792.

MURAUCHI, S., LUDWIG, W. J., DEN, N. *et al.* 1973. Structure of the Sulu Sea and the Celebes Sea. *Journal of Geophysical Research*, **78**, 3437–3447.

PAULL, C., BOROWSKI, W., BLACK, N. & THE SHIPBOARD SCIENTIFIC PARTY LEG 164. 1996. Marine gas hydrate inventory: Preliminary results of ODP Leg 164 and implications for gas venting and slumping associated with the Blake Ridge gas hydrate field. *In*: HENRIET, J.-P. & MIENERT, J. (eds) *First Master Workshop on Gas Hydrates: Relevance to World Margin Stability and Climate Change*, University of Gent, Belgium, 81–93.

RANGIN, C., SILVER, E., VON BREYMANN, M. *et al.* 1990. *Proceedings of the Ocean Drilling Program, Initial Results*. College Station, TX. Ocean Drilling Program, **124**.

SHIPLEY, T., HOUSTON, M., BUFFLER, R. *et al.* 1979. Seismic evidence for widespread possible gas hydrate horizons on continental slopes and rises. *AAPG Bulletin*, **63**, 2204–2213.

TANER, M. T. & KÖHLER, F. 1969. Velocity spectra – digital computer derivation and applications of velocity functions. *Geophysics*, **34**, 859–881.

WEISSEL, L. K. 1980. Evidence for Eocene oceanic crust in the Celebes Basin. *In*: HAYES, D. (ed.) *The Tectonic and Geologic Evolution of Southeast Asian Seas and Islands*. American Geophysical Union, Geophysical Monograph Series, **23**, 37–47.

WOOD, W., STOFFA, P. & SHIPLEY, T. 1994. Quantitative detection of methane hydrate through high resolution seismic velocity analysis. *Journal of Geophysical Research*, **99**, 9681–9695.

YUAN, T., SPENCE, G. & HYNDMAN, R. 1994. Seismic velocity increase and inferred porosities in the accretionary wedge sediments at the Cascadia margin. *Journal of Geophysical Research*, **99**, 4413–4427.

——, HYNDMAN, R., SPENCE, G. & DESMONS, B. 1996. Seismic velocity increase and deep-sea gas hydrate concentration above a bottom-simulating reflector on the northern Cascadia continental slope. *Journal of Geophysical Research*, **101**, 13, 655–13, 671.

The geothermal field of the North Sulawesi accretionary wedge and a model on BSR migration in unstable depositional environments

G. DELISLE, H. BEIERSDORF, S. NEBEN & D. STEINMANN

Bundesanstalt für Geowissenschaften und Rohstoffe (BGR), Postfach 51 01 53, D-30631 Hannover, Germany

Abstract: The distribution of heat flow in the North Sulawesi accretionary wedge was derived from the depths of a bottom simulating reflector (BSR) and nine *in situ* heat flow measurements. The values obtained by both types of measurements agree reasonably well and suggest high heat flow of the order of 70–100 mW m^{-2} near the deformation front and a systematic decrease to 30 mW m^{-2} landwards. In addition, we have tested on the basis of one numerical model the likelihood of the BSR being in a position of complete thermal equilibrium with the surrounding rocks. The model describes the rate by which a BSR layer re-equilibrates after a thermal disturbance at the sea floor (sediment slumping). The result suggests that the BSR does not regain complete thermal equilibrium after slumping in the course of several 10^5 years.

The Celebes Sea continental margin of the North Arm of Sulawesi is characterized by an accretionary wedge over the subducting Celebes Sea micro-plate (Hamilton 1979). During the R/V *Sonne* cruise SO98 GIGICS in 1994 by BGR, Hannover (Germany) this tectonic unit was surveyed with regard to its topography, internal structure, hemipelagic sediment cover, terrestrial heat flow, and gravity and magnetic fields. Multichannel reflection seismic profiling along four lines over the accretionary wedge and more or less perpendicular to the deformation front and the Sulawesi Trench (Fig.1) revealed a distinct bottom simulating reflector (BSR) within the wedge, which is considered to indicate the base of the gas hydrate zone (for more details and illustrations see Neben *et al.* 1998). In this study we assumed that the thickness of the layer with gas hydrates (i.e. distance between BSR and sea floor) is an expression of the temperature conditions within the accretionary wedge. Therefore, we mapped the BSR depth below sea floor, using the four reflection seismic profiles, derived heat flow values from this set of data and, in addition, measured heat flow with a marine heat flow probe within the wedge-draping hemipelagic sediments.

There are sources of uncertainty for both sets of data: the heat flow value from the marine heat flow probe might include a shallow convective heat transport component, otherwise only identifiable by measurements in deeper boreholes. The BSR-derived heat flow relies on the correct BSR depth determination and – in the absence of boreholes – on the choice of an appropriate value for the thermal conductivity of the gas hydrate zone. An additional uncertainty in the BSR-derived heat flow values is the response time of the lower BSR boundary to induced thermal disturbances such as, for example, a slumping event downslope of the accretionary wedge. This paper attempts to develop an approximate picture of the thermal field of the North Sulawesi accretionary wedge in view of these uncertainties.

Data acquisition

The depth of the BSR on all four profiles was calculated using a P-wave velocity of 1900 m s^{-1}. An average interval velocity for the interval above the BSR was determined using the rms-resp. stacking velocities from the processing sequence and the Dix formula (Dix 1955). The heat flow was measured at nine stations on reflection seismic profile SO98–028 (Figs 1 and 2, and Table 1) using a marine heat flow probe consisting of a steel lance with six thermistor outriggers. Electric signals between the recording unit onboard and the probe were transmitted via a coaxial conductor cable.

In the heat flow data calculation based on BSR depths the following assumptions were made for the gas hydrate layer:

- a linear temperature gradient within the gas hydrate zone (the latter starts at least beyond the penetration depth of the piston corer used for sampling, i.e.13.5 m);
- a uniform thermal conductivity of 1.2 W m^{-1} K^{-1} (see e.g. fig. 2 in Fisher & Hounslow 1990);
- a uniform lithostatic pressure gradient within the sediment column of 2.2 MPa per 100 m.

A lithostatic pressure gradient was chosen as the gas hydrate layer must be considered as a

DELISLE, G., BEIERSDORF, H., NEBEN, S. & STEINMANN, D. 1998. The geothermal field of the North Sulawesi accretionary wedge and a model on BSR migration in unstable depositional environments. *In:* HENRIET, J.-P. & MIENERT, J. (eds) *Gas Hydrates: Relevance to World Margin Stability and Climate Change*. Geological Society, London, Special Publications, **137**, 267–274.

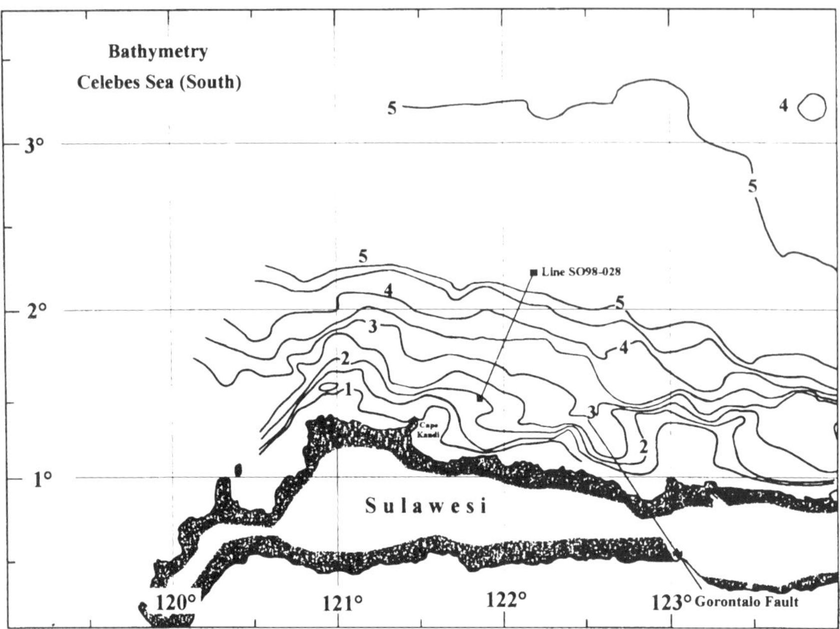

Fig. 1. Bathymetry of the Celebes Sea (water depth in km) and location of the seismic line SO98-028.

low permeability layer (see Kvenvolden 1994). The water pressure in the compacting sediments below the layer should rise to near-lithostatic pressures which are, as is commonly observed (Moore & Vrolijk 1992), the cause for vigorous fluid flow toward accretionary fronts.

The sea bottom heat flow measurements were derived as follows. The bottom-water temperature was determined through a function derived from water depth vs temperature plots using temperatures measured with our heat flow probe in the water column. The temperatures at the BSR were derived from the stability field of methane hydrate compiled by Sloan (1990).

The *in situ* thermal conductivities shown in Table 1 are based on temperature decay measured after heating of the sediment. Heating and temperature increase were tracked in parallel at each thermistor for 8 min. Slight variations in the applied thermal energy during heating, in turn, led to uncertainties in the determination of the conductivity values. However, the values are close to the average thermal conductivity (λ) of 0.85 W m^{-1}K^{-1} of hemipelagic mud near the sea floor (see Langseth *et al.* 1990). In comparison, higher λ-values around 1.2 W m^{-1} K^{-1} (see above) were assumed to be representative for the more compacted sediments of the gas hydrate layer below the uppermost sea bottom sediments (see Fisher & Hounslow 1990).

The seismic line SO98–028 (Figs 1 and 2) between a point near shore (1°27.97′N; 121°53.74′E) to a point above the abyssal plain (2°12.38′N; 122°6.92′E) was chosen for measuring heat flow data with a marine heat flow probe as developed by Von Herzen (Von Herzen & Maxwell 1959). The heat flow values (q) were calculated from the temperature and thermal conductivity data using the thermal resistance method (Bullard 1939; see also Chapman *et al.* 1984). Heat flow is determined from the temperature and the thermal resistance of the sediments where

$$q = (T - T_0)/R(z),$$

with T_0 as the temperature at the sea floor and $R(z)$ as the thermal resistance defined by:

$$R(z) = \Sigma\, d(z)/\lambda(z),$$

where $d(z)$ is the interval of measurement and $\lambda(z)$ is the thermal conductivity of sediments as function of depth (z).

Nine measurements were carried out which are summarized in Table 1.

Heat flow data map

All heat flow values derived from BSR depths were used to construct isolines fitted by hand

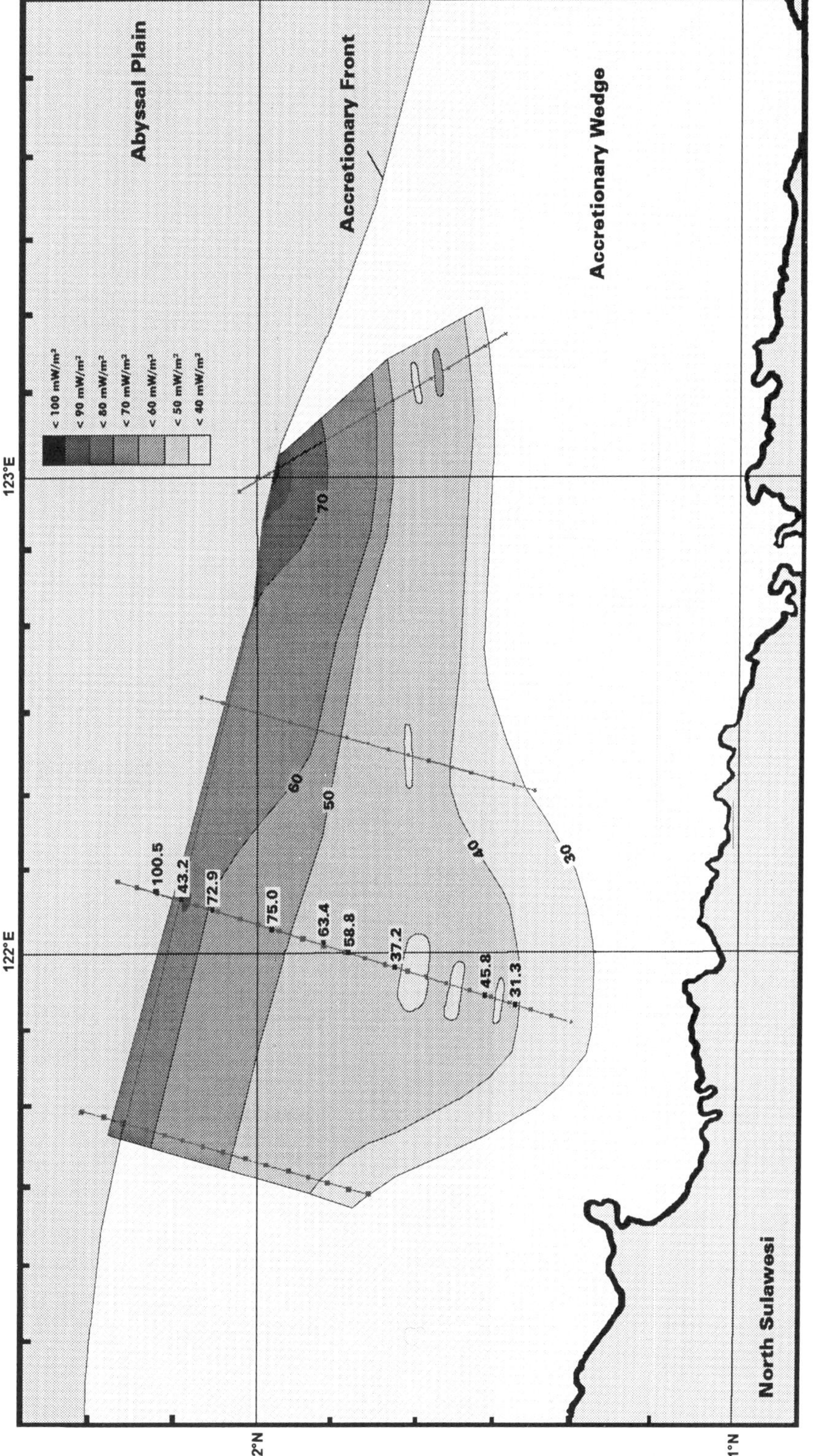

Fig. 2. Heat flow distribution of the North Sulawesi accretionary wedge based on BSR-depth and nine heat flow measurements with a marine heat flow probe along seismic line SO98-028. Measured heat flow values are given in mW m^{-2}. Heat flow decreases in the typical fashion for accretionary wedges from high values near the accretionary front to lower values landwards.

Table 1. *Summary of all heat flow measurements off the North Sulawesi coast with the BGR marine heat flow probe*

Station	Coordinates	Sediment depth (m)	Heat flow (q) (mW m^{-2})	Mean value (q) (mW m^{-2})	Standard deviation (mW m^{-2})
HF 2	2°12.38′N				
	122°06.92′E	0.00–0.65	97.5		
		0.00–1.95	105.8		
		0.00–2.60	98.2	100.5	3.76
HF 26	2°09.93′N				
	122°06.14′E	0.00–1.95	32.5		
		0.00–2.60	42.3		
		0.00–3.25	54.8	43.2	9.126
HF 25	2°06.76′N				
	122°05.23′E	0.00–1.95	54.4		
		0.00–2.60	88.2		
		0.00–3.25	76.0	72.9	13.975
HF 23	1°59.17′N				
	122°03.02′E	0.00–1.95	79.0		
		0.00–2.60	68.8		
		0.00–3.25	77.2	75.0	4.445
HF 20	1°52.84′N				
	122°01.06′E	0.00–1.95	66.4		
		0.00–2.60	59.1		
		0.00–3.25	64.7	63.4	3.119
HF 18	1°49.88′N				
	122°00.23′E	0.00–1.95	54.0		
		0.00–2.60	59.3		
		0.00–3.25	63.1	58.8	3.732
HF 16	1°43.88′N				
	121°58.41′E	0.00–1.95	40.8		
		0.00–3.25	33.6	37.2	3.6
HF 8	1°30.74′N				
	121°54.50′E	0.00–1.95	48.0		
		0.00–2.60	46.3		
		0.00–3.25	43.1	45.8	2.031
HF 12	1°27.97′N				
	121°53.74′E	0.00–1.95	32.3		
		0.00–2.60	32.6		
		0.00–3.25	28.9	31.3	1.678

(Fig. 2). The map demonstrates very clearly the general trend of elevated heat flow near the accretionary front and decreasing heat flow upslope. Heat flow at the accretionary front appears to be higher in the eastern part of the wedge. The heat flow values measured *in situ* are included as well. These values in general correspond well with the BSR-based values. The *in situ* measurements, however, show an approximately 30% higher heat flow in a 30 km wide zone immediately south of the deformation front compared to the BSR-derived values. This possibly indicates convective heat transport by upward-migrating fluids through pathways across the gas hydrate layer heating the sediments near the sea floor. In this case, high-temperature gradients would exist in the topmost layers followed by low temperature gradients below. The BSR-based heat flow values would then have to be lower, as by definition they do not include a convective heat transport component.

The high heat flow of 100.5 mW m^{-2} at Station HF 2 (located seaward of the deformation front) may not reflect the intrinsic heat flow of the Celebes Sea crust, because the crust is too old to explain such high heat flow. Radiometric K–Ar age determination on an ocean island-type basalt, dredged during our cruise, has produced an age for the southern Celebes Sea crust of at least 45.2 ± 1.2 Ma (F. Henjes-Kunst pers. comm.), thus reconciling the age of 42 Ma given by Weissel (1980) and Rangin *et al.* (1990). An oceanic crust of this age, in turn, would have been cooled down to lower heat flow values between 70 and 75 mW m^{-2} (Parsons

& Sclater 1977). We assume this high heat flow to be caused by upwelling fluids being expelled towards the north by the compacting accretionary wedge.

How stable is the BSR? – a numerical model

Irrespective of the reasonable fit between the BSR-based and needle probe heat flow values, it appears to be worthwhile to investigate the probability of the BSR position to reflect the true heat flow. In the case of a sudden event, i.e. slumping of sea sediments downslope of the accretionary wedge, the thermal boundary conditions are suddenly changed and the BSR has to re-adjust its position. A thermal cold wave penetrates into the gas hydrate zone and forces with time the BSR to migrate to deeper levels. Release of latent heat during the formation of new gas hydrate impedes the downward migration over time.

The time-dependent repositioning of the BSR was calculated using the following model configuration and the theoretical approach outlined in Carslaw & Jaeger (1959):

Heat flow q at $t = 0$	$40\,\mathrm{mW\,m^{-2}}$
Constant thermal conductivity λ of the gas hydrate layer and the sediments below	$1.2\,\mathrm{W\,m^{-1}\,K^{-1}}$
Thickness of removed sediment pile	50 m
Temperature at sea floor	3°C
Initial thickness of frozen gas hydrate layer	690 m

The temperature field in the sediments was calculated according to:

$$q = \lambda \operatorname{grad} T. \quad (1)$$

The re-equilibration of the temperature field after the slump event is calculated according to

$$\delta T/\delta t = \lambda/\rho c \operatorname{div} (\operatorname{grad} T) \quad (2)$$

and

$$\lambda(\mathrm{d}T_1/\mathrm{d}z - \mathrm{d}T_2/\mathrm{d}z) = L\,\mathrm{d}X/\mathrm{d}t \quad (3)$$

with $L = 1 \times 10^8\,\mathrm{Ws\,m^{-3}}$ (equivalent to 30% gas hydrate per $1\,\mathrm{m^3}$ rock volume); $\rho c = 2.2 \times 10^6\,\mathrm{Ws\ m^{-3}\ K^{-1}}$; $\mathrm{d}X/\mathrm{d}t$ = rate of movement of the phase change boundary in $\mathrm{m\,s^{-1}}$. Equation (3) describes the rate by which the BSR migrates toward the new equilibrium position. $\mathrm{d}T_1/\mathrm{d}z$ represents the thermal gradient above and $\mathrm{d}T_2/\mathrm{d}z$ the gradient below the phase boundary.

The time-dependent change of the temperature field has been calculated using a finite difference-scheme. The space from the sea bed down to a depth of 800 m has been sub-divided into small elements to facilitate the inclusion of heat sources by the liberation of latent heat at the phase boundary. The spacing is then slowly increased to a total depth of 13 800 m to allow for an accurate approximation of the thermal field in the system. The amount of latent heat attainable in each element (which depends on the amount of gas hydrate which is assumed to have formed within) is calculated in a first step. During each time iteration for the recalculation of the temperature field each element is tested to determine whether, for physical reasons, latent heat is liberated. As long as latent heat is liberated, the temperature of the corresponding element is kept at the phase boundary. The rate by which latent heat is conducted away is given by equation (3). The temperature field in and around an element observes equation (2) again as soon as all latent heat from an element is liberated.

The result of the calculation (Fig. 3a and b) shows that the repositioning of the BSR is a very slow process. The repositioning of the BSR starts only after the cold wave has reached the initial BSR position, after 6000 years. The downward migration of the BSR is greatly impeded by the release of large quantities of latent heat, which can be conducted away only very slowly by the minute difference between the slightly larger T-gradient within the hydrate zone and the T-gradient in the sediments below ($\mathrm{d}T_1/\mathrm{d}z - \mathrm{d}T_2/\mathrm{d}z$). The T-gradient in the hydrate zone depends on the cooling process due to the temperature drop induced by slumping of the top sediments. Under usual circumstances this temperature drop will be of the order of 1–2°C at the newly formed sea floor. The difference in T-gradients will be smaller the thicker the initial hydrate zone is. According to the numerical model presented here, only a 34 m thick (from 690 to 724 m thickness) fresh gas hydrate layer will have developed 500 000 years after the slumping event below the phase boundary at time = 0. During this time any estimation of heat flow based on the depth of the BSR will be in error initially by $<7\%$ and for a long time of the order of 5%. Theoretically, it will take infinite time to reach the final hydrate layer thickness again of 690 m because the difference between $\mathrm{d}T_1/\mathrm{d}z$ and $\mathrm{d}T_2/\mathrm{d}z$, which controls the conductive transport of latent heat away from the phase boundary, decreases with time to 0.

The case of a thickening gas hydrate layer due to a continuous high sediment accumulation rate or redepositioning of a slumped mass is slightly different, as the new gas hydrates grow well

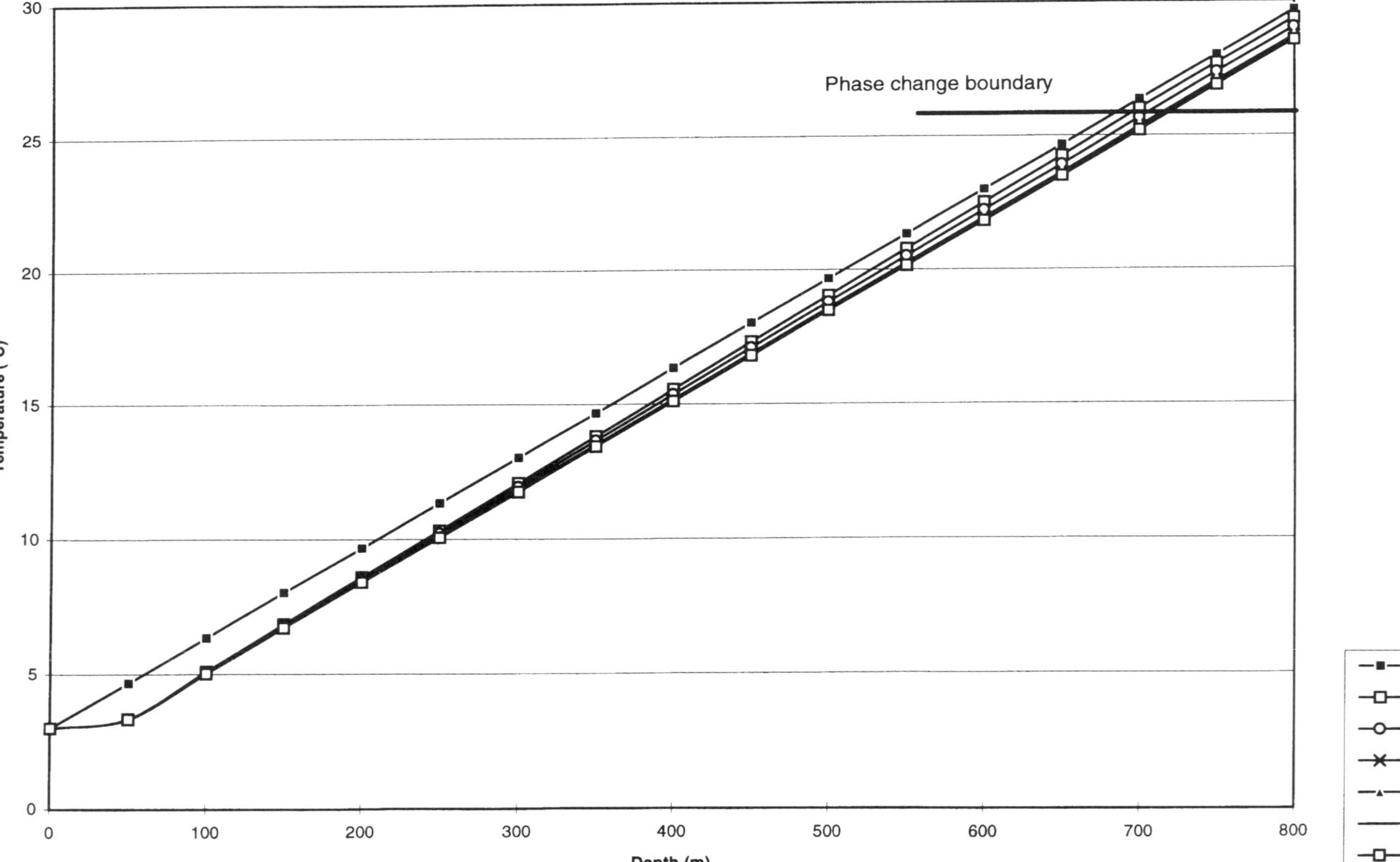

Fig. 3. (**a**) Calculated thermal re-equilibaration of a BSR layer after a 50 m thick pile of top sediments slumped away. The resulting thermal disturbance and re-equilibration, including the downward migration of the BSR with time is shown for 100 000 year intervals.

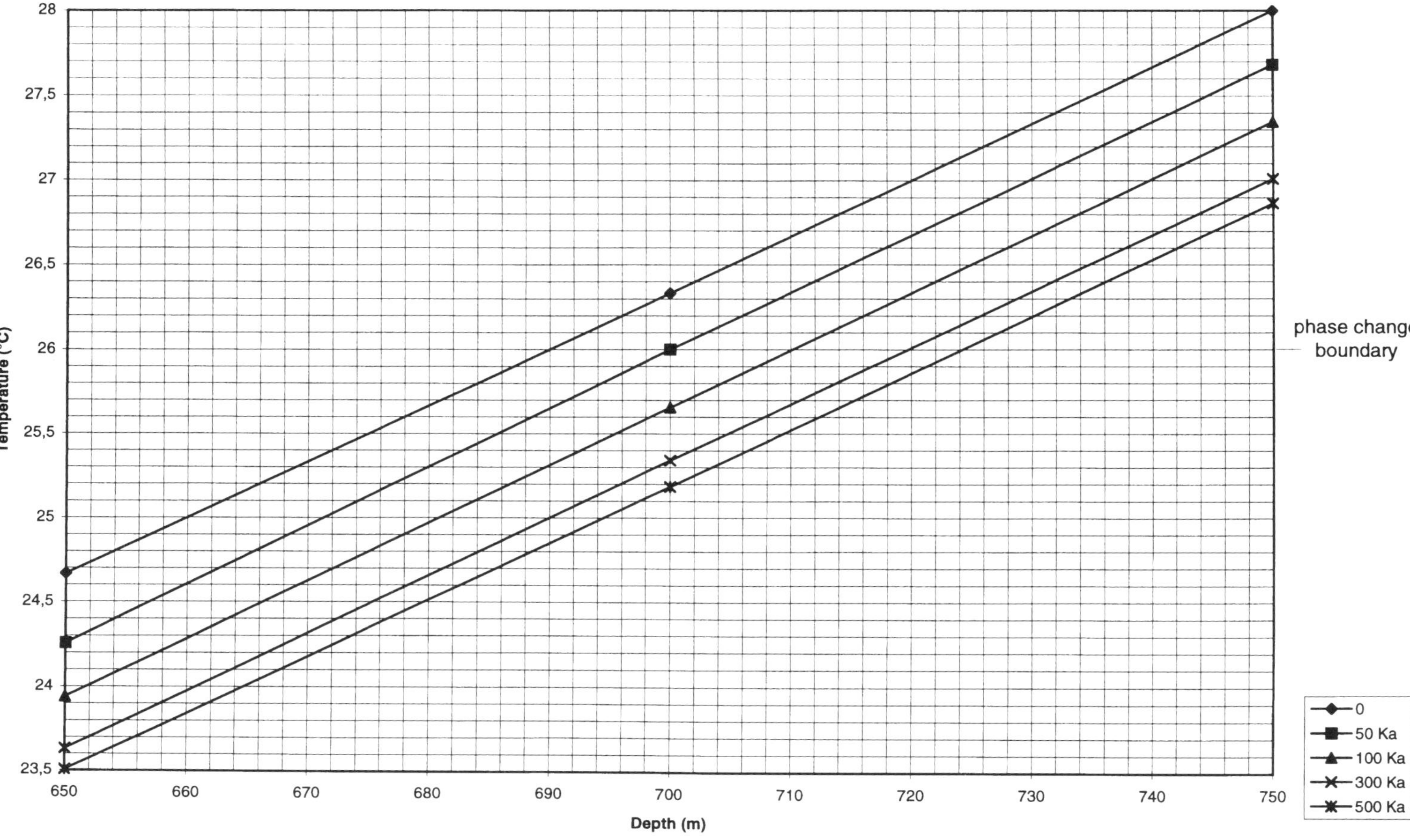

Fig. 3. (b) Same as (a) showing more detail for the depth interval from 650 to 750 m.

within the limits of the hydrate stability field while gas hydrates are slowly dissolved at the bottom. The layer, from which latent heat of hydrate formation is released, acts like a local heat source. The thermal gradient to the bottom waters, which maintain a constant temperature, is increased, allowing (also due to the short distance of the layer to the sea floor) an effective conductive heat transfer from the heat source to the sink. Nevertheless, in response the BSR must migrate upwards with time due to the thermal re-equilibration of the hydrate layer proceeding from top to bottom. This process will be dealt with in more detail elsewhere.

Conclusions

The accretionary wedge north of Sulawesi shows a geothermal behaviour typical for many similar complexes. A high heat flow of between 70 and 100 mW m^{-2} exists near the deformation front. Upslope, the heat flow decreases systematically towards values around 30 mWm^{-2}. The accretionary front of the western part of the North Sulawesi accretionary wedge seems to be cooler than the eastern part.

The repositioning of a BSR after rapid erosion or a slope failure is an extremely slow process. Erosion and slope failures are common in tectonically active accretionary complexes or sedimented continental margins with contour current contact, and BSRs there may not always be in full thermal equilibrium. Therefore, heat flow values derived from BSR depths in these environments are likely to be too high because the BSR still has to migrate downward.

In contrast, if sediment is rapidly piled up, for example after deposition of slump masses, a BSR would have to migrate upward with a comparable slow rate as above. The thermal effect of the newly added relatively cold sediment would again be felt at the existing phase change boundary after considerable time. A complete thermal re-equilibrium over a short time span can be ruled out. The error of heat flow values derived from present BSR depths in areas, where thermal equilibrium is lacking, may be of the order of 5%.

The Bundesministerium für Bildung, Forschung und Technologie (BMBF), Germany, provided the funding of the cruise. We thankfully acknowledge this support. We thank the Government of the Republic of Indonesia and the Agency for the Assessment and Application of Technology, Jakarta, for their permission to work in their Exclusive Economic Zone and for their assistance in the shipboard work. The helpful remarks of two anonymous reviewers are gratefully acknowledged.

References

BULLARD, E. C. 1939. Heat flow in South Africa. *Proceedings of the Royal Society of London, Series A*, **173**, 474–502.

CHAPMAN, D. S., HOWELL, J. & SAASS, J. H. 1984. A note on drillhole depths required for reliable heat flow determinations. *Tectonophysics*, **103**, 11–18.

CARSLAW, H. S. & JAEGER, J. C. 1959. *Conduction of Heat in Solids*. Oxford University Press, New York.

DIX, C. H. 1955. Seismic velocities from surface measurements. *Geophysics*, **20**, 68–86.

FISHER, A.T. & HOUNSLOW, M. W. 1990. Transient fluid flow through the toe of the Barbados Accretionary Complex: constraints from Ocean Drilling Program Leg 110 heat flow studies and simple models. *Journal of Geophysical Research*, **95**, 8845–8858.

HAMILTON, W. 1979. *Tectonics of the Indonesian Region*. US Geological Survey Professional Paper, **1078**.

KVENVOLDEN, K. A. 1994. Natural gas hydrate occurrence and issues. *In*: SLOAN, E., JR, HAPPEL, J. & HNATOW, M. (eds) *International Conference on Natural Gas Hydrates*. Annals of the New York Academy of Sciences, **715**, 232–246.

LANGSETH, M. , WESTBROOK, G. & HOBART, M. 1990. Contrasting geothermal regimes of the Barbados Accretionary complex. *Journal of Geophysical Research*, **95**, 8829–8843.

MOORE, J. & VROLIJK, P. 1992. Fluids in accretionary prisms. *Review of Geophysics*, **30**(2), 113–136.

NEBEN, S., HINZ, K. & BEIERSDORF, H. 1998. Reflection characteristics, depth and geographical distribution of bottom simulating reflectors within the accretionary wedge of Sulawesi. *This volume*.

PARSONS, B. & SCLATER, J.G. 1977. An analysis of the variation of ocean floor bathymetry and heat flow with age. *Journal of Geophysical Research*, **82**, 803–827.

RANGIN, C., SILVER, E., VON BREYMANN, M. *et al.* 1990. *Proceedings of the Ocean Drilling Program, Initial Reports*, College Station, TX. Ocean Drilling Program, **124**.

SLOAN, E. D. JR. 1990. *Clathrate Hydrates of Natural Gases*. Marcel Dekker, New York.

VON HERZEN, R. & MAXWELL, A. 1959. The measurement of thermal conductivity of deep-sea sediments by a needle-probe method. *Journal of Geophysical Research*, **64**, 1557–1563.

WEISSEL, J. K. 1980. Evidence for oceanic crust in the Celebes Basin. *In*: HAYES, E. (ed.) *The Tectonic and Geologic Evolution of Southeast Asian Seas and Islands*. American Geophysical Union. Geophysical Monograph Series, **23**, 37–47.

Gas hydrates along the northeastern Atlantic margin: possible hydrate-bound margin instabilities and possible release of methane

J. MIENERT[1], J. POSEWANG[2] & M. BAUMANN[1]

[1] *GEOMAR, Research Centre for Marine Geosciences, Wischhofstrasse 1–3, D24148 Kiel, Germany*

[2] *SFB 313, Christian-Albrechts-University, Olshausenstrasse 40, D24118 Kiel, Germany*

Abstract: The presence of gas hydrates and free gas in oceanic sediments along the north-eastern European Margin is documented in high-frequency near-vertical and wide-angle seismic reflection data. Shallow-water and deep-water gas hydrate instabilities can cause free gas to escape from oceanic sediments. Particularly, methane from shallow-water gas hydrate destabilization may then get transferred from the sediments into the water column, and eventually into the atmosphere. Deep-water gas hydrates are coincident with areas and depths of slope failures in continental margin sediments. Comparisons between seismicity and the potential hydrate distributions suggest a correlation between hydrate instability and margin instabilities along the north-eastern Atlantic Margin.

Submarine gas hydrates are believed to be common in continental margin sediments because seismic reflection data and sampling have documented their existence along passive and active margins (Kvenvolden & Barnard 1983). Direct measurements and sampling have been carried out on only a few occasions. Dickens *et al.* (1997) showed ODP Leg 164 results from the western US Atlantic Margin (Blake Ridge) and reported that direct measurements of *in situ* methane abundances stored as gas hydrate and free gas in a sediment sequence contain approximately 35 Gt of carbon. Important in this case is that the amount of gas estimated by geophysical measurements is lower by a factor of 3. The discrepancy is mainly due to the fact that the amount of free gas directly measured beneath the gas hydrate zone is significantly larger than that estimated from the direct geophysical approach (Dickens *et al.* 1997). This result shows that, although hydrates and the free gas zone clearly store immense amounts of carbon, the total estimates of methane carbon in gas hydrates remain largely underestimated and speculative.

Quantitative information about the distribution and concentration of gas hydrates and free gas in oceanic sediments is still poorly constrained. They are mainly gained from seismic reflection surveys (e.g. Shipley *et al.* 1979; Andreassen *et al.* 1990; Miller *et al.* 1991; Hyndman & Spence 1992; Eicken & Hinz 1993; Katzmann *et al.* 1994; Minshull *et al.* 1994; Mienert 1994; Mienert *et al.* 1994; Bobsien 1995; Mienert & Posewang 1996; Posewang 1997). The Norwegian Margin may be one of the target areas for future studies because gas hydrates have also been suggested as a future energy source (e.g. Kvenvolden 1993; Gornitz & Fung 1994; Max & Lowrie 1996). In addition, gas hydrates may have a significant effect on global climatic change (e.g. Kvenvolden 1988, 1993; MacDonald 1990; Lammers *et al.* 1995b) and they may cause hazards such as sediment failure on the sea floor (e.g. Kayen & Lee 1991; Kvenvolden 1993).

The available data from a major slide on the Mid-Norwegian Margin suggests that the Storegga Slide (Fig. 1) consists of three major events with additional minor or secondary associated slides (Bugge *et al.* 1988). The first Storegga Slide was the largest and created a slide scar with a 290 km wide headwall. The slide extends down the continental slope and into the abyssal plain for a distance of more than 800 km. The sediments that were removed are presumed to have been the relatively soft, mainly Plio-Quaternary clays, now found downslope as acoustically transparent deposits with small-scale surface roughness. The second and third slides cut down up to 200 m deeper into more consolidated sediments and cut a new headwall some 5–8 km further back into the shelf edge. The deposits have a characteristically blocky appearance on seismic profiles and include some huge, largely unbroken sediment slabs. The first slide is dated at 30 000–50 000 years BP and displaced about 3900 km^3 of material, while the second and third slide occurred in near succession, about 6000–8000 years BP, and involved about 1700 km^3 of sediment. A thick (more than 6 m) fine-grained turbidite in the abyssal plain is related to the second slide. In some places, its thickness is more than 20 m. There is also evidence for a tsunami

MIENERT, J., POSEWANG, J. & BAUMANN, M. 1998. Gas hydrates along the northeastern Atlantic margin: possible hydrate-bound margin instabilities and possible release of methane. *In:* HENRIET, J.-P. & MIENERT, J. (eds) *Gas Hydrates: Relevance to World Margin Stability and Climate Change*. Geological Society, London, Special Publications, **137**, 275–291.

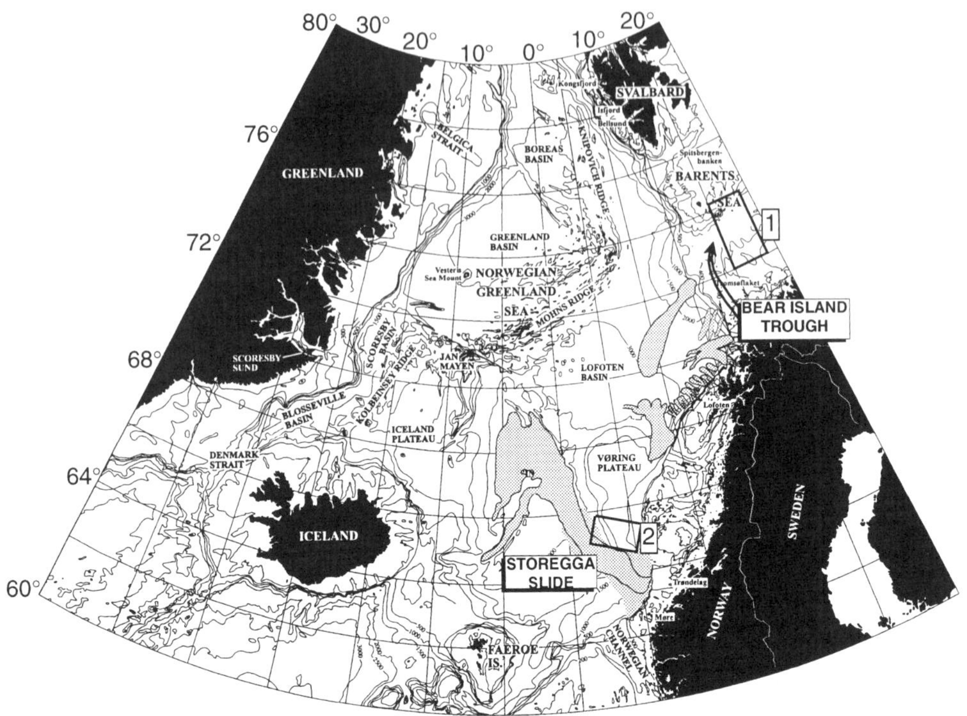

Fig. 1. The Norwegian–Greenland Sea continental margins showing a concentration of large slides on the Norwegian Margin (sources: Vorren *et al.* in press; e.g. Bugge *et al.* 1987; Laberg & Vorren 1993; Dowdeswell *et al.* 1996).

which accompanied the second slide, leaving dated deposits on the coasts of Scotland and Norway (Dawson *et al.* 1988).

The north-eastern Atlantic Continental Margin has been subject to significant mass movements in the past (Fig. 1). Earthquakes are cited as the cause for slides occurring off the Norwegian coast (Bugge *et al.* 1987). Earthquakes in conjunction with the presence of weakened sediment zones due to gas hydrate destabilization may represent a seismic hazard along the north-eastern European Margin. Based on different geological and geophysical data sets, a possible link between the instability of gas hydrates and the north-eastern Atlantic Margin instability is presented. Clathrate instability in the sea floor may also be the origin of gas escape in the water column. The transfer of methane from the oceanosphere to the atmosphere is discussed using the example of a deep-water gas escape field at the Mid-Norwegian Margin and a shallow-water escape field at the Barents Sea (Fig. 1) which can be considered representative of the north-eastern Atlantic Margin (Lammers *et al.* 1995*a*).

Methods

The calculation of compressional wave (P-wave) velocities in marine sediments is one of the ways to detect and estimate free gas and gas hydrate zones through the sediment column. The high-frequency ocean bottom seismometers (HF-OBS), originally developed at the SFB 313 of the Christian-Albrechts-University in Kiel (Bobsien & Mienert 1994; Bobsien 1995; Mienert *et al.* 1995), are specially designed and constructed to take accurate *in situ* measurements of the vertical and lateral changes in compressional wave velocity within the upper 400 m of the sea floor (Fig. 2), the region where the major sliding processes may occur (e.g. Storegga Slide) (Bugge 1983; Bugge *et al.* 1987).

Several types of sources were used to cover a broad range of source frequencies, thus widening the band of the resolution for the inferred seismic structure. The following sources were used at the same locations: pinger (source frequency: 3.5 kHz), deep-tow boomer (0.2–3.5 kHz) and airgun (50–200 Hz). The obtained acoustic penetration depths were 100 m (pinger; Bobsien

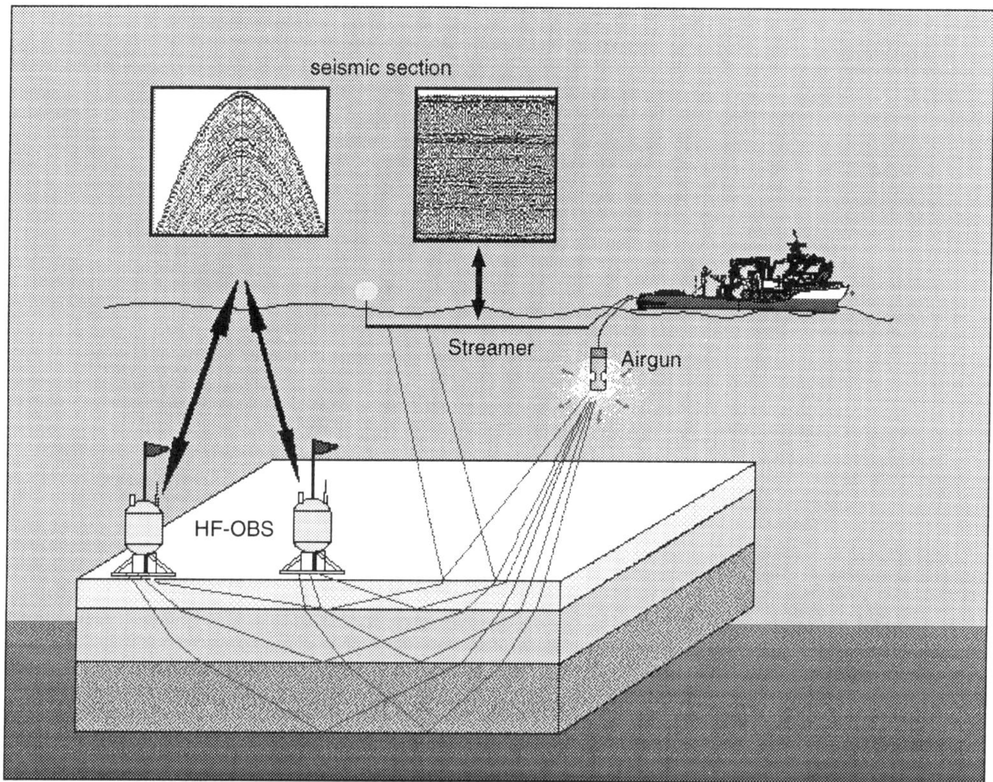

Fig. 2. Schematic diagram of the high-frequency near-vertical and wide-angle seismic reflection experiments to study gas hydrates and free gas zones in sediments below the sea floor. The source (2 l airgun) and six-channel streamer are towed behind the ship at a constant distance and speed. The position of the recording platform (HF-OBS) remains stationary on the sea floor. As a result a reflection hyperbola is recorded.

1995), approximately 250 m (deep-tow boomer; Bobsien 1995) and approximately 500 m (airgun; Posewang & Mienert 1996*a*, *b*). Sedimentary layers and P-wave velocities could be vertically resolved within a few metres and with an accuracy of much less than 50 m s^{-1}.

In addition, the upper sediment column and sea floor were studied with high-resolution methods. A Parasound echo-sounder, which is a sediment sub-bottom profiler, was used. It works with two simultaneous primary frequencies (a fixed frequency of 18 kHz and a variable frequency of 18–23.5 kHz). Through the parametric effect in water, a secondary frequency is produced in the range of 2.5–5.5 kHz. In ocean sediments, a penetration of up to 100 m is possible. Analogue plots are available for preliminary interpretation on board. Digital data are preserved on magnetic tape so that details can be studied with post-processing techniques (Spieß & Breitzke 1992). A deep-tow boomer provided by the British Geological Survey (BGS) was used and towed at depths down to 800 m. It enabled a resolution of more than 1 m and acoustic penetrations of up to 300 ms (approximately 150 m). Furthermore, an ORE side-scan sonar was deployed to image geological structures of the sea floor (Mienert 1994).

The R/V *Meteor* Hydrosweep system, a multibeam echo sounder system, was also used. It operates at a frequency of 50 kHz. The combined use of airgun, Parasound, deep-tow boomer, Hydrosweep systems and side-scan sonar provides detailed imaging of sea-floor morphology and seismic units in the study area, which would not have been achieved by using one system only.

Hydrate distribution based on geophysical signatures

Owing to the physical properties of gas hydrates and gas-bearing sediments, and the limitations inherent to different geological and geophysical methods used to map them, it is difficult to quantify their distribution and concentration. Generally, the gas hydrate zone and the free gas zone directly underneath creates a high-amplitude reflector, which is commonly referred to as the bottom simulating reflector (BSR). The BSR parallels the sea-floor reflection and its polarity is reversed relative to the sea-floor reflection due to a negative impedance contrast (e.g. Hyndman & Spence 1992; Posewang & Mienert 1996*a*). In continental margin sediments, the BSR appears as a sharp reflector with variable amplitude behaviour and character (Minshull *et al.* 1994). In addition, BSRs often cut across the dominant seismic strata (Shipley *et al.* 1979; Katzman *et al.* 1994; Mienert & Posewang 1996*a*; Andreassen & Hansen submitted) (Fig. 3). BSRs might either arise from an impedance contrast between high-velocity, partially hydrated sediments and water-saturated sediments, or from a contrast with partially gas-saturated sediments (Minshull *et al.* 1994). The compressional wave (P-wave) velocity changes are related to the concentration of gas and gas hydrates (cf. Field & Kvenvolden 1985, Hyndman & Spence 1992; Katzman *et al.* 1994; Minshull *et al.* 1994; Andreassen *et al.* 1995, and references therein). The presence of gas hydrates in porous sediments depends on the prevailing pressure and temperature conditions. Gas hydrates may form in water depths greater than 150 m in arctic waters and 610 m in subtropical waters, but this may vary with temperature, pore-water salinity and the content of other gases.

The first deep-sea gas hydrate site in the northeastern European continental margin inferred from seismic surveys was found at Storegga (Bugge *et al.* 1987) at 1200 m water depth and approximately 300 m below the sea floor (mbsf), and the first shallow-water gas hydrate in the Barents Sea at 345 m water depth and 180 mbsf (Andreassen *et al.* 1990) (Fig. 1).

Shallow-water gas hydrates and sea-floor craters at the Barents Sea: implications for marine methane releases

A shallow seismic reflector showing anomalous high amplitude and cutting through dipping layers was observed on a high-resolution multichannel seismic profile, as well as on conventional deep seismic lines in water depth of 345 m (Andreassen *et al.* 1990). The seismic

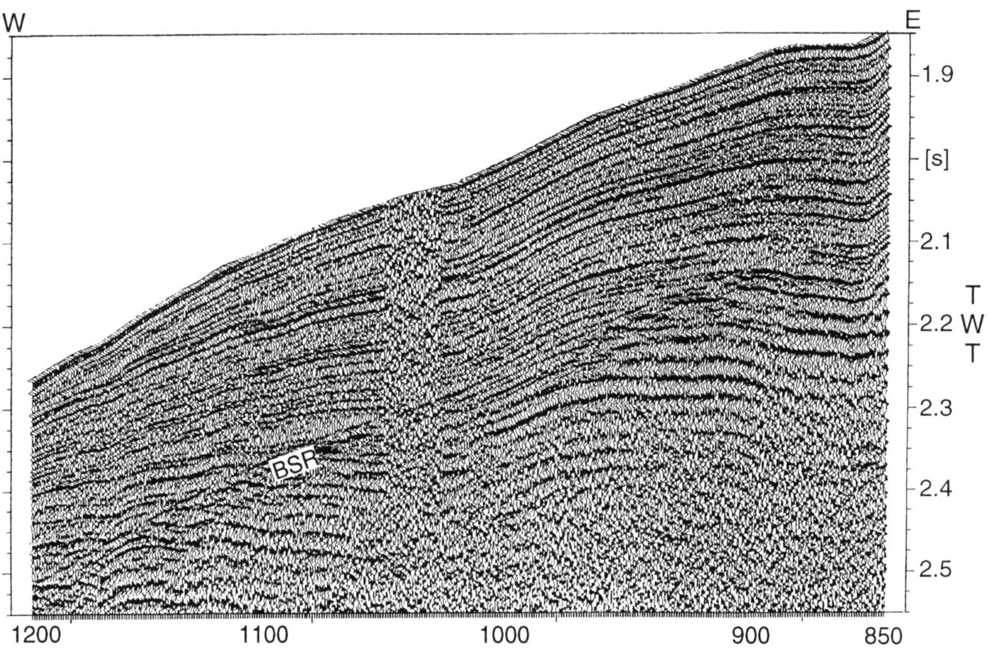

Fig. 3. Example of a high-amplitude BSR on the continental slope of northwest Spitsbergen cutting across the dominant seismic strata (from Posewang 1997). The BSR occurs at 0.25 s TWT bsf. The data are filtered (50–180 Hz) and recorded with a six-channel streamer and a 2 l airgun (Mienert 1994).

anomaly at 180 mbsf is due to the reflection from the base of gas-hydrated sediment which acts as a trap for the underlying free gas (Andreassen *et al.* 1990). Seismic reflection lines from this area document a patchy distribution of BSRs indicating gas hydrates and free gas at about 72°15′N–20°00′E. The BSR segments have various extents and distributions varying in diameter from a few kilometres to tens of kilometres. Short high-amplitude segments with variable extent are often grouped together immediately or close above the deeper faults, indicating that some faults may act as migration pathways from hydrocarbon reservoirs (Andreassen *et al.* 1990). No distinct sea-floor gas escape features such as pockmarks have been observed in this area. If gas hydrates become unstable and release the gas, one would expect an overpressure in sediments and, finally, a release of the gas from the sea floor into the oceanosphere, leaving behind pockmarks as fingerprints of the gas release processes. However, pockmark craters exist in the Barents Sea in the Bøjrnøyrenna area (Fig. 4), which is northeast of the gas hydrate field.

The area is part of a major trough in the west-central Barents Sea in water depths of 320–340 m at 74°54′N–27°34′E. It was first discovered by Solheim & Elverhøi (1993). The pockmark field extending over 35 km^2 east of the Bear Island was revisited during a RV *METEOR* cruise (Suess & Altenbach 1992) and was found to be covered by more than 30 craters between 300 and 700 m in diameter and up to 28 m deep (Hovland & Mienert 1992; Lammers *et al.* 1995*a*) (Fig. 4a). It is speculated that a strong sub-surface reflector, which rises to about 8 mbsf, relates either to a normal lithological change, gas-charged sediments or gas hydrates (Hovland & Mienert 1992; Hempel pers. comm.). The pockmark craters are clearly carving into the Quaternary sediments and the Triassic sandstone below. The craters have probably been created by explosive gas eruptions (Fig. 4b), which indicate a cluster of events in space (Solheim & Elverhøi 1993). Whether the events occurred at the same time remains unclear, but it is most likely that the events took place after the last deglaciation, i.e. 15000 years BP after the decay of the Barents Sea ice sheet. First, this would decrease the ice load and decrease the pressure on sediments, and thus increase the hydrate destabilization from the bottom upward. An enhancement of the free gas beneath the gas hydrate zone, and a migration of the hydrate stability zone (HSZ) from deeper to shallower levels leading to a thinning of the HSZ, would follow. As a result, there is a build-up of pressure and of free gas underneath the hydrate stability zone. Second, because bottom-water temperatures also have a profound effect on the gas hydrate stability (Dickens *et al.* 1995) a temperature increase within the Norwegian current at about 7000 years BP (Weinelt 1993) may also have shifted the gas hydrate in the upper sediment column from stable to unstable conditions (Fig. 5). Based on the phase diagram, an increase of 1°C would be necessary to destabilize the gas hydrate (Fig. 5), which is within the frame of palaeotemperature changes. Third, as there is a high potential for thermogenic hydrocarbons in the area (Johansen *et al.* 1993) and only a small amount of Cenozoic sediment accumulation occurred, the potential for hydrates of biogenic origin is low. Thus, shallow-water gas hydrates of the Barents Sea should be mainly of thermogenic origin. Their distribution should depend on geological structures such as deep fault zones. In addition to temperature and pressure changes, only a small excursion in the thermogenic gas composition may have shifted the hydrate–gas phase boundary (Fig. 5). A decrease of heavier carbons such as ethane and propane would result in a destabilization at lower temperatures, shifting the phase boundary to the left (Fig. 5). However, palaeotemperature changes have been documented for the Norwegian Margin (Sarnthein *et al.* 1995) which support our gas hydrate temperature stability hypothesis.

Anomalous high concentrations of methane in the shelf waters around the craters suggest that a strong methane source near this area is still active today (Lammers *et al.* 1995*a*). The vertical methane concentration decreases towards the sea surface due to biota using this energy pool, thus reducing the methane concentration within the water column by about 98% between 300 m depth and the sea surface; 80–85% of the methane is removed within the lower 50% of the water column (Lammers *et al.* 1995*a*). The removal processes within the water column are a dominating factor in the methane cycle within shallow-marine environments. The efficiency of the removal processes seems to be a function of water depth, i.e. the deeper the water column, the smaller the methane flux into the atmosphere (cf. Lammers & Suess 1994; for sea–air methane flux). Even though the flux of methane into the atmosphere is minimal due to the shape of the methane–depth distribution, the flux from the Barents Sea is in the upper range of the presently estimated global marine methane release. The flux increases when rough weather conditions produce strong vertical mixing during autumn and winter, i.e. the flux is seasonal (cf. Lammers *et al.* 1995*b*).

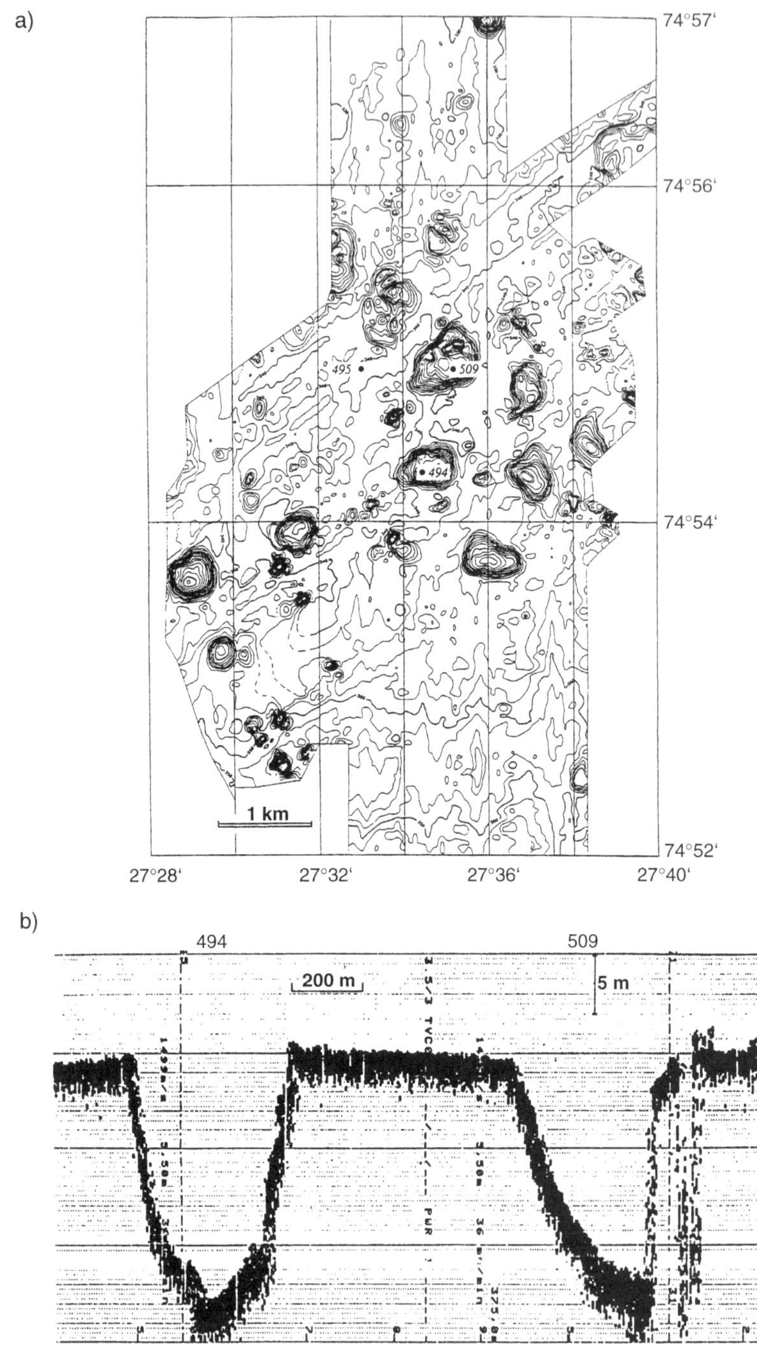

Fig. 4. (**a**) Sea-bed topography in the northern Bear Island trough obtained from the multi-beam echo sounder Hydrosweep (see work area 1 in Fig. 1) (Hovland & Mienert 1992; Lammers *et al.* 1995*a*). Large and small pockmark craters were observed in the area. (**b**) Parasound profile across the pockmark craters at stations 494 and 509 shows approximately 26 m deep and 400 m wide craters, but also larger features exist up to 28 m deep and 700 m wide.

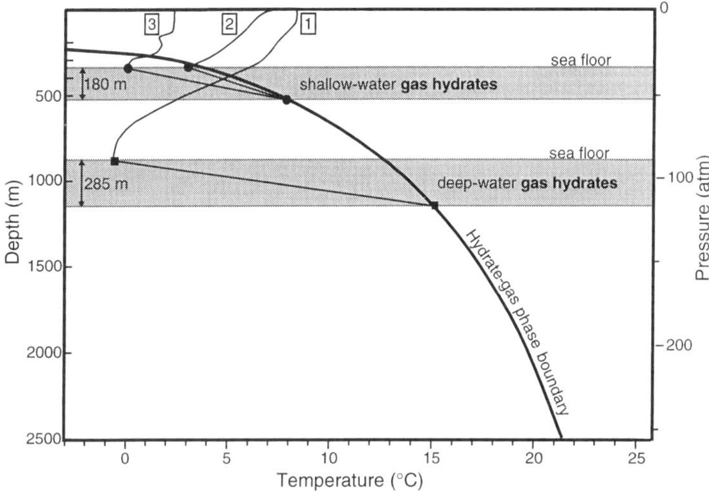

Fig. 5. Phase diagram of the present methane hydrate stability zone for shallow-water (Bear Island trough) and deep-water (Mid-Norwegian Margin) gas hydrates. The grey zone represents the present hydrate stability zone (HSZ). Bottom-water temperatures at 345 m water depth in the Bear Island trough and at 900 m water depth at the Mid-Norwegian Margin are based on CTD measurements. CTD station 1 is on the Mid-Norwegian slope, CTD station 2 is on the Barents Sea slope and CTD station 3 is within the pockmark field. The temperature of station 2 indicates warm northward flowing waters of the Norwegian current, i.e. Atlantic water masses. The geothermal gradient is 3°C per 100 m in Barents Sea sediments and 5.5°C per 100 m in sediments north of the Storegga Slide (Andreassen *et al.* 1990; Posewang 1997). While a bottom-water temperature increase of only 1°C of warm northward-flowing Atlantic waters and a spill over into the Barents Sea would completely destabilize the shallow water gas hydrates, a major temperature increase of 13°C would be necessary for the deep-water sites of the Mid-Norwegian Margin.

Controlling parameters for the methane distribution in the water column are removal (i.e. microbial oxidation), solution in the water column and transports (water mixing). The surface conditions are responsible for the gas exchanges between the water and the atmosphere, but the methane source remains the most important factor in the methane budget at the water–air boundary (Lammers *et al.* 1995*a*, and references therein). The Barents Sea methane anomaly can be considered as an example for many other submarine sedimentary sources of either thermogenic or biogenic methane in the northern shelf areas. From the studies by Lammers *et al.* (1995*a*, *b*), local and temporal pockmark activity seems to strongly determine the budget of the marine methane-related carbon cycle. The contribution of northern latitude shelf areas to the reported seasonal variations is still unclear and more data from long-term investigations of these anomalous methane regions in shallow and deep water should shed light on the methane flux processes from the lithosphere through the oceanosphere to the atmosphere (Fig. 6).

Deep-water gas hydrates and pockmarks at the Mid-Norwegian Margin: potential for marine gas releases

A coincidence of dissociation of gas hydrates and slope failures may exist at the Norwegian continental margin at Storegga, where Bugge *et al.* (1987) first documented a BSR in seismic reflection profiles. The Storegga Slide is located on the outer shelf and continental slope off Norway. Bugge and his co-workers have studied the area mainly within the slide complex (Bugge 1983; Jansen *et al.* 1987; Bugge *et al.* 1988). It is one of the world's largest submarine slides, having moved a total of 5600 km^3 of sediment with a thickness of up to 450 m from an area with a size comparable to that of mainland Scotland. The undisturbed sequences of the northern rim of the Storegga Slide were studied as part of the European North Atlantic Margin (ENAM) project, funded by the European Commission, and the SFB 313, funded by the German Science Foundation (DFG). The study concentrated on water depths between 400 and 1500 m. Several techniques were used to assess the distribution

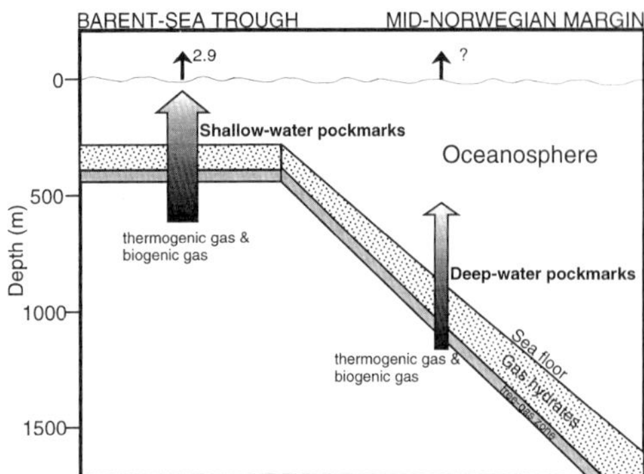

Fig. 6. Schematic diagram of methane release into the oceanosphere and atmosphere in the two work areas (Fig. 1). The values may indicate a 2.9×10^4 g CH_4 km^{-2} $year^{-1}$ net flux of methane from the Bear Island Trough site (Lammers *et al.* 1995*a*), and a generalized estimated flux from the sea floor of 13 g CH_4 m^{-2} $year^{-1}$ based on shallow hydrocarbon-rich areas (Hovland & Judd 1992).

and effects of free gas and gas hydrates in the slope sediments.

Several examples of possible connections between gas hydrate boundaries, submarine slides and slump surfaces exist from ocean margin sites (e.g. McIver 1977; Kayen & Lee 1991). Thus, a detailed high-resolution seismic study of BSRs, including side-scan sonar studies along the northern rim of the Storegga Slide, were carried out (Bobsien & Mienert 1994; Mienert 1994; Bobsien 1995; Posewang & Mienert 1996*a*, *b*; Posewang 1997) (Fig. 1). The experiments aimed at mapping their spatial distribution and their acoustic character, i.e. depths and thicknesses of free gas and gas hydrate zones, and gas or fluid escape features on the sea floor.

Airgun data collected at the northern rim of the Storegga Slide (water depth 700–1500 m) show a patchy distribution of gas hydrates and free gas zones (Fig. 7a and b). Relatively high P-wave velocities of 1900 m s^{-1} in the gas hydrate zones were inferred from the HF-OBS data (Fig. 8a and b). Water depth ranges between 850 and 870 m, and signal penetration reaches 0.6 s TWT. Two high-amplitude reflections occur, the first at 0.125 s TWT and the second at 0.375 s TWT (Fig. 7a). The seismic horizon at 0.375 s TWT shows a varying amplitude, where the BSR ends abruptly at shot point 890 (marked a in Fig. 7b), and is evident again at shot point 920. A similar feature exists from shot points 970–1000 (Fig. 7b). Such broken-up high-amplitude reflections are also documented from the shallow-water gas hydrates (Andreassen & Hansen 1996) where the areas with high-amplitude reflectors occur above faults and free gas is migrating to the gas hydrate stability zone along these faults. However, at Storegga it may document the gas or fluid escaping through the gas hydrate stability zone.

A phase reversal occurs at 0.375 s TWT indicating a velocity inversion (Posewang 1997). It corresponds to a low-velocity zone of 10–30 m in thickness, and it provides evidence for a free-gas zone characterized by velocities between 1300 and 1400 m s^{-1}, i.e. below the speed of sound in sea water (1500 m s^{-1}) (Fig. 8). The velocity–depth structure inferred from the HF-OBS data shows the existence of several alternating zones of high and low velocities (Fig. 8), which might point towards small-scale vertical variations in gas composition within gas hydrate reservoirs or a migration of the stability field into a new equilibrium. An outstanding observation is the first evidence for a possible double BSR in the velocity profile (Fig. 8). The seismic reflection data and the HF-OBS data document the fact that the BSR probably does correspond to a simple first-order impedance boundary, but it also has to be considered as more complex structure, both vertically and laterally (cf. also Minshull *et al.* 1994). Detailed travel-time modelling and frequency studies further showed that the free gas-bearing sediment layers below the BSR act as a low-pass filter on the seismic signal (Posewang 1997). Above the BSR, frequencies vary with values up to 170 Hz, in contrast to prevailing frequencies of less than 80 Hz below the BSR (Posewang 1997). Such shifts towards lower frequencies are also observed in zones below gassy sands (e.g. Taner *et al.* 1979).

A detailed side-scan survey running parallel to the northern rim of the Storegga Slide showed two major pockmark fields, which concentrate in water depths at approximately 700 m and 900 m, respectively (Fig. 9). They provide evidence of gas escaping from the sea floor. While individual pockmarks cluster in the vicinity of the two fields, the pockmark population reaches a minimum at approximately 800 m (Fig. 9). In addition, airgun (Fig. 7) and deep-tow boomer (Fig. 10) records showed clear evidence for subvertical horizons of acoustic ‘wipe-out’ zones, which will be called pockmarks, vents and mud volcanoes, and which may document various development stages of pressure build-up.

Pockmarks (Fig. 10a) are shallow depressions in the sea bed which are thought to be formed by the escape of gas or fluid (e.g. Hovland & Judd 1988). This process may be gradual or explosive, and leads to the formation of a depression at the sea bed (Fig. 10a). Most frequently, the pocka)

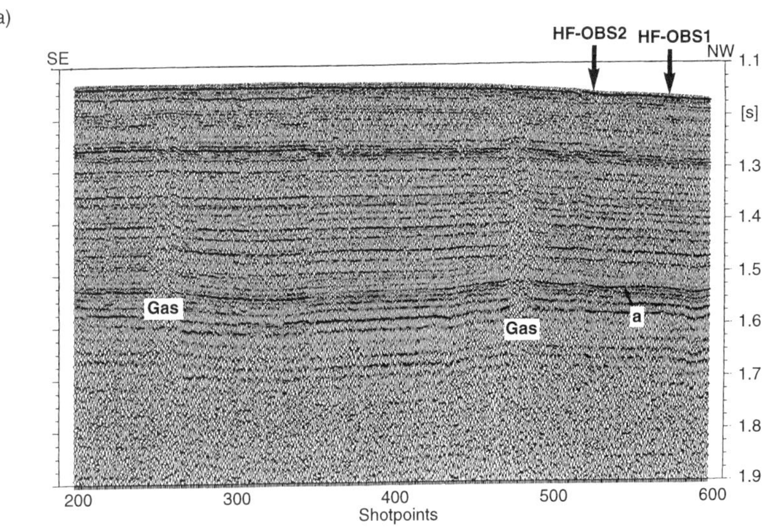

b)

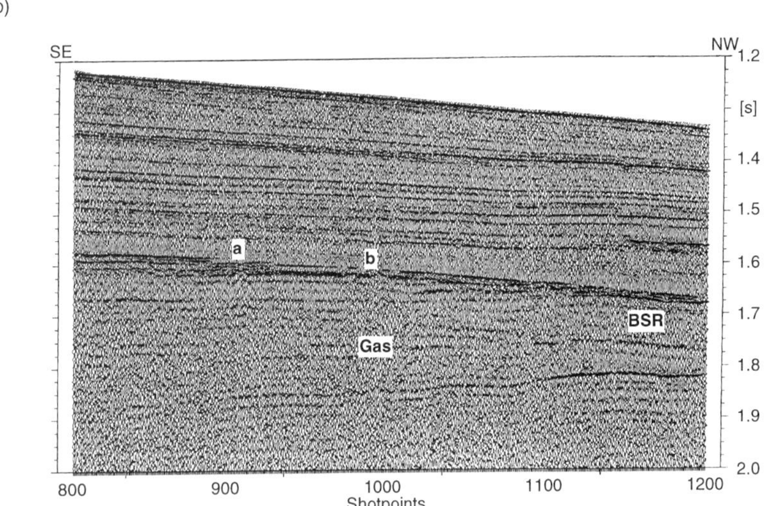

Fig. 7. (**a**) Seismic section of a seismic profile recorded with a six-channel streamer using a 2 l airgun. High-amplitude reflectors occur at 0.125 s and 0.375 s TWT. The lower horizon (a) shows a phase reversal compared to the sea floor. The HF-OBS positions are marked by arrows. Vertical acoustic wipe-out zones may indicate gas/fluid escape zones. (**b**) The BSR occurs at 0.35 s TWT and is clearly cutting the parallel reflections from sediment layers. Its amplitude varies distinctly and sometimes (a and b) vanishes completely. An acoustic blanking is observed directly above the BSR. Vertical acoustic transparent zones may indicate free gas escape zones through the hydrate stability zone.

marks are V-shaped, either with or without levees, and they measure several metres in a vertical section and several tens of metres in cross-section.

Vents are wide acoustic transparent features which extend to the sea bottom (Fig. 10b). Hence, in both cases it is very likely that gas has escaped out of the sedimentary sequence into the water column. In some instances, the location of a former pressure dome can be recognized (Fig. 10c), indicated by seismically transparent columns. The escape route above these pressure domes is commonly narrower than the width of the pressure dome itself.

Mud volcanoes are only recorded in one

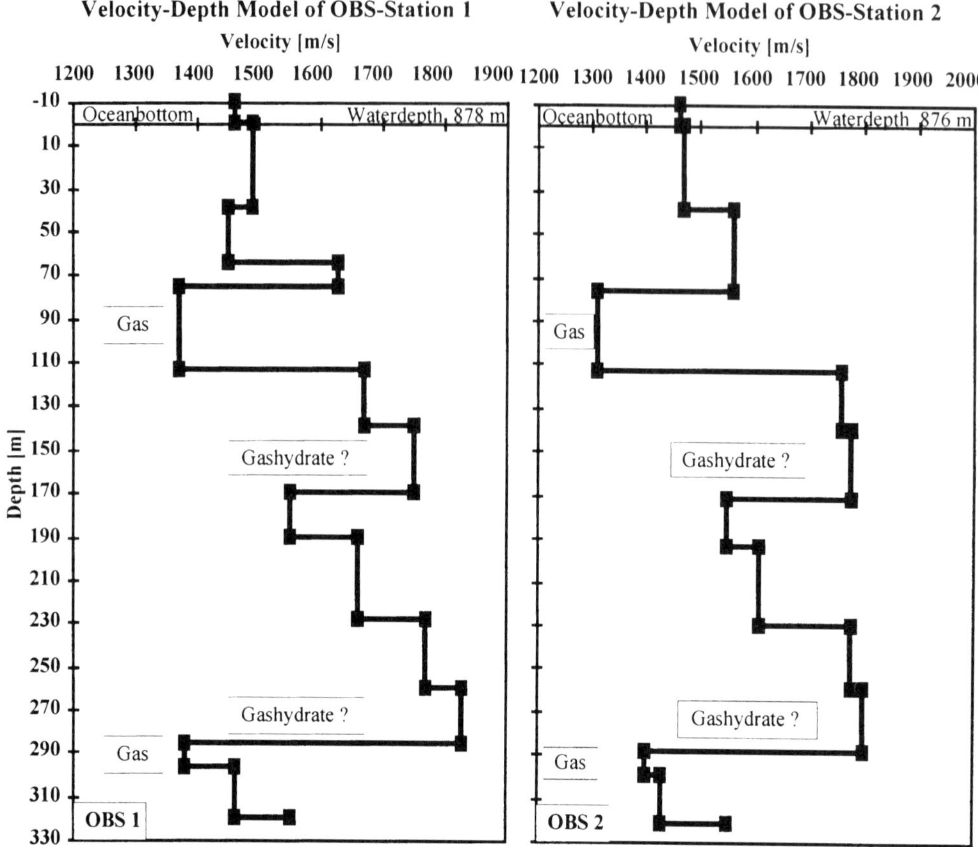

Fig. 8. (**left**) V_P-depth model for HF-OBS station 1 (see Fig. 7a) shows two distinct low-velocity zones well below the speed of sound in water (< 1460 m s^{-1}). They occur at the depth of the high-amplitude reflectors (Fig. 10a). The low-velocity zones can be explained by free gas zones. The high-velocity zone (> 1800 m s^{-1}) should indicate gas hydrates. (**Right**) V_P-depth model for HF-OBS station 2 (see Fig. 7a) shows two distinct low-velocity zones well below the speed of sound in water (< 1460 m s^{-1}) which occur at the depth of the high-amplitude reflectors (Fig. 7a). They can be explained by free gas zones. The high-velocity zones (1800 m s^{-1}) should indicate gas hydrates.

instance (Fig. 10d). The feature is similar to a pockmark except that it has a weakly defined topographic rim on the profile and upturned marginal reflectors, suggesting forceful movement through the sediment (Fig. 10d). It is the largest feature found in the area and approximately 300 m in diameter and 5 m in depth.

The features defined here appear to confirm the presence of fluid/gas escaping through the sea floor in an area where gas hydrates are documented beneath the sea floor. Some pockmarks do not appear to have been active for some time, as they are buried. In other cases, the fluid/gas escape pathways may not reach the surface. However, there is much evidence for the presence of gas hydrate and free gas in the sediments to the north of the slide edge. If this is an area of gas hydrate dissociation, then gas escape features such as the observed ones should be most common.

Earthquake distribution based on historical and instrumental data

The Norwegian Continental Shelf (NCS) is a tectonic part of the Eurasian plate, with a seismicity level well below that of many plate margin areas. There are, however, examples of large intraplate earthquakes in the Norwegian Continental Shelf and zones of weakness that may be potential locations of future earthquakes (cf. Bungum & Selnes 1988).

During the Tertiary, the north-eastern Eur-

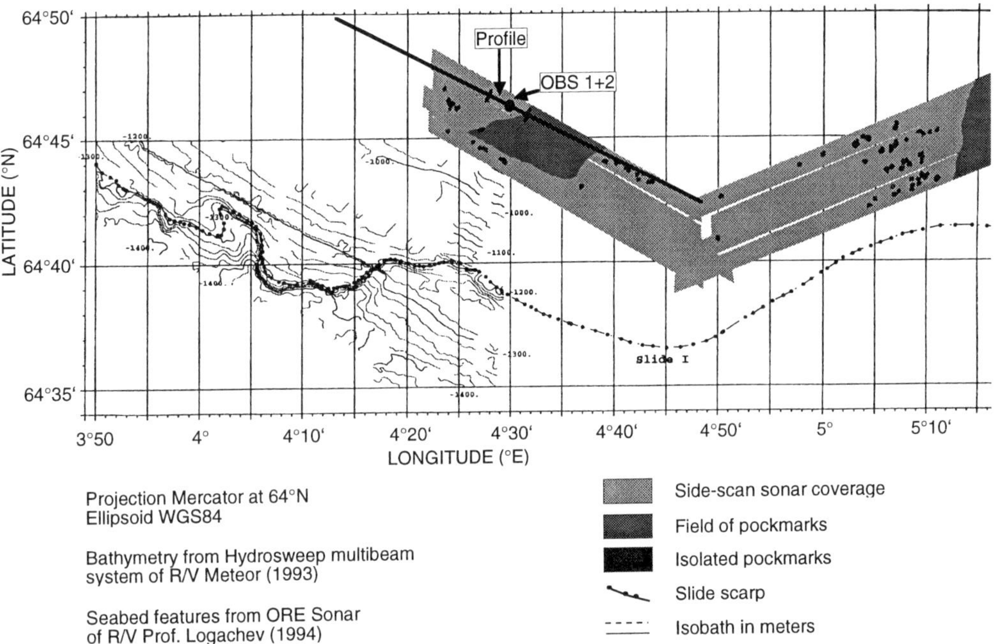

Fig. 9. Interpretation map of side-scan sonar and hydrosweep tracks at the northern rim of Storegga Slide (sources: ENAM project data, processed at IFREMER, see work area 2 in Fig. 1). The positions of the seismic line (Fig. 7) and HF-OBS (Fig. 8) stations are marked. A cluster of pockmarks occur close to two densely populated pockmark fields at 700 and 900 m water depths.

opean Continental Margin was dominated by tectonics of subsidence, inversion and uplift, which have been operating for long periods in parallel in different regions on the NCS. Earthquakes large enough to cause considerable damage to structures took place in the NCS. The largest historical earthquakes, which have been sufficiently documented to allow magnitude assessments, are the 1759 Kattegat ($\mathbf{M}_s$ 6.5), the 1819 Nordland ($\mathbf{M}_s$ 5.8), the 1866 Western Norway ($\mathbf{M}_s$ 5.7), the 1894 Lofoten ($\mathbf{M}_s$ 5.4) and the 1904 Oslofjord ($\mathbf{M}_s$ 5.4) earthquakes. The first earthquake found in historic records in Scandinavia dates back to 1073 (Bungum & Selnes 1988).

The epicentral distribution of historical earthquakes from Muir Wood & Woo (1988) is based on a new historical catalogue. The seismicity of the Norwegian Continental Shelf obtained from instrumentally recorded earthquakes is depicted in Fig. 11. Analyses of the seismological data give a seismicity pattern which correlates well with both regional and local geological features (Bungum 1988). A clear zone of significant seismic activity along the west coast of Norway and on the continental margin off Norway can be detected. On a regional scale, the seismic activity along the NCS shows a more or less continuous level of activity from north to south. The Storegga Slide area is, seismically, a particularly active area (Figs 1 and 11).

The main seismic sources in the Norwegian Continental Margin are earthquakes released by earth movements along faults. On the NCS, faults are often difficult to identify because movements along the faults are relatively small and often not easily seen from controlled seismic source data. In addition, there are relatively few earthquakes and the accuracy in the location of earthquakes is limited. The study by Bungum & Selnes (1988), however, has identified major fault systems which have experienced large movements in the past and, therefore, also must be considered as possible sources of earthquakes in the future, which may trigger mass movements from the continental slope to the deep sea.

Possible link between continental margin earthquake activities and hydrate instabilities

The Norwegian Margin has undergone an extensive sliding activity in the past (Fig. 1). A 35–40 km wide slide occurred about 200 000 years ago off the Bjørnøy Channel. The three largest historically known slides are at Storegga and were possibly triggered by earthquakes (Bugge

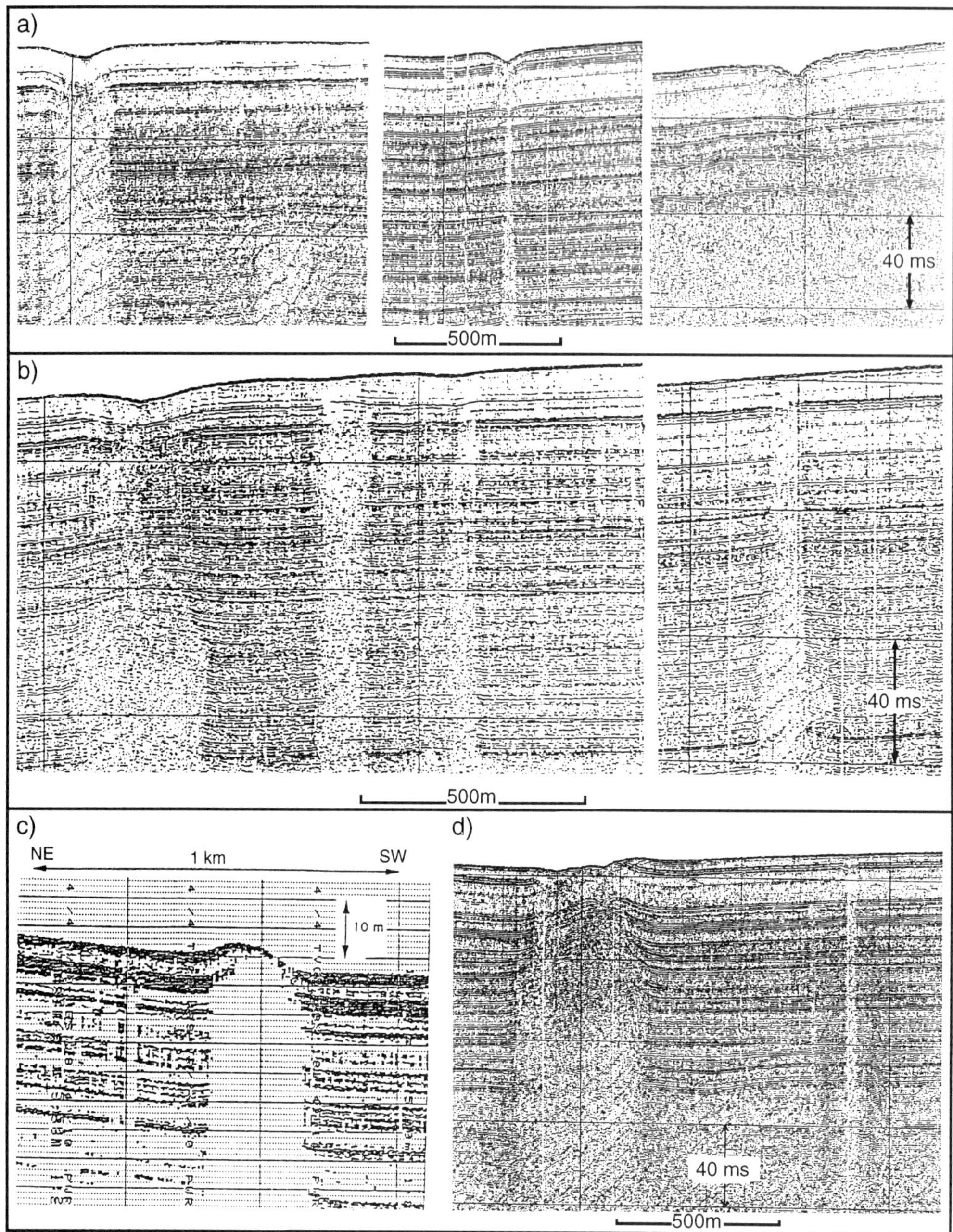

Fig. 10. (**a**) Deep-tow boomer profile showing pockmarks which are less than 5 m deep and up to 200 m wide. Vertical acoustic wipe-out zones evidently give hints to fluid/gas escape features. (**b**) Vents which sometimes show an upward thinning of the acoustic wipe-out zones (left) and, thus, the so-called chimney. (**c**) The parasound profile indicates a doming of a gas vent on the sea floor at a water depth of 735 m. (**d**) The 'Mud volcano' shows an upward turning of reflectors at the outer border of the acoustic wipe-out zone and a rimmed pockmark at its surface.

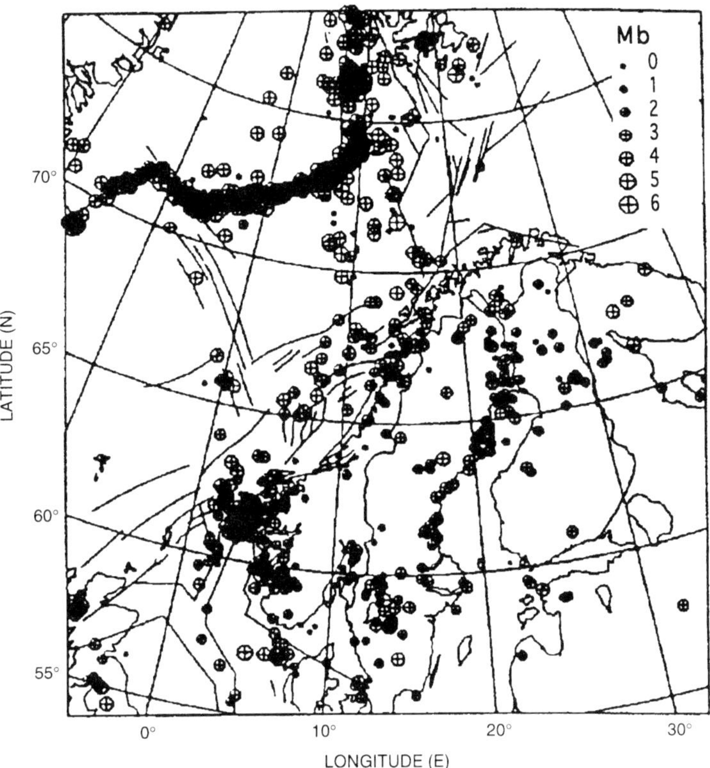

Fig. 11. Instrumentally recorded earthquakes between 1955 and 1987. Symbol size corresponds to the magnitudes (either m_b or ML) (from Bungum 1988).

1983; Bugge *et al.* 1987). Among the most frequently suggested triggering mechanisms for submarine slides are earthquakes and gas hydrates. Several factors may cause submarine slides. The stability of slopes is given by the relation between the resisting forces and the driving forces. The main gravity forces may be aided by earthquake accelerations, oscillatory motions imposed by waves, overloading from sedimentation or upslope slide depositions, ice weight and ice-induced forces in glaciated areas, and gas and internal current water motion in the sediments (Bugge 1983). The resisting forces are based on the shear strength along potential gliding planes. The effective shear strength may be reduced, for instance, by an increase in pore pressure. Excess in pore pressure can be caused by wave or earthquake loading, rapid sedimentation (underconsolidated sediments), by the rapid lowering of sea level, gas generation and gas decomposition (Bugge 1983).

Gas hydrates are dependant on certain temperature and pressure conditions, and earthquakes may disturb this gas hydrate equilibrium. The gas hydrate may then start to decompose. The consequent release of gas and water volume would exceed the volume previously occupied by the gas hydrate, so that the internal pressure would drastically rise if large volumes of hydrate are decomposed. The combination of the gas hydrate decomposition and additional external factors, such as sea-level lowering or earthquake loading, may be sufficient to cause sediment to collapse and to liquefy over large areas. Gas hydrates, free gas and a double BSR were documented in the Storegga area where gas hydrates may occur at the same depths as the glide planes (Fig. 12). Hence, gas hydrate destabilization in the sediments probably caused liquefaction and earthquake loading may have triggered the slides.

Implications for the marine methane cycle

Given that the Storegga Slide is such a major geological feature so close to the coast of Europe,

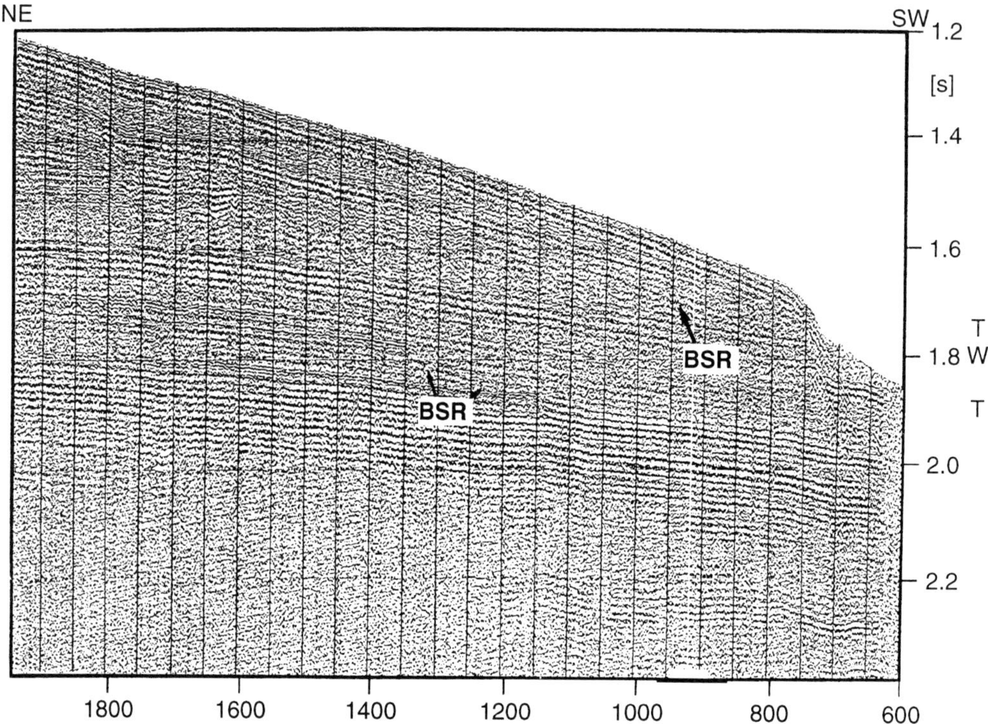

Fig. 12. Seismic reflection profile analogue recorded only with a single-channel streamer. The sound source was a 2 l airgun. Two BSRs are observed: one at 0.1 s TWT, and the deeper one at 0.35 s TWT. The shallow BSR occurs at the same depth as the glide plane of Storegga at approximately 75 m depth.

survey coverage of it is surprisingly limited. More details have been discovered over a very small part of the slide, and already it is clear that the gas hydrate dynamics documented in a double BSR and fluid/gas escape features are more complex than hitherto proposed from other areas. Much further work is necessary, in particular the following should be considered: the seismicity distribution along the margin (Bungum & Selnes 1988) seems to coincide with the distribution of highest potential for gas hydrates proposed by Andreassen & Hansen (1996). Hence, future studies should concentrate on combined geochemical and geophysical approaches.

Methane is one of the most effective greenhouse gases and undergoes significant global variations (Rasmussen & Khalil 1981). Even though the cycling of methane and the study of its global budget clearly indicate a dominant influence from terrestrial sources and a minor marine contribution, the marine methane sources and sinks have yet to be fully quantified. Lammers *et al.* (1995*a*) studied, in detail, the local impact of a methane source near Bear Island (Barents Sea; Fig. 1), one of the largest known shallow-water marine methane escape zones. This site is typical for the heterogeneous distribution of methane in water of the northern continental shelf and is probably an important natural source of methane for the present atmosphere.

There appears to be significant earthquake activity in the Storegga area. Immediately outside the northern margin of the Storegga Slide scar, gas-escape features have been observed in water depths between 600 and 1300 m. It is believed that there is free gas and also a gas hydrate layer parallel to the sea floor and cutting across bedding planes (seen as a Bottom simulating reflector) at a sub-bottom depth of 300 m. However, not only is the rapid release of methane during major sediment slide events in coincidence with gas hydrate dissociation an open question (Fig. 13), but also the present release of methane from the deep-water sites.

Conclusions and outlook

In the north-eastern European Continental Margin and the adjacent deep-ocean regions,

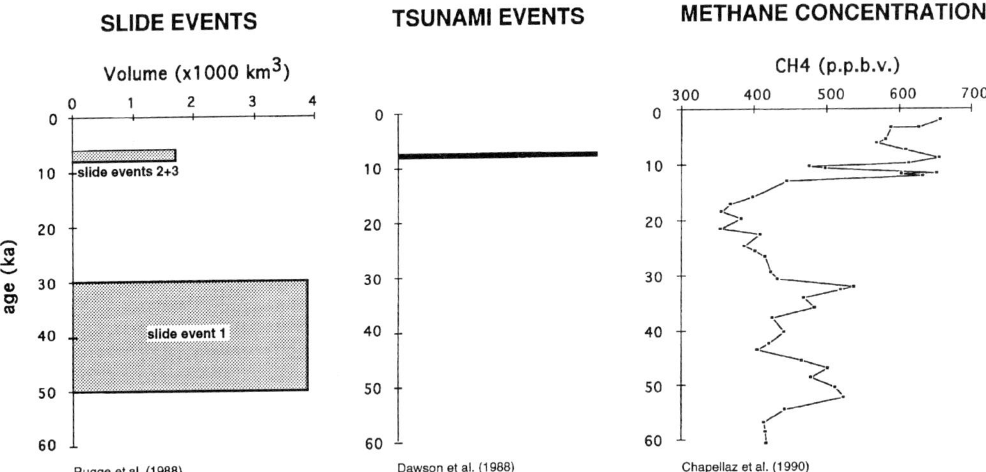

Fig. 13. Comparison between slide events, tsunamis and methane concentrations in the atmosphere (sources: Bugge *et al.* 1988; Dawson *et al.* 1988; Chappellaz *et al.* 1990).

shallow- and deep-water gas hydrates could be identified from observed bottom simulating reflectors on seismic records (e.g. Bugge *et al.* 1987; Mienert 1994; Andreassen *et al.* 1995; Posewang & Mienert 1996*a*, *b*). The concentration of hydrates present above the BSR and free gas in the underlying sediments is still the subject of controversial debates (Miller *et al.* 1991; Hyndman & Spence 1992; Katzman *et al.* 1994; MacKay *et al.* 1994; Minshull *et al.* 1994). There is a strong need to quantitatively constrain the models with independent drilling and sampling in order to better understand how gas hydrates cluster in sediments and how saturation affects the elastic properties of the sediments.

The presence of gas hydrates in the pore space of sediment formations implies a significant change in their petrophysical properties (shear strength, internal friction and elastic constants) and their hydraulic characteristics (porosity and permeability). These alterations affect the margins stability angle and the efficiency of fluid transport in pore space (variations of pore pressure). During the past years, the importance of gas hydrates in terms of ocean sediment stability has been recognized. For a better understanding of the relations between gas escape zones, gas hydrate equilibrium and margin instabilities, the variations at escape zones and the gas hydrates will have to be studied in detail in terms of geological, geochemical and geotechnical behaviour. The relationship between seismicity, activity at gas escape zones and continental margin instabilities have yet to be quantified. The analysis of world-wide distribution and the total amount of gas hydrates (cf. Kvenvolden 1993; Gornitz & Fung 1994) will have to be analysed in relation to their impact on the climate.

The HF-OBH developments and high-frequency seismic studies performed at SFB 313 and GEOMAR were supported by the ENAM programme, funded by the European Commission Mast II and the Deutsche Forschungsgemeinschaft. We are grateful to H. Beese for his technical support during the HF-OBS system deployments. We thank U. Brennwald, M. Schwartz and M. Wilken for assistance in preparing the final draft of this paper.

References

ANDREASSEN, K. & HANSEN, T. 199?. Inferred gas hydrates offshore Norway and Svalbard. *Norsk Geologisk Tidsskrift* (submitted).

——, HART, P. E. & GRANTZ, A. 1995. Seismic studies of a bottom simulating reflection related to gas hydrate beneath the continental margin of the Beaufort Sea. *Journal of Geophysical Research*, **100**, 12 659–12 673.

——, HOGSTAD, K. & BERTEUSSEN, K. A. 1990. Gas hydrate in the southern Barents Sea, indicated by a shallow seismic anomaly. *First Break*, **8**(6), 235–245.

BOBSIEN, M. 1995. *Entwicklung und Einsatz eines Hoch-Frequenz-Ozean-Boden-Hydrophons zur hochauflösenden Bestimmung von Kompressionswellengeschwindig-keiten in Sedimenten.* PhD thesis, Christian-Albrechts-University of Kiel, Germany.

—— & MIENERT, J. 1994. Geoacoustic hydrophone-pinger experiments. *In*: SUESS, E., KREMLING, K. & MIENERT, J. (eds) *Nordatlantik 1993, Cruise*

No. 26. Meteor Report. University of Hamburg, **94**(4).

BUGGE, T. 1983. *Submarine Slides on the Norwegian Continental Margin, With Special Emphasis on the Storegga Area.* Continental Shelf Institute, Norway, **110**.

——, BELDERSON, R. H. & KENYON, N. H. 1988. The Storegga slide. *Philosophical Transactions of the Royal Society of London*, **A325**, 357–388.

——, BEFRING, S., BELDERSON, R. H., EIDVIN, T. *et al.* 1987. A giant three-stage submarine slide off Norway. *Geo-Marine Letters*, **7**, 191–198.

BUNGUM, H. 1988. Instrumentally analysed seismicity of the Norwegian Continental Shelf. *In*: BUNGUM, H. & SELNES, P. (eds) *Earthquake Loading on the Norwegian Continental Shelf (ELOCS)*. Summary Report, Norwegian Geotechnical Institute, 25–26.

—— & SELNES, P. B. 1988. *The Joint Industrial Research and Development Program 'Earthquake Loading on the Norwegian Continental Shelf' (ELOCS)*. Summary Report, Norwegian Geotechnical Institute.

CHAPPELLAZ, J., BARNOLA, J. M., RAYNAUD, D., KORTKEVICH, Y. S. & LORIUS, C. 1990. Ice-core record of atmospheric methane over the past 160,000 years. *Nature*, **345**, 127–131.

DAWSON, A., LONG, D. & SMITH, D. E. 1988. The Storegga Slides: Evidence from eastern Scotland for a possible tsunami. *Marine Geology*, **82**, 271–276.

DICKENS, G. R., O'NEILL, J., REA, D. & OWEN, R. 1995. Dissociation of oceanic methane hydrate as a cause of the carbon isotope excursion at the end of the Paleocene. *Paleoceanography*, **10**(6), 965–971.

——, PAULL, C., WALLACE, P. & THE ODP LEG 164 PARTY 1997. Direct measurement of *in situ* methane quantities in a large gas-hydrate reservoir. *Nature*, **385**, 426–428.

DOWDESWELL, J., KENYON, N., ELVERHOI, A., LABERG, J., HOLLENDER, F.-J., MIENERT, J. & SIEGERT, M. 1996. Large-scale sedimentation on the glacier-influenced Polar North Atlantic margins: Long-range side-scan sonar evidence. *Geophysical Research Letters*, **23–24**, 3535–3538.

EIKEN, O. & HINZ, K. 1993. Contourites in the Fram Strait. *Sedimentary Geology*, **82**, 15–32.

FIELD, M. & KVENVOLDEN, K. 1985. Gas hydrates on the northern California continental margin. *Geology*, **13**, 517–520.

GORNITZ, V. & FUNG, I. 1994. Potential distribution of methane hydrates in the world's oceans. *Global Biogeochemical Cycles*, **8**, 335–347.

HOVLAND, M. & JUDD, A. 1988. *Seabed Pockmarks and Seepages*. Graham and Trotman, London.

—— & JUDD, A. 1992. The global production of methane from shallow submarine sources. *Continental Shelf Research*, **12**, 1231–1238.

—— & MIENERT, J. 1992. Parasound profiling and Hydrosweep mapping of shallow gas reservoirs on the Barents Shelf and the Vring Plateau. *In*: SUESS, E. & ALTENBACH, A. (eds) *Europäisches Nordmeer, Reise 17*. Meteor-Berichte, University of Hamburg, **92**(3), 164.

HYNDMAN, R. & SPENCE, G. 1992. A seismic study of methane hydrate marine bottom simulating reflectors. *Journal of Geophysical Research*, **97**, 6683–6698.

JANSEN, E., BEFRING, S., BUGGE, T. *et al.* 1987. Large submarine slides on the Norwegian continental margin: sediments, transport, and timing. *Marine Geology*, **78**, 77–107.

JOHANSEN, S., OSTISTY, B., BIRKELAND, Ø. *et al.* 1993. Hydrocarbon potential in the Barents Sea region: play distribution and potential. *In*: VORREN, T., BERGSAGER, E. *et al.* (eds) *Arctic Geology and Petroleum Potential.* NPF Special Publication, Volume 2. Elsevier, Amsterdam, **2**, 273–320.

KAYEN, R. & LEE, H. 1991. Pleistocene slope instability of gas hydrate-laden sediment on the Beaufort Sea margin. *Marine Geotechnology*, **10**, 125–141.

KATZMAN, R., HOLBROOK, W. & PAULL, C. 1994. Combined vertical-incidence and wide-angle seismic study of a gas hydrate zone, Blake Ridge. *Journal of Geophysical Re*search, **99**, 17975–17995.

KVENVOLDEN, K. 1988. Methane hydrate and global climate. *Global Biogeochemical Cycles*, **2**, 221–229.

—— 1993. Gas hydrates – Geological perspective and global change. *Review of Geophysics*, **31**(2), 173–187.

——, & BARNARD, L. 1983. Hydrates of natural gas in continental margins. *In*: WATKINS, J. & DRAKE, C. (eds). The American Association of Petroleum Geologists, Memoir, 631–640.

LABERG, J. & VORREN, T. 1993. A late Pleistocene submarine slide on the bear Island Trough Mouth Fan. *Geo-Marine Letters*, **13**, 227–234.

LAMMERS, S. & SUESS, E. 1994. An improved headspace analysis method for methane in seawater. *Marine Chemistry*, **47**, 115–125.

——, SUESS, E. & HOVLAND, M. 1995a. A large methane plume east of Bear Island (Barents Sea): implications for the marine methane cycle. *Geologische Rundschau*, **84**, 59–66.

——, SUESS, E., MANSUROV, M. & ANIKIEV, V. 1995*b*. Variations of atmospheric methane supply from the Sea of Okhotsk induced by the seasonal ice cover. *Global Biogeochemical Cycle*, **9**, 351–358.

MACDONALD, G. 1990. Role of methane clatharates in past and future climates. *Climatic Change*, **16**, 247–281.

MACKAY, M., JARRAD, R., WESTBROOK, G., HYNDMAN, R. & THE SHIPBOARD SCIENTIFIC PARTY OF ODP LEG 146. 1994. Origin of the bottom simulating reflector: Geophysical evidence from the Cascadia accretionary prism. *Geology*, **22**, 459–462.

MAX, M. & LOWRIE, A. 1996. Oceanic methane hydrates: a "frontier" gas resource. *Journal of Petroloeum Geology*, **19**(1), 41–56.

MCIVER, R. 1977. Hydrates of natural gas-important agent in geological processes. Abstracts, *Geological Society of America* Abstracts with Programs, **9**, 1089–1090.

MIENERT, J. 1994. *Cruise Reports: R/V* Livonia *Cruise 92 – GLORIA Studies of the East Greenland Continental Margin Between 70° and 80°N; R/V* Poseidon *PO 200/10 – European North Atlantic Margin: Sediment Pathways, Processes, and Fluxes; R/V* Akademic Aleksandr Karpinskiy –

Gas Hydrates on the Northern European Continental Margin. GEOMAR Report, **30**.

—— & POSEWANG, J. 1996. High-frequency OBH deployments to determine compressional wave velocities in gas hydrate-cemented sediments of continental margins. (Abstract.). *In*: HENRIET, J.-P. & MIENERT, J. (eds) *First Master Workshop on Gas Hydrates: Relevance to World Margin Stability and Climatic Change*, University of Gent, Belgium.

——, BOBSIEN, M., POSEWANG, J. & BELOUSSOV, M. 1994. Bottom simulating reflectors on the northern European continental margin: evidence for gas hydrates. *In: Third International Conference on Gas in Marine Sediments*, NIOZ, Texel, the Netherlands.

——, BOBSIEN, M. & POSEWANG, J. 1995. High-frequency ocean bottom hydrophone (HF-OBH) deployments on the sea floor: Investigations of gassy sediments on the northern European continental margin. *In*: MONATGNER, J.-P. & LANCELOT, Y. (eds) *Multidisciplinary Observations on the Deep Seafloor*. Marseille, France.

——, KENYON, N. , THIEDE, J. & HOLLENDER, F. 1993. Polar continental margins: Studies off East Greenland. *EOS, Transactions of the American Geophysical Union*, **74**(20), 225–236.

MILLER, J., LEE, M. & VON HUENE, R. 1991. An analysis of a seismic reflection from the base of a gas hydrate zone, offshore Peru. *AAPG Bulletin*, **75**, 910–924.

MINSHULL, T., SINGH, S. & WESTBROOK, G. 1994. Seismic velocity structure at a gas hydrate reflector, offshore western Colombia, from waveform inversion. *Journal of Geophysical Research*, **99**, 4715–4734.

MUIR WOOD, R. & WOO, G. 1988. The historical seismicity of the Norwegian Continental Shelf. *In*: BUNGUM, H. & SELNES, P. (eds) *Earthquake Loading on the Norwegian Continental Shelf (ELOCS)*. Summary Report, Norwegian Geotechnical Institute, 24–25.

POSEWANG, J. 1997. *Nachweis von Gashydraten und freiem Gas in den Sedimenten des nordwesteuropäischen Kontinentalabhangs mit hochauflösenden reflexiosseismischen Methoden und HF-OBS-Daten*. PhD thesis, Christian-Albrechts-University of Kiel, Germany.

—— & MIENERT, J. 1996a. High-resolution investigations of gas hydrates and free gas in sediments of the continental margin of Svalbard. (Abstract.). *In*: HENRIET, J.-P. & MIENERT, J. (eds) *First Master Workshop on Gas Hydrates: Relevance to World Margin Stability and Climatic Change*, University of Gent, Belgium.

—— & —— 1996*b*. Gas hydrates and free gas in sediments of the Norwegian continental margin: results from wide-angle seismic. (Abstract.). *In*: HENRIET, J.-P. & MIENERT, J. (eds) *First Master Workshop on Gas Hydrates: Relevance to World Margin Stability and Climatic Change*, University of Gent, Belgium.

RASMUSSEN, R.& KHALIL, M.1981. Increase in the concentration of atmospheric methane. *Atmospheric Environment*, **15**, 883–886.

SARNTHEIN, M., JANSEN, E., WEINELT, M. *et al.* 1995. Variations in Atlantic surface ocean paleoceanography, 50°-80°N: A time-slice record of the last 30,000 years. *Paleoceanography*, **10**(6), 1063–1094.

SHIPLEY, T, HOUSTON, M., BUFFLER, R. *et al.* 1979. Seismic evidence for widespread possible gas hydrate horizons on continental slopes and rises. *AAPG Bulletin*, **63**, 2204–2213.

SOLHEIM, A. & ELVERHØI, A. 1993. A pockmark field in the central Barents Sea. *Geo-Marine Letters*, **13**, 235–243.

SPIEß, V. & BREITZKE, M. 1992. Seismic signature of the narrow beam parasound echosounder: analysis of analog and digital echograms. *In*: WEYDERT, M. (ed.) *European Conference on Underwater Acoustics*. Elsevier Applied Science, London, 489–492.

SUESS, E. & ALTENBACH, A. 1992. Europäisches Nordmeer, Reise 17. *METEOR-Berichte*, University of Hamburg, **92**(3).

TANER, M., KOEHLER, F. & SHERIFF, R. 1979. Complex seismic trace analysis. *Geophysics*, **44**, 1041–1063.

VORREN, T.O., LABERG, J.S., BLAUME, F., MIENERT, J., RUMOHR, J. & WERNER, F., in press. The Norwegian-Greenland Sea continental margins: morphology and late Quaternary sedimentary processes and environment. *In: 'The PONAM Book', Part III, Sedimenary Processes and Products*, Ch. 9, Springer, Berlin.

WEINELT, M. 1993. *Veränderungen der Oberflächenzirkulation im Europäischen Nordmeer während der letzten 60.000 Jahre – Hinweise aus stabilen Isotopen*. Berichte Sonderforschungsbereich 313, Universität Kiel, **41**, 106.

Evidence for faulting related to dissociation of gas hydrate and release of methane off the southeastern United States

W. P. DILLON, W. W. DANFORTH, D. R. HUTCHINSON, R. M. DRURY, M. H. TAYLOR & J. S. BOOTH

US Geological Survey, Woods Hole, MA 02543, USA

Abstract: An irregular, faulted, collapse depression about 38 × 18 km in extent is located on the crest of the Blake Ridge offshore from the south-eastern United States. Faults disrupt the sea floor and terminate or sole out about 40–500 m below the sea floor at the base of the gas hydrate stable zone, which is identified from the location of the bottom simulating reflection (BSR). Normal faults are common but reverse faults and folds also are widespread. Folds commonly convert upward into faults. Sediment diapirs and deposits of sediments that were erupted onto the sea floor are also present. Sea-floor depressions at faults may represent locations of liquid/gas vents. The collapse was probably caused by overpressures and by the decoupling of the overlying sediments by gassy muds that existed just beneath the zone of gas hydrate stability.

Collapse of gas hydrate-bearing sedimentary deposits at the sea floor may be the primary process that releases methane from the hydrate reservoir to the atmosphere. On the continental slopes and rises this release is likely to be associated with landslides (Carpenter 1981; Paull *et al.* 1991). However, at one site off the south-eastern United States (Fig. 1), which we have studied,

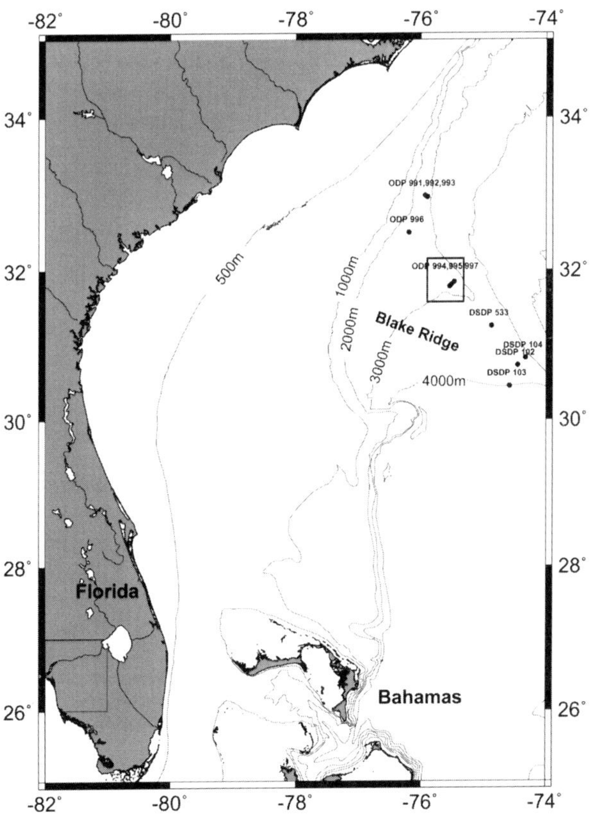

Fig. 1. Location of the study area (square), and the Deep Sea Drilling Project (DSDP) and Ocean Drilling Program (ODP) drill sites.

DILLON, W. P., DANFORTH, W. W., HUTCHINSON, D. R, DRURY, R. M., TAYLOR, M. H. & BOOTH, J. S. 1998. Evidence for faulting related to dissociation of gas hydrate and release of methane off the southeastern United States. *In:* HENRIET, J.-P. & MIENERT, J. (eds) *Gas Hydrates: Relevance to World Margin Stability and Climate Change*. Geological Society, London, Special Publications, **137**, 293–302.

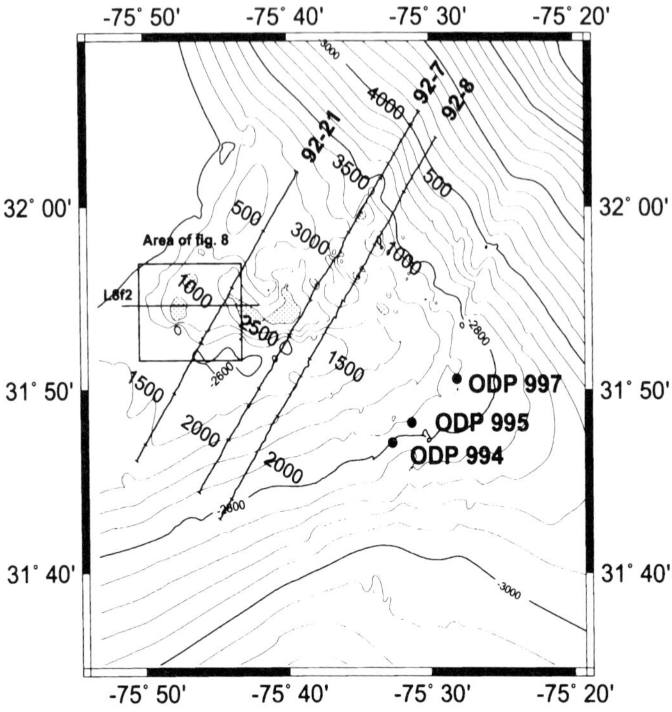

Fig. 2. Detailed bathymetry of the crest of the Blake Ridge in the study area. Locations of seismic profiles shown in this paper are indicated by lines (profiles shown in Figs 4, 5, 6 and 9) and location of the side-scan sonar image (Fig. 8) is shown by the box. Bottoms of depressions are shown by a stippled pattern.

collapse and probable release of methane has occurred with no landsliding, probably because the collapse took place at a location of gentle slopes at the crest of the Blake Ridge. This situation provides an opportunity to study processes involved, while avoiding the complications of mass movement. The Blake Ridge was chosen for several recent drill sites on Ocean Drilling Program Leg 164 (Paull *et al.* 1996), largely because it appears to have the greatest concentration of gas hydrate on the continental margin of the eastern United States (Tucholke *et al.* 1977; Shipley *et al.* 1979; Kvenvolden & Barnard 1983; Dillon *et al.* 1994).

Methods

Various seismic/acoustic surveying approaches were used to characterize sea-floor morphology and sedimentary structure in a grid of 4-km spaced seismic profiles, covering a 4800 km^2 area on the crest of the Blake Ridge. Moderately small, pneumatic seismic sound sources provided adequate power to penetrate below the region of gas hydrate stability, while maintaining the best possible resolution. Sources used were a 160 in^3 (2.62 l) airgun and a generator/injector (GI) gun. In the latter, the generator chamber (which generates the primary signal) was 105 in^3 (1.72 l) and the injector chamber (which controls bubble pulsing), was also 105 in^3. Swept-frequency ('chirp') side-scan sonar (26–33 kHz) and sub-bottom reflection (2–7 kHz) data, both obtained from a deep-towed multisensor system ('fish'), were used to analyse the characteristics of the surface and near-surface sediments (upper 60 m) within the region containing the collapse structure.

Blake Ridge morphology

The Blake Ridge (Fig. 1) is a broad, generally smooth sedimentary accretionary ridge – a deep-sea sediment drift deposit – that is accreting at the site of interaction of major ocean currents (Markl & Bryan 1983; Mountain & Tucholke

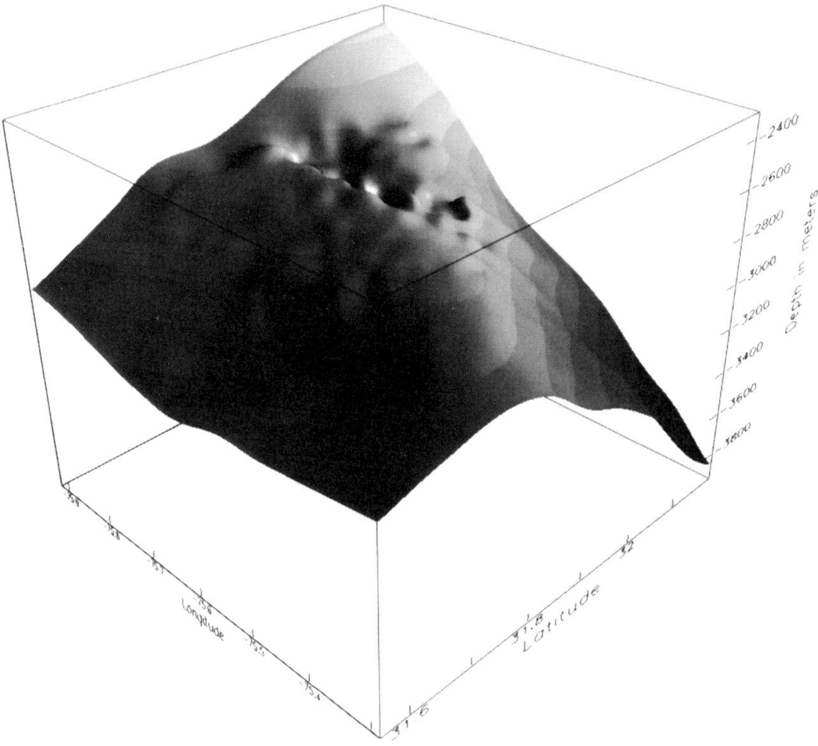

Fig. 3. Perspective image of sea-floor morphology of the crest of the Blake Ridge in the study area (area of Fig. 2).

1985; McCave & Tucholke 1986; Dillon & Popenoe 1988). Ridge sedimentation was initiated near the middle of Oligocene time after an episode of ocean-floor erosion and the ridge has evolved by accretion on its southern flank and erosion on its northern flank (Tucholke & Mountain 1986). The surface of the ridge is generally smooth except for minor differential erosion features on its northern (eroding) flank (EEZ-Scan 87 Scientific Staff 1991), but, at the crest of the ridge at about 32°N, a rough topographic depression covers an area of about 35 × 18 km. Detailed bathymetry of the area was generated from available echo-sounder profiles (Fig. 2); a perspective image of the ridge was produced from these data (Fig. 3).

Structure of the Blake Ridge collapse depression

The depression at the crest of the Blake Ridge is clearly shown by seismic reflection profiles to be a structural collapse (Fig. 4). Faulting took place in a surface layer of sediment about 0.5–0.6 s thick (about 400–500 m). Our seismic profiles indicate that these faults consistently extend from the sea floor to near the base of the gas hydrate stability zone. The base of the gas hydrate stability zone is assumed to be marked by the bottom simulating reflection (BSR, note Fig. 4), which is considered to mark the acoustic contrast between hydrate-bearing sediments above and gas-bearing sediments below (Dillon & Paull 1983).

The structure displayed in Fig. 4 includes normal faults with throws of as much as 150 m, which bound downdropped and rotated blocks. Seemingly, this structure might have been formed by extension of the surface layer by about 12% across the collapse feature, but it is clear from the regional structure that the ridge has not been deformed by any such overall extension. Our regional seismic survey shows that there has been no outward movement of the ridge flanks and that no landsliding has taken place. On some profiles, the results of compres-

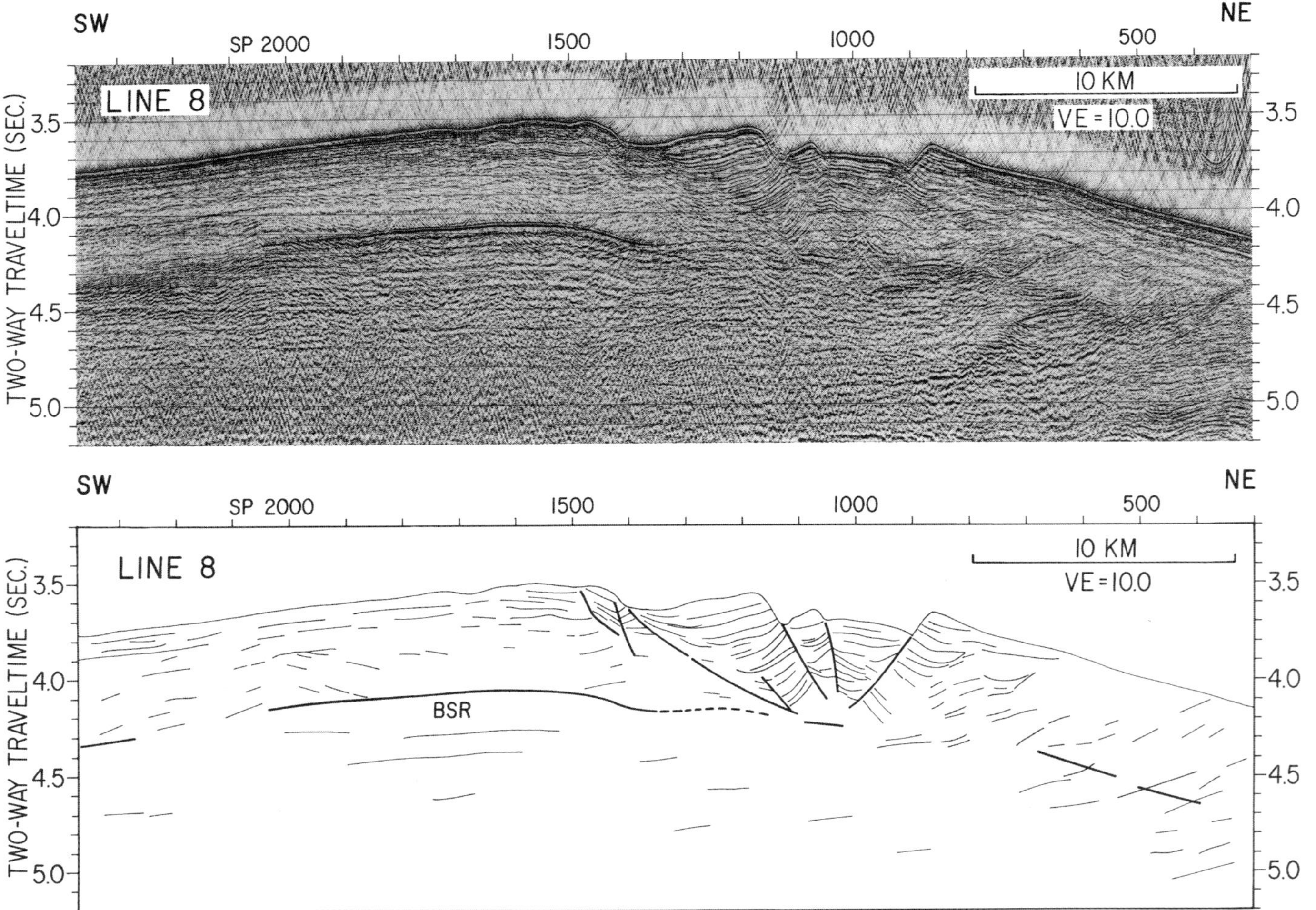

Fig. 4. Airgun seismic reflection profile 8. Location shown in Fig. 2. BSR indicates bottom simulating reflection. Processing of seismic data for this figure and Figs 5 and 6 included minimum phase predictive deconvolution (operator length of 220 ms with a prediction lag of 4 ms), frequency filtering and constant finite difference velocity time migration (1500 m s^{-1}).

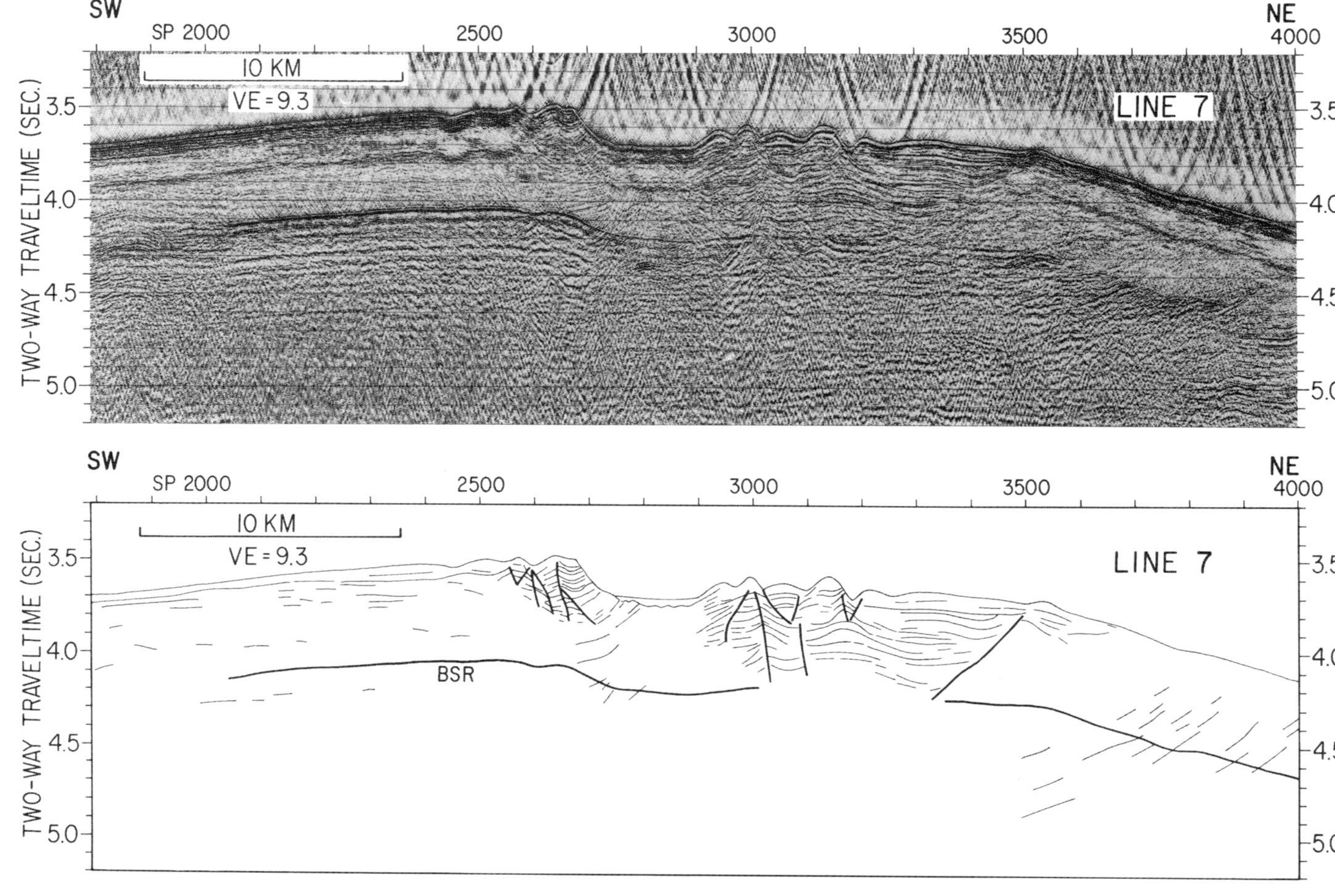

Fig. 5. Airgun seismic reflection profile 7. Location shown in Fig. 2. BSR indicates bottom simulating reflection.

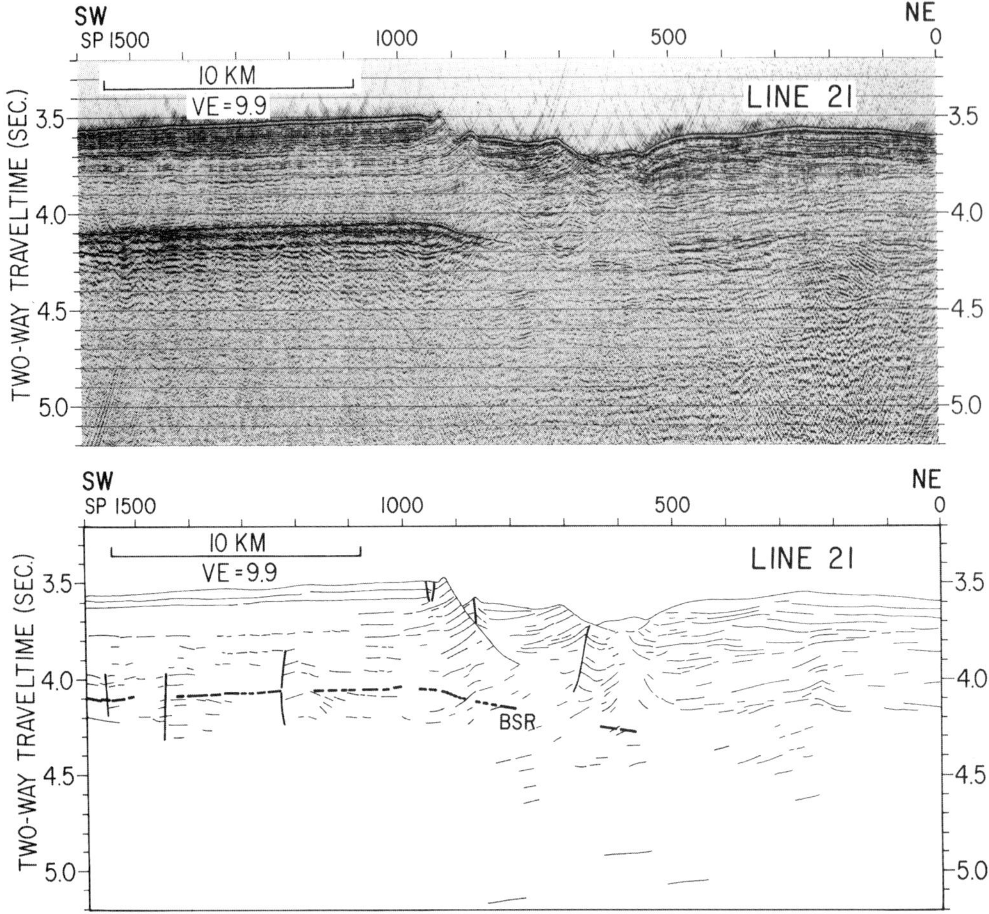

Fig. 6. Airgun seismic reflection profile 21. Location shown in Fig. 2. BSR indicates bottom simulating reflection.

sive movement are apparent; the fault blocks have been forced up and shortening appears within the blocks. The profile of Fig. 5 shows such shortening (note shotpoints 3000–3200) resulting in folding and thrusting. In some cases, folds increase in amplitude upward and convert into faults, a common structural pattern in this area.

Mobilization of sediment is indicated in Fig. 6 (near shotpoint 600). The structure suggests that a sediment diapir arose from near the base of the zone of hydrate stability at the BSR, and that mud was extruded onto the sea floor. A high-resolution profile that was run simultaneously with the seismic profile of Fig. 6 (Fig. 7, the seismic source was a hull-mounted 3.5 kHz transducer) appears to confirm that the older sea floor extends beneath the extruded sediments that were derived from the diapiric intrusion.

Surface expression of faults

Side-scan sonar images disclose that most of the surface traces of the faults are oriented approximately northwest, parallel to the Blake Ridge axis (normal to the direction of the profiles shown in Figs. 4–7). However, near the western limit of the collapse structure, the faults curve around to a northerly orientation as shown by the side-scan sonar image (Fig. 8). Bright stripes (strong reflections) along the faces of the fault scarps delineate outcropping strata (Fig. 8). A high-resolution 'chirp' seismic profile (Fig. 9) taken from the deep-towed fish shows the detailed structure of these east-facing scarps. Note that most of the sea floor is coated by a conformable layer of draped recent sediments. This layer seems to be absent in the depressions at the bases of the major fault scarps (arrows on

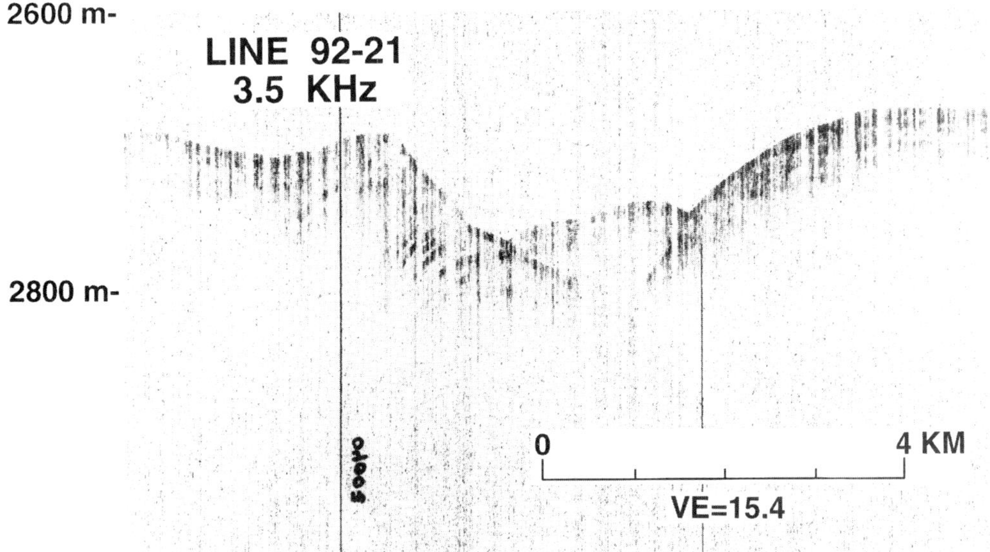

Fig. 7. High-resolution, ship-mounted 3.5 kHz transducer profile across inferred erupted mud deposit that was interpreted on profile 21 near shotpoint 600 (Fig. 6).

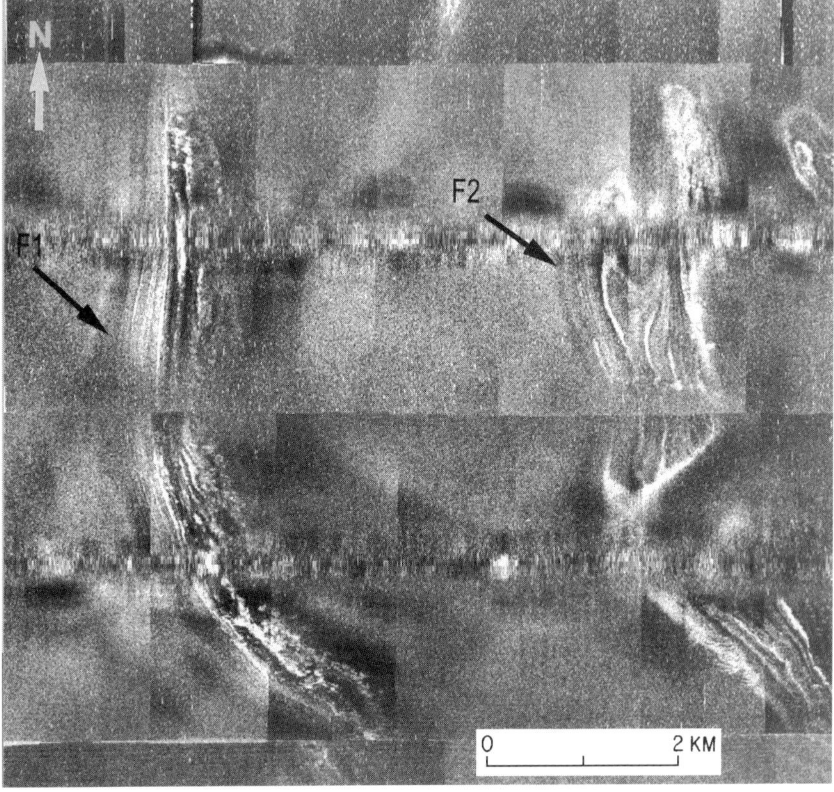

Fig. 8. Deep-towed side-scan image of north-trending fault escarpments at the west end of the collapse structure. Location of image shown on Fig. 2. Crests of fault escarpments are near points of arrows F1 and F2; the east-facing escarpments are marked by parallel bright (strong) reflections that represent outcropping strata.

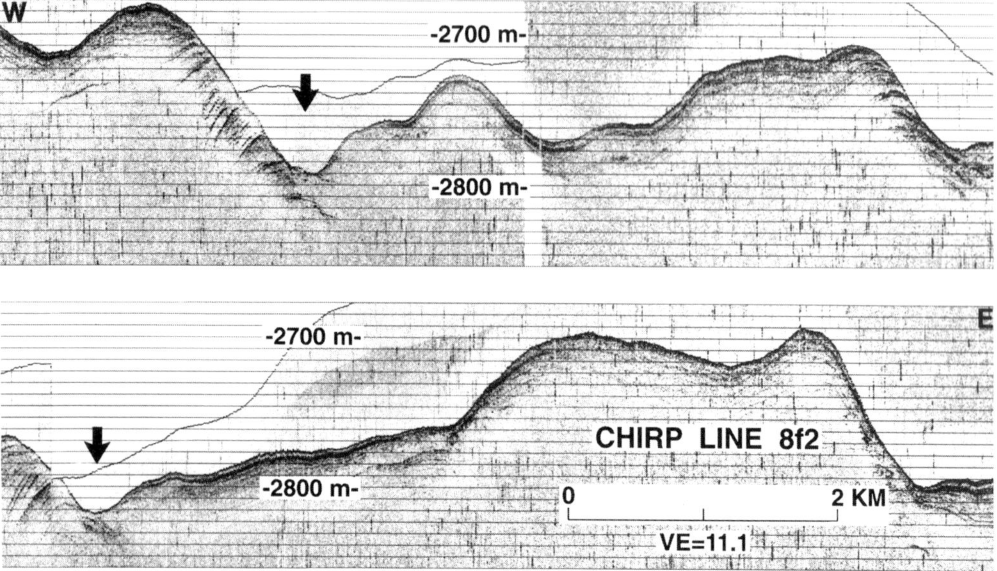

Fig. 9. Swept-frequency ('chirp') high-resolution seismic reflection profile 8f2. Arrows indicate depressions at the bases of the major fault scarps that lack sediment fill, even though most of the profile displays a drape of recent (probably Holocene) sediment.

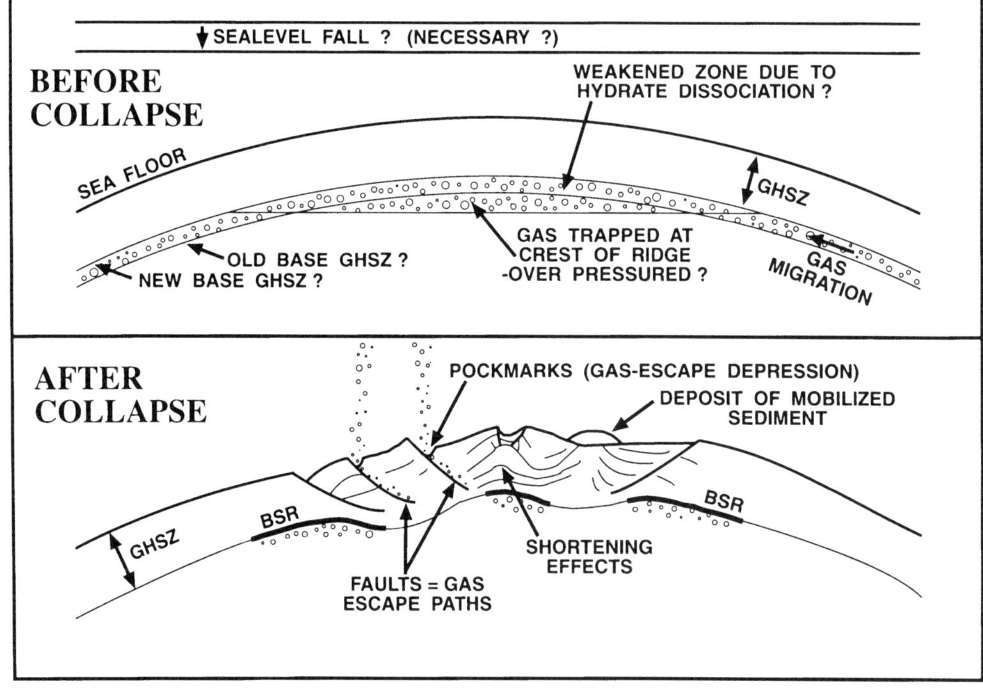

Fig. 10. Conceptual drawing of the Blake Ridge section before and after faulting and collapse. GHSZ is the gas hydrate stability zone. BSR is the bottom simulating reflection, which is considered to mark the base of gas hydrate phase stability.

Fig. 9), an absence that is surprising as a depression would be expected to accumulate greater thicknesses of sediments than adjacent highs. Therefore, some process such as current scour, fluid venting or phenomena related to hydrate formation at the sediment surface would be required to prevent deposition in the depressions. These depressions may be comparable to pockmarks associated with gas escape in other seafloor environments (Hovland & Judd 1988).

Summary and conclusions

A collapse depression exists at the crest of the Blake Ridge where gas hydrate is inferred to be concentrated in the sediments on the basis of seismic profiles (Dillon *et al.* 1994, 1996) and drilling results (Paull *et al.* 1996). The depression results from complex faulting, but involves no overall extension. The region within the collapse structure contains normal faults, folds, folds that convert upward into reverse faults, sediment diapirs, and extruded, mobilized sediment (Fig. 10). Many of the faults extend down to the base of the gas hydrate stability zone, but apparently not below that depth.

We infer that the collapse depression was caused by excess pore pressure within a gas trap formed at the crest of the ridge and sealed by the hydrate-bearing layer of sediments of the gas hydrate stability zone. The crest of the ridge is the location where trapping of gas would be expected, and the crest-centred collapse probably defines the region of overpressure at the time of faulting. A dominant factor in creating the structures probably was the conversion of sediments just below the base of gas hydrate stability into an easily mobilized, gassy mud. This process allowed the formation of mud diapirs and the extrusion of mobilized, presumably fluid-rich, sediment. It also allowed independent movement of blocks of semi-lithified sediment within the gas hydrate stable zone.

Remaining questions (Fig. 10) include: Were the fault blocks partially cemented by hydrate? Would the excess pore pressure generated from hydrate dissociation be enough to cause the convulsion at the Blake Ridge crest? Was the activity triggered by a lowering of sea level and attendant decrease in pressure that caused breakdown of hydrate and release of gas? Is a sea-level fall necessary, or could such an event take place occasionally without pressure changes due to thermal effects caused by the insulation effect of sediment accumulation, or caused by changes in deep-ocean temperatures? Other profiles suggest the presence of similar structures that are now buried, so how frequently does this process happen? Is venting of fluids, including gas, continuing at present?

We wish to acknowledge the US Office of Naval Research and Naval Research Laboratory for partial support of our 1995 cruises, and particularly J. Gettrust and M. Rowe of NRL. We thank the scientists, technicians, and ships officers and crew of the R/V *Cape Hatteras* during cruises CH 15–91 (1991), CH 12–92 (1992), CH 17–95 and CH 18–95 (1995), and especially T. O'Brien of the US Geological Survey. The paper was reviewed by C. W. Poag and H. J. Knebel of the US Geological Survey.

References

CARPENTER, G. 1981. Coincident slump/clathrate complexes on the U.S. Atlantic continental slope. *Geo-Marine Letters*, **1**, 29–32.

DILLON, W. & PAULL, C. 1983. Marine gas hydrates – II: Geophysical Evidence. *In*: COX, J. (ed.) *Natural Gas Hydrates – Properties, Occurrences and Recovery*. Butterworth, Boston, MA, 73–90.

—— & POPENOE, P. 1988. The Blake Plateau and Carolina Trough. *In*: SHERIDAN, R. & GROW, J. (eds) *The Atlantic Continental Margin. US* Geological Society of America, The Geology of North America, **I(2)**, 291–328.

——, HUTCHINSON, D. & DRURY, R. 1996. Seismic reflection profiles on the Blake Ridge near Sites 994, 995, and 997. *In*: PAULL, C., MATSUMOTO, R., WALLACE, P. *et al.* (eds) *Proceedings of the Ocean Drilling Project, Initial Reports*, College Station, TX. Ocean Drilling Program, **164**, 47–56.

——, LEE, M. & COLEMAN, D. 1994. Identification of marine hydrates in situ and their distribution off the Atlantic coast of the United States. *In*: SLOAN, E., JR, HAPPEL, J. & HNATOW, M. (eds) *International Conference on Natural Gas Hydrates*. Annals of the New York Academy of Sciences, **715**, 364–380.

EEZ-SCAN 87 SCIENTIFIC STAFF. 1991. *Atlas of the U.S. Exclusive Economic Zone, Atlantic Continental Margin*. US Geological Survey, Miscellaneous Investigations Series **I-2054**.

HOVLAND, M. & JUDD, A. 1988. *Seabed Pockmarks and Seepages, Impact on Geology, Biology and the Marine Environment*. Graham and Trotman, London.

KVENVOLDEN, K. & BARNARD, L. 1983. Gas hydrates of the Blake Outer Ridge, Site 533. Deep Sea Drilling Project Leg 76. *In:* SHERIDAN, R., GRADSTEIN F. *et al.* (eds) *Initial Reports of the Deep Sea Drilling Project*, Volume 76. US Government Printing Office, Washington, DC, 353–365.

MARKL, R. G. & BRYAN, G. M. 1983. Stratigraphic evolution of Blake Outer Ridge. *AAPG Bulletin*, **67**, 666–683.

MCCAVE, I. & TUCHOLKE, B. 1986. Deep current-controlled sedimentation in the western North Atlantic. *In:* VOGT, P. & TUCHOLKE, B. (eds) *The*

Geology of North America. The Western North Atlantic Region. Geological Society of America, Special Publication, **M**, 451–468.

MOUNTAIN, G. & TUCHOLKE, E. 1985. Mesozoic and Cenozoic geology of the U.S. Atlantic continental slope and rise. *In*: POAG, C. (ed.) *Geologic Evolution of the United States Atlantic Margin*. Van Nostrand Reinhold, New York, 293–341.

PAULL, C., USSLER, W., III & DILLON W. 1991. Is the extent of glaciation limited by marine gas hydrates? *Geophysical Research Letters*, **18**, 432–434.

——, MATSUMOTO, R., WALLACE, P. J. *et al.* 1996. *Proceedings of the Ocean Drilling Program*, Initial Reports. College Station, TX. Ocean Drilling Program, **164**.

SHIPLEY, T., HOUSTON, M., BUFFLER, R. *et al.* 1979. Seismic evidence for widespread possible gas hydrate horizons on continental slopes and rises. *AAPG Bulletin*, **63**(12), 2204–2213.

TUCHOLKE, B. & MOUNTAIN, G. 1986. Tertiary paleoceanography of the western North Atlantic Ocean. *In:* VOGT, P. & TUCHOLKE, B. (eds) *The Geology of North America. The Western North Atlantic Region*. Geological Society of America Special Publication, **M**, 631–650.

——, BRYAN, G. & EWING, J. 1977. Gas-hydrate horizons detected in seismic-profiler data from the western North Atlantic. *AAPG Bulletin*, **61**, 698–707.

Natural gas hydrates: searching for the long-term climatic and slope-stability records

B. U. HAQ

National Science Foundation, Division of Ocean Sciences, Arlington, Virginia, USA

Abstract: The recent revival of interest in gas hydrates has grown from the awareness that they may play significant roles in several global and regional processes, including global carbon recycling, rapid climate change through emission of methane from marine sediments into the atmosphere, and as a cause for massive transport of sediments and structural changes on the continental slope. Their estimated large volumes are also considered to be a potential resource for future exploitation. Here the long-term record of gas hydrate behaviour is examined in terms of climatic, oceanographic, stratigraphic and slope-stability changes of the past. Various intriguing hypotheses recently offered to explain the observed record also highlight the numerous unknowns and our relative ignorance about these widespread features of the continental margins. The paper identifies some of these gaps and poses several relevant questions that will need to be answered before we can appreciate their full potential and cause–effect relationship of hydrate dissociation with climate change and slope stability.

Natural gas hydrates (composed largely of methane and water) occur in marine sediment on the continental margins under the dual conditions of high hydrostatic pressure and low ambient temperature at the sediment–water interface. There has been a great revival of interest in methane hydrates in recent years, both as a natural resource because of their estimated large quantities and because the methane stored in them may be sensitive to changes in sea level and bottom–water temperature.

Global estimates of the methane stored in gas hydrates vary widely, but even the relatively conservative estimate of Kvenvolden (10^4 Gt of methane carbon) exceeds estimates of organic carbon from all other sources with the exception of dispersed carbon in the lithosphere, and is approximately double the estimate of carbon from known fossil fuel sources (Kvenvolden 1988). This may, therefore, represent a prolific, although as yet unproven, source of energy for the future. Methane hydrates have not been considered a viable resource by industry so far because they occur mostly on the continental slope and rise, economically too distant from the consumer markets. This is not likely to remain so in the future when plentiful hydrocarbons on land and near-shore reach their nadir. In addition, hydrate dissociation has implications for health, safety and the environment. Uncontrolled drilling into shallow hydrates and the free gas below can lead to blowouts and sinking of rigs, loss of life and pollution. Thus, better estimates and characterization of hydrate reservoirs will not only lead to a better appreciation of their resource potential, but also to improved understanding of the effects of their mass dissociation.

Ice-core palaeorecords (see e.g. Jouzel *et al.* 1993) suggest that climatic warming occurs in tandem with rapid increase in atmospheric methane. This evidence has been used to suggest that catastrophic release of the greenhouse methane gas into the atmosphere from gas hydrate sources may be a causal factor for abrupt climate change. If a eustatic fall is glacially forced, the eventual addition of large quantities of methane from low-latitude clathrate fields in response to lowered hydrostatic pressure can provide a negative feedback to glacial cooling, that can lead to the reversal of the course of glaciation. Once climate in high latitudes ameliorates, additional methane released from the near-surface sources of shallow shelf and the permafrost could provide a positive feedback to the warming trend, further aiding in the relatively rapid termination of the glacial cycle. These ideas remain speculative, but such scenarios could have important implications for global climatic change if the continuing climatic amelioration trends lead to further warming of the higher latitudes by even a few degrees.

It has also been suggested that when the breakdown of gas hydrate occurs in response to reduced hydrostatic pressure or a rise in bottom-water temperatures it first causes dissociation of the solid hydrate at its base, creating a zone of reduced sediment strength that is prone to faulting and block slumping. Weakening of mechanical strength of sediments leading to megaslumps may be an important first-order mechanism for mass transport on continental

HAQ, B. U. 1998. Natural gas hydrates: searching for the long-term climatic and slope-stability records. *In:* HENRIET, J.-P. & MIENERT, J. (eds) *Gas Hydrates: Relevance to World Margin Stability and Climate Change.* Geological Society, London, Special Publications, **137**, 303–318.

margins. Many major slump features found in the stratigraphic record of the continental slope may have had their origin in such mass movements initiated by gas hydrate decomposition.

In our present state of the knowledge of gas hydrates, however, the proper evaluation of these causal relationships and potentials is severely hampered by a lack of meaningful global estimates of the volume of trapped gas and the quality of gas hydrate reservoirs.

Estimating the global distribution of gas hydrates

The conditions for the stability of gas hydrate (low bottom-water temperature and high pressure) are theoretically met over a high percentage of the floor of the continental slope and rise where water depth exceeds 250 m. The prerequisite of gas saturation in the pore waters, however, may be a limiting factor for the actual formation of hydrate. The depth and thickness of the gas hydrate stability zone in the sediment is also a function of the geothermal gradient and the activity of the pore waters (see Dickens and Quinby-Hunt 1997 for details). Thus, the potential gas hydrate zone may extend from the sediment–water interface down to 2000 m below the sea floor. In the high-latitudes seas, where the ocean bottom-water temperatures are sufficiently low, hydrates occur in marine sediments at shallower depths (at ~200 m). On land at these latitudes they are known to exist near the surface in association with the permafrost. At present the permafrost is distributed over as much as 20% of the land area in the northern Hemisphere, and may thus represent a substantial potential source of methane emissions.

Rapidly deposited sediments with high biogenic content are more amenable to the genesis of large quantities of methane by bacterial alteration of the buried organic matter. It has been suggested that relatively high gas content in pore waters is required to form a solid hydrate (Kvenvolden & Barnard 1983). This assertion has been confirmed by recent results of Ocean Drilling Program (ODP) drilling on Blake Ridge, offshore the US East Coast, which indicate that solid hydrate occurs only in association with pore waters that are saturated with gas (see e.g. Dickens *et al.* 1997). The volume of the solid hydrate based on direct measurements was estimated to be between 0 and 9% of the pore space in the hydrate stability zone (190 and 450 m sub-bottom) (Dickens *et al.* 1997), and between 5 and 7% based on vertical seismic profile velocity data (Holbrook *et al.* 1996).

The most widely employed method of detecting gas hydrates is through the presence of the so-called bottom simulating reflectors (BSRs) on seismic reflection profiles, which, when present, delineate the base of the hydrate stability zone (Fig. 1). BSRs have been suspected to indicate the presence of clathrate and significant gas accumulation at least since the 1970s (Markl *et al.* 1970). Blake Outer Ridge was the first place where BSRs were observed and drilled through to reveal the presence of high methane content (Hollister *et al.* 1972; Bryan 1974; Tucholke *et al.* 1977). Shipley *et al.* (1979) published a series of profiles with BSRs ascribable to gas hydrates from the slope and rise areas of North, Central and South American passive and active margins.

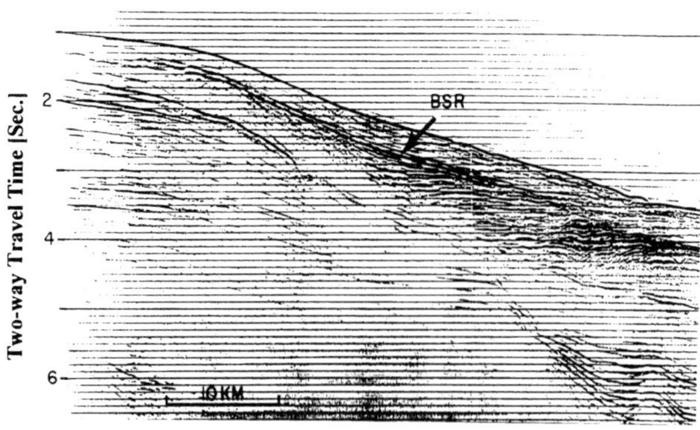

Fig. 1. Seismic reflection profile over a portion of the Blake Outer Ridge, offshore US East Coast, showing the prominent bottom simulating reflector (BSR) at ~4.25 s. The BSR cuts across local bedding and marks the reflective boundary between hydrate-cemented above and free gas below (USGS Line 32, after Dillon 1991).

Since then BSRs have been mapped on many continental margins of the world, including offshore Antarctica (see Kvenvolden 1993, 1998). These reflectors may cut across primary regional bedding, and local structure (folds and faults) often controls the upward migration of gas. Steeper geometries encourage upward migration and gas expulsion where a BSR might intersect the sea floor.

It has been suggested that significant quantities of free gas may be present below the hydrates. This provides the velocity contrast and, thus, the presence of free gas is essential for a BSR show on the seismic profiles. These reflectors commonly have half the amplitude and opposite polarity relative to the the sea floor. The sediment velocity change at the BSR marks a diagenetic boundary, often cutting across local depositional surfaces, and a transition between gas hydrate-cemented sediments above and water-filled, uncemented and gas-charged, sediments below (Dillon & Paull 1983; see also Bangs *et al.* 1993; Katzman *et al.* 1994; MacKay *et al.* 1994 for further discussion of the nature of BSRs). Earlier it had been assumed that BSRs always mark the base of the hydrate zone, but ODP Leg 164 drilling on Blake Ridge proved that hydrates also exist in areas with no discernible BSR at the base, the only difference here being the lack of free gas below the base (Paull *et al.* 1997). From direct measurements of methane in cores free gas below the hydrate zone was estimated to occupy up to 12% of the pore space, significantly higher than previous estimates (1–2%) (Dickens *et al.* 1997). Holbrook *et al.* (1996) used vertical seismic profiling data from the same section to interpret a smaller amount of free gas (occupying ~1% of the porosity) beneath the hydrate stability zone. This dual study on the same section illustrates the fallibility of estimates based on velocity data alone and underscores the need for direct measurements (Dickens *et al.* 1997).

Reduction of reflection amplitude (blanking) may be another seismic characteristic that can be employed to infer the presence of gas hydrates. Blanking occurs when the presence of clathrate increases the velocity of sediment through the introduction of higher velocity material in pore spaces producing a 'blanking' effect for the sedimentary section. Zones of increased reflection amplitude within the gas hydrate stability window may thus suggest areas of dissociation and loss of gas hydrate (Lee *et al.* 1993; Dillon *et al.* 1998), or areas where the solid hydrate did not accumulate due to lack of gas saturation. Holbrook *et al.* (1996), however, contend that the low reflectance above the BSR is more likely to be caused by lithological homogeneity of the sediment rather than hydrate cementation.

BSRs have been mapped over wide areas of the continental margins of the world, but clathrates have only been sampled a few times (see Kvenvolden 1993). Although new information has now begun to accumulate from recent ODP drilling in hydrate fields (e.g. Legs 141, 146 and 164), the general lack of ground-truthing means that the estimates of gas trapped in the hydrate zones, or the free gas below them, remain largely speculative in most places. Blake Ridge (drilled during ODP Leg 164) remains the only reservoir where estimates of the volume of hydrate are available from seismic reflection profiling, vertical seismic profiling and direct measurements on cores obtained through the hydrate stability zone and below (Holbrook *et al.* 1996; Dickens *et al.* 1997; Paull *et al.* 1997). The *in situ* measurements indicate some 35 Gt of methane carbon tied up on Blake Ridge (equal to ~7% of the total terrestrial biota) (Dickens *et al.* 1997).

Several important questions about BSRs remain to be answered or better documented. These include: (1) Do BSRs on continental margins always mark the transition between solid phase and free gas? (What about other diagenetic boundaries at similar depths?) (2) Is the distribution of the hydrate mostly patchy or can it be homogenous under some circumstances? (3) How large are the pools of free methane trapped below the BSRs? The last question is especially relevant if gas hydrates also function as permeability barriers for significant natural gas reservoirs. Geophysical and direct measurements on sediments, similar to those obtained on Blake Ridge through drilling, need to be attempted on at least several representative margins and settings before a more meaningful picture of the nature and the size of the gas hydrate reservoir will emerge. ODP results from the Blake Ridge so far indicate that the distribution of the hydrate is highly heterogeneous with respect to depth and there is more free gas below the BSR than previously suspected. However, it is not clear whether the Blake Ridge is an anomalous situation or more characteristic of the gas hydrate reservoirs in general.

Well-log response is another method of detecting and estimating gas hydrate in a region. High electrical resistivity, short acoustic transit time relative to water and significant release of gas during drilling (see Fig. 2) have been shown to characterize the presence of gas hydrate layers (Collett *et al.* 1988; see also Bangs *et al.* 1993; MacKay *et al.* 1994). In a study of the North Slope of Alaska, Collett (1993) described the

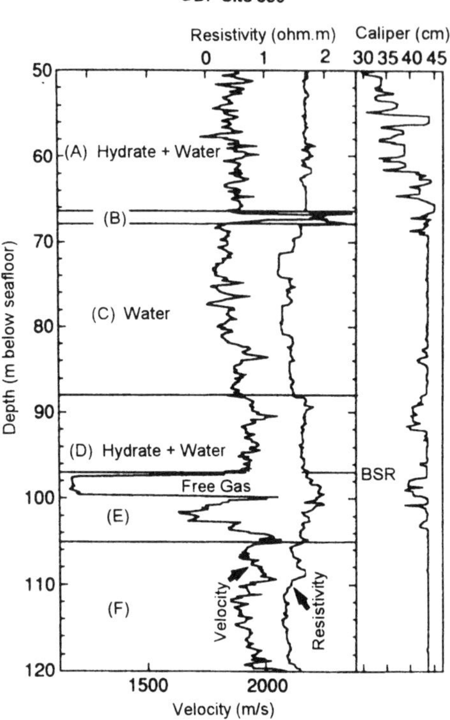

Fig. 2. Well-log response of hydrate-bearing zone at ODP Site 859 near Chile triple junction. Two hydrate-rich layers (A and D) and a 7 m thick layer of free gas zone below the BSR were interpreted from resistivity and velocity data (after Bangs *et al.* 1993).

regional distribution of hydrate based on 50 wells from Prudhoe Bay and Kuparuk River areas, ground-truthed through the actual recovery of clathrate in one well. Regionally the hydrate stability zone is known to be extensive beneath the coastal plain and is over 1000 m thick in the Prudhoe Bay area. The hydrate is considered to have formed during the last glacial cycle. Geochemically, it is composed of shallow biogenic methane mixed with deeper thermogenic methane, sourced upward through the same regional faults through which large volumes of heavy oil had migrated upward. The gas in the clathrate was estimated to be twice the volume of the conventional gas in the Prudhoe Bay field. Another significant finding was that the hydrate largely occurs within distinct layers of laterally continuous sandstone and conglomerate (Collett 1993). This also implies that hydrate may preferentially accumulate in sediments with greater porosities and, thus, occur only in discrete layers rather than uniformly through the sedimentary column of the hydrate stability zone.

At ODP sites near Chile triple junction, drilled and logged during Leg 141, Bangs *et al.* (1993) modelled the wave form and amplitude vs offset of the BSR, and compared it to the *in situ* observations and physical properties. At Site 859 the BSR is interpreted to be generated from a 7 m thick layer containing free gas rather than from a discrete surface. Based on velocity values two intervals above the BSR are interpreted to contain hydrate in pore spaces. The upper interval contains some 10% hydrate while in the lower layer ~18% of the pore space is estimated to be occupied by hydrate (Fig. 2). At Site 860 a 12 m interval below the BSR with low V_p is estimated to contain free gas in small quantities (~1%). During ODP Leg 146, off Cascadia Margin, logs and vertical seismic profiles from two sites also indicate reduced seismic velocities below the BSR indicating the presence of 1–5% free gas (MacKay *et al.* 1994). Normal velocities above the BSR are considered to be indicative of only small amounts of hydrate in the hydrate stability zone. Thus, even small volumes of free gas can dramatically lower the velocity – as little as 1–3% gas can reduce the normal sediment velocity of 1900–1700 m s^{-1} to as low as 1400 m s^{-1} – additional free gas produces very little change (Domenico 1976; MacKay *et al.* 1994).

Reduction in pore water chlorinity within the gas hydrate stability zone can also indicate dissociation of solid hydrate and therefore its recent presence. Chloride anomalies in pore waters occur when during its genesis the hydrate crystallizes and expels NaCl ions (Hesse & Harrison 1981). Surrounding pore water initially become more saline, but over time the salts are advected or diffused away. In the hydrate chlorinity is reduced from an average of 19.8% (in sea water) to between 3.2 and 0.5% (in hydrate). Thus, while drilling freshening of pore water (as hydrate melts) may be another useful signal for the recognition of the presence of hydrate (Kvenvolden, 1998). At ODP Site 859 ~ 5–15% fresher than sea water chloride levels indicate a dissociation of the hydrate (Bangs *et al.* 1993). Such chloride anomalies within the gas hydrate stability zone depths have also been observed in the sites drilled on the Blake Ridge (Paull *et al.* 1997).

Other indications of the presence of gas hydrates at depth may include gas-escape features on land and on the sea floor. Gas is often vented to the surface from the hydrated layers through structural faults, or as dissolved gas through advective venting. Such gas-escape structures have been mapped widely, including the Gulf of Mexico, the Carolina Margin of the

US East Coast, the North Slope of Alaska, among others (see e.g. Brooks *et al.* 1986; Prior *et al.* 1989; Collett 1993; Paull *et al.* 1995). In the Gulf of Mexico hydrates actually breach the surface and crop out on the sea floor (MacDonald *et al.* 1994).

Global estimates of the total methane trapped in gas hydrate reservoirs (both in the hydrate-stability zone and beneath it) based on BSRs and other indicators vary wildly. For example, according to various estimates the Arctic permafrost holds anywhere between 7.6 and 18 000 Gt of methane carbon, while marine sediments, extrapolated globally, hold between 1700 and 4 100 000 Gt of methane carbon (Kvenvolden, 1988). These estimates are based on many assumptions, for example one estimate considers only those places where at least 1% or more organic matter has been accumulating, with 0.5 km average thickness of the hydrate zone, an average 50% porosity of which hydrate occupies 10% of the pore space, etc. (Kvenvolden & Claypool 1988). Another estimate is based on 2–4% porosity, with clathrate occupying only 1% of the pore space, etc. (MacDonald 1990). Although one is forced to make such assumptions in our present state of the knowledge about gas hydrate reservoirs, such premises are largely conjectural. For example, assumptions of porosities on which these estimates are based could be considered unrealistic. Marine sediment, when first deposited, may have porosities of over 90% and, depending on the lithology, this may be reduced to 30% or less as the sediment is buried and compacted over time. Similarly, rates of organic productivity vary significantly on different margins and an assumption of 1% average may not be meaningful. Another problem in estimating hydrate abundance is the fact that we do not know which stability curve is appropriate (e.g. methane–hydrate–ice, methane–hydrate—sea water or methane–ice-sea water equilibrium curve) to use for wider estimates. In addition, not knowing how much free gas is trapped beneath the hydrates also adds to the considerable uncertainty in estimating the total gas reservoirs.

The problem of obtaining a more accurate global estimate of the methane sequestered in the clathrate reservoirs remains the most significant challenge, and the resolution of all other issues in gas hydrate research is dependent on it. Another vital question, especially for climatic implications, is to ascertain the mode of expulsion of methane from the hydrate. How and how much of the gas escapes from the hydrate zone, and how much of it is dissolved in the water column vs escaping into the atmosphere? Although changes in temperature and pressure regime are obviously among the forcing mechanisms, most of the emission scenarios remain speculative and untested at this time. In a steady state much of the methane diffusing from marine sediments (> 99%) is thought to be oxidized in the surficial sediment and the water column (Cranston 1996). What happens to the significant volume of gas released catastrophically from the hydrates? Does most of it make it to the atmosphere (which would be necessary for climatic scenarios discussed below) or is most of it oxidized in the water column?

Gas hydrates and rapid climate change

The temperature–pressure relationship necessary for the stability of gas hydrate implies that any major change in either of these controlling factors will tend to alter the zone of hydrate stability. For example, a significant drop in eustatic sea level will reduce the hydrostatic pressure on the slope and rise, altering the temperature–pressure regime, leading to destabilization of the gas hydrates. McIver (1977) was first to recognize the potential causal relationship between gas hydrates and submarine slumps. Paull *et al.* (1991) offered a scenario of sea-level fall associated with the Pleistocene glaciation, leading to gas hydrate instability and major slumping on the continental margins. It has been suggested that a sea-level drop of ~ 120 m during the last glacial maximum reduced hydrostatic pressure sufficiently to raise the lower limit of gas hydrates by about 20 m (assuming constant *in situ* temperatures) (Dillon & Paull 1983). The ensuing destabilization created a zone of weakness where sedimentary failure could take place, leading to major slumps world-wide. These authors ascribe the occurrence of common Pleistocene slumps on the sea floor to this catastrophic mechanism. Slumps have been identified in young sediments of widely separated margins of the world, including the Gulf of Mexico, the US Atlantic Margin, the Bering Sea, the North Sea, offshore West Africa and other areas (see e.g. Bugge *et al.* 1987; Collett *et al.* 1990; Koyen & Lee 1991; Kvenvolden 1993; Booth *et al.* 1994; Paull *et al.* 1996).

Submarine slumping also could be accompanied by the liberation of methane trapped below the level of the slump, injecting a significant amount of this greenhouse gas into the atmosphere. The emission of methane could increase with the frequency of slumps as glaciation progressed. This could eventually trigger a negative feedback to advancing glaciation once methane emissions increased over a threshold level, leading to termination of the glacial cycle. Thus, there may be a built-in terminator to glaciation, via the gas hydrate connection. Paull *et al.* (1991) attribute the abrupt nature of Pleistocene glacial terminations to such a process.

The glacially forced sea-level lowering leading to slumping and release of methane, providing negative feedback to glaciation, can at first function effectively only in the lower latitudes. At higher latitudes glacially induced freezing would tend to delay the negative feedback effect, but once deglaciation begins even a relatively small increase in atmospheric temperature of the higher latitudes could cause further release of methane from near-surface sources, providing a positive feedback to warming. Nisbet (1990) suggested that a small triggering event and liberation of one or more Arctic gas pools could initiate massive release of methane from the permafrost. The strong positive feedback could provide increased emissions of methane and accelerated warming. Nisbet ascribed the abrupt nature of the termination of the Younger Dryas (some 10 000 years ago) to such an event and suggested that gas hydrates may play a dominant role, more important than ocean degassing, in recharging the biosphere with carbon dioxide at the end of a glaciation. It is, however, not clear whether a single gas-escape event from the Arctic gas pools in itself is sufficient to cause a 'domino effect' of further methane release as envisioned by Nisbet.

The 'negative–positive feedback loop'

The palaeoclimatic record of the recent past, e.g. ice-core records of the past 200 ka, clearly show gradual decrease in atmospheric carbon dioxide and methane at the onset of glaciations (Jouzel *et al.* 1993). Deglaciations, on the other hand, tend to be relatively abrupt and are associated with rapid increases in methane and carbon dioxide (Fig. 3). Glaciation is believed to be initiated by Milankovitch orbital forcing, a mechanism that also can explain the broad variations in glacial cycles, but not the relatively abrupt terminations. Oceanic degassing of carbon dioxide alone cannot explain the relatively rapid switch from glacials to interglacials (Nisbet 1990). To explain the combined effect of glacially induced sea-level lowering leading to hydrate instability, and low-latitude warming leading to higher-latitude

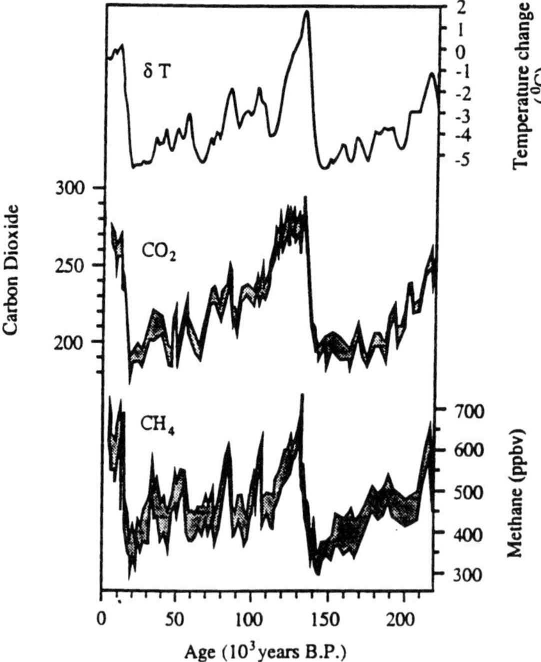

Fig. 3. Ice-core record from the Antarctic Vostock ice core showing parallel variations in methane, carbon dioxide and temperature over the past 220 ka (modified after Jouzel *et al.* 1993).

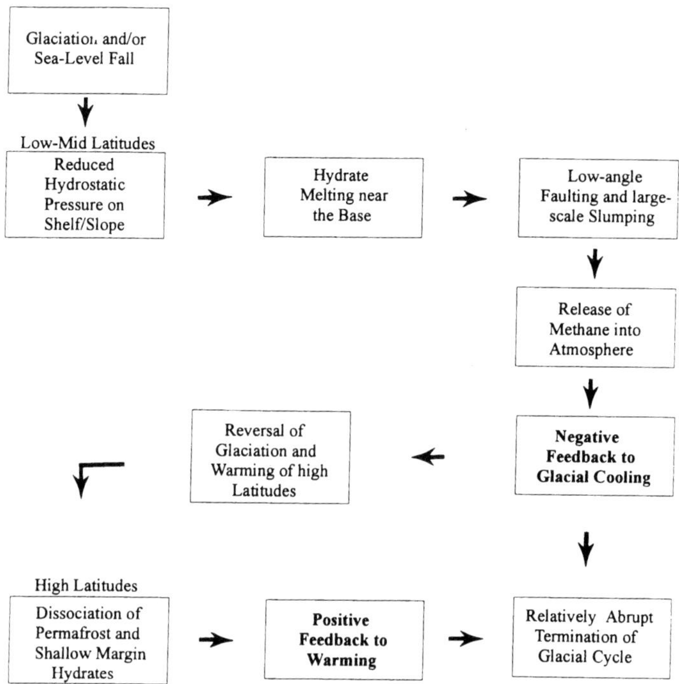

Fig. 4. The negative–positive feedback loop model of sea-level fall, hydrate decomposition and climate change (reversal of glaciation and rapid warming) through methane release in the low and high latitudes (after Haq 1993).

methane release, Haq (1993) extended these notions into a 'negative–positive feedback' model (see Fig. 4) (a similar scenario was also proposed by Kvenvolden 1993).

The initially delayed effect of sea-level fall in the high latitudes compared to low latitudes constitute a negative–positive feedback loop that could be an effective mechanism for explaining the rapid warmings at the end of glacial cycles (the Dansgaard–Oeschger events) in the late Quaternary, the transitions occurring often only on decadal to centennial time scales (see e.g. Bond & Lotti 1995). It is conceivable that a combined effect of lowstand-induced slumping and methane emissions in low latitudes triggers a negative feedback to glaciation as suggested by Paull *et al* (1991), and the ensuing positive feedback from ocean carbon dioxide degassing and warming in higher latitudes and further release of methane from near-surface sources as envisioned by Nisbet (1990). The former would help force a reversal of the glacial cycle and the latter could reinforce the trend resulting in apparent rapid warming observed at the end of the glacial cycle. The present-day estimate of methane in the atmosphere is a relatively small 3.6 Gt of carbon (Kvenvolden 1988). Thus, even a relatively modest quantity of methane released from the vast hydrate reservoir could conceivably double or triple the atmospheric methane content and cause increased greenhouse warming lasting a few decades or more, if significant methane emissions continue.

A study of the stable isotopic record of the Santa Barbara Basin by Kennett *et al.* (1996) based on recently drilled ODP sites has revealed rapid warmings in the late Quaternary that are synchronous with warmings associated with Dansgaard–Oeschger (D–O) events in the Greenland ice record. Greenland ice cores indicate that the D–O events are often characterized by rapid warming followed by slower cooling. Transitions from glacial to interglacial modes are relatively rapid, lasting only a few decades. Kennett *et al.* (1996) suggest that the energy needed for these rapid warmings probably came from methane hydrate dissociation. Relatively large excursions of $\delta^{13}C$ (up to 5‰) in benthic foraminifera are associated with the D–O events. However, during several brief intervals the planktonics also show large negative shifts in $\delta^{13}C$ (up to 2.5‰), implying that the entire water column may have experienced rapid ^{12}C enrichment. One plausible mechanism for these changes may be the release of methane from clathrates during the interstadials. Thus, abrupt warmings at the onset of D–O events may have been forced by dissociation of gas hydrates modulated by temperature changes in overlying intermediate waters (Kennett *et al.* 1996).

In spite of the attractiveness of this scenario, the problem of explaining the cause and effect remains. Did the intermediate water temperature rise first (if so, then how?) which led to gas hydrate dissociation and eventual rapid warmings? Or did the sea-level drops during glacials cause hydrate to degas and lead to rapid warming both of intermediate waters, the sea surface and the atmosphere?

For the optimal functioning of the negative–positive feedback model, methane would have to be constantly replenished from new and larger sources during the switchover (Haq 1993). Although as a greenhouse gas methane is nearly 10 times as effective as carbon dioxide (see Table 1), its residence time in the atmosphere is relatively short (on the order of 15 years), after which it reacts with the hydroxyl (OH) radical and oxidizes to carbon dioxide and water (Lashof & Ahuja 1990). Carbon dioxide accounts for up to 65% of the contribution to greenhouse warming in the atmosphere. The atmospheric retention of carbon dioxide is somewhat more complex than methane because it is readily transferred to other reservoirs, such as oceans and the biota, from which it can re-enter the atmosphere. Lashof & Ahuja (1990) estimate an effective average residence time of *c.* 230 years for carbon dioxide. These retention times are short enough that for a cumulative impact of methane and carbon dioxide through the negative–positive feedback loop to be effective methane levels would have to be continuously sustained from gas hydrate and permafrost sources. The feedback loop (methane from continental margins and permafrost clathrate sources, and carbon dioxide from degassing of the ocean water) would close when a threshold is reached where sea level is once again high enough that it can stabilize the remaining hydrates and encourage the genesis of new ones. Alternatively, slowing of the rate of warming may also slow down or cease further catastrophic release of methane from hydrate sources and close the loop (Dickens pers. comm. 1997).

In this connection the modelling experiments of Thorpe *et al.* (1996, 1998) are of relevance. These authors modelled the release of methane during climatic amelioration as indicated by ice-core results to test its potential effects on glacial climate and came to the conclusion that methane alone could not trigger deglaciation. However, a combination of methane release, increased carbon dioxide, heat transport by the oceans, coupled with high climatic sensitivity can produce climatic changes of the observed magnitude (3–5°C).

Several unresolved problems remain with this model. The negative–positive feedback loop (Haq 1993) assumes a certain amount of time lag between events as they shift from lower to higher latitudes, but the duration of the lag remains unresolved, although a short duration (on decadal to centennial time scales) is implied by the ice-core records. At present the resolution of the ice-core record is not refined enough to understand the sequence of leads and lags in such a scenario (Raynaud *et al.* 1996, 1998) which makes it difficult to resolve whether the apparent parallel changes in carbon dioxide + methane vs the temperature inferred from ice cores are a cause or an effect relationship or occur truly in tandem.

Thus, several issues need to be resolved for the validity of these scenarios: (1) The issue of time lead and lag between various forcing functions. How long does it take for the low-latitude hydrate fields to start dissociating after a sea-level fall?

Table 1. *Global warming potential of some greenhouse gases (after Lashof & Ahuja 1990)*

Gas	Atmospheric residence time (years)	Greenhouse potential	
		Molar basis	Weight basis
Carbon dioxide	230	1.0	1.0
Carbon monoxide*	(21)	(1.4)	(2.2)
Methane	14.4	3.7	10
Nitrous oxide	160	180	180
CFC-11	60	4000	1300
CFC-12	120	10 000	3700

* Carbon monoxide is not a greenhouse gas in itself but its presence enhances the residence time of methane in the atmosphere by destroying the OH radical.

How long do the higher latitudes lag behind the low latitudes in their response to the warming trend begun in the lower latitudes? (2) Then there is the issue of the methane-forced atmospheric warming and how it might affect temperature rise in intermediate and deep waters, as well as the time lags associated with this pathway? Is the warming of deep waters a cause of methane release or an effect? (3) Also, how does a methane release event from the hydrate cause additional release events? As noted earlier, knowing how methane is actually released from the hydrate after dissociation into the water column and the atmosphere is another prerequisite, without which the questions about time leads and lags will largely remain unanswered.

Other modelling results play down the role of methane release from hydrate sources. Harvey & Huang (1995) modelled heat transfer and methane destabilization processes in oceanic sediments in a coupled atmosphere–ocean model with various input assumptions and anthropogenic emission scenarios. They found the hydrate dissociation effects to be smaller than the effects of increased carbon dioxide emissions by human activity. In a worst-case scenario global warming increased by 10–25% more with clathrate destabilization than without. However, these models did not take into account the associated free gas beneath the hydrate zone that may play an additional and significant role as well. It is obvious from drilling results on Blake Ridge that a large volume of free methane are readily available for transfer without requiring dissociation (Paull *et al.* 1997).

The long-term record of climate change

Are there any clues in the longer term geological record where cause and/or effect can be ascribed to gas hydrates? One potential clue for the release of significant volumes of methane into the ocean waters is the changes in $\delta^{13}C$ composition of the carbon reservoir. The $\delta^{13}C$ of methane in hydrates averages approximately −60‰ (Kvenvolden 1993), perhaps the lightest (most enriched in ^{12}C) carbon anywhere in the Earth system. Dickens *et al.* (1995) have argued that massive methane release from hydrate sources is the most likely mechanism for the pronounced input of carbon greatly enriched in ^{12}C during a period of rapid bottom-water warming. Thus, major negative shifts in $\delta^{13}C$ that occur together with increases in benthic temperature (bottom-water warming) or sea-level fall events (reducing hydrostatic pressure) may provide plausible clues to past behaviour of gas hydrates.

Dickens *et al.* (1995) explain a prominent excursion in global carbonate and organic matter $\delta^{13}C$ during the latest Palaeocene peak warming as a consequence of such hydrate dissociation due to rapid warming of the bottom waters. The Late Palaeocene – Early Eocene was a period of peak warming, and overall the warmest interval in the Cenozoic when latitudinal thermal gradients were greatly reduced (Haq 1984). In the latest Palaeocene, bottom-water temperature also increased rapidly (in less than 10 ka) by as much as 4°C, with a coincident excursion of about −2 to −3‰ in $\delta^{13}C$ of all carbon reservoirs in the global carbon cycle (Kennett & Stott 1991; Koch *et al.* 1992; Bralower *et al.* 1995). Dickens *et al.* (1997) suggest that this rapid excursion cannot be explained by conventional mechanisms (increased volcanic emissions of carbon dioxide, changes in oceanic circulation and/or terrestrial and marine productivity, etc.). However, a rapid warming of bottom waters from 11 to 15°C could abruptly alter the sediment thermal gradients leading to methane release from gas hydrates. Increased flux of methane into the ocean–atmosphere system and its subsequent oxidation is considered sufficient to explain the ~ -2.5 ‰ excursion in $\delta^{13}C$ in the inorganic carbon reservoir. Adding large quantities of carbon dioxide to the ocean should also increase its acidity leading to elevation of the lysocline and greater carbonate dissolution. Although there is some indication of increased carbonate dissolution in the latest Palaeocene, its distribution and magnitude are unclear. Dickens *et al.* (1997) hold explosive volcanism and rapid release of carbon dioxide and changes in the sources of bottom water during this time as a plausible triggering mechanism for the peak warming leading to hydrate dissociation. But they wonder what could also be responsible for terminating this sequence of events relatively rapidly over some 200 ka. Several short-lived sea-level rise events in the earliest Eocene (Haq *et al.* 1988), which were most probably unrelated to the gas hydrate events, may offer a potential forcing mechanism that would put an end to further hydrate dissociation and would instead work to generate and stabilize gas hydrates.

The eustatic record of the Late Palaeocene–Early Eocene could offer further clues for the behaviour of the gas hydrates and their contribution to the overall peak warm period of this interval (Haq *et al.* 1988). The longer term trend shows a rising sea level through the latest Palaeocene and early Eocene, but there are several shorter term sea-level drops throughout this period and one prominent drop straddling the

Palaeocene–Eocene boundary (which could be an additional forcing component to hydrate dissociation, not considered by Dickens *et al.* 1995). Early Eocene is particularly replete with high-frequency sea-level drops of several tens of metres. Could these events have contributed to the instability of gas hydrates adding significant quantities of methane to the atmosphere and maintaining the general warming of the period? If the Dickens *et al.* scenario is true for the latest Palaeocene, could it not also work for the pre-icehouse Palaeogene? These ideas seem testable if detailed faunal and isotopic data for the interval in question were available with at least the same kind of resolution as that obtained for the latest Palaeocene interval by Kennett & Stott (1991). Although such higher resolution data are not available as yet, other Palaeogene candidates for sea-level fall events associated with prominent shifts in $\delta^{13}C$ include an event near the Eocene–Oligocene boundary and another in the mid-Oligocene (36 Ma and 30 Ma events of Haq *et al.* 1988) where significant shifts in $\delta^{13}C$ have also been recorded in marine isotopic record offshore Brazil (Abreu & Haddad in press). A closer scrutiny may reveal more such correlations.

Gas hydrates and continental slope stability

The above discussion implies that past sea-level lowerings, especially major eustatic falls, may have been accompanied by massive slumping along the margins. Slumping need not be forced by gravity forced failure of unconsolidated sediment alone, but development of zones of weakness within the sedimentary column after the melting of gas hydrate may also contribute to significant mass movements and rearrangement of sediment bodies on the continental margins.

Gravity driven mass transport of sediment on the continental slope can be classed into three broadly defined categories (Shanmugam *et al.* 1995) (see Fig. 5). Large slides occur on planar glide surfaces. They may have no interval deformation and may survive with the bedding intact. Slumps occur as coherent rotational mass movements on a concave-up glide planes and contain internal deformation of the slumped mass. The third end-member is the debris flow that occurs as an incoherent viscous mass in which intergrannular movement produces a chaotic end product. If such a mass of debris flow is fluidized due to increased water content during its movement it can turn into a turbidity current. Thus, debris flows may be associated with sheet-like turbiditic deposits at their distal ends.

Decomposition of gas hydrates and weakening of the mechanical strength of sediments that encourages failure along low-angle slopes may produce more slides and some slumps rather than debris flows. Examples are provided by the Carolina Trough area where slump features are associated with numerous faults that sole

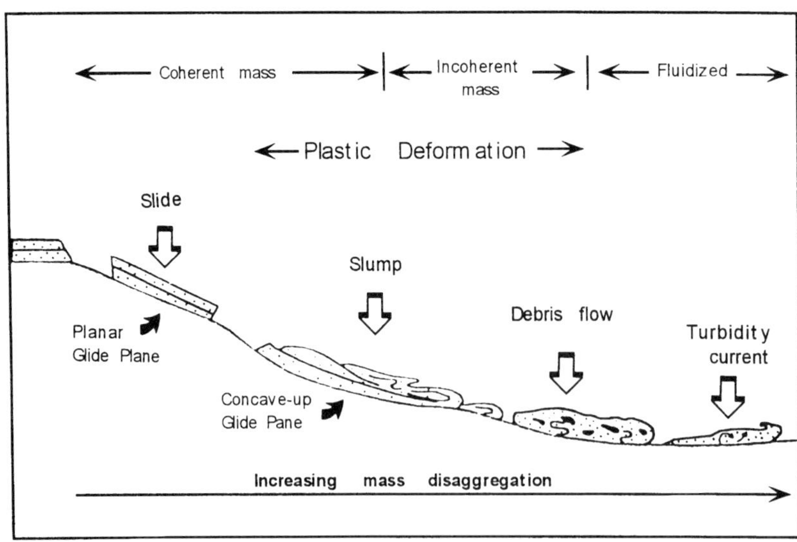

Fig. 5. Schematic of various end-members of gravity driven sediment transport processes on the continental slope. Slope failures caused by gas hydrate dissociation and weakening of sediment strength are more likely to occur as slides and slumps and less likely to be associated with distal turbidites (see text) (modified after Shanmugan *et al.* 1995).

out at or above the BSR levels (Paull *et al.* 1989). Earthquakes, on the other hand, are more likely to produce slumps and debris flows associated with more random mass failure of the slope. Thus, slope and foot-of-the-slope slide and slump features caused by hydrate dissociation are less likely to be associated with distal turbidites.

A relatively recent series of giant slides and slumps off the Norwegian continental margin (Storregga) have been ascribed to earthquakes and gas hydrate dissociation (Bugge *et al.* 1987; Jansen *et al.* 1987). The Storregga composite slump scar is 290 km long with a maximum runout distance of 800 km. The total volume of the slumped material involving Tertiary and Quaternary sediments is estimated to be 5600 km^3. The mass failure took place in three recognizable phases (first prior to 30 ka, and the other two between 6 and 8 ka). Distal parts show remnants of debris flows and fluidized turbidity deposits. Large slabs can be 200 m thick and up to 30 km across, that slid at low angle slopes of only 0.3°. At least the second slump is considered to have been caused by gas hydrate dissociation (Bugge *et al.* 1987).

Summerhayes *et al.* (1979) documented large slides and slumps of probable Pleistocene age off the south-western African coast that displaced over 250 km^3 of sediments from the lower slope to upper rise and a runout distance of about 250 km. The displacement affected only the upper few tens of metres of sediment that slid on a concave-up glide plane leaving behind large scars, and were apparently not accompanied by turbidity currents. Similar mass movements are commonly observed on the African west coast seismic profiles that displace several hundred cubic kilometres of sediments in each failure event. High productivity and rapid generation of biogenic methane is a characteristic of this margin. The authors held slope failure by earthquakes and gas hydrate dissociation as likely mechanisms for triggering these events. The lack of turbidity deposits at the foot of the slumps could be a clue to their gas hydrate dissociation-related origin.

Recent ODP drilling on the Amazon Fan has also revealed several mass-transported deposits that are obviously linked to catastrophic failure of the continental slope of Brazil. These deposits occur over 10 km^3 and can be up to 200 m thick. Maslin *et al.* (in press) observe that the contained benthic assemblages point to an origin at the slope depths some 150 km away and 1500 m above the present position where the masses came to rest. At least three phases of mass failure are recognized occurring near sea-level lowstands (between 42 and 45 ka, at ~ 33 ka, and between 14 and 17 ka). Because the slumps occur near sea-level drops, Maslin *et al.* ascribe loading of unconsolidated sediments and gas hydrate destabilization as the plausible triggering mechanisms for these mass failures.

This linkage between slumped features and gas hydrate dissociation, however, remains mostly conjectural at the present. To be able to convincingly distinguish between normal sediment-loaded, gravity-forced, failure of unconsolidated sediment from failure triggered by gas hydrate decomposition with objective criteria remains a challenge.

Timing of the gas hydrate development

When did the gas hydrates first develop in the geological past? The specific low-temperature–high-pressure requirement for the stability of gas hydrates suggests that they may have existed at least since the latest Eocene, the timing of the first development of the oceanic psychrosphere and cold bottom waters. Prior to that bottom waters in the world ocean are inferred to have been relatively warm even in the higher latitudes (see Haq 1984, for a discussion). This raises questions about the presence of hydrates in pre-psychrospheric times. Does this imply that gas hydrates are a post-psychrospheric phenomena? Or could hydrostatic pressure alone have maintained the clathrate stability?

According to the gas hydrate stability window (Kvenvolden & Barnard 1983) it is apparent that bottom-water temperature need not approach zero, but instead the geothermal gradient within the sediments and the hydrostatic pressure above would be more critical for clathrate stability. Theoretically clathrates could exist on the slope and rise when bottom waters approach those estimated for late Cretaceous and Palaeogene (~7–15°C), although they would occur deeper within the sedimentary column and the stability window would be relatively thinner. Dickens *et al.* (1995) estimated a depth of ~ 900 m below sea level for the hydrate stability zone in the late Palaeocene. If the bottom waters were to warm up to 22°C only then would most margins of the world be free of gas hydrate accumulation (Fig. 6). The implied smaller stability zone during warm bottom-water regimes, however, does not necessarily mean an overall reduced methane reservoir, as it also follows that the sub-hydrate free gas zone could be larger making up for the hydrate deficiency.

For the pre-Oligocene there is little evidence of large ice caps, and the mechanism for short-term

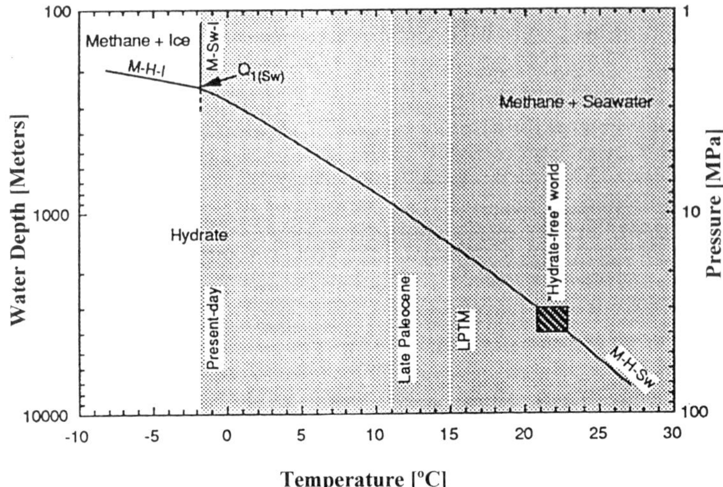

Fig. 6. Dickens *et al.*'s (1995) methane hydrate temperature–pressure phase diagram for a methane-sea water system. The late Palaeocene bottom-water temperature is assumed to be 11°C and hydrate are estimated to exist in sediments with water depth exceeding 920 m (as opposed to ~250 m depth at −1.5°C today). During the latest Palaeocene thermal maximum (LPTM) the bottom temperature is interpreted to have risen to *c.* 15°C, lowering the depth of hydrate stability zone to ~1460 m. For continental margins shallower than 4000 m the hydrate-free world would occur when bottom-water temperature reached at about 22°C (after Dickens *et al.* 1995).

sea-level changes remains uncertain. And yet, the Mesozoic–Early Cenozoic eustatic history is replete with major sea-level falls of 100 m or more that are comparable in magnitude, if not in frequency, to glacially induced eustatic changes of the Late Neogene (see Haq *et al.* 1988). If gas hydrates existed in the pre-Oligocene, major sea-level falls would imply that hydrate dissociation may have contributed significantly to climate change and shallow-seated tectonics along continental margins. Such massive methane emissions, however, should also be accompanied by prominent $\delta^{13}C$ excursions, as exemplified by the terminal Palaeocene climatic maximum (Kennett & Stott 1991; Dickens *et al.* 1995).

The long-term record of slope failure

Is there geological evidence of increased frequency of slumping associated with major falls in the pre-Oligocene sea level that can be ascribed to gas hydrate breakdown? As a test case we could look for such evidence in a careful study of buried canyons, slumps and erosion along the New Jersey Margin of the US East Coast based on seismic data. It has been obvious that sediment accumulation and preservation patterns along this margin (as along any other continental margin) resulted from a complex interaction of sea-level changes, fluctuating sediment supply rates and regional subsidence, and bottom current flow along the margins and in the basin. But perhaps destabilization and movement of sediment wedges caused by gas hydrate breakdown during lowstand times were also significant.

In a seismic study of the New Jersey Margin, Mountain (1987) documented four periods of Palaeogene slope failure, slumping and infilling along the slope, i.e. near the Cretaceous–Tertiary boundary, at the Palaeocene–Eocene boundary, at the top of Lower Eocene and in the Middle Eocene. In addition, Mountain & Tucholke (1985) have shown a widespread unconformity near the Eocene–Oligocene boundary which wiped out much of the Oligocene stratigraphic record. These unconformities have little or no shallow-water debris resting on them. The events near the Palaeocene–Eocene boundary and at the top of Lower Eocene also are associated with major slumps. In the latter, a megaslump which is compositionally similar to enclosing sediments seems to have travelled a few kilometres downslope to its present position. Slope failure related to episodic collapse of the underlying Mesozoic carbonate margin was suggested as a mechanism for these hiatal surfaces (Mountain & Tucholke 1985; Mountain 1987). A puzzle for the authors was the presence of onlapping units at the foot of the slope (slumps) and coeval unconformities that indicate shelfal erosion and transportation of sediments at the same time as slope failure.

All of the events documented on the New Jersey Margin occur close to major sea-level lowstands (see Haq *et al.* 1988). In particular, the events near the Palaeocene–Eocene boundary, at the top of Lower Eocene and another in the Middle Miocene are associated with major slumps. Some slump blocks maintain their original bedding. Mountain (1987) ascribed slope detachment to be triggered by diagenesis and/or local faulting. However, the process may be more readily explained in terms of gas hydrate destabilization, following lowered base levels, and reduced hydrostatic pressure on the shelf and slope. This would also explain the puzzle of apparent coeval shelfal and slope erosion as during lowstands the sub-areally exposed shelf would be prone to extensive erosion while the slope will suffer from accelerated slumping due to hydrate dissociation.

Another example of a slope scour and associated sea-floor unconformity that could be attributed to gas hydrate destabilization is provided by Angstadt *et al.* (1983) from the Gulf of Mexico. Seismic data and cores from two drill sites provide a precise age for the missing section at the unconformity. The rough surface of the unconformity was also interpreted as due to channellized flow. The prominent hiatus is centred on a global sea-level fall event near the Middle–Late Eocene boundary at 39.5 Ma (Haq *et al.* 1988). Angstadt *et al.* (1983) ascribed the event to slope instability and mass wasting, perhaps caused by shoreline retreat and intensified current activity. These authors go on to speculate that a perhaps a meteoritic impact may have been a forcing factor for the high-magnitude sea-level changes. Nevertheless, decomposition of the gas hydrate on the slope of Florida Escarpment following a major base-level drop at 39.5 Ma and ensuing mass wasting is a simpler and more plausible scenario for this and similar events elsewhere, especially if a connection with sea-level lowering can be established.

For methane emissions related to sea-level fall to have been effective agents of global climate change over the longer term (on the order of million years) in the non-glacial world, methane must be replenished continuously over much of this period. This implies that a step-wise sea-level fall (i.e. with numerous fall events) would be more conducive to a sustained climatic change – shorter sea-level drops would cause continued and increasingly frequent slumps and release of methane, leading to greater climatic modification.

These ideas are largely conjectural at present and will need further testing. A re-examination of seismic and stratigraphic data along continental margins for evidence of normal and growth faults and major slumping/sliding within gas hydrate field depths and associated negative $\delta^{13}C$ excursions could point to causal linkages between gas hydrates and sedimentary tectonic processes hitherto unaccounted for.

Gas hydrates reservoir characteristics

Gas hydrates are composed predominantly of methane, a clean burning fuel. Because the hydrate structure concentrates 164 times more methane in the same space as free gas, the clathrate reservoirs may represent a major, although as yet unproven, energy resource. However, if gas hydrates are to be exploited as a resource better understanding of the nature of the hydrate reservoir and the volume of methane trapped in them (including the free gas underneath) needs to be gained. Good empirical data that could lead to insights into reservoir quality and exploitibility are rare at the present time.

As mentioned earlier, first direct measurements of *in situ* methane in sediments with clathrates and free gas were made only recently during ODP Leg 164 (Paull *et al.* 1997; Dickens *et al.* 1997). Although actual recovered samples of hydrate were rare, except at Site 996, these results indicate that large quantities of methane (estimated to be ~ 15 Gt of carbon) are stored in the clathrates on Blake Ridge, and even more free gas below the solid hydrates. In the hydrate stability depth (450–190 m below the sea floor) 0–9% of the pore space may be occupied by solid hydrate (consisting of 98.5% methane and remainder is largely carbon dioxide). Although the gas hydrate is distributed mostly as finely disseminated, solid hydrate bodies up to 30 cm thick occur in this interval. Presence of solid clathrate is associated with high resistivity and sonic velocity well-log response and coincident with chloride anomalies (Paull *et al.* 1997).

Below the hydrate stability depth pore spaces are saturated with free methane gas. The total free gas volume on Blake Ridge may be significantly larger than that inferred from seismic data (up to 12% compared to 1–2% estimated from seismic data) (Dickens *et al.* 1997; Holbrook *et al.* 1996). Acoustic velocity within the gas hydrate stability zone (above the BSR) is similar to that in the overlying, non-hydrates sediments (on average $2000\,m\,s^{-1}$), but just below the BSR boundary it may be as low as $1200\,m\,s^{-1}$ in the free gas zone. This explains why there are no BSR shows in areas where there is no free gas below the hydrate. The free gas zone is also heterogeneous in distribution. There is a thin (~25 m) zone beneath the BSR in which gas volumes may reach a high of 12% and sonic velocities may drop to their lowest. Below this zone gas volumes decrease and velocities increase rapidly (Paull *et al.* 1997; Dickens *et al.* 1997). A single core was obtained with a pressure core sampler below the hydrate zone at site 994 (at a depth similar to just below the BSR depth of Sites 995 and 997). The amount of methane in this core plots on the saturation curve for methane and sea water, which is also consistent with the idea that no BSR is apparent in the absence of free gas below the hydrate zone (Dickens pers. comm. 1997).

Concluding remarks

It is very likely that gas hydrates have been an important factor in reshaping the structure of the continental margins. As potential agents of faulting, slumping and block sliding, they may have played a very significant role in modifying stratigraphic patterns, particularly during the lowstand phases of the eustatic cycles. However, the role of gas hydrate breakdown in the rearrangement of sediment wedges is obviously a complex issue. Some of the ideas concerning this role have been touched upon here, but many major questions remain unanswered, underscoring the need for considerably more research on this issue of fundamental importance to sedimentary geology. The role of gas hydrate as a source of significant emissions of greenhouse gas in global change scenarios, or their contribution as a major potential source of carbon in considerations of global carbon cycle, is even more controversial and can only be resolved with more detailed studies of hydrated intervals in conjunction with high-resolution studies of the ice cores, with as great a resolution as 50 years (Raynaud *et al.* 1993, 1998).

A better understanding of gas hydrates may well show their considerable role in controlling continental margin stratigraphy and shallow structure, as well as in global climatic change, and through it as agents of biotic evolution. A systematic search for evidence of slumping and normal and growth faulting associated with clathrate destabilization needs to be carried out using existing seismic and stratigraphic data on the continental margins. A much greater effort is warranted to unravel the enigma of the gas hydrate, which may also prove to be an enormous untapped energy resource for the future, both as a direct source of natural gas and as potential stratigraphic seals for the large pools of free gas below.

The author thanks J. Dickens and M. Maslin for reviewing this article that led to substantial improvement of the text. The views expressed here are those of the author and do not necessarily reflect the views of National Science Foundation.

References

ABREU, V. S. & HADDAD, G. A. 1998. *Glacioeustatic Fluctuations: The Mechanism Linking Stable Isotope Events and Sequence Stratigraphy From Early Oligocene to Middle Miocene*. Society of Economic Paleontologists and Mineralogists, Special Publication, **57** (in press).

ANGSTADT, D., AUSTIN, J., JR & BUFFLER, R. 1983. Deep Sea erosional unconformity in the southeastern Gulf of Mexico. *Geology*, **11**, 215–218.

BANGS, N., SAWYER, D. & GOLOVCHENKOVA, X. 1993. Free gas at the base of gas hydrate zone in the vicinity of the Chile triple junction. *Geology*, **21**, 905–908.

BOND, G. C. & LOTTI, R. 1995. Iceberg discharges into the North Atlantic on millennial time scales during the last glaciation. *Science*, **267**, 1005–1010.

BOOTH, J., WINTERS, W. & DILLON, W. 1994. Circumstantial evidence of gas hydrate and slope failure associations on the US Atlantic continental margin. *Annals of the New York Academy of Sciences*, **75**, 487–489.

BRALOWER, T., ZACHOS, J., THOMAS, E. *et al.* 1995. Late Paleocene to Eocene paleoceanography of the equatorial Pacific Ocean. ODP Site 865, Allison Guyot. *Paleoceanography*, **10**, 841–865.

BROOKS, J., BRYANT, H., KENNICUT, W., MANN, T. & MCDONALD, T. 1986. Association of gas hydrates and oil seepage in the Gulf of Mexico. *Organic Geochemistry*, **10**, 221.

BRYAN, G. M. 1974. *In situ* indicators of Gas hydrates. *In: Natural Gas in Marine Sediments*. Marine Science, **3**, 299–308.

BUGGE, T., BEFRING, S., BELDERSON, R. H. *et al.* 1987. A giant three-stage submarine slide off Norway. *Geo-Marine Letters*, **7**, 191–198.

COLLETT, T. S. 1993. Natural gas hydrates of the Prudhoe Bay and Kuparuk River Area, North Slope, Alaska. *AAPG Bulletin*, **77**, 793–812.

——, KVENVOLDEN, K. & MAGOON, L. 1990. Characterization of hydrocarbon gas within the stratigraphic interval of gas hydrate stability on the North Slope of Alaska. *Applied Geochemistry*, **5**, 279–287.

——, BIRD, K., KVENVOLDEN, K. & MAGOON, L. 1988. *Geologic Interrelations Relative to Gas Hydrate Within the North Slope of Alaska*. US Geological Survey Open File Report, **88–389**.

CRANSTON, R. E. 1996. Marine sediments as a source of atmospheric methane (Abstracts). *In*: HENRIET, J.-P. & MIENERT, J. (eds) *First Master Conference on*

Gas Hydrates: Relevance to World Margin Stability and Climatic Change, University of Gent, Belgium, 2.

DICKENS, G. & QUINBY-HUNT, M. 1997. Methane hydrate stability in pore water: a simple theoretical approach for geophysical applications. *Journal of Geophysical Research*, **102**, 773–783.

——, O'NEIL, J., REA, D. & OWEN, R. 1995. Dissociation of oceanic methane hydrate as a cause of the carbon isotope excursion at the end of the Paleocene. *Paleoceanography*, **10**, 965–971.

——, PAULL, C., WALLACE, P. & ODP LEG 164 SCIENTIFIC PARTY 1997. Direct measurement of *in situ* methane quantities in a large gas hydrate reservoir. *Nature*, **385**, 426–428.

DILLON, W. P. 1991. *Seismic Interpretation of Gas Hydrates in the Blake Ridge Area*. Transections of Natural Gas Research and Development Contractors Review Meeting, US Department of Energy.

—— & PAULL, C. 1983. Marine gas hydrates II: Geophysical evidence. *In:* COX, L. L. (ed.) *Natural Gas Hydrates, Properties, Occurrence, Recovery*. Butterworth, Woburn, MA, 73–90.

——, DANFORTH, W. W., HUTCHINSON, D. R., DRURY, R. M., TAYLOR, M. H. & BOOTH, J. S. 1998. Evidence for faulting related to dissociation of gas hydrate and release of methane off the south-eastern United States. *This volume*.

DOMENICO, S. N. 1976. Effect of brine–gas mixture on velocity in unconsolidated sand reservoir. *Geophysics*, **41**, 882–894.

HAQ, B. U. 1984. Paleoceanography – A synoptic overview of 200 million years of ocean history. *In*: HAQ, B. & MILLIMAN, J. (eds) *Marine Geology and Oceanography of Arabian Sea and Coastal Pakistan*. Van Nostrand Reinhold, New York, 201–231.

—— 1993. *Deep Sea Response to Eustatic Change and the Significance of Gas Hydrates for Continental Margin Stratigraphy*. International Association of Sedimentologists, Special Publication, **18**, 93–106.

——, HARDENBOL, J. & VAIL, P. R. 1988. *Mesozoic and Cenozoic Chronostratigraphy and Eustatic Cycles*. Society of Economic Paleontologists and Mineralogists, Special Publication, **42**, 71–108.

HARVEY, L. & HUANG, Z. 1995. Evaluation of potential impact of methane clathrate destabilization on future global warming. *Journal of Geophysical Research*, **100**, 2905–2926.

HESSE, R. & HARRISON, W. 1981. Gas hydrates (clathrates) causing pore-water freshening and oxygen isotope fractionation in deep-water sedimentary sections of terrigenous continental margins. *Earth and Planetary Science Letters*, **55**, 453–562.

HOLBROOK, W., HOSKINS, H., WOOD, W., STEPHEN, R., LIZARRALDE, D. & LEG 164 SCIENCE PARTY 1996. Methane hydrate and free gas on the Blake Ridge from vertical seismic profiling. *Science*, **273**, 1840–1843.

HOLLISTER, C. D., EWING, J., HABIB, D. *et al.* 1972. *Initial Reports of the Deep Sea Drilling Project*, Volume 11. US Government Printing Office, Washington, DC.

JANSEN, E., BEFRING, S., BUGGE, T. *et al.* 1987. Large submarine slides on the Norwegian continental margin: sediments, transport and timing. *Marine Geology*, **78**, 77–107.

JOUZEL, J. *et al.* 1993. Extending the Vostock ice-core record of paleoclimate to the penultimate glacial period. *Nature*, **364**, 407–412.

KATZMAN, R., HOLBROOK, W. & PAULL, C. 1994. A combined vertical-incidence and wide-angle seismic study of gas hydrate zone, Blake Ridge. *Journal of Geophysical Research*, **99**, 17 975–17 995.

KAYEN, R. E. & LEE, H. J. 1991. Pleistocene slope instability of gas hydrate laden sediment on the Beaufort Sea margin. *Marine Geotechnology*, **10**, 125–141.

KENNETT, J., STOTT, L. 1991. Abrupt deep sea warming, paleoceanographic changes and benthic extinctions at the end of Palaeocene. *Nature*, **353**, 319–322.

——, HENDY, I. & BEHL, R. 1996. Late Quaternary foraminiferal carbon isotope record of Santa Barabara Basin: Implications for rapid climate change. (Abstract). *In: Annual Meeting of the American Geophysical Union*, San Francisco, F.292.

KOCH, P., ZACHOS, J. & GINGERRICH, P. 1992. Correlation between isotope records in marine and continental carbon reservoirs near the Paleocene–Eocene boundary. *Nature*, **358**, 319–322.

KVENVOLDEN, K. A. 1988. Methane hydrates – A major reservoir of carbon in shallow geosphere. *Chemical Geology*, **72**, 41–51.

—— 1993. Gas hydrates – Geological perceptive and global change. *Reviews of Geophysics*, **31**, 173–187.

—— 1998. A primer on the geological occurrence of gas hydrate. *This volume*.

—— & BARNARD, L. A. 1983. Hydrates of natural gas in continental margins. *AAPG Bulletin*, **34**, 631–640.

—— & CLAYPOOL, G. E. 1988. *Gas Hydrates in Oceanic Sediments*. United States Geological Survey Open File Report, **88–216**.

LASHOF, D. & AHUJA, D. 1990. Relative contribution of greenhouse gas emissions to global warming. *Nature*, **344**, 529–531.

LEE, M., HUTCHINSON, D., DILLON, W. *et al.* 1993. Method of estimating the amount of in situ gas hydrates in deep marine sediments. *Marine and Petroleum Geology*, **10**, 493–506.

MACDONALD, G. T. 1990. The future of methane as an energy resource. *Annual Review of Energy*, **15**, 5–83.

MACDONALD, I. R., GUINASSO, N., SASSEN, R. *et al.* 1994. Gas hydrate that breaches the seafloor on the continental slope of the Gulf of Mexico. *Geology*, **22**, 699–702.

MACKAY, M. E., JARRAD, R. D., WESTBROOK, G. K., HYNDMAN, R. E. & LEG 146 SHIPBOARD SCIENTIFIC PARTY. 1994. Origin of bottom simulating reflectors: Geophysical evidence from the Cascadia accretionary prism. *Geology*, **22**, 459–462.

MARKL, R., BRYAN, G. & EWING, J. 1970. Structure of the Blake–Bahama Outer Ridge. *Journal of Geophysical Research*, **75**, 4539–4555.

MASLIN, M., MIKKELSEN, N. & HAQ, B. U. 1998. Se level controlled catastrophic failures of the Amazon Fan. *Nature* (in press).

MCIVER, R. 1977. Hydrates of natural gas – Important agent in geological processes. *Geological Society of America, Abstracts*, **9**, 1989–1990.

——. 1982. Role of naturally occurring gas hydrates in sediment transport. *AAPG Bulletin*, **66**, 789–792.

MOUNTAIN, G. S. 1987. *Cenozoic Margin Construction and Destruction Offshore New Jersey. Cushman Foundation for Foraminiferal Research, Special Publication*, **24**, 57–83.

—— & TUCHOLKE, B. E. 1985. Mesozoic and Cenozoic geology of the US Atlantic continental slope and rise. *In*: POAG, W. (ed.) *Geologic Evolution of the United States Atlantic Margin*. Van Nostrand Reinhold, New York, 293–341.

NISBET, E. G. 1990. The end of ice age. *Canadian Journal of Earth Sciences*, **27**, 148–157.

PAULL, C., USSLER, W., III & DILLON, W. 1991. Is the extent of glaciation limited by marine gas hydrates. *Geophysical Research Letters*, **18**, 432–434.

——, BUELOW, W., USSLER, W., III, & BOROWSKI, W. 1996. Increased continental-margin slumping frequency during sea-level lowstands above gas hydrate-bearing sediments. *Geology*, **24**, 143–146.

——, USSLER, W., III, BOROWSKI, W. & SPIESS, F. 1995. Methane-rich plumes on the Carolina continental rise: Associations with gas hydrates. *Geology*, **23**, 89–92.

——, SCHMUCK, E., CHANTON, J., MANNHEIM, F., & BRALOWER, T. 1989. Carolina Trough diapirs: Salt or shale? *EOS, Transactions of the American Geophysical Union, Abstracts*, **70**, 370.

——, MATSUMOTO, R., WALLACE, P. *et al.* 1997. *Proceedings of the Ocean Drilling Project, Initial Reports*, College Station, TX. Ocean Drilling Project, **164**.

PRIOR, D., DOYLE, E. H. & KALUZA, M. J. 1989. Evidence for sediment eruption on deep sea floor, Gulf of Mexico. *Science*, **234**, 517–519.

RAYNAUD, D., CHAPPELLAZ, J. & BLÜNIER, T. 1996. Ice core record of atmospheric methane changes: Relevance to climatic changes and possible gas hydrate sources. (Abstract). *In*: HENRIET, J.-P. & MIENERT, J. (eds) *First Master Conference on Gas Hydrates: Relevance to World Margin Stability and Climatic Change*, University of Gent, Belgium, 148–155.

——, CHAPPELLAZ, J. & BLÜNIER, T. 1998. Ice-core record of atmospheric methane changes: relevance to climatic changes and possible gas hydrate sources. *This volume*.

——, JOUZEL, J., BRANOLA, J.-M., CHAPPELLAZ, J., DELMAS, R. J. & LORIUS, C. 1993. The ice core record of greenhouse gases. *Science*, **259**, 926–934.

SHANMUGAM, G., BLOCH, R., MITCHELL, S. M. *et al.* 1995. Basin-floor fans in the North Sea: Sequence stratigraphic models vs. sedimentary facies. *AAPG Bulletin*, **79**, 477–512.

SHIPLEY, T. H., HOSUTON, M. H., BUFFLER, R. *et al.* 1979. Seismic evidence for widespread possible gas hydrate horizons on continental slopes and rises. *AAPG Bulletin*, **63**, 2204–2213.

SUMMERHAYES, C., BORHOLD, B. & EMBLEY, R. 1979. Surficial slides and slumps on the continental slope and rise of South West Africa: A reconnaissance study. *Marine Geology*, **31**, 265–277.

THORPE, R. B., PYLE, J. A. & NISBET, E. G. 1998. What does the ice-core record imply concerning the maximum climatic impact of possible gas hydrate release at Termination 1A? *This volume*.

——, LAW, K., BEKKI, S., PYLE, J. & NISBET, E. 1996. Is a CH_4 – hydrate driven deglaciation consistent with the available evidence? (Abs.). *In*: HENRIET, J.-P. & MIENERT, J. (eds) *First Master Conference on Gas Hydrates: Relevance to World Margin Stability and Climatic Change*, University of Gent, Belgium, 4.

TUCHOLKE, B., BRYAN, G. & EWING, J. 1977. Gas hydrate horizons detected in seismic profiler data from western North Atlantic. *AAPG Bulletin*, **61**, 698–707.

What does the ice-core record imply concerning the maximum climatic impact of possible gas hydrate release at Termination 1A?

R. B. THORPE[1], J. A. PYLE[1] & E.G. NISBET[2]

[1]*Centre for Atmospheric Science at the University Chemical Laboratories, Lensfield Road, Cambridge, UK*

[2]*Department of Geology, Royal Holloway and Bedford New College, University of London, Egham, Surrey, UK*

Abstract: The rapid climate changes at the terminations marking the end of the last glaciation are poorly understood. CH_4 hydrate decomposition has been suggested as a possible trigger of deglaciation. This would imply that there was a large pulse of atmospheric CH_4 contemporaneous with Termination lA. This implication is tested here in a two-stage process using an adapted version of the Cambridge 2-D model. Firstly, the ice-core record was used to place an upper limit on the magnitude of possible CH_4 release at Termination 1A. Secondly, the climatic impact of this realistic maximum (RM) pulse was estimated and compared with the actual climate change occurring at deglaciation. It was found that the maximum pulse that could readily be squared with the ice-core record was around 4000 Tg. Its climatic impact was modelled for a range of scenarios covering climate sensitivities from 1.5 to 5.2°C, and involving possible changes in atmospheric CO_2 and latitudinal heat transport. The direct radiative effects of this pulse were too small to account for the deglaciation alone, but, for certain combinations of CH_4, CO_2, and heat transport changes, coupled with high climate sensitivity, it was possible to simulate changes of the same magnitude as those observed.

The realization that large stores of CH_4 might be stored in the form of gas hydrates (Trofimuk *et al.* 1977; Dobryin *et al.* 1981; Kvenvolden 1988; Kvenvolden & Claypool 1988), and the fact that the stability of these deposits is strongly dependent upon temperature and pressure led to the suggestion that outgassing from CH_4-hydrates could have been a causal factor in the last deglaciation (MacDonald 1990). Specifically, Nisbet (1990, 1992) argued that: 'The last deglaciation was triggered by the thermal forcing of climate due to a sudden and catastrophic release of a very large amount of methane'.

His rationale for advancing this hypothesis (which is here termed the Catastrophic Hydrate Release (CHR) hypothesis) was four-fold.

- Methane has an atmospheric lifetime of the order of 10 years, comparable to the time scale of the climate change implied by the ice core record at Terminations lA and lB (Dansgaard *et al.* 1989; Alley *et al.* 1993).
- Methane is a strong greenhouse gas.
- There exists large and potentially unstable stores of methane in marine sediments and permafrost.
- Geological evidence suggests that methane bearing deposits could have been disturbed at around the time of the last deglaciation, liberating maybe 4–8 Gt ($1 Gt = 10^{15}$ g) of methane.

Investigation methodology

In order for the CHR mechanism to be considered a valid possibility as an explanation (either partial or full) of the last deglaciation, it must provide a strong greenhouse forcing immediately prior to the commencement of deglaciation at Termination lA, implying that greatly enhanced levels of atrnospheric CH_4 were co-incident with Termination lA. Consequently, the following are all required.

- The release mechanism must be both rapid and must make geophysical sense.
- The CH_4 must not be oxidized before it can reach the atmosphere and hence can influence the climate.
- The enhanced levels of atmospheric CH_4 necessary to produce a significant climatic impact must not be inconsistent with the ice-core record.

The first two issues are not considered further in this study, where it is assumed that the sudden release of large amounts of CH_4 is possible, and that this CH_4 would reach the atmosphere essentially intact. Instead, this study concentrates solely on the third issue. This is a potential problem for the CHR hypothesis because, when the ice-core record of atmospheric CH_4 is studied, no massive spike of CH_4 is apparent. The non-appearance of this predicted spike allows a con-

THORPE, R. B., PYLE, J. A. & NISBET, E.G. 1998. What does the ice-core record imply concerning the maximum climatic impact of possible gas hydrate release at Termination 1A?. *In:* HENRIET, J.-P. & MIENERT, J. (eds) *Gas Hydrates: Relevance to World Margin Stability and Climate Change*. Geological Society, London, Special Publications, **137**, 319–326.

straint to be placed upon the size of any possible CH_4 outgassing event at the end of the last deglaciation, and consequently can be used to test the CHR hypothesis. The methodology is summarized below.

- The CHR hypothesis implies a sharply enhanced atmospheric CH_4 concentration coincident with the commencement of deglaciation which is not apparent in the ice-core record.
- The non-appearance of CH_4 at deglaciation allows a realistic upper limit to be placed upon the possible size of the hypothetical CH_4 pulse. Here this is defined as being that pulse of a magnitude such that its chance of detection in the ice core record is 50%, and is termed the 'realistic maximum' (RM) pulse.
- Model predictions of the climate change due to the hypothetical RM CH_4 pulse are then compared with the observed temperature shift at Termination 1A (the commencement of deglaciation).

Modelling considerations

A thorough model investigation of the CHR hypothesis implies the following requirements.

- Global model coverage, although the latitudinal/longitudinal resolution of the coverage will not necessarily have to be that high.
- A full chemical scheme capable of accurate representation of CH_4 chemistry.
- A dynamical scheme adequate to allow modelling of the planet-wide advection of CH_4.
- A radiation scheme adequate to allow modelling of radiative forcing due to CH_4, O_3, CO_2, and H_2O throughout the lower atmosphere.
- A high level of computational efficiency that will allow model runs of several decades of simulated time to be accomplished quickly.

The choice of the Cambridge 2-D model with its full chemistry, interactive radiative treatment of climatically important trace gases in the stratosphere and troposphere (Zhong & Haigh 1995), and adequate dynamics, represents an 'optimum compromise' as far as the above requirements are concerned (see Table 1). A three-dimensional model, although more desirable in theory, would at present run too slowly to be practical.

Establishing the realistic maximum pulse

As suggested earlier, the 'realistic maximum' (RM) pulse was defined to be a pulse of magnitude such that its chance of detection in the GRIP core record was 50%. The experimental procedure that was used to establish the RM pulse is summarized below.

- The LGM budget for CH_4 was established, and run to approximate equilibrium over a period of six years (Thorpe 1996).
- A pulse of 4000 Tg of CH_4 was injected into the lowest model level at a latitude of ~66°N.

Table 1. *The Cambridge 2-D Model (see Law & Pyle (1993) for details)*

Model Domain	Altitude: 0–60 km (~3.5 km scale height) Latitude: 90°S–90°N (9.47° resolution)
Dynamics	Classical Eulerian latitude/altitude *K*s from Luther (1973) with K_{ZZ}s from Hidaglo & Crutzen (1977)
Radiation	Solar heating (O_2, O_3) Near infra-red heating (O_3, CO_2, H_2O) Long-wave cooling (H_2O, CO_2, CH_4, N_2O, O_3)
Water vapour	Interactive scheme based on Harwood & Pyle (1980)
Chemistry	Continuity equations for e.g. O_x, NO_x, ClO_x, HO_x, HNO_3, HNO_4, N_2O_5, H_2O_2, C_2H_4, C_2H_6, n-C_4H_{10}, PAN
Source gases	CH_4, CO, NO_x, C_2H_4, CFCs, N_2O,C_2H_6, n-C_4H_{10}
Dry deposition	Depends upon f (land, sea, ice cover) O_3, NO_2, HNO_3, CO, PAN, H_2O_2, HCHO, CH_3OOH, CH_3CHO, RO_2
Emissions	CH_4, CO, NO_X, C_2H_4,C_2H_6, n-C_4H_{10} Vary with latitude/season

- The subsequent evolution of modelled atmospheric CH_4, OH, O_3, H_2O, etc., was followed over a period of 110 years.

Atmospheric CH_4 concentrations in the lowest model level at 66°N and 85°S were converted into synthetic ice-core signals by averaging daily values using the diffusion-only approach of Schwander *et al.* (1993). The exact period over which the atmospheric CH_4 was averaged so as to create the synthetic core signal (the synthetic ice-core averaging time) was a tunable parameter in the experiment, and was allowed to vary from 10 to 1000 years. The exact value pertaining to the GRIP core at the LGM is not known with certainty, but might be expected to lie between 15 and 100 years (Raynaud *et al.* 1996).

- The synthetic ice-core records were compared with the observed record at deglaciation. It was assumed that once the synthetic core signal breached the 440 ppbv level, it was a detectable event vis-à-vis the GRIP core record (see Thorpe *et al.* 1996 for details). The length of time for which the synthetic core signal remained above the 440 ppbv detection threshold (termed the 'detection window length') was then compared with the time between successive measurements of CH_4 for the GRIP core at LGM (the sampling interval).
- Assuming a sampling interval of 150 years between successive measurements of atmospheric CH_4 in the GRIP core at LGM (Raynaud *et al.* 1996), the RM pulse was then simply that pulse of a magnitude sufficient to give a synthetic ice-core signal with a detection window length of exactly 150/2 or 75 years.

Figure 1 shows a plot of the length of the detectable signal (relative to the 440 ppbv threshold) against the synthetic ice-core averaging time for the standard chemical scenario (CH_4 lifetime of 11.5 years) and a low NO_x scenario (CH_4 lifetime of 22 years). It can be seen that if the sampling interval were 300 years, detection of the signal would not be guaranteed for any core averaging time. For the GRIP core sampling interval of 150 years (Raynaud *et al.* 1996), detection of the signal would be more likely, but is still not guaranteed. However, if the sampling interval were cut to 50 years, then detection of a 4000 Tg pulse would be guaranteed. The bottom line is that a 1 year resolution is not required to detect this pulse; a 50 year sampling interval would suffice.

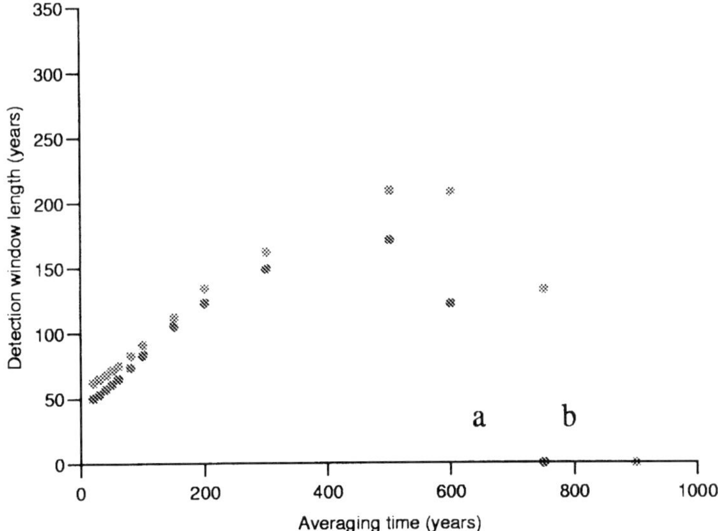

Fig. 1. The variation of maximum detection window length with averaging time for (**a**) standard and (**b**) low NO_x cases. As expected, low NO_x leads to longer detection windows, especially if the core averaging time is long (500+ years). However, the detection window length remains below the sampling interval for the GRIP core at LGM (averaging time 15–100 years), and the conclusion that detection of the CH_4 pulse is not inevitable remains unaltered.

It can be seen from Fig. 1 that the length of the detection window is a strong function of the synthetic core averaging time. This is because short-period averaging (a couple of decades) concentrates the ice-core signal as a short intense spike that resembles the atmospheric signal and allows the event to slip undetected between two successive measurements of CH_4, whereas averaging over a couple of centuries results in a broad spike that is relatively easy to detect. (A very long averaging time, for example 1000 years, smears out the signal to such an extent that it is swallowed up by background noise and cannot be detected – see far right of Fig. 1, where the detection window shrinks to zero.) The implication of this relationship between detection window length and synthetic core averaging time (AT) is that the magnitude of the realistic maximum pulse (the one with the 50% chance of detection) is also very sensitive to the choice of AT. In fact, for the likely AT appropriate to the GRIP core of 15–100 years (Raynaud *et al.* 1996; Chappellaz pers. comm.), the suggested 'detection window length' (Thorpe *et al.* 1996) for a 4000 Tg CH_4 release event is around 75 years and, therefore, the chance of detecting a pulse of this magnitude is around 50%. For this reason, a pulse of exactly 4000 Tg was taken to be the 'realistic maximum' RM pulse as suggested by the GRIP record.

Climate change experiments

The predicted climatic impact of the RM pulse is a function of the model's so-called 'climate sensitivity'; the model climate response to a radiative forcing perturbation (e.g. an increase in greenhouse gases) of standard size. It is conventional to represent this parameter in terms of a global mean temperature response to a doubling of atmospheric CO_2 from today's level. The 'true' value appropriate to the Earth's climate system is only poorly constrained, and so it is best to present model predictions of climate change in terms of a range of possible climate sensitivities, rather than on focusing on any single model run as providing 'the answer'. Here, the climatic impact of the RM pulse was estimated using three versions of the Cambridge 2-D model (including an interactive surface that incorporates temperature–albedo and water vapour–temperature feedback (Thorpe 1996)) with climate sensitivities of 1.55, 2.26 and 5.15°C. This range of climate sensitivities represents a measure of our uncertainty regarding the 'true' climate sensitivity, and approximates to the range of 1.5–4.5°C as suggested by IPCC (1990, 1992). Whilst the calculations using a climate sensitivity of 2.26°C might loosely be thought of as the 'best guess', with the other simulations being valid as possible upper and lower limits, it is not impossible that the 'true' climate sensitivity, and hence the climatic impact of the hypothetical RM pulse, could lie outside of this range. Three specific scenarios were considered.

Experiment P 'Pulse'

Instantaneous release of 4000 Tg CH_4 at 66°N.

Experiment C 'North Atlantic Conveyor'

Pulse release as experiment P, with 25% increase in efficiency of latitudinal heat transport phased in over 1000 days when global mean temperature exceeds 282.3 K (0.25 K above starting value).

Experiment O 'CO_2 outgassing'

As experiment C, with CO_2 increasing linearly from 200 to 280 ppmv over a period of 40 years, commencing when global mean temperature exceeds 282.3 K (as before). It is important to bear in mind that these should be considered as sensitivity studies only, not as rigorous attempts to model the last deglaciation.

Results for the basic experiment P are shown in Fig. 2. The results suggest that the potential contribution of the RM pulse to the glacial/interglacial temperature change ($\Delta TG/I$) of 3.5–5.5°C lies somewhere between 4 and 15%. This implies that whilst CHR could be significant in the context of deglaciation, it could *not* be either the sole, or even the main, cause of deglaciation. Other factors over and above the temperature–albedo and water-vapour–temperature feedback modelled in this study are also required.

The purpose of experiment C was to consider whether the RM pulse, coupled with changes in the efficiency of cross-latitudinal heat transport of 25%, could account for the deglaciation. The modelled changes in heat transport can crudely be thought of as representing changes in the North Atlantic thermohaline circulation that may have occurred during deglaciation (Broecker 1994; Sarnthein *et al.* 1994; Keigwin & Jones 1994; Kennett & Ingram 1995). The results are shown in Fig. 3. The considerable impact of the change in cross-latitudinal heat transport is apparent, with the resulting climate change increasing some four-fold to between 17

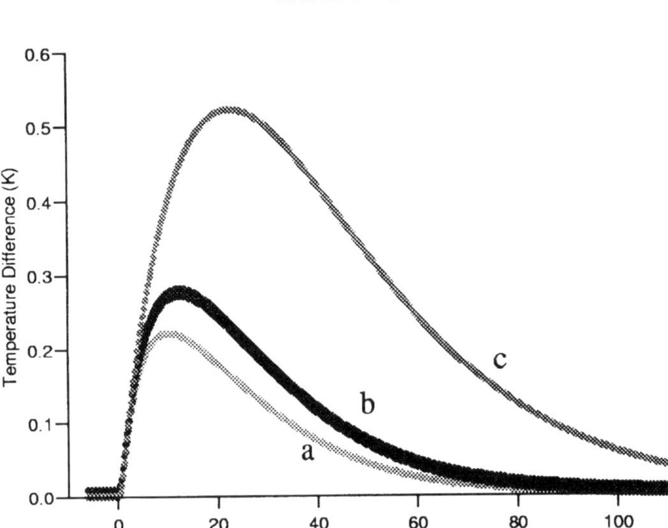

Fig. 2. The time evolution of the global mean temperature relative to the control run in experiment P (CH_4 pulse only) for climate sensitivities of (**a**) 1.55°C, (**b**) 2.26°C and (**c**) 5.15°C. Case (b), denoted by the thick line, is the standard scenario representing the 'best guess'.

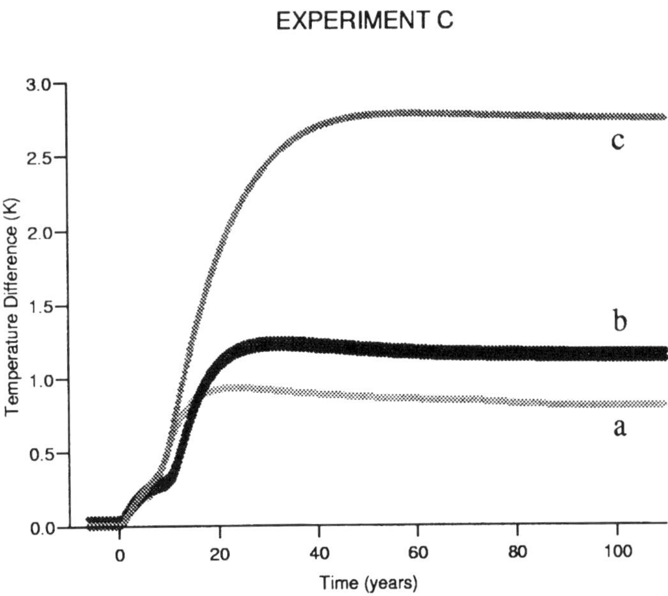

Fig. 3. The time evolution of the global mean temperature relative to the control run in experiment C (CH_4 pulse + conveyor) for climate sensitivities of (**a**) 1.55°C, (**b**) 2.26°C and (**c**) 5.15°C. Case (b), denoted by the thick line, is the standard scenario representing the 'best guess'.

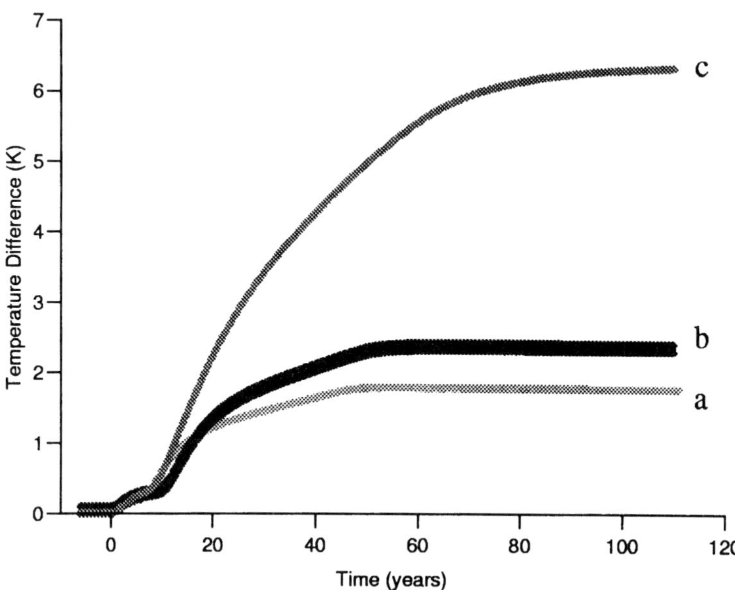

Fig. 4. The time evolution of the global mean temperature relative to the control run in experiment O (CH_4 pulse + conveyor + CO_2) for climate sensitivities of (**a**) 1.55°C, (**b**) 2.26°C and (**c**) 5.15°C. Case (b), denoted by the thick line, is the standard scenario representing the 'best guess'.

and 79% of ΔT G/I. However, even on the most favourable assumptions it is not possible to account for the whole of ΔT G/I, and additional factors must also been invoked.

The deglaciation was approximately contemporaneous with large changes in atmospheric CO_2 from around 200 to 280 ppmv (Leuenberger *et al.* 1992; Raynaud *et al.* 1993) which would have acted as a powerful amplifier of any climate change. This was modelled in experiment O. The results for this experiment are shown in Fig. 4. It can be seen that the impact of the CO_2 is highly significant, roughly doubling the climatic response seen in experiment C. In this case, changes of the magnitude of ΔT G/I can be simulated given favourable assumptions about the overall climate sensitivity.

Discussion of experimental results

The results of the three experiments are displayed in Table 2, with some conclusions summarized below.

- The ice-core record is not inconsistent with the CHR hypothesis, but it does imply an upper limit of around 4000 Tg for the magnitude of any realistic hypothetical CH_4 pulse.

Table 2. *Summary of results for the climate change experiments*

Experiment	Description	Peak climate change ΔT (K)	Climate change as % of ΔT G/I*
P	CH_4	0.22–0.52	4–15
C	CH_4 + conveyor	0.93–2.78	17–79
O	CH_4 + conveyor + CO_2	1.79–6.35	33–181

* ΔT G/I = temperature shift between glacial and interglacial, expressed in terms of global mean temperature. Assumed to lie between 3.5 and 5.5 K.

- A 4000 Tg pulse can only account for 4–15% of the temperature shift at deglaciation, and thus cannot represent the sole cause of deglaciation.
- If it is assumed that the hypothetical CH_4 pulse could have triggered changes in heat transport and CO_2, then it is possible to *rationalize* the course of the last deglaciation, if climate sensitivity is additionally assumed to be high.

However, it should always be borne in mind that the experiments P, C and O are just sensitivity studies. The representation of latitudinal heat transport (in experiments C and O) is of necessity a crude one in our two-dimensional latitude/altitude model. Many uncertainties are involved, and the 'successful' simulation of climate change of the required magnitude should in no circumstances be taken as proof that CHR did actually occur.

Conclusions

The purpose of this study was to consider whether the Catastrophic Hydrate Release (CHR) hypothesis was consistent with the evidence available from the ice-core record. The ice-core record suggests an upper limit to an outgassing event at the commencement of deglaciation of around 4000 Tg. It was seen that a pulse of this magnitude cannot cause deglaciation on its own, but if it resulted in changes in oceanic circulation, CO_2, etc., it might then be possible to account for ΔT G/I.

How might this question be answered more decisively in the future? It was seen that if the GRIP core sampling interval could be cut to around 50 years, then it would be possible to detect or to refute the existence of a pulse of 4000 Tg, making it possible to determine the truth or falsity of the CHR hypothesis.

So the final verdict must be that it is not possible to determine whether the CHR hypothesis is true or false at present using the approach outlined in this presentation. However, if the sampling interval for the ice core record of CH_4 at Termination 1A could be cut to say 50 years, then a decisive test of the CHR hypothesis ought to be possible.

This work was funded by the NERC. Thanks are expressed to E. Wolff for his help and constructive advice. The Centre for Atmospheric Science is a joint initiative of the Departments of Applied Mathematics and Theoretical Physics, and the Department of Chemistry. This modelling study is a contribution to the UGAMP project.

References

Alley, R., Meese, D., Shuman, C. *et al.* 1993. Abrupt increase in the Greenland snow accumulation at the end of the Younger Dryas event. *Nature,* **362**, 527–529.

Broecker, W. S. 1994. An unstable superconveyor. *Nature,* **367**, 414–415.

Dansgaard, W., White, J. & Johnsen, S. 1989. The abrupt termination of the Younger Dryas. *Nature,* **339**, 532–534.

Dobrynin, V., Korotajev, Y. & Plyuschev, D. 1981. Gas hydrates, a possible energy source. *In*: Meyer, R. & Olson, J. (eds) *Long-term Energy Resources.* Pitman, Boston, MA., 727–729.

Kvenvolden, K. 1988. Methane hydrates: a major source of carbon in the shallow geosphere? *Chemical Geology,* **71**, 41–51.

—— & Claypool, G. 1988. *Gas Hydrates in Oceanic Sediment.* US Geological Survey Open File Report, **88–216**.

Harwood, R. & Pyle, J. 1980. The dynamical behaviour of a two dimensional model of the stratosphere. *Quarterly Journal of the Royal Meteorological Society of London,* **106**, 395–420.

Hidaglo, H. & Crutzen, P. 1977. The tropospheric and stratospheric composition perturbed by NO_X emissions by high-altitude aircraft. *Journal of Geophysical Research,* **82**, 5833–5866.

IPCC. 1990. *Climate Change: The IPCC Scientific Assessment*, Houghton, J., Jenkins, G. & Ephraums, J. (eds) Cambridge University Press, Cambridge.

——. 1992. *Climate Change 1992, The Supplementary Report to the IPCC Scientific Assessment*, Houghton, J., Callender, S. & Varney, S. (eds). Cambridge University Press, Cambridge.

Keigwin, L. & Jones, G. 1994. Western North Atlantic evidence for millennial scale changes in ocean circulation and climate. *Journal of Geophysical Research,* **99**(C6), 12 397–12 410.

Kennett, J. & Ingram, B. 1995. A 20,000 year record of ocean circulation and climate change from the Santa Barbara Basin. *Nature,* **377**, 510–514.

Law, K. S. & Pyle, J. A. 1993. Modeling trace gas budgets in the troposphere, 1, ozone and odd nitrogen. *Journal of Geophysical Research,* **98**, 18,377–18,400.

Leuenberger, M., Siegenthaler, U. & Langway, C. C. 1992. Carbon isotope composition of atmospheric CO_2 during the last ice age from an Antarctic ice core. *Nature,* **357**, 488–490.

Luther, F. M. 1973. *Monthly Mean Values of Eddy Diffusion Coefficients in the Lower Stratosphere.* AIAA/AMS Conference Paper, Denver, Colorado, **73–49**.

MacDonald, G. J. 1990. Role of methane clathrates in past and future climates. *Climatic Change,* **16**(3), 247–281.

Nisbet, E. G. 1990. The end of the ice age. *Canadian Journal of Earth Sciences*, **27**(1), 148–157.

—— 1992. Sources of atmospheric CH_4 in early post-glacial time. *Journal of Geophysical Research,* **97**(D12), 12,859–12,867.

RAYNAUD, D., CHAPPELLAZ, J. & BLUNIER, T. 1996. Ice core record of atmospheric methane changes: relevance to climatic changes and possible gas hydrate sources. *In:* HENRIET, J.-P. & MIENERT, J. (eds) *First Master Workshop: Gas Hydrates, Relevance to World Margin Stability and Climatic Change*, University of Gent, Belgium, 148–155.

——, JOUZEL, J., BARNOLA, J. M., CHAPPELLAZ, J., DELMAS, R. J. & LORIUS, C. 1993. The ice record of greenhouse gases. *Science*, **259**, 926–934.

SARNTHEIN, M., WINN, K., JUNG, S. *et al.* 1994. Changes in East Atlantic deep-water circulation over the last 30,000 years – 8 time slice reconstructions. *Paleoceanography*, **9**(2), 209–267.

SCHWANDER, J., BARNOLA, J. M., ANDRIE, C. *et al.* 1993. The age of the air in the firn and the ice at Summit, Greenland. *Journal of Geophysical Research*, **98**(D2), 2831–2838

THORPE, R. B. 1996. *Can Methane Driven Deglaciation Provide a Plausible Account of the End of the Last Ice Age?* PhD Thesis, University of Cambridge, UK.

——, BEKKI, S., LAW, K. S., PYLE, J. A. & NISBET, E. 1996. Is methane-driven deglaciation consistent with the ice core record? *Journal of Geophysical Research*, **101**, 28 627–28 635.

TROFIMUK, A., CHERSKIY, N. & TSAREV, V. 1977. The role of continental glaciation and hydrate formation on petroleum occurrences. *In*: MEYER, R. (ed.) *Future Supply of Nature-made Petroleum and Gas*. Pergamon, New York, 919–926.

ZHONG, W. & HAIGH, J. 1995. Improved broad-band emissivity parameterisation for water vapour cooling rate calculations. *Journal of Atmospheric Sciences*, **52**, 124–138.

Ice-core record of atmospheric methane changes: relevance to climatic changes and possible gas hydrate sources

D. RAYNAUD[1], J. CHAPPELLAZ[1] & T. BLÜNIER[2]

[1]*Laboratoire de Glaciologie et de Géophysique de l'Environnement du CNRS, Saint-Martin-d'Héres cedex, France*

[2]*Climate and Environmental Physics, Physics Institute, University of Bern, Sidlerstrasse 5, Bern, CH-3012 Switzerland*

Abstract: The Antarctic and Greenland ice contains an almost direct record of past atmospheric CH_4. The record over the last 200 years reveals a spectacular 150% increase of the CH_4 atmospheric mixing ratio since pre-industrial times. At the scale of a glacial–interglacial cycle the record shows a remarkable correlation with climatic changes, with high (low) CH_4 levels during warm (cold) periods. A striking feature of the glacial–interglacial CH_4 record is the presence of large and abrupt (at the scale of a century or less) changes during the last glaciation and glacial–interglacial transition. The classical interpretation for the origin of CH_4 changes prior to the industrial era involves mainly the wetland source. In the context of gas hydrates the question is to know whether the past ice-core record contains fingerprints of catastrophic hydrate release (CHR). We currently conclude that the available record shows no evidence for CHR but additional ice-core analyses are necessary to reach a more definitive conclusion.

The Antarctic ice-core record of CH_4 now covers two glacial–interglacial cycles and will be extended in the future to several other cycles. These climatic cycles experienced important changes in the extent and shape of the huge ice sheets of the northern hemisphere as well as major global climatic changes. Such changes have affected the global CH_4 cycle as demonstrated by the ice-core record. A question of direct interest in the frame of the First Master Workshop on Gas Hydrates Is: is there evidence arising from the ice-core record of possible releases of CH_4 to the atmosphere from gas hydrate destabilization in ocean sediments or in the permafrost area?

The first section of this paper deals with the atmospheric significance of the methane record from ice cores. We then describe the major trends of the record for different time scales. Finally, we summarize the source mechanisms classically considered when interpreting the ice-core record and how likely it was that the methane clathrates played a role in the atmospheric CH_4 evolution over a glacial–interglacial cycle.

Atmospheric significance of the CH_4 record from ice cores

The snow deposited in areas free of summer melting at the surface of the ice sheets is compressed and sintered as a result of water vapour diffusion and plastic deformation of the solid H_2O grains under the weight of subsequently fallen snow. During this stage, called firn, of the transformation of snow into ice, the air found in the open pores becomes trapped in bubbles which close off during the last step of the sintering process, mainly in the bottom part of the firn layer. When all pores are finally closed off, the resulting material, called ice, contains 90% solid water and 10% of air by volume.

The porous firn layer on the surface of the ice sheet is continuously exchanging air with the overlying atmosphere, the firn–air is a mixture of air masses that were last in contact with the atmosphere at different times. Air, which gets enclosed in bubbles typically about 50–100 m below the surface (Schwander & Stauffer 1984), thus has a mean age which is younger than the age of the surrounding ice.

Because of the mixing of the air in the firn, which is dominated by molecular diffusion, and the gradual enclosure process of the air bubbles in the ice in formation, the air in the bubbles does not have a discrete age but rather an age distribution. The width of the age distribution is dominated by the occlusion process at low accumulation sites and by the mixing process in the firn at high accumulation sites (Schwander 1996). It varies, under present-day conditions, from about 600 years (low accumulation sites) to about 15 years (high accumulation sites).

It is necessary to ascertain the extent to which the CH_4 concentration in the air extracted from

RAYNAUD, D., CHAPPELLAZ, J. & BLÜNIER, T. 1998. Ice-core record of atmospheric methane changes: relevance to climatic changes and possible gas hydrate sources. *In:* HENRIET, J.-P. & MIENERT, J. (eds) *Gas Hydrates: Relevance to World Margin Stability and Climate Change*. Geological Society, London, Special Publications, **137**, 327–331.

the ice is the same as the original atmospheric methane concentration prevailing at the time when the air was trapped. There are potentially several parameters that could make the gas record measured in ice samples different to the real atmospheric concentrations, and a thorough discussion of these parameters with regard to the methane record can be found in Raynaud & Chappellaz (1993). According to current knowledge, the initial atmospheric CH_4 signal is generally well preserved (within a few per cent) in ice. An important clue for the preservation is the good connection between the CH_4 record obtained from ice-core measurements for the recent period and the atmospheric record measured since 1978 (Etheridge *et al.* 1992).

The glacial–interglacial atmospheric CH_4 record

Description of the record

The analysis of the air trapped in the polar ice at the time of its formation near the surface of the ice sheets has provided a large amount of information concerning the past history of the background levels of CH_4 atmospheric mixing ratios. The results have brought the current anthropogenic changes into a wide temporal perspective by providing (for a review see Raynaud *et al.* 1993):

- an extension towards past direct atmospheric background records covering the few last decades and showing a steady increase in the atmospheric CH_4 mixing ratios since the end of pre-industrial times (roughly 150 years ago) of about 150%;
- a CH_4 record over a full glacial–interglacial cycle (approximately the last 150 000 years) showing a remarkable correlation with climatic changes, with the highest mixing ratio levels (except for the last 200 years!) equivalent to the pre-industrial values (about 700 ppbv) during the warm interglacials and the lowest levels close to 350 ppbv during the coldest parts of the glaciations.

The ice-core record of methane has been more recently been extended back into the past. The correlation between CH_4 and low-frequency climate changes over the last climatic cycle has been tested on the preceding cycle (Jouzel *et al.* 1993). The overall covariation for the extended part of the record is similar to that observed for the last glacial–interglacial cycle, highlighting the strong interaction between methane and climate at such time scales. It is admitted that the flux changes between the different reservoirs (atmosphere, continents and oceans) result from climatic variations. However, the modifications in atmospheric mixing ratios of the greenhouse gases induce an initial radiative forcing which, together with the fast climatic feedbacks (water vapour, clouds, sea ice), will amplify the climatic changes. A recent comparison between CO_2 and CH_4 records with other palaeoclimatic records (Sowers & Bender 1995) confirms the sequential relationship with changes in greenhouse trace-gases and temperature at high southern latitudes preceding those of northern temperature and continental ice volume during glacial–interglacial transitions.

Also, recently an effort has been concentrated on the analysis at high resolution of methane in the Greenland ice (Chappellaz *et al.* 1993*b*; Blünier *et al.* 1995; Brook *et al.* 1996). The results show surprising Holocene variations and abrupt changes during the glacial periods. The Holocene (the present interglacial period initiated 11 500 years ago) climate has been fairly uniform with conditions generally close to those prevailing today. We now have a complete CH_4 Holocene record from the Greenland ice core (Blünier *et al.* 1995) indicating a wide oscillation throughout the entire period (amplitude of about 140 ppbv). This oscillation starts during the early Holocene at levels close to the pre-industrial value and reaches minimum values during the middle Holocene (Fig. 1). These CH_4 changes are mainly attributed to modifications in wetland extents at different latitudinal bands in response to changes in precipitation patterns. Another striking feature of the Greenland CH_4 record is the sharp change observed around 8.2 ka (see Fig. 1).

By the term 'abrupt changes' we refer to drastic changes having occurred during the past over time intervals of a few centuries or less, i.e. similar to the recent environmental changes attributed to the anthropogenic activities. Examples of such changes are the Dansgaard–Oeschger (D–O) events during the last glaciation or the Younger Dryas, whose climatic signature is found in the isotopic Greenland ice record, as illustrated in Fig. 1. The question arises as to how global these events were and whether or not they were accompanied by changes in atmospheric greenhouse trace-gases. There is clearly one abrupt CH_4 change that is recorded both in the Antarctic and Greenland ice: the Younger Dryas oscillation during the last deglaciation. Furthermore, the recent careful study of the deglaciation and the glacial period revealed remarkable synchronous changes between CH_4

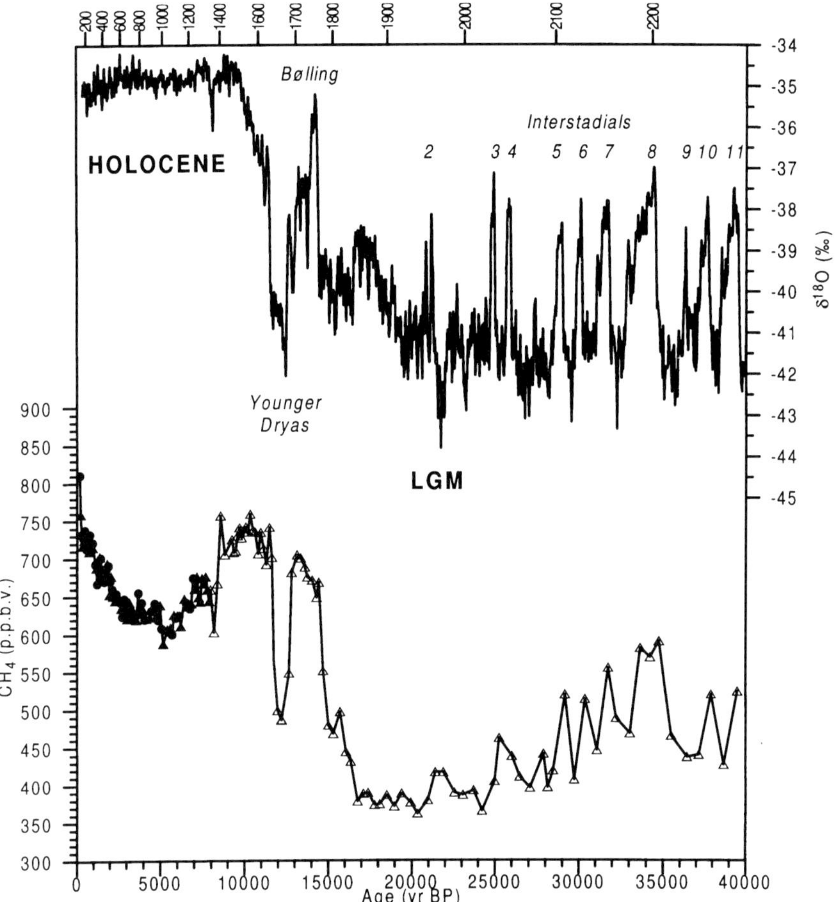

Fig. 1. GRIP ice-core profiles for the last 40 ka. Bottom: CH_4 record, adapted from Chappelaz *et al.* (1993*b*) and Blünier *et al.* (1995). Top: ice isotopic record (D180) adapted from Dansgaard *et al.* (1993). The $\delta^{18}O$ record reflects the climatic variations over Greenland. The significant climatic periods or events are noted by name. The numbering corresponds to the rapid warm interstadials, known as Dansgaard–Oeschger events.

and the Greenland climate during the D–O events, suggesting that these events were at least interhemispheric in their extent (Chappellaz *et al.* 1993*b*; Brook *et al.* 1996).

Sources mechanisms: relevance to climate changes and gas hydrates

The classical interpretation of atmospheric methane changes prior to the industrial period mainly involves modifications in wetland extents at various latitudes under different climatic conditions; other natural sources such as ruminants, termites, ocean and destabilization of methane hydrates may also have played a role, but they are considered as less important (Chappellaz *et al.* 1993*a*). Changes in the strength of the atmospheric sink (the oxidative capacity of the atmosphere) are also believed to have affected to a lesser extent the atmospheric methane variations.

Concerning the destabilization of gas hydrates in ocean sediments or permafrost area, it should be pointed out that important climatic and environmental changes occurred during the last glacial–interglacial cycles that may have influenced this methane reservoir. For instance, the last glacial maximum (LGM) corresponds to a period during which the sea level reached a minimum, on average at about 120 m below present-day sea level. The fall in sea level may have destabilized deep-sea gas hydrates which decomposed with decreasing pressure and then, via marine slumps, released methane to the sea water and possibly the atmosphere. It has even been proposed that such release would have made the deglaciation the methane trigger (Paull *et al.* 1991, and references therein).

On the other hand, the glacial–interglacial warming and the retreat of the Laurentide ice sheet during the deglaciation and the early Holocene could have been the cause of CH_4 releases to the atmosphere from permafrost hydrates either by pressure reduction and/or by temperature increase (directly through the atmospheric warming or induced by flooding of the permafrost as a consequence of increasing sea level); in the case of thermal destabilization of the hydrates, it can take several thousands of years for the initial surface thermal wave to reach the base of the hydrate layer (Nisbet 1992).

The ice-core record of methane begins to be well documented over the LGM, the last deglaciation and the early Holocene (see section *Description of the record* above). This enables us to investigate the atmospheric CH_4–gas hydrate aspect. We should first mention that the time resolution of the record may be a limiting factor in this investigation. Modelling the atmospheric methane change, consecutive to a dramatic hydrate release, and its fingerprint in ice cores has recently been performed (Thorpe 1994). The results indicate that, in the case of the Greenland record examined above, a sampling resolution of about 50 years would be necessary to detect each potential events. As the current resolution for the critical time period (LGM–early Holocene) is on average about 150 years, we could have lost the fingerprints of some of the hypothetical dramatic hydrate releases. An effort should be made in the future to increase the time resolution of the record corresponding to this critical time interval. Nevertheless, there are already strong indications from the record available.

The first question arising is: can the abrupt changes observed in the methane record be explained by dramatic releases of methane due to the destabilization of gas hydrates in oceanic sediments or permafrost?

The rapid variations of methane discovered in the LGM part of the Greenland ice record and associated with Dansgaard–Oeschger events (Chappellaz *et al.* 1993*b*; Brook *et al.* 1996) present a morphology which may be compatible with huge releases of CH_4 into the atmosphere originating from gas hydrate destabilization. There are nevertheless two major reasons against this interpretation:

- the durations of the events are too large compared to the simulated ice-core signal for catastrophic hydrate release (Thorpe 1994);
- the sea level shows a marked minimum during the LGM and we would expect a maximum of catastrophic events during that period; in fact, the rapid variations are associated with most of the Dansgaard–Oeschger events not only during the glacial maximum but all through the ice age over the past 110 000 years (Brook *et al.* 1996).

During the deglaciation the methane ice-core record exhibits two abrupt and important increases (Fig. 1). The first one appears to be associated with the drastic Bolling warming recorded in the Greenland ice cores, and the second one with the abrupt warming characterizing the end of the Younger Dryas. These sharp CH_4 increases cannot be easily explained by a catastrophic event as the constant methane high concentrations following their abrupt increases (Chappellaz *et al.* 1993*b*), (Fig. 1) would require a regular occurrence of such catastrophic events. Furthermore, the Younger Dryas methane signal has to be considered as a full oscillation starting by a marked decrease, the most straightforward explanation of which is the direct response of wetlands to the Younger Dryas climatic fluctuation. A similar explanation applies for the Holocene oscillation at about 8.2 ka.

Finally, we consider a second relevant question: are the phase relationships observed in the palaeorecord between CH_4, CO_2 and other climatic indicators compatible with a CH_4 release due to the destabilization of gas hydrates in oceanic sediments or permafrost? Nisbet (1992) interprets the CH_4 changes, depicted in the ice record for the transition from glacial conditions to the Holocene period, as the initial cause for the warming. He hypothesizes that the lower sea level during the LGM caused methane hydrates to destabilize. The resulting methane atmospheric concentration increase of several ppmv would have initiated the glacial–interglacial warming and a chain of climatic events during the deglaciation. The Nisbet scenario suggests that the initial warming concentrated in the Arctic. This and part of the following sequence of events described in the scenario are not supported by the phase relationships existing between southern and northern palaeoclimatic indicators (see for instance: Raynaud & Siegenthaler 1993; Sowers & Bender 1995). In other words, such a climatic scenario based on the initial instability of gas hydrates is, at least partly, in contradiction with the lead and lag sequence described by the palaeorecord.

Conclusions

The ice-core record provides the most direct reconstruction of global changes in atmospheric methane concentrations over the last glacial–interglacial cycles. A careful examination of this record should in principle enable us to search for the fingerprints of catastrophic hydrate releases (CHR) during the past and to check the validity of scenarios involving the climatic role of these catastrophic events. The available record shows no evidence of CHR.

A more definitive conclusion concerning the hydrate question can be brought about by additional ice-core analyses. In particular, an effort should be made to increase the time resolution in the critical parts of the record. Furthermore, measurements of the isotopic composition of CH_4 ($^{13}C/^{12}C$ and D/H) would help to identify the importance of the various sources. However, such isotopic measurements on air trapped in ice cores remains a technical challenge (Craig *et al.* 1988).

Furthermore, comparison with other palaeorecords does not support the sequence of environmental and climatic changes involved in scenarios where CHR and the corresponding abrupt and huge CH_4 releases to the atmosphere are driving the glacial–interglacial warming.

References

BLÜNIER, T., CHAPPELLAZ, J. A., SCHWANDER, J., STAUFFER B. & RAYNAUD, D. 1995. Variations in the atmospheric methane concentration during the Holocene epoch. *Nature*, **374**, 46–49.

BROOK E. J., SOWERS T. & ORCHADO, J. 1996. Rapid variations in atmospheric methane concentration during the past 110,000 years. *Science*, **273**, 1087–1091.

CHAPPELLAZ, J. A., FUNG, I. Y. & THOMPSON, A. M. 1993*a*. The atmospheric CH_4 increase since the last glacial maximum – (1) Source estimates. *Tellus*, **45B**, 228–241.

——, BLUENIER , T., RAYNAUD, D. *et al.* 1993*b*. Synchronous changes in atmospheric CH_4 and Greenland climate between 40 and 8 kyr BP. *Nature*, **366**, 443–445.

CRAIG, H., CHOU, C. C., WELHAN, J. A., STEVENS, C. M. & ENGELKEMEIR, A. 1988. The isotopic composition of methane in polar ice cores. *Science*, **242**, 1535–1539.

DANSGAARD, W. *et al.* 1993. Evidence for general instability of past climate from a 250-kyr ice-core record. *Nature*, **364**, 218–220.

ETHERIDGE, D., PEARMAN, G. I. & FRASER, P. J. 1992. Changes in tropospheric methane between 1841 and 1978 from a high accumulation-rate Antarctic ice core. *Tellus*, **44B**, 282–294.

JOUZEL, J. *et al.* 1993. Extending the Vostok ice-core record of paleoclimate to the penultimate glacial period. *Nature*, **364**, 407–412.

NISBET, E. G. 1992. Sources of atmospheric CH_4 in early postglacial time. *Journal of Geophysical Research*, **97**, 12 859–12 867.

PAULL, C. K., USSLER, W., III & DILLON, W. P. 1991. Is the extent of glaciation limited by marine gas-hydrates? *Geophysical Research Letters*, **18**, 432–434.

RAYNAUD, D. & CHAPPELLAZ, J. A. 1993. The record of atmospheric methane. *In*: KHALIL, M. (ed.) *Atmospheric Methane: Sources, Sinks, and Role in Global Change*. NATO ASI Series I, **13**, 38–61.

—— & SIEGENTHALER, U. 1993. Role of trace gases: the problem of lead and lag. *In*: EDDY, J. & OESCHGER, H. (eds) *Global Changes in the Perspective of the Past (Dahlem Conference)*. Wiley, Chichester, 173–188.

——, JOUZEL, J., BARNOLA, J-M., CHAPPELLAZ, J. A., DELMAS, R. J. & LORIUS, C. 1993. The ice core record of greenhouse gases. *Science*, **259**, 926–934.

ROBBINS, R., CAVANAGH, L., SALAS, L. & ROBINSON, E. 1973. Analysis of ancient atmospheres. *Journal of Geophysical Research*, **78**, 5341–5344.

SCHWANDER, J. 1996. Gas diffusion in firn. *In*: WOLF, E. & BALES, R. (eds) *Chemical Exchange Between the Atmosphere and Polar Snow*. NATO ASI Series I, **43**, 527–540.

—— & STAUFFER, B. 1984. Age difference between polar ice and the air trapped in its bubbles. *Nature*, **311**, 45–47.

SOWERS, T. & BENDER, M. 1995. Climate records covering the last deglaciation. *Science*, **269**, 210–214.

THORPE, R. B. 1994. *Modelling the Effect of Atmospheric Chemistry Upon Climate on Geological Timescales*. Thesis, University of Cambridge.

Index

Page numbers in *italics* refer to Figures or Tables